STEIN'S
Refresher
Mathematics

NEW Edition

Edwin I. Stein

ALLYN AND BACON, INC.
Newton, Massachusetts

Rockleigh, NJ   Atlanta   Warrensburg, MO   Dallas   Rancho Cordova, CA
London   Sydney   Toronto

Editors: Andrew P. Mastronardi
  Sylvia Gelb
Designer: Volney Croswell
Art Director: L. Christopher Valente
Production Designers: Carol H. Rose
  Shay J. Mayer
Cover Designer: John Martucci
Photo Researcher: Susan VanEtten
Technical Artists: Anco/Boston Inc.
  Roget Ink.
Preparation Services Manager: Martha E. Ballentine
Senior Buyer: Roger E. Powers

ISBN 0-205-08165-7

Printed in the United States of America

8 9 93 92 91 90

Contents

UNIT 1 Basic Skills in Arithmetic   2

CHAPTER 2 DECIMALS   65

UNIT 2 Problem Solving and Consumer Applications 242

CHAPTER 10 CUSTOMARY SYSTEM, TIME, AND TEMPERATURE   383

CHAPTER 11 GRAPHS, STATISTICS, AND PROBABILITY   407

UNIT 4 Geometry 426

CHAPTER 12 LINES, ANGLES, AND TRIANGLES 431

CHAPTER 13 CONSTRUCTIONS 469

CHAPTER 14 INDIRECT MEASUREMENT 485

CHAPTER 17 EQUATIONS, FORMULAS, AND INEQUALITIES 567

CHAPTER 18 GRAPHING IN THE NUMBER PLANE 599

Overview

REFRESHER MATHEMATICS is a comprehensive basal text in general mathematics, including arithmetic skills, consumer topics, metric and customary units of measure, graphs, statistics, probability, geometry, and algebra. The text also includes a thorough step-by-step approach to the development of problem-solving skills, career applications, and computer and calculator activities.

REFRESHER MATHEMATICS features a specially designed program for teaching arithmetic skills and problem-solving strategies. The materials are so organized that diagnostic and prescriptive teaching techniques may easily be used and are recommended. Students, by a simple testing procedure in the Lesson, are directed quickly to specific assignments based on individual needs. No students need waste time drilling on exercises and problems they already can solve.

Each lesson contains completely worked-out model problems highlighted with color to indicate procedural steps toward finding the solution. Items in the *Diagnostic Test* that follows are closely calibrated in difficulty and are keyed to sets of *Related Practice Exercises*. These *Diagnostic Tests* segment the components of each skill and are a time-saver for teachers when planning for varying levels of ability. Lessons end with a *Skill Achievement Test* to indicate if additional instruction and practice are necessary.

REFRESHER MATHEMATICS is the ideal textbook for any review program in basic mathematics and problem solving. Each Unit has an *Inventory Test* for locating specific mathematical deficiencies and directing students to the appropriate lessons for help.

The testing program consists of *Unit Inventory Tests* and *Achievement Tests, Chapter Reviews* and *Competency Check Tests, Skill Achievement Tests,* and maintenance tests called *Refresh Your Skills.*

The carefully planned, flexible organization of **REFRESHER MATHEMATICS** allows for individual differences and at the same time provides maximum and minimum materials for class, home, and optional assignments.

The author acknowledges the assistance of his wife Elaine, and the contributions made by Marilyn Lieberman, Mathematics Coordinator, Meadowbrook School, Meadowbrook, PA; Charlotte Jaffe, Director of Gifted Education, Clementon School, Clementon NJ, and Peter J. Jaffe.

UNIT 1

Basic Skills in Arithmetic

Inventory Test I (BASIC SKILLS)

1-1 **1.** Read the numeral 95,237 or write it in words.

1-2 **2.** Write the numeral naming: Eight hundred seven thousand, forty-eight

1-3 **3.** Round 6,849 to the nearest hundred.

1-4 **4.** Add: 578
946
+ 825

1-5 **5.** Subtract: 9,856
− 3,682

1-6 **6.** Multiply: 59×83

1-7 **7.** Divide: $34\overline{)918}$

1-8 **8.** Name the greatest common factor of 54 and 72.

2-1 **9.** Read the numeral .375 or write it in words.

2-2 **10.** Write as a decimal numeral: Ninety-six hundredths

2-3 **11.** Find correct to the nearest tenth: 6.79

2-4 **12.** Add: $.39 + 2.5$

2-5 **13.** Subtract: $9 − 4.6$

2-6 **14.** Which is less: .7 or .65?

2-7 **15.** Multiply: $5.4 \times .13$

2-8 **16.** Divide: $.2\overline{)6.8}$

2-9 **17.** What decimal part of 25 is 16?

2-10 **18.** .3 of what number is 12?

2-11 **19.** Multiply: 100×35

2-12 **20.** Divide: $4.9 \div 10$

2-13 **21.** Write a complete numeral for 2.7 *million*.

2-14 **22.** Write the shortened name for 16,300,000 in millions.

3-1 **23.** Express $\frac{10}{25}$ in lowest terms.

3-2 **24.** Rewrite $\frac{11}{8}$ as a mixed number.

3-3 **25.** Raise to higher terms: $\frac{3}{5} = \frac{?}{20}$

3-4 **26.** Are $\frac{12}{15}$ and $\frac{28}{35}$ equivalent fractions?

3-5 **27.** Which is greater: $\frac{5}{12}$ or $\frac{5}{8}$?

3-6 **28.** Find the least common multiple of 12 and 16.

3-7 **29.** Find the least common denominator of the fractions: $\frac{1}{2}$ and $\frac{2}{3}$

3-8 **30.** Add: $4\frac{2}{5}$
$+ 7\frac{3}{10}$

3-9 **31.** Subtract: $5\frac{3}{4}$
$- 2\frac{1}{3}$

3-10 **32.** Rewrite $4\frac{1}{2}$ as an improper fraction.

3-11 **33.** Multiply: $\frac{2}{3} \times \frac{4}{5}$

3-12 **34.** Divide: $2\frac{1}{4} \div \frac{1}{2}$

3-13 **35.** 8 is what part of 20?

3-14 **36.** $\frac{1}{4}$ of what number is 36?

3-15 **37.** Write $\frac{3}{4}$ as a decimal.

3-16 **38.** Write .8 as a fraction.

3-17 **39.** Find the ratio of 25 to 75.

3-18 **40.** Solve and check: $\frac{n}{12} = \frac{27}{36}$

4-1 **41.** Write *seventeen-hundredths* as a percent, as a decimal, and as a fraction.

4-2 **42.** Write 9% as a decimal.

4-3 **43.** Write .12 as a percent.

4-4 **44.** Write 40% as a fraction.

4-5 **45.** Write $\frac{9}{10}$ as a percent.

4-6 **46.** Find 25% of 93.

4-7 **47.** What percent of 60 is 45?

4-8 **48.** 80% of what number is 72?

Inventory Test II (BASIC SKILLS)

1-1 **1.** Read the numeral 163,000,987 or write it in words.

1-2 **2.** Write the numeral naming: Seventeen billion, four hundred seven million

1-3 **3.** Round 18,268,395 to the nearest hundred thousand.

1-4 **4.** Find the sum of 4,132; 375; 81,580; 69; and 5,364.

1-5 **5.** Subtract 259 from 3,120.

1-6 **6.** Multiply: 736 × 6,080   **1-7** **7.** Divide: $693\overline{)29,799}$

1-8 **8.** Name the greatest common factor for 24, 36, and 96.

2-1 **9.** Read the numeral .0839 or write it in words.

2-2 **10.** Write as a decimal numeral: Two hundred and seventy-two thousandths

2-3 **11.** Find correct to the nearest cent: $8.3674

2-4 **12.** Add: 6.8 + .24 + 16   **2-5** **13.** Subtract: 18 − 1.5

2-6 **14.** Which is less: 1.05 or .247?   **2-7** **15.** Find .75 of 2.60

2-8 **16.** Divide: $3.6\overline{)51.12}$ **2-9** **17.** 54 is what decimal part of 72?

2-10 **18.** 9 is .5 of what number?

2-11 **19.** Multiply: 100 × .09

2-12 **20.** Divide: 576.5 ÷ 1,000

2-13 **21.** Write a complete numeral for 9.14 *million*.

2-14 **22.** Write the shortened name for 75,280,000,000 in billions.

3-1 **23.** Express $\frac{52}{64}$ in lowest terms.   **3-2** **24.** Rewrite $\frac{8}{3}$ as a mixed number.

3-3 **25.** Express $\frac{3}{4}$ as a fraction with the denominator 32.

3-4 **26.** Are $\frac{15}{35}$ and $\frac{21}{42}$ equivalent fractions?   **3-5** **27.** Which is less: $\frac{3}{4}$ or $\frac{7}{10}$?

3-6 **28.** Find the least common multiple of 10, 15, and 18.

3-7 **29.** Find the least common denominator of the fractions: $\frac{1}{6}$ and $\frac{3}{8}$.

3-8 **30.** Add: $2\frac{3}{4} + 6\frac{5}{8}$ **3-9** **31.** Subtract: $\begin{array}{r} 8 \\ -\ 3\frac{2}{5} \\ \hline \end{array}$

3-10 **32.** Rewrite $1\frac{5}{6}$ as an improper fraction.

3-11 **33.** Find $\frac{7}{12}$ of 84.   **3-12** **34.** Divide: $7\frac{1}{2} \div 3$

3-13 **35.** What part of 50 is 35?   **3-14** **36.** $\frac{3}{8}$ of what number is 27?

3-15 **37.** Write $\frac{7}{8}$ as a decimal.   **3-16** **38.** Write 4.75 as a mixed number.

3-17 **39.** Find the ratio of 96 to 18.

3-18 **40.** Solve and check: $\frac{16}{x} = \frac{80}{15}$

4-1 **41.** Write *seven-hundredths* as a percent, as a decimal, and as a fraction.

4-2 **42.** Write 30% as a decimal.   **4-3** **43.** Write .08 as a percent.

4-4 **44.** Write 12% as a fraction.   **4-5** **45.** Write $\frac{27}{36}$ as a percent.

4-6 **46.** Find 8% of $8.27 (nearest cent).   **4-7** **47.** 16 is what percent of 12?

4-8 **48.** 42 is 7% of what number?

5

Inventory Test III (BASIC SKILLS)

1-1   **1.** Read the numeral 5,493,268,003,751 or write it in words.

1-2   **2.** Write the numeral naming: Forty-three million, nine hundred five thousand, twenty-six

1-3   **3.** Round 839,599,696 to the nearest million.

1-4   **4.** Add: 583,289
698,785
835,869
549,397
897,298

1-5   **5.** Subtract:
3,210,050
1,839,096

1-6   **6.** Multiply: 6,905
9,087

1-7   **7.** Divide:
5,022,702 ÷ 5,937

1-8   **8.** Name the greatest common factor of 45, 75, and 120.

2-1   **9.** Read the numeral 42.09 or write it in words.

2-2   **10.** Write as a decimal numeral: Sixty and nine ten-thousandths

2-3   **11.** Find correct to the nearest cent: $68.5349

2-4   **12.** Add: 3.24 + 53.7 + .938

2-5   **13.** Subtract: 4.2 − .351

2-6   **14.** Is the statement 7.5 > .750 true?

2-7   **15.** Multiply: .013 × .07

2-8   **16.** Divide: $.05\overline{)6}$

2-9   **17.** 91 is what decimal part of 104?

2-10   **18.** 120 is .625 of what number?

2-11   **19.** Multiply: 100 × 90.235

2-12   **20.** Divide: 6,340.7 ÷ 10,000

2-13   **21.** Write a complete numeral for 10.25 *billion*.

2-14   **22.** Write a shortened name for 1,839,000,000,000 in trillions.

3-1   **23.** Express $\frac{48}{108}$ in simplest terms.

3-2   **24.** Rewrite $\frac{29}{9}$ as a mixed number.

3-3   **25.** Express $\frac{7}{12}$ as a fraction with denominator 60.

3-4   **26.** Are $\frac{42}{63}$ and $\frac{34}{51}$ equivalent fractions?

3-5   **27.** Is the statement $\frac{7}{9} < \frac{10}{13}$ true?

3-6   **28.** Find the least common multiple of 8, 18, and 24.

3-7   **29.** Find the least common denominator of the fractions: $\frac{1}{10}$, $\frac{1}{4}$, and $\frac{1}{6}$.

3-8   **30.** Add: $1\frac{2}{3} + 5\frac{7}{12} + 3\frac{1}{6}$

3-9   **31.** Subtract: $14\frac{1}{3} - 9\frac{11}{16}$

3-10   **32.** Rewrite $6\frac{5}{7}$ as an improper fraction.

3-11   **33.** Multiply: $2\frac{3}{4} \times 8\frac{4}{5}$

3-12   **34.** Divide: $6\frac{2}{3} \div 1\frac{3}{5}$

3-13   **35.** What part of 192 is 54?

3-14   **36.** 220 is $\frac{11}{16}$ of what number?

3-15   **37.** Write $\frac{21}{32}$ as a decimal (2 places).

3-16   **38.** Write $.16\frac{2}{3}$ as a fraction.

3-17   **39.** Find the ratio of 27 to 72.

3-18   **40.** Solve and check: $\frac{48}{10} = \frac{n}{25}$

4-1   **41.** Write *one-fourth hundredth* as a percent, as a decimal, and as a fraction.

4-2   **42.** Write 103% as a decimal.

4-3   **43.** Write .0875 as a percent.

4-4   **44.** Write $233\frac{1}{3}\%$ as a mixed number.

4-5   **45.** Write $\frac{27}{45}$ as a percent.

4-6   **46.** Find $100\frac{1}{2}\%$ of $9,000.

4-7   **47.** $.56 is what percent of $.48?

4-8   **48.** $83\frac{1}{3}\%$ of what number is 30?

6

CHAPTER 1 Whole Numbers

Reading Numerals
Naming Whole Numbers

TRILLIONS			BILLIONS			MILLIONS			THOUSANDS			ONES		
hundred trillions	ten trillions	trillions	hundred billions	ten billions	billions	hundred millions	ten millions	millions	hundred thousands	ten thousands	thousands	hundreds	tens	ones
		4	2	6	0	7	5	9	2	4	1	3	7	8

PROCEDURE

1 Separate 4260759241378 into periods.

4 2 6 0 7 5 9 2 4 1 3 7 8
periods

Start from the right and mark off groups of three digits.

4,260,759,241,378 or 4 260 759 241 378

Place commas or spaces between each period.

ANSWER: 4,260,759,241,378 or 4 260 759 241 378

2 Separate 8176 into periods.

8,176   8 176

ANSWER: 8,176 or 8 176

Four-digit numerals may also be written without either a space or comma.

---

3 Read 476,923.

Read each period of digits. Then, state the name of the period.

4 7 6 , 9 2 3

four hundred seventy-six thousand   nine hundred twenty-three

ANSWER: Four hundred seventy-six thousand, nine hundred twenty-three.

4 Read 34,008,290.

3 4 , 0 0 8 , 2 9 0

thirty-four million   eight thousand   two hundred ninety

ANSWER: Thirty-four million, eight thousand, two hundred ninety.

5 Read 9,100 in two ways.

9 , 1 0 0   9 1 0 0

nine thousand   one hundred   ninety-one hundred

ANSWER: Nine thousand, one hundred or Ninety-one hundred.

6 Read 2 540 000 250.

When all digits of a period are zero, do not name the period.

$$2\,540\,000\,250$$

two billion   five hundred forty million   two hundred fifty

ANSWER: Two billion, five hundred forty million, two hundred fifty.

Vocabulary

A **period** is a group of three places in the number scale such as thousands, millions, etc.

The **Whole Numbers** are the numbers 0, 1, 2, 3, 4, 5, 6, 7, 8, 9, 10, and so on without end.

The whole numbers beginning with 1 that are used in counting are called **Counting** or **Natural Numbers**.

DIAGNOSTIC TEST

Separate the following numerals into proper periods:

1 By using commas: 29740049265040   **2** By using spaces: 817604009000

Read, or write in words, each of the following:

3 68	**4** 582	**5** 4,975	**6** 5,003	**7** 10,500
8 823,659	**9** 4,900,000	**10** 7,847,000	**11** 3,582,942	**12** 18,467,125
13 995,028,000	**14** 62,669,004,716	**15** 904,060,381,269,008	**16** 17500000000000	
17 740 509	**18** 63 086 493	**19** 218 904 226 007	**20** 5 693 169 040 000	

21 Read in two ways: 5300

RELATED PRACTICE EXERCISES

Separate the following numerals into proper periods:

1 By using commas:
   1. 793058   **2.** 96400127   **3.** 592473085604   **4.** 8335000000000000

2 By using spaces:
   1. 61437   **2.** 483050095   **3.** 216038823700000   **4.** 8903250000000000000

Read, or write in words, each of the following:

3 **1.** 96	**2.** 47	**3.** 84	**4.** 79
4 **1.** 426	**2.** 784	**3.** 905	**4.** 380
5 **1.** 8,278	**2.** 3,926	**3.** 4,206	**4.** 7,959
6 **1.** 6,004	**2.** 4,045	**3.** 9,080	**4.** 3,007

7 **1.** 27,432   **2.** 20,730   **3.** 45,063   **4.** 81,896
8 **1.** 150,000   **2.** 291,429   **3.** 900,281   **4.** 674,364
9 **1.** 3,000,000   **2.** 8,000,000   **3.** 2,500,000   **4.** 7,600,000
10 **1.** 9,250,000   **2.** 3,746,000   **3.** 2,006,000   **4.** 5,670,000
11 **1.** 7,122,843   **2.** 8,900,527   **3.** 5,416,084   **4.** 6,004,009
12 **1.** 32,429,784   **2.** 57,105,933   **3.** 96,863,848   **4.** 25,000,975
13 **1.** 700,000,000   **2.** 241,849,000   **3.** 928,376,500   **4.** 843,297,861
14 **1.** 5,000,000,000   **2.** 6,450,000,000   **3.** 93,884,526,767   **4.** 842,699,019,508
15 **1.** 9,000,000,000,000   **2.** 23,587,914,060,000
    3. 139,005,407,862,100   **4.** 768,841,253,196,369
16 **1.** 7593000   **2.** 82100476   **3.** 360005430000   **4.** 67800000000000
17 **1.** 28 900   **2.** 983 225   **3.** 7 003   **4.** 455 607
18 **1.** 6 582 000   **2.** 34 809 500   **3.** 803 578 632   **4.** 59 200 048
19 **1.** 75 000 000 000   **2.** 8 497 500 000   **3.** 93 674 317 505   **4.** 245 390 006 700
20 **1.** 9 000 000 000 000   **2.** 141 325 913 006 200
    3. 2 607 559 040 172   **4.** 53 482 080 000 000
21 Read in two ways:
    1. 6900   **2.** 2800   **3.** 8700   **4.** 7400

SKILL ACHIEVEMENT TEST

1. Separate 59320718400 into periods.   **2.** Read, or write in words, 4600 in two ways.
Read, or write in words: **3.** 925,030   **4.** 86,209,473   **5.** 7,083,006,400

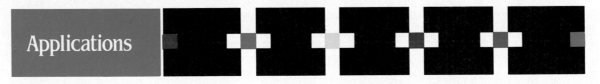

Applications

Read, or write in words, each of the numerals that appear in the following:

1. The Pacific Ocean has an area of 165,246,320 square kilometers; Atlantic, 82,441,560 square kilometers; Indian, 73,442,755 square kilometers; and Arctic, 14,090,110 square kilometers.

2. The total number of United States coins minted during the year was 13,377,659,747 valued at $769,466,209 while the total money in circulation was valued at $137,244,000,000.

3. Recently Shanghai had a population of 10,820,000; Tokyo, 8,349,000; New York, 7,895,543; Moscow, 8,203,000; and Bombay, 8,202,759.

4. The trading volume on the New York Stock Exchange one day was 49,802,130 shares making the total volume so far for the year 2,921,686,626 shares.

ROMAN NUMERALS

The ancient Romans used capital letters to write numerals. Today Roman numerals are still used on clock and watch faces, for the display of dates, and for numbering in outlines.

The Roman number symbols and their values are shown at the right.

Roman numerals are formed by writing from left to right as a sum, first the symbol for the greatest possible value followed by symbols of decreasing values.

The symbols I, X, C, or M may be used as many as three times in succession.

When one of the symbols I, X, or C precedes a Roman number symbol of greater value, its value is subtracted from the greater value.

Symbol	Value
I	1
V	5
X	10
L	50
C	100
D	500
M	1000

IV = 5 − 1 = 4	XL = 50 − 10 = 40	CD = 500 − 100 = 400
IX = 10 − 1 = 9	XC = 100 − 10 = 90	CM = 1,000 − 100 = 900

The symbols V, L, and D never precede a symbol of greater value and are never used in succession.

EXAMPLE 1: What number does MDCCXLVIII represent?

MDCCXLVIII = M + D + C + C + XL + V + III
$$\downarrow \qquad \downarrow \qquad \downarrow \qquad \downarrow \qquad \downarrow \qquad \downarrow \quad \downarrow$$
1,000   500   100   100   (50−10)   5   3 = 1,748

ANSWER: 1,748

EXAMPLE 2: Write the Roman numeral representing 287.

287 = 200 + 50 + 30 + 5 + 2
$$\qquad \downarrow \quad \downarrow \quad \downarrow \quad \downarrow \quad \downarrow$$
= CC + L + XXX + V + II = CCLXXXVII

ANSWER: CCLXXXVII

EXERCISES

What number does each of the following Roman numerals represent?

1. XXXVI   **2.** CLXV   **3.** MDCLIX   **4.** MCCLIV   **5.** MCMXCIII   **6.** MMXLVII

Write a Roman numeral for each of the following numbers:

7. 19   **8.** 299   **9.** 847   **10.** 1492   **11.** 1776   **12.** 2015

13. A cornerstone is marked MCMXIX. What date does it represent?

What number is represented by the Roman numeral in each of the following:

14. Page VII   **15.** Unit XLVI   **16.** Item CXC

In making an outline, what Roman numeral should you use following:

17. XIII   **18.** XXIV   **19.** XCIX   **20.** XLVIII

DIED
·JANU
ARY·X
XVII·
MDCC
CXCV
IIII

1-2 Writing Numerals
Naming Large Whole Numbers

PROCEDURE

1 Write the numeral naming: Seven million, five hundred four thousand, twenty-nine.

Write each period of digits. Then, use a comma or space to represent the period.

seven million ⎡five hundred four⎤ thousand ⎡twenty-nine⎤
7 , 504 , 029

ANSWER: 7,504,029 or 7 504 029

2 Write the numeral naming: Twelve billion, nine hundred twenty-six million, eight hundred thousand.

twelve billion ⎡nine hundred twenty-six⎤ million ⎡eight hundred⎤ thousand
12 , 926 , 800 , ⎡000⎤

ANSWER: 12,926,800,000 or 12 926 800 000

Write zeros to complete the numeral.

3 Write the numeral naming: One trillion, eight hundred forty million, five hundred sixteen thousand, seventy-one.

one trillion ⎡eight hundred forty⎤ million ⎡five hundred sixteen⎤ thousand ⎡seventy-one⎤
1 , ⎡000,⎤ 840 , 516 , 071

NO billions

ANSWER: 1,000,840,516,071 or 1 000 840 516 071

DIAGNOSTIC TEST

Write, using commas, the numeral naming each of the following:

1 Six hundred fifty-two

2 Five hundred nineteen thousand, eight hundred forty-seven

3 Thirty million, four hundred five thousand, eighteen

4 Seven billion, nine hundred thirty million, two hundred thousand

5 Ten trillion, twenty-one billion, three hundred million, three hundred fifty

6 Write, using spaces, the numeral: Four hundred sixty million

RELATED PRACTICE EXERCISES

Write, using commas, the numeral that names each of the following:

1 1. Four hundred
3. Seven hundred nine

2. Eight hundred thirty-three
4. Two hundred forty-eight

2 1. Four thousand, six hundred ten
3. Seven hundred eighty-one thousand

2. Ninety thousand, fifty
4. Five hundred sixteen thousand, one hundred seventy-three

3 1. Three million, nine hundred thousand

2. Eighty-five million, six thousand, two hundred forty

3. Six hundred eight million, one hundred fifty thousand, ninety two

4. Nine hundred fourteen million, seven hundred thousand, fifty

4 1. Forty-eight billion, eighteen million, seven thousand

2. Two hundred billion, sixty-two thousand, nine hundred

3. Eight hundred seventeen billion, four hundred million, seventy-six thousand

4. Seven billion, five hundred four million, seven hundred thousand, four hundred fifty-nine

5 1. Four trillion, nine hundred fifty billion

2. Eleven trillion, eight hundred billion, two hundred thirty-six million

3. Three hundred trillion, forty billion, five million, eighty-nine thousand, two hundred one

4. One trillion, six hundred eighty-two billion, four hundred seven million, sixteen thousand

6 Write, using spaces, the numeral that names each of the following:
1. Fourteen thousand, six hundred
2. Three million, eighty thousand, nine hundred twenty-three
3. Nineteen billion, seventy-five million, six hundred forty thousand
4. Eight trillion, one hundred billion, fifty million

SKILL ACHIEVEMENT TEST

Write, using commas, the numeral naming:

1. Twenty-five thousand, three hundred eighty-seven
2. Six hundred three million, four hundred seventy-one thousand, two hundred forty-nine
3. Twelve billion, sixty million, five hundred thousand
4. Nine trillion, six hundred billion, seventy-two million, two hundred fifty thousand
5. Write, using spaces, the numeral naming: Ten billion, eight hundred thousand

Place Value

In the numeral 85,269,713,504, what digit is in the:

1. thousands' place?   **2.** hundreds' place?
3. millions' place?   **4.** ten millions' place?
5. billions' place?   **6.** hundred thousands' place?

7. Complete the following:

9,437 = _____thousands _____hundreds _____tens _____ones

= _____hundreds _____tens _____ones

= _____tens _____ones

= _____ones

Name the place of the underlined digit in each of the following:

8. 649,536   **9.** 538,247
10. 19,406,589   **11.** 2,735,962,814,000

Find the value of the underlined digit in each of the following:

12. 95,076   **13.** 846,709
14. 283,914,500   **15.** 96,827,543,000

What new group is formed by 1 more than:

16. 9,999?   **17.** 999,999?
18. 99,999,999?   **19.** 9,999,999,999?

20. Name the greatest possible 12-place whole number.

21. Name the least possible 7-place whole number.

Write the numeral that names (1) the greatest possible number, and (2) the least possible number, using the digits:

22. 5, 2, 9   **23.** 6, 3, 0, 7
24. 4, 0, 5, 0, 8, 2   **25.** 3, 1, 9, 0, 1, 6, 8

26. Arrange the following numbers according to size, writing the numeral for the greatest number first:

759,368   759,638   759,863   759,836

27. Arrange the following numbers according to size, writing the numeral for the least number first:

3,582,745   3,578,542   3,587,425   3,578,524

SHORTENED NAMES FOR LARGE WHOLE NUMBERS

Newspapers and magazines often express large whole numbers in a shortened form. For example, you might read that, "The population of China is about 1 billion," or that "This year the state budget is up $35 million."

$35 million is a *shortened name* for $35,000,000. It is read, "Thirty-five million dollars."

To write the complete numeral for a number written in shortened form, replace the period name by the appropriate number of zeros.

Period Name	Representation with Digits
thousand	___, 000 (3 zeros)
million	___, 000,000 (6 zeros)
billion	___, 000,000,000 (9 zeros)
trillion	___, 000,000,000,000 (12 zeros)

EXERCISES

Write the complete numeral for each of the numbers in the following:

1. Money-fund assets fell $75 million last week.
2. It is estimated that the Federal budget deficit this year will total about $180 billion.
3. Natural gas deposits estimated at 3 trillion cubic feet were discovered in the North Sea.

4.
United States Foreign Trade
(In millions of dollars)

	Exports	Imports
Western Hemisphere	$67,312	$84,467
Europe	63,664	53,413
Asia	64,822	85,170
Oceania	5,700	3,131
Africa	10,271	17,770

5.
Balance Sheet
(Dollars in thousands)

Assets		Liabilities and Stockholder's Equity	
Cash	$25,360	Notes Payable	$ 73,418
Accounts Receivable	194,525	Accounts Payable	138,734
Inventories	321,837	Other Liabilities	41,583
Property and Equipment	273,548	Long-term Debt	210,895
Other Assets	69,856	Stockholder's Equity	420,496
Total	$885,126	Total	$885,126

1-3 Rounding Whole Numbers

PROCEDURE

1 Round 48,628 to the nearest thousand.

48,628   Locate the place to be rounded.

↖ thousands

48,628   If the digit to the right is 5 or more, round to the
                                                        next higher digit
↖ 6 is greater than 5.

49,000   Replace digits to the right of the place to be
                                                        rounded with zeros
ANSWER: 49,000

2 Round 26,385,000 to the nearest million.

↙ millions

26,385,000

↙ 3 is less than 5.

26,385,000   If the digit to the right is less than 5, write the
26,000,000   same digit

ANSWER: 26,000,000

3 Round 6,430,859,627 to the nearest:
Ten:
ANSWER: 6,430,859,630
Hundred:
ANSWER: 6,430,859,600
Thousand:
ANSWER: 6,430,860,000 ←————————— Sometimes rounding to two different places can
Ten thousand: give the same result.
ANSWER: 6,430,860,000 ←
Hundred thousand:
ANSWER: 6,430,900,000 ←————————— Since digit to the right was 5, round to next
Million: higher digit.
ANSWER: 6,431,000,000
Ten million:
ANSWER: 6,430,000,000
Hundred million:
ANSWER: 6,400,000,000
Billion:
ANSWER: 6,000,000,000

DIAGNOSTIC TEST

Round each of the following numbers to the nearest

Ten:	**1** 57	**2** 394	
Hundred:	**3** 5,626	**4** 890	
Thousand:	**5** 4,501	**6** 63,289	
Ten thousand:	**7** 71,989	**8** 149,000	
Hundred thousand:	**9** 243,762	**10** 1,650,059	
Million:	**11** 3,728,409	**12** 41,371,000	
Billion:	**13** 24,492,568,200	**14** 170,964,135,000	
Trillion:	**15** 1,602,373,068,040	**16** 43,089,967,000,000	

RELATED PRACTICE EXERCISES

Round each of the following numbers to the
Nearest ten:

1 **1.** 28	**2.** 97	**3.** 416	**4.** 2,539
2 **1.** 31	**2.** 152	**3.** 683	**4.** 3,844

Nearest hundred:

3 **1.** 230	**2.** 947	**3.** 3,814	**4.** 5,408
4 **1.** 551	**2.** 7,863	**3.** 9,092	**4.** 15,654

Nearest thousand:

5 **1.** 8,726	**2.** 24,539	**3.** 57,603	**4.** 10,954
6 **1.** 9,485	**2.** 5,094	**3.** 40,159	**4.** 878,208

Nearest ten thousand:

7 **1.** 94,872	**2.** 80,056	**3.** 142,625	**4.** 6,531,009
8 **1.** 56,000	**2.** 37,160	**3.** 295,928	**4.** 7,849,323

Nearest hundred thousand:

9 **1.** 345,946	**2.** 1,510,214	**3.** 3,835,837	**4.** 22,403,541
10 **1.** 560,500	**2.** 2,375,908	**3.** 5,481,481	**4.** 19,792,624

Nearest million:

11 **1.** 4,500,000	**2.** 9,742,820	**3.** 8,906,437	**4.** 84,827,466
12 **1.** 7,398,250	**2.** 14,487,095	**3.** 462,264,128	**4.** 8,250,175,000

Nearest billion:

13 **1.** 6,150,000,000   **2.** 39,076,524,692   **3.** 253,264,800,000   **4.** 598,337,060,489

14 **1.** 8,832,000,000   **2.** 21,584,427,000   **3.** 67,940,839,148   **4.** 908,671,595,286

Nearest trillion:

15 **1.** 5,592,000,000,000   **2.** 46,860,813,289,400
    3. 130,914,325,674,149   **4.** 815,637,453,116,588

16 **1.** 7,459,540,000,000   **2.** 6,187,969,486,725
    3. 95,019,831,645,103   **4.** 604,231,993,584,270

SKILL ACHIEVEMENT TEST

Round each of the following to the nearest:

1. Thousand: 25,309   **2.** Hundred: 13,982

3. Million: 8,174,060   **4.** Billion: 26,840,000,000

5. Ten: 967   **6.** Hundred thousand: 3,774,000

7. Thousand: 466,298

8. Ten thousand: 16,009,217

9. Million: 19,389,000

10. Trillion: 8,597,300,000,000

Enrichment

Decimal System of Numeration

A system of numeration is a method of naming numbers by writing numerals.

In the system of numeration called the decimal system, ten number symbols, 0, 1, 2, 3, 4, 5, 6, 7, 8, and 9, are used to represent all numbers. There is no single number symbol in the notation system for the number ten or for numbers greater than ten. The numerals representing numbers greater than nine are formed by writing two or more number symbols next to each other in different positions or places.

The decimal system is built on the base, ten. The base of a system of numeration is the number it takes in any one place to make 1 in the next higher place. In the decimal system it takes ten in any one place to make 1 in the next higher place. It takes 10 ones to make 1 ten, 10 tens to make 1 hundred, 10 hundreds to make 1 thousand, and so forth.

READING AND WRITING SHORTENED LISTS OF NUMERALS

Long lists of numbers that follow a pattern can be written as shortened lists using three dots. The three dots indicate that the numbers continue in the same pattern.

To write in shortened form an unending list of numbers that follow a pattern:

(1) Write the first few numerals to show the pattern.
(2) Place three dots after the last listed numeral.

EXAMPLE: Write the list of numerals indicating:

Zero, seven, fourteen, twenty-one, and so on, without end.
ANSWER: 0,7,14,21,...

To write a list of numbers that follow a pattern but with too many numbers to list:

(1) Write the first few numerals to show the pattern.
(2) Place three dots before the numeral naming the last number.

EXAMPLE: Write the list of numerals indicating:

One, three, five, seven, and so on, up to and including twenty-nine.
ANSWER: 1,3,5,7,...,29

EXERCISES

Read, or write in words, each of the following:

1. 0,1,2,3,...
2. 1,3,5,7,...
3. 0,5,10,15,...
4. 0,2,4,6,...
5. 2,3,5,7,...
6. 1,4,9,16,...
7. 0,1,2,3,...,25
8. 0,4,8,12,...,64
9. 1,3,5,7,...,61
10. 0,10,20,30,...,200
11. 0,8,16,24,...,96
12. 2,4,6,8,...,50

Write the list of numerals indicating:

13. Zero, three, six, nine, and so on, without end.
14. Zero, six, twelve, eighteen, and so on, up to and including seventy-two.
15. Two, three, five, seven, and so on, up to and including eighty-nine.
16. Zero, twenty, forty, sixty, and so on, without end.
17. One, two, three, four, and so on, up to and including one hundred.
18. Three, nine, twenty-seven, eight-one, and so on, without end.

1-4 Addition of Whole Numbers

PROCEDURE

1 Add: 684 + 1,292

$$\begin{array}{r} 684 \\ +\ 1,292 \\ \hline 6 \end{array}$$ addends

Arrange in columns. Add each column starting from the right.

$$\begin{array}{r} \overset{1}{6}84 \\ +\ 1,292 \\ \hline 1,976 \end{array}$$ ← 8 + 9 = 17   ← sum

Check:
$$\begin{array}{r} 684 \\ +\ 1,292 \\ \hline 1,976 \end{array}$$ ↑

Regroup when necessary.
17 tens = 1 hundred   7 tens.

ANSWER: 1,976

Check by adding up.

2 Find the sum: 2,484; 337; 4,069; 42

$$\begin{array}{r} \overset{22}{2,484} \\ 337 \\ 4,069 \\ +\ \ \ \ 42 \\ \hline 6,932 \end{array}$$

Check:
$$\begin{array}{r} 2,484 \\ 337 \\ 4,069 \\ +\ \ \ \ 42 \\ \hline 6,932 \end{array}$$ ↑

22 ones = 2 tens   2 ones.
23 tens = 2 hundreds   3 tens.

ANSWER: 6,932

---

3 Estimate the sum: 742,908 + 67,484 + 85,367

$$\begin{array}{rcl} 742,908 & \to & 740,000 \\ 67,484 & \to & 70,000 \\ 85,367 & \to & 90,000 \\ \hline & & 900,000 \end{array}$$

Round the numbers to the same place.
Usually, this is the greatest place in the smallest number.
Add.

ANSWER: 900,000

---

4 Solve the equation: 847 + 9,269 = n

$$\begin{array}{r} \overset{11}{\underset{1}{8}47} \\ +\ 9,269 \\ \hline 10,116 \end{array}$$

Check:
$$\begin{array}{r} 847 \\ +\ 9,269 \\ \hline 10,116 \end{array}$$ ↑

To find the value of n, add the given numbers.

ANSWER: 10,116

Vocabulary

An **addend** is a number that you add.
The **sum** is the answer in addition.

DIAGNOSTIC TEST

Add and check:

1 32
45

2 56
17

3 67
59

4 9
5
8

5 15
26
87

6 6
2
4
9

7 25
94
84
39

8 5
7
6
8
9

9 34
90
23
65
57

10 389
459

11 3,592
2,738

12 86,056
44,598

13 435
599
796

14 3,598
6,487
5,739

15 24,673
12,762
37,857

16 556
479
628
493

17 3,962
6,109
2,854
6,875

18 79,459
68,417
75,388
91,754

19 258
584
845
207
396

20 4,973
9,282
3,970
2,639
9,789

21 83,625
94,774
87,146
32,753
64,838

22 9,651
78
83,795
206
5,184
16,745

23 1,483,297
3,915,485
20,500
372,173
5,283,094
7,848
608,556

24 Add as indicated:
3,845 + 928 + 63,847 + 795 + 1,356

25 Find the sum of:
2,381; 967; 29; 8,406; 750

26 Select the nearest given estimate for:
29,625 + 8,370 + 41,598
a. 75,000 **b.** 80,000 **c.** 85,000

27 Solve: 726 + 93 = n

RELATED PRACTICE EXERCISES

Add and check:

1
1. 34
25
2. 67
32
3. 20
58
4. 63
26

2
1. 59
8
2. 37
57
3. 59
24
4. 28
62

3
1. 48
94
2. 24
76
3. 83
60
4. 95
99

4
1. 7
5
8
2. 9
7
7
3. 4
6
9
4. 9
8
9

5
1. 25
50
83
2. 86
92
58
3. 65
7
94
4. 34
79
88

6
1. 3
5
7
4
2. 8
4
9
6
3. 4
0
9
8
4. 7
6
2
9

7

1.	**2.**	**3.**	**4.**
82	45	94	86
43	79	36	59
92	63	49	97
30	42	78	95

8

1.	**2.**	**3.**	**4.**
5	8	8	9
1	6	9	9
3	9	4	6
9	4	7	8
2	7	8	9

9

1.	**2.**	**3.**	**4.**
12	73	91	76
26	9	30	89
21	24	54	97
13	8	79	88
22	87	68	53

10

1.	**2.**	**3.**	**4.**
231	267	974	608
457	385	566	96

11

1.	**2.**	**3.**	**4.**
6,142	2,358	5,705	7,478
3,756	4,135	986	9,757

12

1.	**2.**	**3.**	**4.**
37,454	946	35,878	21,973
41,345	30,306	89,646	92,027

13

1.	**2.**	**3.**	**4.**
582	779	926	639
143	698	15	852
214	579	658	543

14

1.	**2.**	**3.**	**4.**
4,214	4,043	8,588	6,957
1,152	3,285	765	9,799
3,403	9,772	89	7,186

15

1.	**2.**	**3.**	**4.**
13,581	45,496	68,432	73,858
61,132	27,383	257	59,576
15,273	31,576	9,746	38,643

16

1.	**2.**	**3.**	**4.**
637	458	682	39
131	173	745	896
528	649	384	9
300	326	973	372

17

1.	**2.**	**3.**	**4.**
2,105	4,966	7,581	6,479
3,216	3,859	470	8,642
1,143	5,573	28	9,736
2,319	8,426	6	6,857

18

1.	**2.**	**3.**	**4.**
34,832	56,834	42,763	6,445
11,913	67,509	96,833	95,214
22,741	31,915	47,485	87
10,420	47,628	92,739	1,756

19

1.	**2.**	**3.**	**4.**
129	289	646	7
325	769	53	226
356	427	8	589
218	892	24	36
497	976	593	328

20

1.	**2.**	**3.**	**4.**
1,673	2,765	7,328	6,494
2,191	4,497	478	3,849
1,814	2,086	9,663	8,589
1,271	2,199	98	6,355
2,353	7,385	8,976	2,914

21

1.	**2.**	**3.**	**4.**
15,924	45,618	75	96,545
24,872	4,973	5,491	32,678
29,644	87	67,987	49,924
18,873	564	446	85,397
15,692	9	51,743	26,785

22

1. 6	**2.** 16	**3.** 529	**4.** 6,583
5	8	63	3,476
2	29	784	4,285
8	9	8	3,842
4	65	29	5,273
2	22	685	2,646

5. 95,329	**6.** 82,541	**7.** 4,953	**8.** 642,566
458	31,796	58,275	255,887
7,509	28,865	87	974,464
26	49,496	893,574	850,952
84,531	22,738	64,705	527,327
727	57,633	9,268	741,969

23

1. 4,591	**2.** 46,398	**3.** 698,523	**4.** 4,369,565
8,216	127	34,769	7,436,859
9,137	3,874	178	3,050,762
1,615	58,462	9,874	4,598,083
2,293	95	968,452	6,452,136
3,748	4,267	789,347	7,587,327
5,157	67,839	59,791	3,967,009

24 Add as indicated:

1. 8 + 5 + 6 + 4 + 9 **2.** 69 + 75 + 29 + 82 + 51

3. 47 + 419 + 72 + 5 + 27 **4.** 328 + 139 + 426

5. 886 + 529 + 793 + 438 + 927 + 572 + 106

6. 69,483 + 24,752 + 80,599 + 2,843

7. 14,208 + 932 + 5,919 + 28 + 36,545 + 15,970

8. 37,442 + 42,596 + 807 + 6,638 + 52,800 + 51 + 9,046

25 Find the sum of:

1. 86; 153; 128; 249 **2.** 425; 836; 595; 26; 84

3. 200; 175; 335; 95; 480; 45 **4.** 731; 453; 846; 924; 877; 382; 502

5. 1,469; 8,355; 2,076; 3,416; 123; 4,125 **6.** 8,500; 250; 960; 63,075; 407; 2,155

7. 24,057; 6,799; 4,276; 9,341; 8,223; 37,165 **8.** 83; 9; 472; 1,857; 343; 8,158; 48,494

26 For each of the following, select the nearest given estimate:

1. 927 + 596 + 487	**a.** 1,000	**b.** 1,500	**c.** 2,000
2. 6,380 + 5,162 + 8,609	**a.** 20,000	**b.** 25,000	**c.** 10,000
3. 7,859 + 235 + 376	**a.** 8,000	**b.** 9,000	**c.** 12,000
4. 67,618 + 8,953	**a.** 140,000	**b.** 89,000	**c.** 77,000

27 Solve by performing the indicated operation:

1. 18 + 35 = ? **2.** 653 + 989 + 864 = ▓ **3.** 1,528 + 9,396 = n

4. ? = 59 + 427 + 76 **5.** ▓ = 3,855 + 2,962 **6.** n = 35,849 + 966 + 6,758

SKILL ACHIEVEMENT TEST

Add:

1. 359
697
486
932

2. 3,426
998
7,657
374
5,868

3. 15,905
63,287
49,569
32,878
89,389

4. Add as indicated: 943 + 8,107 + 62,498 + 9,965

5. Find the sum of: 3,586; 7,459; 68,426; 9,817; 59,649

6. For 48,693 + 6,381, select your nearest estimate: **a.** 10,000 **b.** 50,000 **c.** 55,000

7. Solve by performing the indicated operation: 475 + 928 = n

First estimate the sum, then add the given numbers:

8. 24,856 + 5,392

9. 41,484 + 93,542 + 57,969

10. 86,527 + 6,603 + 21,749 + 3,458

Enrichment

Word Names for Numbers

In most countries throughout the world, the number symbols 0, 1, 2, 3, 4, 5, 6, 7, 8, and 9 are commonly used. People from various countries say and write names for these symbols differently because of their respective languages.

The table shows how the numbers one to ten are written in several languages.

	English	French	Spanish	Italian	German
1	one	un	uno	uno	eins
2	two	deux	dos	due	zwei
3	three	trois	tres	tre	drei
4	four	quatre	cuatro	quattro	vier
5	five	cinq	cinco	cinque	fünf
6	six	six	seis	sei	sechs
7	seven	sept	siete	sette	sieben
8	eight	huit	ocho	otto	acht
9	nine	neuf	nueve	nove	neun
10	ten	dix	diez	dieci	zehn

EVEN AND ODD NUMBERS

Whole numbers may be separated into even and odd numbers.

Any whole number that can be divided exactly by two (2) is called an even number. Zero is considered an even number.

The numerals for even numbers always end in 0, 2, 4, 6, or 8.

Any whole number that cannot be divided exactly by two (2) is called an odd number.

The numerals for odd numbers always end in 1, 3, 5, 7, or 9.

EXERCISES

1. Which of the following are even numbers? Odd numbers?
30 72 49 97 386 565 908 2,183

2. Write all the one-digit numerals that name even whole numbers.

3. Write the numerals that name all the odd numbers greater than 94 and less than 106.

Is the sum an odd number or an even number when we add:

4. Two even numbers? Illustrate.   **5.** Two odd numbers? Illustrate.

6. An odd number and an even number? Illustrate.

Is the product an odd number or an even number when we multiply:

7. Two even numbers? Illustrate.   **8.** Two odd numbers? Illustrate.

9. An odd number and an even number? Illustrate.

10. Is one more than any whole number an even number?

11. Is two more than any whole number an even number?

12. Is two times any whole number an odd number or an even number?

BUILDING PROBLEM-SOLVING SKILLS

PROBLEM: What is the total enrollment at the County High School if there are 1,136 students in the 9th grade, 1,067 students in the 10th grade, 894 students in the 11th grade, and 742 students in the 12th grade?

1 READ the problem carefully.

a. Find the **given facts.**   9th grade: 1,136 students   10th grade: 1,067 students
                               11th grade: 894 students   12th grade: 742 students

b. Find the **question asked.** What is the *total* enrollment?

2 PLAN how to solve the problem.

a. Choose the **operation needed.**
   The word *total* with *unequal* quantities indicates addition.

b. **Think or write out your plan** relating the given facts to the question asked.
   The sum of 1,136 students, 1,067 students, 894 students, and 742 students is equal to the total number of students.

c. Express as an **equation.**
   $1{,}136 + 1{,}067 + 894 + 742 = n$

3 SOLVE the problem.

a. **Estimate the answer.**
   $1{,}136 \rightarrow 1{,}100$
   $1{,}067 \rightarrow 1{,}100$
   $\ \ \ 894 \rightarrow \ \ \ 900$
   $\ \ \ 742 \rightarrow \ \ \ 700$
   3,800 students, estimate

b. **Solution.**
   $\ \ \ 1{,}136$
   $\ \ \ 1{,}067$
   $\ \ \ \ \ \ 894$
   $+ \ \ \ 742$
   3,839 students

26

4 CHECK

a. **Check the accuracy** of your arithmetic.

$$
\begin{array}{r}
1,136 \;\uparrow \\
1,067 \;\mid \\
894 \;\mid \\
+\quad 742 \;\mid \\
\hline
3,839 \;\checkmark
\end{array}
$$

b. **Compare the answer to the estimate.** The answer 3,839 students compares reasonably with the estimate, 3,800 students.

ANSWER: The total enrollment is 3,839 students.

PRACTICE PROBLEMS

1. What was the total attendance at the 4 school league football games played by Central High School if the first game was witnessed by 2,972 spectators, the second by 2,684, the third by 3,189, and the last game by 2,708 spectators?

2. How many farms are in the East North Central States (Ohio, Indiana, Illinois, Michigan, Wisconsin) if there are 111,322 farms in Ohio, 101,479 farms in Indiana, 123,565 in Illinois, 77,944 in Michigan, and 98,973 in Wisconsin?

3. How many seats are there altogether at the county baseball stadium if there are 978 box seats, 19,564 general admission seats, and 3,825 bleacher seats?

4. The township public library has 2,309 fiction books, 1,894 nonfiction books, 195 reference books, and 275 magazines. How many books and magazines are there in all?

Select the letter corresponding to your answer.

5. What is the population of New England if Maine has a population of 1,124,660; New Hampshire, 920,610; Vermont, 511,456; Massachusetts, 5,737,037; Rhode Island, 947,154; and Connecticut, 3,107,576?
  a. 12,347,493 people **b.** 11,348,493 people
  c. 12,834,493 people **d.** Answer not given

6. How many calories are contained in the following meal: vegetable soup, 86; roast lamb, 175; fresh peas, 66; boiled potato, 117; two slices of white bread, 134; apple, 81; glass of milk, 170?
  a. 819 calories **b.** 739 calories
  c. 929 calories **d.** 829 calories

Refresh Your Skills

The numerals in boxes indicate Lessons where help may be found.

1. Write 6,862,925 as a word statement. `1-1`

2. Round 129,484,056 to the nearest million. `1-3`

First estimate, then add the given numbers. `1-4`

3.	**4.**	**5.**
16834	92,153	38,669
32605	8,449	81,767
57428	376	63,898
45887	9,849	89,948
84065	75,698	56,887

1-5 Subtraction of Whole Numbers

PROCEDURE

1 Subtract: $8,627 - 1,482$

$$8,627 \leftarrow \text{minuend}$$
$$- 1,482 \leftarrow \text{subtrahend}$$

$$\overset{512}{8,6\cancel{2}7}$$
$$- 1,48\cancel{2} \quad \text{8 is greater than 2.}$$
$$7,145 \leftarrow \text{difference}$$

Check: $1,482$
       $+ 7,145$
        $8,627$ ✔

ANSWER: 7,145

Arrange in columns. Subtract, starting from the right.

Regroup when necessary.

6 hundreds   2 tens = 5 hundreds   12 tens.

Check by adding.

---

2 From 94,500 subtract 93,264.

$$\overset{4\,9\,10}{94,500}$$
$$- 93,264$$
$$1,236$$

Check: $93,264$
       $+ 1,236$
        $94,500$ ✔

ANSWER: 1,236

5 hundreds = 4 hundreds   9 tens   10 ones.

---

3 Find the difference: $702 - 469$

$$702$$
$$- 469$$
$$233$$

Check: 469
       $+ 233$
        702 ✔

ANSWER: 233

4 Take 279 from 3,856.

$$3856$$
$$- 279$$
$$3577$$

Check: 279
       $+ 3577$
        3856 ✔

ANSWER: 3,577

---

5 Estimate the difference: $53,108 - 7,865$

$$53,108 \rightarrow 53,000$$
$$7,865 \rightarrow 8,000$$
$$45,000$$

ANSWER: 45,000

Round the numbers to the same place. Usually, this is the greatest place in the smaller number. Subtract.

---

6 Solve the equation: $925 - 467 = n$

$$925$$
$$- 467$$
$$458$$

Check: 467
       $+ 458$
        925 ✔

ANSWER: 458

To find the value of n, subtract the given numbers.

28

Vocabulary

The **minuend** is the number from which you subtract.
The **subtrahend** is the number you subtract.
The **difference** is the answer in subtraction.

DIAGNOSTIC TEST

Subtract and check:

1 28
  5

2 31
  6

3 76
 14

4 87
 68

5 32
 12

6 59
 51

7 50
 29

8 496
 266

9 562
 228

10 643
 367

11 465
 359

12 800
 698

13 604
 589

14 255
  37

15 4,965
 1,842

16 6,841
 2,327

17 7,653
 4,725

18 3,962
 1,487

19 5,000
 3,792

20 8,429
   64

21 85,924
 23,713

22 92,045
 21,924

23 68,247
 44,395

24 89,463
 15,684

25 70,571
 39,782

26 35,823
 19,341

27 45,936
    798

28 845,094
 384,276

29 4,575,000
 1,395,463

30 56,070 − 984

31 98,370 − 84,697

32 From 8,463 subtract 579.

33 Take 3,582 from 9,348.

34 Find the difference between 17,947 and 13,799.

35 Select the nearest given estimate for 50,943 − 18,499.
a. 20,000   **b.** 30,000   **c.** 40,000

36 Solve: $1,305 - 498 = n$

RELATED PRACTICE EXERCISES

Subtract and check:

1
1. 59
   4
2. 84
   2
3. 96
   3
4. 69
   5

2
1. 23
   8
2. 94
   6
3. 47
   9
4. 52
   3

3
1. 85
  61
2. 98
  26
3. 49
  15
4. 57
  23

4
1. 36
  19
2. 64
  25
3. 93
  58
4. 85
  37

5
1. 78
  48
2. 95
  35
3. 29
   9
4. 87
  17

6
1. 37
  34
2. 86
  82
3. 98
  90
4. 79
  72

7
1. 60
  36
2. 30
  23
3. 40
   8
4. 90
  65

8
1. 978
 642
2. 362
 150
3. 549
 325
4. 853
 721

9 1. 891   2. 582   3. 439   4. 918   **10** 1. 842   2. 516   3. 975   4. 923
         379   166   284   627   358   179   587   476

11 1. 428   2. 785   3. 964   4. 545   **12** 1. 609   2. 750   3. 500   4. 600
         218   380   357   138   392   273   307   485

13 1. 972   2. 784   3. 901   4. 735   **14** 1. 253   2. 467   3. 884   4. 573
         965   697   838   686   26   58   9   83

15 1. 8,654   2. 4,296   3. 6,582   4. 7,849
          2,431   3,003   4,172   3,812

16 1. 9,642   2. 2,938   3. 3,463   4. 9,185
          6,425   1,682   2,702   6,345

17 1. 7,194   2. 8,362   3. 5,774   4. 3,680
          3,457   4,554   2,968   2,962

18 1. 9,426   2. 8,335   3. 4,900   4. 5,371
          3,258   6,452   3,825   4,691

19 1. 8,324   2. 8,050   3. 6,000   4. 9,000
          5,975   4,584   2,505   8,427

20 1. 4,823   2. 3,058   3. 1,971   4. 3,000
             8   65   986   479

21 1. 38,762   2. 55,706   3. 45,383   4. 79,648
          26,521   52,304   32,162   36,136

22 1. 45,962   2. 79,614   3. 90,349   4. 52,767
          12,535   48,532   70,281   28,343

23 1. 37,424   2. 56,382   3. 76,200   4. 38,509
          15,258   42,491   41,350   32,754

24 1. 61,847   2. 54,000   3. 18,508   4. 79,406
          25,952   23,475   10,809   25,837

25 1. 97,437   2. 50,583   3. 62,471   4. 90,000
          28,769   48,794   42,587   82,575

26 1. 24,593   2. 57,924   3. 67,500   4. 29,070
          10,638   39,819   59,045   14,924

27 1. 65,925   2. 21,507   3. 86,500   4. 95,374
             386   4,742   95   8,298

28 1. 645,987   2. 705,961   3. 450,000   4. 532,654
          314,265   536,809   328,542   2,867

29   **1.** 8,935,299   **2.** 2,963,475   **3.** 7,694,251   **4.** 4,000,000
         <u>3,411,274</u>   <u>1,724,133</u>   <u>86,206</u>   <u>2,950,966</u>

30   **1.** 25 − 18   **2.** 645 − 382   **3.** 500 − 298   **4.** 4,563 − 3,275
        5. 5,694 − 4,987   **6.** 95,375 − 92,186 **7.** 636,059 − 278,770 **8.** 3,540,000 − 3,269,145

31   **1.** 63 − 8   **2.** 427 − 76   **3.** 3,788 − 909   **4.** 1,500 − 6
        5. 58,625 − 9,775 **6.** 32,743 − 814   **7.** 545,611 − 39,768   **8.** 750,000 − 7,050

32   **1.** From 300 subtract 152.   **2.** From 798 subtract 476.

        3. From 237 subtract 95.   **4.** From 3,489 subtract 2,183.

        5. From 2,000 take 405.   **6.** From 9,153 take 2,846.

        7. From 37,450 take 15,968.   **8.** From 84,073 take 28,536.

33   **1.** Take 63 from 324.   **2.** Take 574 from 1,582.
        3. Take 497 from 520.   **4.** Take 38 from 1,000.
        5. Subtract 633 from 1,981.   **6.** Subtract 200 from 1,582.
        7. Subtract 4,156 from 8,915.   **8.** Subtract 2,059 from 37,685.

34   Find the difference between:

        1. 942 and 368   **2.** 450 and 199
        3. 3,794 and 900   **4.** 45,836 and 21,862
        5. 68,593 and 40,500   **6.** 300,000 and 171,938
        7. 761,350 and 385,025   **8.** 5,000,000 and 2,750,000

35   For each of the following, select the nearest given estimate:

        1. 7,950 − 2,035   **a.** 4,000   **b.** 5,000   **c.** 6,000
        2. 5,831 − 1,928   **a.** 3,000   **b.** 4,000   **c.** 5,000
        3. 91,475 − 39,696 **a.** 50,000 **b.** 60,000 **c.** 70,000
        4. 40,237 − 2,869   **a.** 12,000 **b.** 23,000 **c.** 37,000

36   Solve by performing the indicated operation:

        1. 48 − 29 = ?   **2.** 174 − 87 = ▓   **3.** 2,586 − 1,248 = n **4.** 16,400 − 904 = n
        5. ? = 200 − 63 **6.** ▓ = 4,525 − 1,875 **7.** n = 3,600 − 950   **8.** n = 20,000 − 8,295

SKILL ACHIEVEMENT TEST

Subtract and check:

  1. 843   **2.** 9,407   **3.** 10,000   **4.** 782,651
      <u>695</u>   <u>8,308</u>   <u>994</u>   <u>591,826</u>

  5. Subtract as indicated: 63,412 − 58,057   **6.** From 9,206 take 8,056.
  7. Subtract 3,529 from 24,500.
  8. Find the difference between 79,835 and 78,953.
  9. For 6,147 − 4,922 select the nearest given estimate: **a.** 1,000   **b.** 2,000   **c.** 3,000
10. Solve by performing the indicated operation: 5,030 − 2,709 = n

BUILDING PROBLEM-SOLVING SKILLS

SUBTRACTION INDICATORS

- How many more
- How many fewer
- How much larger
- How much smaller
- Amount of increase
- Amount of decrease
- Difference
- Balance
- How many left over
- How much left over

FLASHBACK

1 READ

2 PLAN

3 SOLVE

4 CHECK

PROBLEM: The seating capacity at the new athletic field is 4,120 persons. The stands at the old field held only 1,984 persons. How many more people can now be seated?

1 **Given facts:** New field seats 4,120 persons. Old field seats 1,984 persons.

Question asked: *How many more* people can now be seated?

2 **Operation needed:** The words *how many more* indicate subtraction.

Think or write out your plan: The difference in seating capacities is equal to the increase in the number of seats.

Equation: $4{,}120 - 1{,}984 = n$

3 **Estimate the answer:** $4{,}120 \rightarrow 4{,}000$
$ 1{,}984 \rightarrow 2{,}000$

$4{,}000 - 2{,}000 = 2{,}000$ seats, estimate

Solution:
$4{,}120 - 1{,}984 = 2{,}136$ seats

ANSWER: 2,136 more people can be seated.

4 **Check the accuracy:**
$$\begin{array}{r} 1{,}984 \\ +\ 2{,}136 \\ \hline 4{,}120 \ \checkmark \end{array}$$

Compare answer to the estimate: The answer 2,136 seats compares reasonably with the estimate, 2,000 seats.

PRACTICE PROBLEMS

1. The enrollment at the Township High School decreased from 3,468 students to 2,849 students. What was the decrease in enrollment?

2. In the student association election, Joan received 1,206 votes and Mike, 978 votes. How many fewer votes did Mike receive?

3. How much larger is the state of Texas with an area of 688,617 square kilometers than the state of Rhode Island with an area of 3,232 square kilometers?

4. The distance from the moon to the earth is 384,393 kilometers and from the sun to the earth is 149,499,812 kilometers. What is the difference in distances?

MIXED PROBLEMS

Solve each of the following problems. For problems 1-4 select the letter corresponding to your answer.

1. A sales person's automobile mileage for the past year was 41,266 kilometers. If 11,291 kilometers represents pleasure driving, how many kilometers was the car driven for business purposes?
   a. 30,874 kilometers   **b.** 29,985 kilometers
   c. 29,875 kilometers   **d.** 29,975 kilometers

2. During the year Mr. Santos bought the following amounts of fuel oil to heat his house: 665 liters, 787 liters, 936 liters, 905 liters, 738 liters, and 946 liters. How many liters of fuel oil did he buy in all?
   a. 4,877 liters   **b.** 4,887 liters
   c. 4,977 liters   **d.** 5,067 liters

3. This year Washington High School presented three performances of their dramatic show. 1,058 persons saw the opening performance. On the succeeding two nights 993 and 1,196 persons attended. What was the total attendance at the school show?
   a. 3,147 persons   **b.** 3,247 persons
   c. 3,437 persons   **d.** Answer not given

4. How many fewer farms are there in the United States if the number of farms decreased from 5,859,169 to 2,730,242?
   a. 3,028,297 farms   **b.** 3,138,927 farms
   c. 3,027,927 farms   **d.** 3,128,927 farms

5. The total attendance for the year at all athletic events at the Amerigo Vespucci High School was 48,547. The attendance for the previous year was 39,498. Find the amount of increase in attendance.

6. Find the total area of the Pacific States if the area of California is 404,975 square kilometers; Oregon, 249,117 square kilometers; and Washington, 172,416 square kilometers.

Refresh Your Skills

The numerals in boxes indicate Lessons where help may be found.
1. Write 208,070,009 as a word statement.   2. Round 7,583,284 to the nearest thousand.
   1-1   1-3

First estimate, then solve each of the following:
3. Add: 45,836   4. Subtract: 21,673   5. Subtract: 965,425
   1-4   29,797   1 5   4,689   1-5   829,517
         75,086
         18,925
         56,498

1-6 Multiplication of Whole Numbers

PROCEDURE

1 Multiply: 641×6

$4 \times 6 = 24$ $\quad 641 \leftarrow$ **multiplicand** ⎱
$\qquad\qquad \underline{\times \; 6} \leftarrow$ **multiplier** ⎰ **factors**
$\qquad\qquad\;\; 3,846 \leftarrow$ **product**

ANSWER: 3,846

Arrange in columns. Multiply, starting from the right.
Regroup when necessary.
24 tens = 2 hundreds   4 tens.

2 Multiply 63 by 32.

$$\begin{array}{r} 63 \\ \underline{\times\ 32} \end{array}$$

$63 \times 2 \rightarrow \quad 126$ ⎱
$63 \times 3 \rightarrow \underline{\ 189}$ ⎰ **partial products**
$\qquad\qquad\quad 2,016$

Check: $\begin{array}{r} 32 \\ \underline{\times\ 63} \\ 96 \\ \underline{192} \\ 2,016 \end{array}$ ✔

Multiply by *each* digit in the multiplier to form partial products.
Place the right-hand digit of each partial product directly under the multiplier.
Add.
Check by interchanging the factors.

ANSWER: 2,016

3 Find the product: 681×205

$\begin{array}{r} 681 \\ \underline{\times\ 205} \\ 3405 \\ \underline{13620} \\ 139,605 \end{array}$ Check: $\begin{array}{r} 205 \\ \underline{\times\ 681} \\ 205 \\ 1640 \\ \underline{1230} \\ 139,605 \end{array}$ ✔

When one partial product is zero, the next partial product may be written on the same line.

ANSWER: 139,605

4 Multiply: $38 \times 4 \times 284$

$38 \times 4 = 152$
$\qquad\quad 152 \times 284 = 43,168$

ANSWER: 43,168

Multiply 2 factors.
Multiply the product by the remaining factor.

---

5 Estimate the product: $842 \times 6,559$

$842 \rightarrow 800$
$6,559 \rightarrow 7,000$
$800 \times 7,000 = 5,600,000$

ANSWER: 5,600,000

Round each factor to its greatest place. (Do not round 1-digit factors.)
Multiply.

---

6 Solve the equation: $6{,}297 \times 2{,}005 = n$

    6,297
  × 2,005
  ──────
    31485
  1259400
  ──────
  12,625,485

ANSWER: 12,625,485

To find the value of n, multiply the given numbers.

Vocabulary

The **multiplicand** is the number you multiply.
The **multiplier** is the number by which you multiply.
A **factor** is any of the numbers used in multiplication to form a product.
The **product** is the answer in multiplication.
A **partial product** is obtained by multiplying the multiplicand by any figure in the multiplier.

DIAGNOSTIC TEST   (multipliers—one digit)

Multiply and check:

1   23
     3

2   72
     4

3   24
     3

4   39
     7

5   231
     2

6   319
     3

7   874
     6

8   1,728
     9

9   34,267
     6

10   8
   2,532

11   60
     3

12   400
     5

13   302
     2

14   5,208
     8

15   3,000
     5

16   4,006
     7

17   50,800
     4

18 6×985

19 Multiply 48 by 9.
20 Find the product: 314×8

21 Select the nearest given estimate for 8×687:   **a.** 5,600   **b.** 4,800   **c.** 48,000
22 Solve: $5 \times 1{,}496 = n$

RELATED PRACTICE EXERCISES

Multiply and check:

1   **1.** 32   **2.** 11   **3.** 43   **4.** 21
      2   8   2   3

2   **1.** 63   **2.** 94   **3.** 52   **4.** 71
      3   2   3   6

3   **1.** 23   **2.** 28   **3.** 19   **4.** 15
      4   3   5   6

4   **1.** 48   **2.** 65   **3.** 64   **4.** 59
      6   8   4   9

5   **1.** 321   **2.** 542   **3.** 823   **4.** 912
      3   2   4   4

6   **1.** 217   **2.** 816   **3.** 191   **4.** 649
      4   5   6   2

7   **1.** 376   **2.** 593   **3.** 938   **4.** 697
      2   7   9   8

8   **1.** 2,143   **2.** 1,728   **3.** 5,914   **4.** 4,789
      2   4   5   9

9 **1.** 24,239 **2.** 49,663 **3.** 39,145 **4.** 84,574   **10** **1.** 7 **2.** 9 **3.** 3 **4.** 6
               4   7   6   8   13   537   2,182   72,589

11 **1.** 20 **2.** 60 **3.** 4 **4.** 50   **12** **1.** 700 **2.** 9 **3.** 200 **4.** 6
              8   5   70   9   3   900   5   800

13 **1.** 104 **2.** 2 **3.** 3,012 **4.** 34,032   **14** **1.** 405 **2.** 1,806 **3.** 8 **4.** 46,084
               2   304   3   2   4   7   9,072   5

15 **1.** 5,280   **2.** 9,500   **3.** 82,000   **4.** 6
               8   7   5   30,000

16 **1.** 2,004 × 2   **2.** 4,008 × 7   **3.** 80,005 × 6   **4.** 90,026 × 9
17 **1.** 3040   **2.** 6,080   **3.** 50,700   **4.** 40,030
               7   5   9   8

18 **1.** 8 × 21   **2.** 4 × 923   **3.** 6 × 709   **4.** 3 × 5,762
        5. 5 × 700   **6.** 8 × 5,030   **7.** 351 × 9   **8.** 605 × 4

19 Multiply:
        1. 26 by 4   **2.** 752 by 7   **3.** 6 by 401   **4.** 3,600 by 5
        5. 5,975 by 8   **6.** 5,000 by 3   **7.** 3,002 by 9   **8.** 16,000 by 5

20 Find the product:
        1. 144 × 9   **2.** 6 × 231   **3.** 5 × 320   **4.** 4 × 1,000
        5. 5,175 × 3   **6.** 9 × 2,446   **7.** 6,080 × 7   **8.** 25,050 × 6

21 For each of the following select the nearest given estimate:
        1. 4 × 79   **a.** 360   **b.** 320   **c.** 300
        2. 8 × 908   **a.** 7,000   **b.** 7,200   **c.** 7,500
        3. 9 × 3,844   **a.** 2,700   **b.** 27,000 **c.** 35,000
        4. 5 × 61,485   **a.** 66,000 **b.** 350,000 **c.** 300,000

22 Solve each of the following equations by performing the indicated operation:
        1. 3 × 93 = ?   **2.** 8 × 506 = ▓   **3.** 7 × 75 = n
        4. ? = 5 × 740   **5.** ▓ = 6 × 4,058   **6.** n = 9 × 8,762

SKILL ACHIEVEMENT TEST

Multiply and check:
1.   583   **2.**   4,078   **3.**   6
       ×   4   ×   9   × 50,708

4. 5 × 86   **5.** 7 × 290   **6.** 8 × 8,005   **7.** Multiply 50,900 by 3.
8. Find the product of 9 and 75,000.
9. For 8 × 7,955, select the nearest estimate: **a.** 64,000   **b.** 60,000   **c.** 56,000
10. Solve: 6 × 496 = n

DIAGNOSTIC TEST (multipliers—two or more digits)

Multiply and check:

1 37
24

2 78
56

3 485
92

4 6,948
89

5 45,847
65

6 36
9,967

7 592
231

8 6,342
358

9 16,959
786

10 4,574
1,728

11 74,686
9,743

12 38,457
75,962

13 8,500
54

14 700
500

15 208
144

16 693
907

17 5,009
69

18 40,603
28

19 8,001
306

20 6,080
705

21 384 × 597

22 36 × 407 × 743

23 Multiply 75 by 49.

24 Find the product: 144 and 24

25 Select the nearest given estimate for 308 × 49:
   a. 1,500 **b.** 15,000 **c.** 150,000

26 Solve: $85 \times 21 = n$

RELATED PRACTICE EXERCISES

Multiply and check:

1 **1.** 23
12
2. 28
25
3. 19
37
4. 53
14

2 **1.** 72
18
2. 93
27
3. 57
68
4. 74
96

3 **1.** 144
23
2. 526
42
3. 967
36
4. 897
88

4 **1.** 4,113
21
2. 6,374
35
3. 8,439
78
4. 1,728
93

5 **1.** 93,153
24
2. 38,642
85
3. 68,459
47
4. 84,696
63

6 **1.** 24
312
2. 53
659
3. 32
2,978
4. 25
68,426

7 **1.** 144
324
2. 975
638
3. 347
231
4. 886
697

8 **1.** 2,786
231
2. 149
9,687
3. 8,327
524
4. 4,784
379

9 **1.** 21,462
344
2. 78,356
492
3. 764
34,687
4. 15,647
989

10 1. 2,467   2. 2,462   3. 8,622   4. 9,675
          1,236   6,374   7,393   8,326

11 1. 12,345   2. 56,397   3. 3,846   4. 96,749
           1,728   8,457   86,798   8,795

12 1. 23,814   2. 49,279   3. 57,625   4. 91,352
          16,523   82,536   34,719   73,858

13 1. 50   2. 300   3. 6,000   4. 3,600   **14** 1. 59   2. 365   3.   640   4. 5,280
            62   27   24   375   30   200   3,000   15,000

15 1. 501   2. 603   3. 803   4. 3,205   **16** 1. 529   2. 69   3. 2,534   4. 3,842
             26   59   144   475   706   608   109   8,067

17 1. 4,009   2. 6,008   3. 9,005   4. 80,006   **18** 1. 2,050   2. 6,080   3. 60,506   4.   329
              87   365   2,183   176   16   53   125   7,070

19 1. 4,003   2. 3,009   3. 7,006   4. 20,058
            205   6,082   7,006   1,009

20 1. 6,080   2. 73,050   3. 46,050   4. 50,903
             203   8,007   2,500   60,704

21 1. 24×35 2. 87×832 3. $56 \times 4,973$ 4. 156×849
        5. $523 \times 6,942$ 6. $709 \times 52,723$ 7. 200×503 8. $4,520 \times 3,006$

22 1. $42 \times 37 \times 23$ 2. $30 \times 605 \times 178$ 3. $264 \times 671 \times 9$
        4. $829 \times 72 \times 485$ 5. $273 \times 100 \times 126$ 6. $3,000 \times 20 \times 50$
        7. $10 \times 6,020 \times 200$ 8. $845 \times 172 \times 674$

23 Multiply:
        1. 32 by 16   2. 50 by 29   3. 18 by 45   4. 296 by 83
        5. 314 by 52   6. 400 by 125   7. 693 by 834   8. 1,265 by 200

24 Find the product:

        1. 56 and 17   2. 253 and 85   3. 72 and 481   4. 953 and 249
        5. 405 and 840   6. 9,700 and 82   7. 762 and 300   8. 5,280 and 487

25 For each of the following select the nearest given estimate:

        1. 33×67 **a.** 1,800   **b.** 2,100   **c.** 2,400
        2. 519×806 **a.** 40,000   **b.** 400,000   **c.** 4,000
        3. $993 \times 5,784$ **a.** 600,000   **b.** 450,000   **c.** 6,000,000
        4. $1,893 \times 725$ **a.** 1,000,000   **b.** 1,200,000   **c.** 1,400,000

26 Solve each of the following equations by performing the indicated operation:

        1. $24 \times 89 = ?$ 2. $72 \times 125 = \blacksquare$ 3. $96 \times 310 = n$
        4. $? = 47 \times 18$ 5. $\blacksquare = 491 \times 163$ 6. $n = 50 \times 845$

SKILL ACHIEVEMENT TEST

Multiply and check:

1. $\begin{array}{r} 98 \\ \times\ 56 \end{array}$ **2.** $\begin{array}{r} 327 \\ \times\ 409 \end{array}$ **3.** $\begin{array}{r} 9{,}056 \\ \times\ \ \ 930 \end{array}$

Multiply as indicated:

4. 96 × 105 **5.** 749 × 328 **6.** 5,004 × 8,060

7. Multiply 4,900 by 85.
8. Find the product of 560 and 3,708.
9. First estimate the product 896 × 318, then multiply the given numbers.
10. Solve: 25 × 600 = n

Enrichment

Properties of Zero

I Zero added to any number is the number. $5 + 0 = 5$ $0 + 8 = 8$
II Zero subtracted from any number is the $6 - 0 = 6$
number.
III The difference between any number and $9 - 9 = 0$
itself is zero.
IV The product of a non-zero number and $3 \times 0 = 0$ $0 \times 8 = 0$
zero is zero.
V When zero is multiplied by zero, the prod- $0 \times 0 = 0$
uct is zero.
VI If the product of two numbers is zero, then
one of the factors is zero or both factors
are zero.
VII If zero is divided by any number other than $0 \div 2 = 0$
zero, the quotient is zero.
VIII In arithmetic the division by zero is excluded.

EXERCISES

Determine the value of each of the following:

1. 4 − 0 **2.** 10 − 10 **3.** 12 + 0 **4.** 0 ÷ 5 **5.** 0 × 0

Which of the following represent the number zero?

6. 15 ÷ 15 **7.** 6 − 6 **8.** 4 × 0 **9.** 1 × 18 **10.** 3 × 0 × 7

BUILDING
PROBLEM-SOLVING
SKILLS

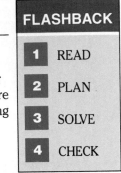

FLASHBACK

1 READ

2 PLAN

3 SOLVE

4 CHECK

PROBLEM: At an average speed of 685 kilometers per hour, how many kilometers does an airplane fly in 6 hours?

1 **Given facts:** Rate of speed is 685 kilometers per hour; Time of flight is 6 hours

Question asked: How many kilometers does the airplane fly?

2 **Operation needed:** The words *how many* with equal quantities indicate multiplication.

Think or write out your plan: The rate of speed times the time of flight is equal to the distance flown (number of kilometers).
Equation: $685 \times 6 = n$

3 **Estimate the answer:**

$685 \rightarrow 700$
$6 \rightarrow 6$
$700 \times 6 = 4,200$ kilometers, estimate

Solution:

$$\begin{array}{r} 685 \\ \times\ \ 6 \\ \hline 4,110 \text{ kilometers} \end{array}$$

ANSWER: The distance is 4,110 kilometers.

4 **Check the accuracy:**

$$\begin{array}{r} 6 \\ \times\ 685 \\ \hline 4,110 \ \checkmark \end{array}$$

Compare answer to the estimate: The answer 4,110 kilometers compares reasonably with the estimate, 4,200 kilometers.

PRACTICE PROBLEMS

1. If a box contains 144 envelopes, how many envelopes will there be in 36 boxes?

2. A merchant purchased 48 computer tapes at $29 each. How much was the total cost?

3. If 1 centimeter represents 12 kilometers, how many kilometers apart are two cities 12 centimeters apart on the chart?

4. Joan's average reading rate is 195 words per minute. How many words can she read in 30 minutes?

MIXED PROBLEMS

Solve each of the following problems. For problems 1-3 select the letter corresponding to your answer.

1. How far can a car go on a tankful of gasoline if it averages 6 kilometers on a liter and the tank holds 80 liters?
   a. 470 kilometers   **b.** 380 kilometers
   c. 480 kilometers   **d.** Answer not given

2. The present enrollment at the Township Junior High School is 849 pupils. There are 675 pupils registered at the senior high. What is the difference in their enrollments?
   a. 185 pupils   **b.** 164 pupils
   c. 174 pupils   **d.** Answer not given

3. Each of the 26 major baseball teams has 25 players on its team roster. How many major league players are there in all?
   a. 640 players   **b.** 650 players
   c. 550 players   **d.** 560 players

4. A 2-story factory building with office and showroom is available for rent. The office has 87 square meters of floor space; showroom, 224 square meters; first floor, 1,986 square meters; and second floor, 1,468 square meters. What is the total floor space of the building?

5. The area of the earth is 510,100,500 square kilometers. If there are 361,149,600 square kilometers of water, how many square kilometers of land are on the earth?

6. Find the winner of a 9-hole golf match (the smaller number of strokes wins):
   Jones, Carson H.S.   4|4|5|6|3|4|5|4|5 =
   Williams, Dalton H.S. 3|5|4|5|4|5|6|5|4 =

Refresh Your Skills

The numerals in boxes indicate Lessons where help may be found.

1. Write 19,207,485,000,000 in words. 1-1

2. Write the numeral naming: Four million, twenty-six thousand, seven hundred ninety-six 1-2

3. Write the complete numeral naming 75 billion. 1-2

4. Round 5,963,499,716 to the nearest million. 1-3

First estimate, then solve each of the following:

5. Add: 1-4
   9,832
   17,496
   50,957
   8,268
   594

6. Subtract: 1-5
   105,000
   98,634

7. Multiply: 1-6
   807
   609

8. Multiply: 1-6
   1,728
   853

EXPONENTS

An **exponent** tells how many times a number called the **base** is used as a factor. It is indicated by a small numeral written to the upper right of the factor (base).

$$5^2 \xleftarrow{} \begin{array}{l}\textbf{exponent}\\\textbf{base}\end{array}$$

5^2 represents $\underline{5} \times \underline{5}$ or the product, 25. It is read "five squared" or "five to the second power."

2^3 represents $\underline{2} \times \underline{2} \times \underline{2}$ or the product, 8. It is read "two cubed" or "two to the third power."

10^4 represents $\underline{10} \times \underline{10} \times \underline{10} \times \underline{10}$ or the product, 10,000. It is read "ten to the fourth power."

3^5 represents $\underline{3} \times \underline{3} \times \underline{3} \times \underline{3} \times \underline{3}$ or the product, 243. It is read "three to the fifth power."

Numbers like 10, 10^2, 10^3 10^4 10^5 10^6, ... are called **powers of 10**.

EXERCISES

Read, or write in words:

1. 3^4 **2.** 7^8 **3.** 2^{11}
4. 10^9 **5.** 6^1 **6.** 11^5
7. 5^3 **8.** 20^2

Write as a numeral:

9. Six to the eighth power   **10.** Nine cubed   **11.** Fifty squared

Name the base and exponent in each:

12. 8^7 **13.** 3^9 **14.** 11^4
15. 6^{11} **16.** 2^8 **17.** 21^{10}
18. 5^{12} **19.** 7^6

Use exponential form to write:

20. $2 \times 2 \times 2 \times 2$
21. $5 \times 5 \times 5 \times 5 \times 5 \times 5 \times 5 \times 5 \times 5 \times 5$
22. Express 36 as a power of 6.
23. Express 32 as a power of 2.

Find the square and cube of each of the following:

24. 6   **25.** 10   **26.** 9
27. 12   **28.** 30   **29.** 25

Find the value of each of the following:

30. 9^2 **31.** 6^4 **32.** 2^8
33. 3^7 **34.** $8^3 + 7^2$ **35.** $5^4 - 1^{10}$

1-7 Division of Whole Numbers

PROCEDURE

1 Divide 978 by 6.

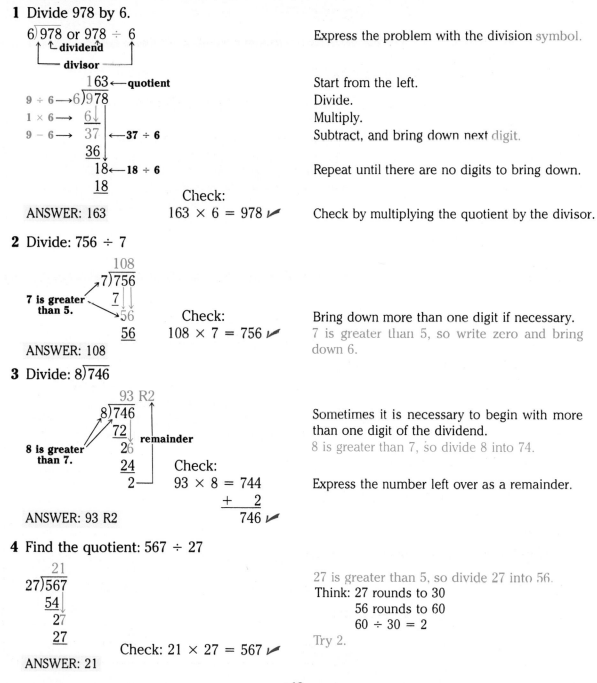

$6\overline{)978}$ or $978 \div 6$

↑ └ **dividend** ↑
└── **divisor** ──

Express the problem with the division symbol.

```
           163 ←quotient
9 ÷ 6 →6)978
1 × 6 →   6↓
9 − 6 →  37  ←37 ÷ 6
         36↓
          18 ←18 ÷ 6
          18
```

ANSWER: 163

Check:
163 × 6 = 978 ✓

Start from the left.
Divide.
Multiply.
Subtract, and bring down next digit.

Repeat until there are no digits to bring down.

Check by multiplying the quotient by the divisor.

2 Divide: 756 ÷ 7

```
          108
        7)756
7 is greater   7↓↓
than 5.       →56
          56
```

ANSWER: 108

Check:
108 × 7 = 756 ✓

Bring down more than one digit if necessary.
7 is greater than 5, so write zero and bring down 6.

3 Divide: 8)746

```
          93 R2
        8)746
          72↓
8 is greater  26  remainder
than 7.       24
          2
```

ANSWER: 93 R2

Check:
93 × 8 = 744
+   2
746 ✓

Sometimes it is necessary to begin with more than one digit of the dividend.
8 is greater than 7, so divide 8 into 74.

Express the number left over as a remainder.

4 Find the quotient: 567 ÷ 27

```
          21
       27)567
          54↓
          27
          27
```

Check: 21 × 27 = 567 ✓

ANSWER: 21

27 is greater than 5, so divide 27 into 56.
Think: 27 rounds to 30
       56 rounds to 60
       60 ÷ 30 = 2
Try 2.

5 Find the quotient: $89\overline{)45,034}$

$$
\begin{array}{r}
506 \text{ R3} \\
89\overline{)45,037} \\
\underline{44\ 5} \\
537 \\
\underline{534} \\
3
\end{array}
$$

Check:
$506 \times 89 = 45,034 + 3 = 45,037$ ✔

ANSWER: 506

89 is greater than 4 and greater than 45, so divide 89 into 450.
Think: 89 rounds to 90
        450 rounds to 450
        $450 \div 90 = 5$
Try 5.
Think: 537 rounds to 540
        $540 \div 90 = 6$
Try 6.

6 Find the quotient and remainder: $15,979 \div 438$

$$
\begin{array}{r}
36 \text{ R211} \\
438\overline{)15,979} \\
\underline{13\ 14} \\
2\ 839 \\
\underline{2\ 628} \\
211
\end{array}
$$

Check: $36 \times 438 = \begin{array}{r}15,768 \\ +211 \\ \hline 15,979 \end{array}$ ✔

ANSWER: 36 R211

438 is greater than 1, greater than 15, greater than 159, so divide 438 into 1,597.
Think: 438 rounds to 400
        1,597 rounds to 1,600
        $1600 \div 400 = 4$
Try 4. $438 \times 4 = 1,752$. The product is too great. Try 3.

Think: 2,839 rounds to 2,800
        $2,800 \div 400 = 7$
Try 7. $438 \times 7 = 3,066$. The product is too great. Try 6.

---

7 Estimate the quotient: $47,928 \div 61$

$$
\begin{array}{l}
61 \rightarrow 60 \\
47,928 \rightarrow 48,000 \\
48,000 \div 60 = 800
\end{array}
$$

ANSWER: 800

Round the divisor to its highest place.
(Do not round 1-digit numbers.)
Round the dividend so that it can be divided exactly by the rounded divisor.
Divide.

---

8 Solve the equation: $70,861 \div 165 = n$

$$
\begin{array}{r}
429 \text{ R76} \\
165\overline{)70,861} \\
\underline{66\ 0} \\
4\ 86 \\
\underline{3\ 30} \\
1\ 561 \\
\underline{1\ 485} \\
76
\end{array}
$$

Check:
$429 \times 165 = \begin{array}{r}70,785 \\ +76 \\ \hline 70,861 \end{array}$ ✔

ANSWER: 429 R76

To find the value of *n*, divide the given numbers.

Vocabulary

The **dividend** is the number you divide.
The **divisor** is the number by which you divide.
The **quotient** is the answer in division.
When the division is not exact, the number left is the **remainder**.

DIAGNOSTIC TEST   (divisors—one digit)

Divide and check:

1 $2\overline{)68}$ **2** $3\overline{)78}$ **3** $2\overline{)846}$ **4** $4\overline{)928}$

5 $6\overline{)834}$ **6** $7\overline{)427}$ **7** $6\overline{)552}$ **8** $4\overline{)8,448}$

9 $3\overline{)8,526}$ **10** $6\overline{)9,852}$ **11** $8\overline{)3,928}$ **12** $4\overline{)96,852}$

13 $7\overline{)36,533}$ **14** $3\overline{)6,900}$ **15** $4\overline{)804}$ **16** $6\overline{)642}$

17 $2\overline{)6,008}$ **18** $5\overline{)5,030}$

Find the quotient and remainder:

19 $7\overline{)156}$ **20** $4\overline{)6,923}$ **21** $6\overline{)935}$ **22** $8\overline{)4,668}$ **23** $7\overline{)843}$

24 Divide as indicated: $8,795 \div 5$ **25** Divide 8,870 by 9.

26 Select the nearest given estimate for $5,375 \div 9$ **a.** 700 **b.** 600 **c.** 500

27 Solve: $553 \div 7 = n$

RELATED PRACTICE EXERCISES

Divide and check:

1 1. $3\overline{)93}$ 2. $2\overline{)46}$ 3. $4\overline{)84}$ 4. $2\overline{)28}$

2 1. $2\overline{)34}$ 2. $3\overline{)87}$ 3. $5\overline{)85}$ 4. $6\overline{)96}$

3 1. $3\overline{)693}$ 2. $2\overline{)682}$ 3. $4\overline{)488}$ 4. $3\overline{)366}$

4 1. $6\overline{)846}$ 2. $4\overline{)496}$ 3. $7\overline{)798}$ 4. $8\overline{)968}$

5 1. $4\overline{)944}$ 2. $5\overline{)745}$ 3. $6\overline{)918}$ 4. $3\overline{)861}$

6 1. $6\overline{)246}$ 2. $9\overline{)729}$ 3. $2\overline{)128}$ 4. $3\overline{)216}$

7 1. $7\overline{)504}$ 2. $4\overline{)348}$ 3. $5\overline{)315}$ 4. $8\overline{)608}$

8 1. $2\overline{)6,448}$ 2. $4\overline{)8,484}$ 3. $3\overline{)9,636}$ 4. $2\overline{)2,486}$

9 1. $5\overline{)8,255}$ 2. $8\overline{)9,928}$ 3. $3\overline{)9,738}$ 4. $4\overline{)9,568}$

10 1. $3\overline{)7,491}$ 2. $4\overline{)9,156}$ 3. $6\overline{)8,748}$ 4. $5\overline{)7,490}$

11 1. $7\overline{)5,243}$ 2. $6\overline{)2,154}$ 3. $8\overline{)6,792}$ 4. $4\overline{)3,392}$

12 1. $3\overline{)69,876}$ 2. $4\overline{)95,860}$ 3. $2\overline{)76,952}$ 4. $6\overline{)85,794}$

13 1. $8\overline{)72,976}$ 2. $5\overline{)37,460}$ 3. $7\overline{)40,992}$ 4. $4\overline{)29,984}$

14 1. $4\overline{)840}$ 2. $2\overline{)9,680}$ 3. $3\overline{)3,600}$ 4. $6\overline{)90,000}$

15 1. $2\overline{)608}$ 2. $4\overline{)8,408}$ 3. $2\overline{)4,082}$ 4. $3\overline{)9,060}$

16 1. $5\overline{)525}$ 2. $8\overline{)8,416}$ 3. $6\overline{)67,218}$ 4. $2\overline{)81,106}$

17 1. $3\overline{)9,003}$ 2. $2\overline{)60,024}$ 3. $4\overline{)40,008}$ 4. $5\overline{)50,005}$

18 1. $3\overline{)6,012}$ 2. $6\overline{)24,036}$ 3. $4\overline{)80,032}$ 4. $2\overline{)10,010}$

Write quotient and remainder:

19 1. $5\overline{)14}$ 2. $3\overline{)553}$ 3. $7\overline{)27,645}$ 4. $4\overline{)98,622}$

20 1. $6\overline{)65}$ 2. $8\overline{)243}$ 3. $3\overline{)8,582}$ 4. $5\overline{)38,403}$

21 1. $9\overline{)35}$ 2. $5\overline{)74}$ 3. $6\overline{)4,657}$ 4. $8\overline{)89,749}$

22 1. $4\overline{)78}$ 2. $6\overline{)940}$ 3. $8\overline{)7,534}$ 4. $6\overline{)59,374}$

23 1. $6\overline{)365}$ 2. $8\overline{)644}$ 3. $3\overline{)812}$ 4. $7\overline{)85,264}$

24 1. $438 \div 3$ 2. $8,694 \div 2$
3. $69,376 \div 9$ 4. $94,500 \div 4$
5. $37,624 \div 8$ 6. $23,583 \div 7$
7. $84,215 \div 5$ 8. $91,862 \div 6$

25 1. Divide 932 by 4.   2. Divide 5,795 by 8.
3. Divide 3,482 by 9.   4. Divide 83,007 by 3.
5. Divide 94,075 by 2.   6. Divide 42,840 by 5.
7. Divide 51,382 by 6.   8. Divide 70,056 by 7.

26 For each of the following, select the nearest given estimate:
1. $584 \div 6$ **a.** 10   **b.** 100   **c.** 1,000
2. $9,105 \div 3$ **a.** 3,000   **b.** 5,000   **c.** 6,000
3. $27,491 \div 5$ **a.** 4,000   **b.** 6,000   **c.** 8,000
4. $40,963 \div 8$ **a.** 2,000   **b.** 4,000   **c.** 5,000

27 Solve each of the following by performing the indicated operation:
1. $69 \div 3 = ?$ 2. $468 \div 9 = $ ▓   3. $8,040 \div 5 = n$
4. $? = 6,894 \div 6$ 5. ▓ $= 9,738 \div 2$ 6. $n = 96,544 \div 7$

SKILL ACHIEVEMENT TEST

Divide and check:
1. $3\overline{)6,993}$ 2. $7\overline{)882}$ 3. $2\overline{)60,180}$

4. $6\overline{)4,584}$ 5. $9\overline{)77,445}$ 6. Write quotient and remainder: $6\overline{)98,405}$

7. Divide as indicated: $42,960 \div 5$ 8. Divide 65,424 by 8.

9. For $478 \div 5$ select the nearest estimate: **a.** 50   **b.** 100   **c.** 200

10. Solve: $5,313 \div 7 = n$

DIAGNOSTIC TEST   (divisors—two or more digits)

Divide and check:

1 $24\overline{)96}$ **2** $38\overline{)228}$ **3** $27\overline{)567}$ **4** $14\overline{)1,022}$

5 $26\overline{)8,164}$ **6** $85\overline{)71,995}$ **7** $48\overline{)61,584}$ **8** $54\overline{)349,272}$

9 $60\overline{)28,980}$ **10** $57\overline{)34,656}$ **11** $36\overline{)230,400}$ **12** $42\overline{)126,252}$

13 $63\overline{)27,044}$ **14** $72\overline{)45,018}$ **15** $144\overline{)864}$ **16** $174\overline{)7,482}$

17 $298\overline{)11,026}$ **18** $257\overline{)32,382}$ **19** $946\overline{)810,722}$ **20** $907\overline{)61,676}$

21 $506\overline{)405,306}$ **22** $843\overline{)552,660}$ **23** $400\overline{)338,000}$ **24** $391\overline{)185,805}$

25 $144\overline{)75,984}$ **26** $1,760\overline{)14,080}$ **27** $1,728\overline{)101,952}$ **28** $8,526\overline{)3,939,012}$

29 $5,280\overline{)23,876,160}$ **30** $8,005\overline{)4,306,690}$ **31** $21,714 \div 231$ **32** Divide 300,960 by 5,280

33 Select the nearest given estimate for: $80,526 \div 92$ **a.** 700   **b.** 800   **c.** 900

34 Solve: $2,184 \div 56 = n$

RELATED PRACTICE EXERCISES

Divide and check:

1   1. $16\overline{)80}$ 2. $23\overline{)92}$ 3. $27\overline{)81}$ 4. $19\overline{)76}$

2   1. $54\overline{)378}$ 2. $48\overline{)432}$ 3. $83\overline{)332}$ 4. $67\overline{)536}$

3   1. $32\overline{)896}$ 2. $48\overline{)672}$ 3. $12\overline{)324}$ 4. $24\overline{)984}$

4   1. $24\overline{)1,824}$ 2. $72\overline{)2,664}$ 3. $95\overline{)6,935}$ 4. $41\overline{)1,845}$

5   1. $37\overline{)4,551}$ 2. $26\overline{)8,918}$ 3. $17\overline{)3,366}$ 4. $52\overline{)9,672}$

6   1. $72\overline{)39,456}$ 2. $85\overline{)41,820}$ 3. $46\overline{)30,038}$ 4. $77\overline{)64,603}$

7 **1.** 29)70,035 **2.** 32)49,984 **3.** 17)48,892 **4.** 25)81,325

8 **1.** 43)224,417 **2.** 54)526,932 **3.** 64)212,416 **4.** 71)138,947

9 **1.** 30)5,670 **2.** 50)37,450 **3.** 70)61,040 **4.** 80)41,840

10 **1.** 36)25,380 **2.** 69)41,607 **3.** 37)29,933 **4.** 83)25,232

11 **1.** 15)3,300 **2.** 57)51,300 **3.** 30)144,000 **4.** 66)574,200

12 **1.** 12)24,048 **2.** 24)120,168 **3.** 69)276,207 **4.** 35)105,140

13 **1.** 56)48,297 **2.** 75)19,366 **3.** 48)36,923 **4.** 64)24,869

14 **1.** 14)2,412 **2.** 52)47,639 **3.** 24)58,819 **4.** 78)651,092

15 **1.** 128)896 **2.** 427)3,416 **3.** 640)1,920 **4.** 569)3,414

16 **1.** 173)8,996 **2.** 144)6,768 **3.** 234)5,382 **4.** 285)9,690

17 **1.** 229)10,992 **2.** 231)13,398 **3.** 929)52,024 **4.** 745)51,405

18 **1.** 152)95,608 **2.** 223)66,008 **3.** 475)75,525 **4.** 337)95,708

19 **1.** 863)650,702 **2.** 629)367,336 **3.** 847)337,953 **4.** 768)433,920

20 **1.** 208)7,696 **2.** 706)55,068 **3.** 405)25,110 **4.** 504)49,392

21 **1.** 106)31,906 **2.** 204)124,032 **3.** 903)367,521 **4.** 502)355,416

22 **1.** 478)152,960 **2.** 640)524,800 **3.** 679)549,990 **4.** 725)304,500

23 **1.** 200)85,200 **2.** 300)290,100 **3.** 700)408,800 **4.** 600)492,000

24 **1.** 692)365,489 **2.** 375)323,467 **3.** 589)433,704 **4.** 866)583,759

25 **1.** 320)149,520 **2.** 231)188,518 **3.** 144)41,088 **4.** 625)294,625

26 **1.** 5,280)42,240 **2.** 2,774)22,192 **3.** 8,304)74,736 **4.** 4,568)22,840

27 **1.** 6,080)255,360 **2.** 9,286)529,302 **3.** 3,974)337,790 **4.** 5,280)153,120

28 **1.** 8,366)3,906,922 **2.** 5,863)2,872,870 **3.** 1,728)1,627,776 **4.** 6,080)1,824,000

29 **1.** 2,240)12,286,400 **2.** 8,267)31,050,852 **3.** 2,794)12,802,108 **4.** 3,246)24,377,460

30 **1.** 7,006)4,581,924 **2.** 6,080)4,572,160 **3.** 3,600)2,109,600 **4.** 3,007)1,202,800

31 **1.** 408 ÷ 24 **2.** 93,940 ÷ 35 **3.** 173,712 ÷ 231 **4.** 401,280 ÷ 5,280

32 **1.** Divide 5,616 by 144. **2.** Divide 13,398 by 231.
3. Divide 153,600 by 320. **4.** Divide 1,487,200 by 1,760.

33 For each of the following select the nearest given estimate:

1. 3,120 ÷ 48 **a.** 40 **b.** 50 **c.** 60
2. 28,086 ÷ 93 **a.** 200 **b.** 300 **c.** 400
3. 71,865 ÷ 89 **a.** 700 **b.** 800 **c.** 900
4. 219,480 ÷ 415 **a.** 50 **b.** 500 **c.** 5,000

34 Solve each of the following by performing the indicated operation:

1. $72 \div 24 = ?$ **2.** $900 \div 50 = $ ▓   **3.** $7,344 \div 34 = n$

4. $? = 4,002 \div 87$ **5.** ▓ $= 8,295 \div 105$ **6.** $n = 216 \div 18$

SKILL ACHIEVEMENT TEST

Divide and check:

1. $29\overline{)667}$ **2.** $43\overline{)301}$ **3.** $48\overline{)74,976}$

4. $144\overline{)10,800}$ **5.** $805\overline{)329,245}$ **6.** Write quotient and remainder:
$82\overline{)37,981}$

7. Divide as indicated: $37,040 \div 40$.   **8.** Divide 342,720 by 2,240.

9. For $24,328 \div 58$ select the nearest estimate:
a. 250   **b.** 2,500   **c.** 400   **10.** Solve: $2,124 \div 36 = n$

Enrichment

Properties of One

I Any number multiplied by one (1) is the number.

II Any number, except zero, divided by itself equals one (1).

III One (1) raised to any power is one (1).

IV When one (1) is added to any whole number, the sum is the next higher whole number.

V When one (1) is added to any even number, the sum is an odd number. When one (1) is added to any odd number, the sum is an even number.

$1 \times 7 = 7 \quad 5 \times 1 = 5$

$60 \div 60 = 1$

$1^5 = 1 \times 1 \times 1 \times 1 \times 1 = 1$

EXERCISES

Determine the value of each of the following:

1. 1×6 **2.** 1^{10} **3.** $3 \div 3$ **4.** 8×1^5 **5.** $1^3 \div 1^2$

Which of the following represent the number 1?

6. $8 - 8$ **7.** 1^6 **8.** $7 - 0$ **9.** $12 \div 12$ **10.** $1 \times 1 \times 1$

BUILDING PROBLEM-SOLVING SKILLS

DIVISION INDICATORS

• How much is each
• How many in each
• Find the average
• Rate per unit

Note: All indicators generally refer to equal quantities.

FLASHBACK

1 READ

2 PLAN

3 SOLVE

4 CHECK

PROBLEM: Pam has taken out a car loan for $5,760. She is going to repay the loan in 36 equal monthly payments. How much is each monthly payment?

1 **Given facts:** Amount of loan is $5,760. Number of equal payments is 36.

Question asked: How much is each monthly payment?

2 **Operation needed:** The words *how much is each* with equal quantities indicate division.

Think or write out your plan: The amount of the loan divided by the number of equal monthly payments is equal to the monthly payment.

Equation: $5,760 \div 36 = n$

3 **Estimate the answer:**

$5,760 \rightarrow \$6,000

36 \rightarrow 40$

$6,000 \div 40 = \$150$, estimate

Solution:

$$36\overline{)5,760} \quad \$160$$
$$\underline{3\ 6}$$
$$2\ 16$$
$$\underline{2\ 16}$$

4 **Check the accuracy:**

$$\begin{array}{r} \$160 \\ \times\ \ 36 \\ \hline 960 \\ 480 \\ \hline \$5,760\ \checkmark \end{array}$$

Compare answer to the estimate: The answer $160 compares reasonably with the estimate, $150.

ANSWER: Each monthly payment is $160.

PRACTICE PROBLEMS

1. The AA Computer Store purchased 18 model F10 printers for $5,382. How much did each printer cost?

2. The total attendance at the Bartram High School for 21 days in March was 19,467. Find the average daily attendance.

3. Ms. Watson earns an annual salary of $18,468. How much does she earn each month?

4. In the last football game of the season, the Hartville High School team gained 168 yards from scrimmage in 28 tries. How many yards gained did the team average per try?

TWO-STEP PROBLEMS
Finding the Missing Fact

FINDING AVERAGES

The average of two or more quantities is equal to the sum of the quantities divided by the number of quantities.

PROBLEM: Maria scored 89, 74, 93, 79, and 70 in five arithmetic tests. Find her average score.

1 **Given facts:** The scores in five tests are: 89, 74, 93, 79, and 70.

Question asked: What is the average score?
    Think: To find the average, first you must find the sum of the scores.

Missing fact: The *sum* of the scores.

2 **Operations needed:** Step 1 The word *sum* indicates addition.
                          Step 2 The words *find the average* indicate division.

Think or write out your plan: The average score is equal to the sum (step 1) of the scores divided by (step 2) the number of scores.

Equation: $(89 + 74 + 93 + 79 + 70) \div 5 = n$

3 **Estimate the answer:**
Step 1   89 + 74 + 93 + 79 + 70
            ↓   ↓   ↓   ↓   ↓
            90 + 70 + 90 + 80 + 70 = 400
                80
Step 2   5)‾400‾   80, estimate
Solution:
Step 1   89 + 74 + 93 + 79 + 70 = 405
                81
Step 2   5)‾405‾

ANSWER: The average score is 81.

4 **Check the accuracy:**
    89
    74   81
    93   × 5
    79   405 ✓
  + 70
    405 ✓

Compare answer to the estimate: The answer 81 compares reasonably with the estimate 80.

PRACTICE PROBLEMS

1. What is the average temperature for the full day if the daytime temperature is 23°C and the nighttime temperature is 15°C?

2. The front four defensive linemen of the school football team weigh 91kg, 95kg, 102kg, and 88kg respectively. What is the average weight of the linemen?

3. Andrew can type an average of 48 words per minute. His report contains 2,136 words. Will he be able to type the report in 45 minutes?

4. Find the average number of points scored per game when the local basketball team scored 97 points in game 1, 89 points in game 2, and 108 points in game 3.

MIXED PROBLEMS

Solve each of the following problems. For problems 1-4 select the letter corresponding to your answer.

1. The deepest place thus far discovered in the world is 10,863 meters in the Marianas Trench near the island of Guam. The greatest depth in the Atlantic Ocean is 9,219 meters. Find the difference in these depths.
   a. 20,082 meters   **b.** 1,054 meters
   c. 1,644 meters   **d.** Answer not given

2. If the car averages 5 kilometers on a liter of gasoline, how many liters of gasoline are needed to drive from Los Angeles to San Francisco, a distance of 680 kilometers?
   a. 142 liters   **b.** 136 liters
   c. 132 liters   **d.** 685 liters

3. What is the total enrollment at the Southwest High School if there are 785 freshmen, 697 sophomores, 648 juniors, and 657 seniors?
   a. 2,697   **b.** 2,767
   c. 2,687   **d.** 2,787

4. How many sheets of writing paper are in 24 packages, each containing 500 sheets?
   a. 524 sheets   **b.** 24,500 sheets
   c. 12,000 sheets   **d.** Answer not given

5. How many people attended the Baltimore-Philadelphia 5-game World Series if 52,204 saw the first game, 52,132 the second, 65,792 the third, 66,947 the fourth, and 67,064 the fifth game?

6. How many more square kilometers of territory did the United States acquire by the Louisiana Purchase of 2,142,427 square kilometers than by the purchase of Alaska with 1,518,776 square kilometers?

7. An airplane flies from Atlanta to Seattle, a distance of 2,690 miles in 5 hours. What is the average speed of the airplane?

8. On her weekly science tests Maria received grades of 87, 79, 92, 64, 88, 75, and 96. Find her average grade in science.

Refresh Your Skills

The numerals in boxes indicate Lessons where help may be found.
1. Round 409,261 to the nearest hundred. [1-3]

First estimate, then solve each of the following:

2. Add: [1-4]
```
    916
  3,827
 54,296
  8,689
 26,778
```

3. Subtract: [1-5]
```
483,251
379,628
```

4. Multiply: [1-6]
```
6,080
  905
```

5. Divide:
```
875)619,500
```
[1-7]

TESTS FOR DIVISIBILITY

When 252 is divided by 3, the remainder is 0.

$$252 \div 3 = 84$$

Thus 3 divides 252 exactly or 252 is divisible by 3.

The following tests may be used to determine whether a given number is divisible by 2, 3, 4, 5, 6, 8, 9, or 10.

Divisor	Test	Examples
2	The number must be even. Its last digit must be 0, 2, 4, 6, or 8.	30 or 154 or 518 or 96 or 372
3	The sum of the digits must be divisible by 3.	For 834: $8 + 3 + 4 = 15$; $15 \div 3 = 5$ For 4,971: $4 + 9 + 7 + 1 = 21$; $21 \div 3 = 7$
4	The number named by the last two digits must be divisible by 4.	For 92,136: $36 \div 4 = 9$ For 7,992: $92 \div 4 = 23$
5	The last digit of the number must be 0 or 5.	430 3,645
6	The number must be even and the sum of its digits must be divisible by 3.	For 9,558: An even number $9 + 5 + 5 + 8 = 27$ $27 \div 3 = 9$
8	The number named by the last three digits must be divisible by 8. Numbers ending in 3 zeros are divisible by 8.	For 39,712: $712 \div 8 = 89$ For 153,000: Ends in 3 zeros
9	The sum of the digits must be divisible by 9.	For 50,382: $5 + 0 + 3 + 8 + 2 = 18$ $18 \div 9 = 2$
10	The last digit of the number must be 0.	80 9,670

EXERCISES

Determine whether the following numbers are divisible:

By 3:	**1.** 417	**2.** 2,853	**3.** 7,415	**4.** 29,538	**5.** 593,618
By 5:	**6.** 251	**7.** 8,970	**8.** 1,565	**9.** 38,006	**10.** 686,400
By 8:	**11.** 4,794	**12.** 1,352	**13.** 86,727	**14.** 75,000	**15.** 827,223
By 2:	**16.** 518	**17.** 6,359	**18.** 4,506	**19.** 83,000	**20.** 179,244
By 9:	**21.** 837	**22.** 5,706	**23.** 89,784	**24.** 40,389	**25.** 779,895
By 4:	**26.** 536	**27.** 9,252	**28.** 1,874	**29.** 92,425	**30.** 954,768
By 6:	**31.** 914	**32.** 8,349	**33.** 6,858	**34.** 27,291	**35.** 906,702
By 10:	**36.** 470	**37.** 3,000	**38.** 20,105	**39.** 90,790	**40.** 813,060

1-8 Factors; Greatest Common Factor (GCF)

PROCEDURE

1 Name all the factors of 24.

$$\begin{array}{r} 24 \\ 1\overline{)24} \end{array}$$ 1 divides 24 exactly, so 1 and 24 are factors.

Divide the given number by 1, 2, 3, etc. until the factors repeat.

$$\begin{array}{r} 12 \\ 2\overline{)24} \end{array}$$ 2 divides 24 exactly, so 2 and 12 are factors.

If the given number is divisible, then the divisor and quotient are a pair of factors.

$$\begin{array}{r} 8 \\ 3\overline{)24} \end{array}$$ 3 divides 24 exactly, so 3 and 8 are factors.

$$\begin{array}{r} 4 \text{ R4} \\ 5\overline{)24} \end{array}$$ 5 does not divide 24 exactly, no factors.

$$\begin{array}{r} 6 \\ 4\overline{)24} \end{array}$$ 4 divides 24 exactly, so 4 and 6 are factors.

$$\begin{array}{r} 4 \\ 6\overline{)24} \end{array}$$ 6 divides 24 exactly but repeats, so stop.

ANSWER: 1, 2, 3, 4, 6, 8, 12, and 24 are factors of 24.

---

2 Name the common factors of 54, 72, and 108.

54: 1 2 3 6 9 18 27 54

72: 1 2 3 4 6 8 9 12 18 24 36 72

108: 1 2 3 4 6 9 12 18 27 36 54 108

List all the factors for each given number.
Identify the factors that are common in each.

ANSWER: 1, 2, 3, 6, 9, and 18 are the common factors.

---

3 Name the greatest common factor (GCF) of 18 and 48.

18: 1 2 3 6 9 18

48: 1 2 3 4 6 8 12 16 24 48

List all the factors for each given number. Then, select the greatest common factor.

Alternate method—prime factorization (see page 57)

$18 = 3 \cdot 3 \cdot 2$
$48 = 3 \cdot 2 \cdot 2 \cdot 2 \cdot 2$ GCF $= 3 \cdot 2 = 6$

Factor each given number as a product of primes.

ANSWER: 6 is the greatest common factor.

The product of the common primes is the GCF

Vocabulary

A **common factor** of two or more whole numbers is any whole number that is a factor of all the given numbers.

The **greatest common factor (GCF)** of two or more whole numbers is the greatest whole number that will divide all the given numbers exactly.

DIAGNOSTIC TEST

1 Is 9 a factor of 63?

2 What is the second factor of a pair of factors when 6 is one factor of 96?

3 Name all the factors of 120. **4** Name the common factors of 56 and 84.

5 Name the common factors of 32, 80, and 128.

6 Name the greatest common factor of 27 and 72.

7 Name the greatest common factor of 60, 75, and 105.

RELATED PRACTICE EXERCISES

1 **1.** Is 6 a factor of 24? **2.** Is 5 a factor of 40?
3. Is 9 a factor of 56? **4.** Is 18 a factor of 90?

2 What is the second factor of a pair of factors when:
1. 7 is one factor of 21? **2.** 8 is one factor of 72?
3. 14 is one factor of 98? **4.** 15 is one factor of 135?

3 Name all the factors of each of the following:
1. 48 **2.** 75 **3.** 56 **4.** 132 **5.** 225 **6.** 400 **7.** 162 **8.** 112

4 Name the common factors of each pair of numbers:
1. 12 and 16 **2.** 63 and 84
3. 32 and 104 **4.** 105 and 150

5 Name the common factors of each group of numbers:
1. 24, 60, and 96 **2.** 39, 65, and 91
3. 36, 81 and 108 **4.** 252, 588, and 420

6 Name the greatest common factor of each pair of numbers:
1. 10 and 16 **2.** 12 and 40
3. 30 and 75 **4.** 52 and 78

7 Name the greatest common factor of each group of numbers:
1. 8, 12, and 18 **2.** 34, 85, and 102
3. 48, 80, and 112 **4.** 84, 144, and 360

SKILL ACHIEVEMENT TEST

1. Is 16 a factor of 192? **2.** Name all the factors of 180.
3. Name the common factors of 72, 90, and 144. **4.** Name the greatest common factor of 28 and 70.
5. Name the greatest common factor of 36, 108, and 156.

PRIME AND COMPOSITE NUMBERS

Whole numbers other than 0 and 1 may be classified as prime or composite numbers.

Any whole number greater than 1 that can be divided exactly by only the number itself and 1 is called a **prime number**.

23 can be divided exactly only by 23 and by 1.

23 is a prime number.

Any whole number greater than 1 that can be divided exactly by at least one whole number other than 1 and the number itself is called a **composite number.**

15 can be divided exactly not only by 15 and by 1

but also by 5 and by 3.

15 is a composite number.

EXERCISES

Which of the following are prime numbers?

1. 14   **2.** 23   **3.** 79   **4.** 51   **5.** 85   **6.** 97

Which of the following are composite numbers?

7. 81   **8.** 18   **9.** 2   **10.** 91   **11.** 53   **12.** 39

Write all the one-digit numerals naming:

13. Prime numbers   **14.** Even prime numbers   **15.** Odd prime numbers

Write the group of numerals naming:

16. All prime numbers greater than 18 and less than 32.

17. All composite numbers less than 70 and greater than 53.

18. Are all even numbers composite numbers? If not, name an even prime number.

19. Are all odd numbers prime numbers? If not, name an odd composite number.

20. Prime numbers that differ by 2 are called **twin primes**. For example 17 and 19 are twin primes. Find four pairs of twin primes between 25 and 75.

Goldbach, a mathematician, guesses that, "Any even number greater than 4 can be expressed as the sum of two odd prime numbers."

For each of the following even numbers, find two odd prime numbers whose sum equals the given numbers:

21. 8   **22.** 12   **23.** 36   **24.** 50   **25.** 108

26. Find an odd number that is the sum of two odd prime numbers.

PRIME FACTORIZATION

A composite number can be expressed as a product of prime numbers.

Factor 18 as a product of prime numbers.

By using a factor tree:

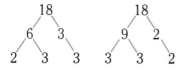

First find two factors of the given number. (Do not use the number 1.) Then continue to factor any factor that is a composite number until only prime factors result.

By division:

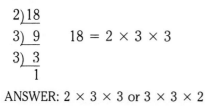

$18 = 2 \times 3 \times 3$

ANSWER: $2 \times 3 \times 3$ or $3 \times 3 \times 2$

Divide the given number and the resulting quotients successively by prime numbers that divide exactly until a quotient of 1 is obtained. The divisors are the prime factors.
Observe in the model how the quotients are brought down in each row.

NOTE:
While each composite number has only one group of prime factors, these factors may be arranged in different orders. As shown above, the complete or prime factorization of 18 is: $2 \times 3 \times 3$
but these factors may be written also as $3 \times 3 \times 2$ or $3 \times 2 \times 3$

EXERCISES

Factor each of the following numbers as a product of prime numbers:

1. 14   **2.** 54   **3.** 72   **4.** 135   **5.** 168   **6.** 600   **7.** 306

Factor each of the following numbers as a product of prime numbers using a factor tree.

8. 20   **9.** 64   **10.** 80   **11.** 48   **12.** 100   **13.** 144   **14.** 960

Career Applications

When merchandise—a radio, jacket, or other item—leaves the factory, it may be taken to a wholesaler's warehouse. Then it may go to a warehouse belonging to a chain of retail stores. Finally, it goes to one of the stores, where it may be sold to a consumer. The workers who keep track of merchandise in its journey from factory to consumer are called clerks. Shipping clerks pack items for shipment and prepare bills. Receiving clerks log in and distribute merchandise and deal with damaged items. Stock clerks count and mark items in a warehouse or store. Most clerks are high school graduates who can do arithmetic well. They get special training on the job. Some clerks use computers.

1. A receiving clerk receives 48 radios on Monday, 105 on Tuesday, and 74 on Wednesday. How many radios have been received?

2. There are 53 refrigerators in a warehouse. A shipping clerk ships 17 of them to a store. How many refrigerators are left in the warehouse?

3. A receiving clerk receives 415 lamps. Of these, 22 are damaged. How many of the lamps are not damaged?

4. A stock clerk counts 31 cases of calculators. There are 12 calculators in each case. How many calculators are there in all?

5. A shipping clerk ships 116 pairs of shoes to one store and 209 pairs to another store. How many pairs are shipped in all?

7. A store had 86 record players before a sale. During the sale, 49 were sold. How many are left?

9. A shipping clerk has an order for 200 chairs. Each truck can hold 40 chairs. How many trucks are needed?

6. A shipping clerk can pack 22 books in a carton. There is an order for 352 books. How many cartons are needed?

8. A receiving clerk receives 5 truckloads of pianos. There are 8 pianos in each truck. How many pianos are there in all?

10. A stock clerk counts basketballs of six different brands. The counts are 62, 41, 113, 27, 89, and 5. What is the total number of basketballs?

11. There are 832 ski jackets in a warehouse. A shipping clerk ships 288 of them to a store. How many ski jackets are left?

Before an order is shipped from a warehouse, a form, similar to the one below, must be completed by the shipping clerk to keep an accurate inventory of stock.

Description	ID Number	Number of Cartons	Number Per Carton	Quantity
Item A	001	12	12	**12.** ?
Item B	002	14	24	**13.** ?
Item C	003	20	10	**14.** ?
Item D	004	22	16	**15.** ?
Item E	005	40	8	**16.** ?
Item F	006	36	18	**17.** ?
TOTALS		**18.** ?		**19.** ?

Help the shipping clerk determine each of the following.

12-17. The quantity of each item

19. Total quantity of items

18. Total number of cartons

20. Number of delivery vans needed, if each van can carry 24 cartons

CHAPTER REVIEW

Vocabulary

addend p. 20
common factor p. 54
counting number p. 9
difference p. 29
dividend p. 45

divisor p. 45
factor p. 35
GCF p. 54
minuend p. 29
multiplicand p. 35

multiplier p. 35
natural number p. 9
partial product p. 35
period p. 9
product p. 35

quotient p. 45
remainder p. 45
subtrahend p. 29
sum p. 20
whole number p. 9

EXERCISES

The numerals in boxes indicate Lessons where help may be found.

Write each of the following numerals in words: 1-1

1. 4,998,281   **2.** 67,050,003   **3.** 2,000,083,000

Write the numeral naming each of the following: 1-2

4. Nine hundred six thousand, twenty-five

5. Seven million, eleven thousand, four hundred seventy-three

6. Sixteen billion, three hundred fifteen million, two hundred fifty thousand

Round: 1-3

7. 275,648 to the nearest hundred   **8.** 19,841,007 to the nearest million

9. 386,248,500 to the nearest hundred thousand   **10.** 7,009,989 to the nearest thousand

Add: 1-4

11. 5,894 **12.** 68,299 **13.** Find the sum of:
 7,579   85,067   462; 8,395; 1,984;
 6,658   3,188   26; 3,572
 8,784   59,083
 9,679   4,348

Subtract: 1-5

14. 8,392   **15.** 25,703
 5,487   9,698

16. 610,324   **17.** 7,800,000
 280,719   5,975,008

18. Subtract 438,794 from 647,600.

Multiply: 1-6

19. 48 **20.** 207 **21.** 896 **22.** 685 by 45
 93   900   759

Divide: 1-7

23. 67)3,953   **24.** 75)29,850

25. 4,609,440 ÷ 5,280 **26.** 561,660 by 138

Find all the factors of each of the following numbers. 1-8

27. 90   **28.** 104   **29.** 210   **30.** 320

Find the common factors of: 1-8

31. 40 and 72   **32.** 18, 27, and 45

Find the greatest common factor of: 1-8

33. 36 and 84   **34.** 81, 108, and 162

COMPETENCY CHECK TEST

The numerals in boxes indicate Lessons where help may be found.

Solve each problem and select the letter corresponding to your answer.

1. Write the numeral naming: Twelve thousand, six hundred forty 1-2
   a. 12,640 **b.** 1264 **c.** 12,600,40 **d.** Answer not given

2. Write 3,680,120 in words. 1-1
   a. Three million, six thousand eighty, one hundred twenty
   b. Three million, sixty eight thousand, one hundred twenty
   c. Three million, six hundred eighty thousand, one hundred twenty
   d. Three billion, six hundred eighty million, one hundred twenty thousand

3. Round 87,493,056 to the nearest hundred thousand. 1-3
   a. 87,000,000 **b.** 87,490,000 **c.** 87,500,000 **d.** 87,000,000,000

4. Add: 5,967 **a.** 33,465
   1-4   8,458 **b.** 33,565
         9,389 **c.** 32,456
         6,554 **d.** Answer
         3,197   not given

5. Subtract: 53,046 **a.** 21,057
   1-5   32,989 **b.** 20,157
                    c. 20,057
                    d. Answer
                         not given

6. Multiply: 798 × 907 1-6
   a. 723,786 **b.** 723,876 **c.** 732,786 **d.** 732,876

7. Divide: 72)43,848 1-7
   a. 69 **b.** 604 **c.** 609 **d.** 64

8. Find the greatest common factor of 28, 56, and 98. 1-8
   a. 14 **b.** 4 **c.** 2 **d.** 7

9. The Amazon River is 6,276 kilometers long and the St. Lawrence is 3,130 kilometers long. How many kilometers longer is the Amazon River? 1-5
   a. 3,246 km **b.** 3,156 km **c.** 3,256 km **d.** 3,146 km

10. A boat sailed from Panama to Jacksonville, a distance of 2,550 kilometers in 75 hours. What speed did it average for the trip? 1-7
    a. 36km/h **b.** 26km/h **c.** 34km/h **d.** Answer not given

11. Last year Joan's mother received $12,350 salary, $4,785 commission, and $495 bonus. Find the total amount of her earnings for the year. 1-4
    a. $17,630 **b.** $17,135 **c.** $18,225 **d.** $18,035

12. A merchant purchased 18 model R6 printers at $379 each. What is the total cost of the printers? 1-6
    a. $6,822 **b.** $7,032 **c.** $6,952 **d.** Answer not given

CUMULATIVE ACHIEVEMENT TEST

The numerals in boxes indicate Lessons where help may be found.

1. Write 4,010,001 in words. [1-1]

2. Write the numeral naming: Three million, seven hundred thousand [1-2]

3. Round 1,290,075,986 to the nearest million. [1-3]

4. Add: 68,426 [1-4]
      99,587
      42,686
      87,967
      95,989

5. Subtract: 724,005
   [1-5]   296,486

6. Multiply: 8,004
  [1-6]   609

7. Divide: 869)7,873,140
  [1-7]

8. Find the greatest common factor of 72, 108, and 180. [1-8]

9. How long will it take a person, driving at a speed of 85 kilometers per hour, to drive a distance of 1,615 kilometers? [1-7]

10. Find the increase in population in the Northeast region of the United States if recently the population was 49,135,283 and ten years earlier it was 49,060,514. [1-5]

11. At a speed of 299,851 kilometers per second, how far does light travel in 1 minute? [1-6]

12. Find the final score of the basketball game: [1-4]

	First Quarter	Second Quarter	Third Quarter	Fourth Quarter	Final Score
Kennedy High School	27	24	19	31	?
Lincoln High School	23	38	25	27	?

Calculator Skills

Use a calculator to check whether the following answers are correct or incorrect.

Add:

1.	**2.**	**3.**	**4.**	**5.**
846	2,984	83,455	91,548	397,426
5,863	5,979	6,908	28,307	586,979
9,596	4,898	56,568	45,279	8,395
688	3,419	9,675	74,486	65,074
2,599	9,382	63,398	29,158	858,987
19,592	26,762	219,004	268,778	1,916,861

Subtract:

6.	**7.**	**8.**	**9.**	**10.**
5,439	81,506	68,042	791,000	926,183
2,148	9,408	38,569	295,044	799,287
3,291	72,098	29,573	505,956	126,896

Multiply:

11. $326 \times 284 = 91,584$

12. $657 \times 469 = 308,133$

13. $5,309 \times 6,058 = 32,161,922$

14. $9,845 \times 7,699 = 75,696,655$

Divide:

15. $29,743 \div 49 = 67$

16. $45,105 \div 485 = 93$

17. $410,892 \div 706 = 592$

18. $785,519 \div 943 = 833$

Enrichment

Order of Operations

Numerical expressions may have more than one operation indicated.

To find the value of (evaluate) these expressions, from left to right, <u>multiply</u> and <u>divide</u> before you <u>add</u> and <u>subtract</u>.

Find the value of each of the following expressions:

1. $12 \times 7 \times 8$ **2.** $8 \times 9 \div 3$ **3.** $48 \div 6 \div 4$

4. $12 - 5 + 7 - 4$ **5.** $4 \times 5 + 9$ **6.** $18 \div 6 + 7$

7. $2 + 3 \times 8$ **8.** $5 + 12 \div 2$ **9.** $20 \div 5 - 3$

10. $7 \times 9 - 40$ **11.** $45 - 30 \div 5$ **12.** $54 - 4 \times 8$

13. $8 \times 9 + 24 \div 2$ **14.** $96 \div 4 - 5 \times 4$ **15.** $12 \times 9 + 6 \times 7$

16. $84 \div 7 + 63 \div 9$ **17.** $6 \times 40 - 30 \div 10 + 8$ **18.** $25 - 54 \div 6 + 7 \times 2$

19. $90 \div 9 + 9 - 3 \times 5$ **20.** $48 - 8 \div 4 + 6 \times 7 - 2$

Computer Activities

The computer program below is written in the BASIC computer language. It is designed to help you practice your addition skills. Enter the program on your computer keyboard exactly as it appears, one line at a time. Press the RETURN or ENTER key at the end of each line.

10 Clearscreen	Clears the screen. (Use appropriate command for *your* microcomputer.)
20 LET X = INT(1000*RND(1))	Formulas which store "random" (unpredictable) numbers in locations X and Y in the computer's memory.
30 LET Y = INT(1000*RND(1))	
40 IF Y > X THEN 30	Guarantees that first number is always greater than or equal to the second.
50 LET Z = X + Y	Computes the sum of X + Y and stores it in Z.
60 PRINT X; "+"; Y	Prints out the problem for you on your computer.
70 INPUT A	Displays a "?" as a way of asking for your answer.
80 IF A = Z THEN PRINT "YES"	Say "YES" if your answer is correct.
90 IF A < > Z THEN PRINT "NO"	Says "NO" if your answer is incorrect.
100 GOTO 20	Sends the computer back to line 20 for another example.

(Note: TRS-80 and IBM users replace RND(1) in lines 20 and 30 with RND(0)).

If you make a typing mistake:
  (a) Use the ← key to back up and make corrections.
     or
  (b) Simply press RETURN or ENTER and retype the line.

EXERCISES

1. Type RUN and press RETURN or ENTER to start your program. Use the program until you have achieved mastery with your addition skills.

2. In BASIC, the following symbols are used for the arithmetic operations:

+ addition	− subtraction	* multiplication

  a) Change the program to help you practice *subtraction*.
     HINT: Only two lines need to be retyped. Look carefully at the program and
        decide *which* two.

  b) Change the program to help you practice *multiplication*. Again, only two lines
     need to be changed.

CHAPTER 2 Decimals

INTRODUCTION TO DECIMALS

Our number scale is extended to the right of the ones place to express parts of one. Since the decimal number scale is based on tens, the value of each place on the number scale is one-tenth of the value of the next place to the left.

   The first place to the right of the ones place has the value of one-tenth of one whole unit. It is indicated as .1 and expresses "tenths." The dot is called the decimal point.

   When a unit is divided into ten equal parts, the size of each equal part is one-tenth and thus is expressed as .1 and read "one-tenth."

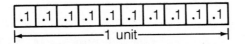

The second place to the right of the ones place expresses hundredths (.01).
The third place expresses thousandths (.001).
The fourth place expresses ten-thousandths (.0001).
The fifth place expresses hundred-thousandths (.00001).
The sixth place expresses millionths (.000001).

$3 \times (1)$	decimal point	$9 \times (.1)$	$6 \times (.01)$	$4 \times (.001)$	$2 \times (.0001)$	$8 \times (.00001)$	$7 \times (.000001)$
3	.	9	6	4	2	8	7

   A decimal is a fractional number that is named by a numeral using place value such as .83 or .0259.

   A mixed decimal is a number containing a whole number and a decimal such as 16.5 where a decimal point is used to separate the whole number from the fractional part.

2-1 Reading Decimals

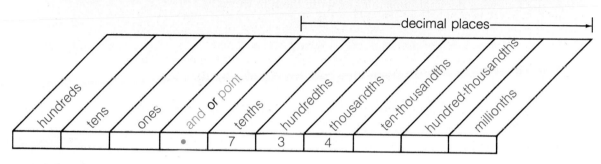

PROCEDURE

1 Read .734

.⌊734⌋

seven hundred thirty-four thousandths

ANSWER: Seven hundred thirty-four thousandths.

Read the digits as for a whole number. Then name the place value of the last digit.

2 Read 0.5291

A zero may be written in the ones' place to prevent the decimal point from being overlooked.

0.⌊5291⌋

five thousand two hundred ninety-one ten-thousandths

ANSWER: Five thousand two hundred ninety-one ten-thousandths.

3 Read 0.005 in two ways.

⌊0.005⌋

five thousandths, or point zero zero five

ANSWER: Five thousandths, or point zero zero five.

Decimals may also be read by naming each digit separately after saying the word point.

4 Read 84.06

⌊84⌋.⌊06⌋

eighty-four and six hundredths

ANSWER: Eighty-four and six hundredths.

For mixed decimals, read the word and for the decimal point.

5 Read 634.368 in two ways.

⌊634.368⌋

six hundred thirty-four and three hundred sixty-eight thousandths,
or six hundred thirty-four point three six eight

ANSWER: Six hundred thirty-four and three hundred sixty-eight thousandths,
or Six hundred thirty-four point three six eight.

DIAGNOSTIC TEST

Read the following decimal numerals (or write them in words):

1 .2   **2** .06   **3** 0.58

4 1.5   **5** 3.73   **6** .004

7 0.076   **8** .289

9 14.708   **10** .0037

11 0.17925   **12** .000456

13 Read (or write) in two ways: 129.4261

14 .836

15 24.78

RELATED PRACTICE EXERCISES

Read the following decimal numerals (or write them in words):

1 1. .8   **2** 1. .03   **3** 1. .24   **4** 1. 1.6
       2. .1   2. .07   2. .85   2. 2.9
       3. 0.5   3. 0.02   3. 0.91   3. 38.5
       4. 0.4   4. 0.08   4. 0.60   4. 126.4

5 1. 2.51   **6** 1. .005   **7** 1. .024   **8** 1. .832
       2. 7.37   2. .008   2. .063   2. .946
       3. 89.03   3. 0.001   3. 0.080   3. 0.253
       4. 248.19   4. 0.007   4. 0.092   4. 0.798

9 1. 6.005   **10** 1. .0007   **11** 1. .00006   **12** 1. .000001
       2. 21.769   2. .0089   2. .00392   2. .000534
       3. 186.528   3. 0.0574   3. 0.09413   3. 0.080076
       4. 200.042   4. .9350   4. .72815   4. 0.175283

13 1. 8.0025
       2. 23.9317
       3. 49.08329
       4. 571.05875

Read (or write in words) in two ways:

14 1. .93   2. .619   **15** 1. 36.89   2. 17.6
       3. 0.2   4. 0.3428   3. 5.925   4. 87.4662

SKILL ACHIEVEMENT TEST

Read (or write in words) each of the following:

1. .09   **2.** 6.8

3. 0.014   **4.** 19.57

5. 0.256   **6.** .1038

7. 0.83421   **8.** 600.005

Enrichment

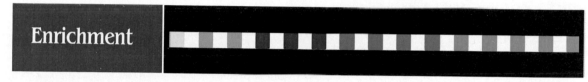

Number Symbols

After people learned to count, they developed number symbols so that they could make permanent records. Clay tablets have been found containing "cuneiform" number symbols which were written more than 5,000 years ago by the Babylonians. The British Museum has in its possession an arithmetic manuscript on a roll of papyrus, said to have been written about 4,000 years ago by Ahmes, an Egyptian.

The Babylonians used repeated wedge-shaped symbols. ▼◀

The Egyptians used different grouping symbols.

| (1)   ∩ (10)   ⊋(100)   ⚘(1000)   ⌐(10,000)

⌐⊃ (100,000) and ⚱ (1,000,000)

The Mayans used combinations of dots and horizontal bars. ⋮̿

The Romans used letter symbols.

I (1)   V (5)   X (10)   L (50)   C (100)   D (500)   M (1,000)

The Chinese used their traditional number symbols.

一 (1)   二 (2)   三 (3)   四 (4)   五 (5)   六 (6)   七 (7)   八 (8)   九 (9)   十 (10)

百 (100)   千 (1000)

The Greeks and Hebrews used the letters of their respective alphabets.

2-2 Writing Decimals

PROCEDURE

1 Write fifty-two hundredths as a decimal.

fifty-two hundredths
↓
.52
↑
hundredths place

ANSWER: .52

Write the digits as for a whole number. Then place the decimal point so that the last digit has the stated place value.

2 Write seven tenths as a decimal.

seven tenths
↓
0.7 or .7

ANSWER: 0.7 or .7

Decimals may also be written with a zero in the ones' place.

3 Write point four nine two as a decimal.

point four nine two
↓   ↓   ↓   ↓
.   4   9   2

ANSWER: .492

Write a decimal point for the word point.

4 Write four hundred and three hundred fifty-three thousandths as a decimal.

four hundred and three hundred fifty-three thousandths
↓   ↓   ↓
400   .   353

ANSWER: 400.353

Write a decimal point for the word and.

5 Write sixty-two point zero four five as a decimal.

sixty-two point zero four five
↓   ↓   ↓   ↓   ↓
62   .   0   4   5

ANSWER: 62.045

6 Write seventy-six and eight ten-thousandths as a decimal.

seventy-six and eight ten-thousandths
↓   ↓   ↓
76   . 000   8
↑
Write 3 zeros.

ANSWER: 76.0008

If necessary, write enough zeros to the right of the decimal point.

DIAGNOSTIC TEST

Write each of the following as a decimal:
1 Three tenths
2 Four and eight tenths
3 Seven hundredths
4 Twenty-five hundredths
5 One hundred nine and eighty-four hundredths
6 Nine thousandths
7 Ninety-four thousandths
8 Five hundred twenty-seven thousandths
9 Seven hundred and ninety-three thousandths
10 Four thousand six hundred thirty-six ten-thousandths
11 Eight hundred forty-two hundred-thousandths
12 Four millionths
13 Two hundred sixty and three hundred forty-seven ten-thousandths
14 Point eight three   15 Five hundred nineteen point nine

RELATED PRACTICE EXERCISES

Write each of the following as a decimal numeral.

1
1. Four tenths
2. Eight tenths
3. Two tenths
4. One tenth

2
1. Six and five tenths
2. Nine and seven tenths
3. Five and one tenth
4. Twenty and six tenths

3
1. Eight hundredths.
2. Two hundredths
3. Four hundredths
4. Five hundredths

4
1. Thirty-six hundredths
2. Fifty-seven hundredths
3. Seventeen hundredths
4. Forty-nine hundredths

5
1. Six and four hundredths
2. Seventy-three and eighteen hundredths
3. Two hundred and five hundredths
4. Four hundred seven and twenty-five hundredths

6
1. Three thousandths
2. Seven thousandths
3. Five thousandths
4. One thousandth

7
1. Sixty-nine thousandths
2. Forty-seven thousandths
3. Sixteen thousandths
4. Eighty-three thousandths

8
1. Two hundred seventy-four thousandths
2. Four hundred thirty-nine thousandths
3. Seven hundred twenty-one thousandths
4. Three hundred six thousandths

9 **1.** Two and seventeen thousandths
  2. Eight hundred and thirty-five thousandths
  3. Thirty-six and two hundred fifty-three thousandths
  4. Nineteen and one hundred twenty-two thousandths

10 **1.** Eight ten-thousandths
  2. Thirty-six ten-thousandths
  3. Four hundred ninety-four ten-thousandths
  4. Three thousand five hundred sixty ten-thousandths

11 **1.** Three hundred-thousandths
  2. Forty-two hundred-thousandths
  3. Four hundred fifty-six hundred-thousandths
  4. Six thousand eight hundred twenty-two hundred-thousandths

12 **1.** Six millionths
  2. Ninety-five millionths
  3. Three hundred seven millionths
  4. Seventy-two thousand one hundred forty-nine millionths

13 **1.** Five hundred and fifty-eight ten-thousandths
  2. Seventy and two thousand five hundred twelve ten-thousandths
  3. Eighty-five and seven hundred thirty-two hundred-thousandths
  4. Six hundred forty-three and sixty-seven millionths

14 **1.** Point seven
  2. Point four nine
  3. Point zero six eight
  4. Point five one zero nine

15 **1.** Sixty-one point two
  2. Zero point six four one
  3. Eight hundred eighty point one six one three
  4. Three thousand ten point seven nine

SKILL ACHIEVEMENT TEST

Write each of the following as a decimal numeral:

1. Seventy-nine thousandths

2. Fifty and four tenths

3. Twelve and ninety-two hundredths

4. One thousand thirty-eight ten-thousandths

2-3 Rounding Decimals

PROCEDURE

1 Round .26 to the nearest tenth.

.26

↑
└─tenths

.26

↑
└─6 is greater than 5.

.3

ANSWER: .3

Locate the place to be rounded.

If the digit to the right is 5 or more, round to the next higher digit. Omit all digits to the right of the place to be rounded.

2 Round .83249 to the nearest thousandth.

.83249

.83249

↑
└─4 is less than 5.

.832

ANSWER: .832

If the digit to the right is less than 5, write the same digit.

Omit all digits to the right of the place to be rounded.

3 Round 15.964 to the nearest tenth.

15.964

15.964

↑
└─Adding 1 to the tenths' place changes the ones' place.

16.0

ANSWER: 16.0

Sometimes the digit to the left of the place to be rounded is increased by one.

4 Round $4.9782 to the nearest cent.

$4.9782

$4.9782

$4.98

ANSWER: $4.98

5 Round 73.568 to the nearest whole number.

73.568

73.568

74

ANSWER: 74

DIAGNOSTIC TEST

Round to:
Nearest tenth:
1 .68
2 5.425

Nearest thousandth:
5 1.9685
6 3.24728

Nearest hundred-thousandth:
9 .005924
10 54.632857

Nearest cent:
13 $2.386

Nearest whole number:
15 57.18

Nearest dollar:
17 $9.84

Nearest hundredth:
3 8.349
4 .5146

Nearest ten-thousandth:
7 .42769
8 26.58634

Nearest millionth:
11 .0000985
12 2.0500302

14 $4.843

16 89.562

18 $136.42

RELATED PRACTICE EXERCISES

Round to:
Nearest tenth:
1 1. .25
2. .87
3. .984
4. 2.39

2 1. .14
2. .32
3. 1.83
4. 5.205

Nearest thousandth:
5 1. .2146
2. .3998
3. 3.0815
4. 16.76876

6 1. .5452
2. .9293
3. 2.0084
4. 24.82509

Nearest hundred-thousandth:
9 1. .000083
2. .005142
3. 1.079324
4. 83.591243

10 1. .000079
2. .030585
3. 2.000467
4. 23.075306

Nearest hundredth:
3 1. .517
2. .308
3. 5.845
4. 15.2263

4 1. .323
2. .934
3. 6.544
4. 8.3025

Nearest ten-thousandth:
7 1. .20585
2. .932473
3. 4.00619
4. 43.020471

8 1. .34143
2. .528129
3. 18.781304
4. 39.56372

Nearest millionth:
11 1. .0000029
2. .0001538
3. .0254985
4. 1.0038296

12 1. .0000052
2. .0000194
3. .0039481
4. 5.0006373

Nearest cent:

13 1. $.267
2. $0.59
3. $1.316
4. $5.8682

14 1. $.643
2. $.834
3. $6.572
4. $14.9615

Nearest whole number:

15 1. 6.3
2. 19.4
3. 2.09
4. 34.28

16 1. 9.6
2. 68.5
3. 100.81
4. 399.704

Nearest dollar:

17 1. $5.69
2. $37.90
3. $420.71
4. $19.53

18 1. $2.46
2. $53.17
3. $269.39
4. $84.25

SKILL ACHIEVEMENT TEST

Round to:

1. Nearest hundredth: .568

2. Nearest tenth: .294

3. Nearest thousandth: 1.6939

4. Nearest millionth: 6.0726142

5. Nearest whole number: 18.7

6. Nearest dollar: $100.83

7. Nearest cent: $4.595

8. Nearest ten-thousandth: .096077

9. Nearest hundred-thousandth: 2.361784

10. Nearest cent: $24.491

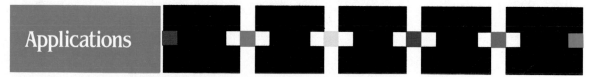

Applications

Round the following conversion factors correct to the nearest hundredth:

1. 1 nautical mile = 1.1515 statute miles
2. 1 meter = 1.0936 yards
3. 1 kilogram = 2.2046 pounds

Round correct to the nearest tenth:

4. 1 centimeter = .3937 inch
5. 1 bushel = 2,150.42 cubic inches
6. 1 cubic inch = 16.3872 cubic centimeters

Round correct to the nearest thousandth:

7. 1 statute mile = .8684 nautical mile
8. 1 square inch = 6.4516 square centimeters
9. 1 liquid quart = .9463 liter

2-4 Addition of Decimals

PROCEDURE

1 Add: .82 + .57

```
 |82 ⟵ addends
+|57 ⟵
 | 9
```

```
   1 ⟵
  .82        .8 + .5 = 1.3
+ .57
 1.39 ⟵ sum
```

Check:
```
   .82 ↑
 + .57 |
  1.39 ✔
```

ANSWER: 1.39

Arrange in columns. Line up the decimal points. Add, starting from the right.

Regroup when necessary.
13 tenths = 1 one   3 tenths

Place the decimal point in the sum directly below the decimal points in the addends. Check by adding up.

2 Add .7 to 3.46

```
   1
  3.46
+  .70
  4.16
```

Check:
```
  3.46 ↑
+  .70 |
  4.16 ✔
```

ANSWER: 4.16

Write ending zeros so both addends have the same number of decimal places.
11 tenths = 1 one   1 tenth

3 Add: 46 + .927

```
  46.000
+   .927
  46.927
```

Check:
```
  46.000 ↑
+   .927 |
  46.927 ✔
```

ANSWER: 46.927

When adding whole numbers and decimals, place a decimal point after each whole number.

4 Add: .26 + .38 + .06

```
  .26
  .38
+ .06
 .70 or .7
```

Check:
```
  .26 ↑
  .38 |
+ .06 |
 .70 or .7 ✔
```

ANSWER: .70 or .7

When a decimal ends in one or more zeros to the right of the decimal point, the zeros may be dropped.

---

5 Find the sum: .523; 4.16; 3; and 2.2

```
   .523
  4.160
  3.000
+ 2.200
  9.883
```

Check:
```
   .523 ↑
  4.160 |
  3.000 |
  2.200 |
  9.883 ✔
```

ANSWER: 9.883

6 Estimate the sum: .71; 9.052; 47.53

$$.71 \rightarrow .7$$
$$9.052 \rightarrow 9.1$$
$$47.53 \rightarrow 47.5$$
$$\overline{ 57.3}$$

Round the numbers to the same place. Usually, this is the greatest place in the smallest number.
Add.

ANSWER: 57.3

---

7 Solve the equation: $20.56 + $14.27 = n

$$\begin{array}{r} \overset{1}{\$20.56} \\ +\ \ 14.27 \\ \hline \$34.83 \end{array}$$

Check: $\begin{array}{r} \$20.56 \\ 14.27 \\ \hline \$34.83 \end{array}$ ✔

To find the value of n, add the given numbers.

ANSWER: $34.83

DIAGNOSTIC TEST

Add:

1
.4
.3
.2

2
.6
.9
.4

3
.02
.01
.03

4
.08
.03
.02
.06

5
.15
.03
.48
.17

6
.52
.43
.69
.74

7
.85
.10
.05

8
1.5
2.1
4.3

9
6.4
8.9
4.4
3.5

10
3.06
4.09
2.08

11
4.57
2.93
4.87

12
7.49
3.5

13
2.86
.7
.12

14
8
.05

15
2.103
4.839
3.542

16
50.48
37.59
23.84

17
326.04
183.75
225.39
491.26

18
19.47
8.46
592.75
74.81
126.78
91.33

19
$.50
.28
.79
.84
.67

20
$42.85
9.74
223.30
54.67
7.98

21 .08 + 1.5

22 .752 + 4.53 + 6

23 $1.43 + $.89 + $5.07 + $.36 + $9.58

24 Find the sum of: 6.4, .976, and 2.87

25 Select the nearest given estimate for:
.493 + .78 + .6
a. 1.5   **b.** 2.0   **c.** 2.5

26 Solve: .13 + .7 = n

RELATED PRACTICE EXERCISES

Add:

1
1. .2
   .6

2. .3
   .2
   .2

3. 0.2
   0.3
   0.1
   0.2

4. .1
   .3
   .2
   .1
   .2

2
1. .9
   .7

2. .5
   .8
   .4

3. 0.4
   0.8
   0.7
   0.4

4. .5
   .7
   .4
   .8
   .3

3
1. .03
   .04

2. .03
   .02
   .03

3. 0.01
   0.03
   0.02
   0.03

4. .03
   .01
   .02
   .01
   .02

4
1. .06
   .09

2. .05
   .05
   .07

3. 0.02
   0.07
   0.04
   0.05

4. .09
   .06
   .08
   .04
   .09

5
1. .68
   .26

2. .57
   .04
   .18

3. 0.12
   0.23
   0.45
   0.16

4. .23
   .18
   .29
   .03
   .25

6
1. .43
   .89

2. .38
   .45
   .22

3. 0.82
   0.59
   0.77
   0.31

4. .27
   .74
   .51
   .93
   .89

7
1. .53
   .47

2. .48
   .26
   .16

3. 0.52
   0.81
   0.49
   0.68

4. .17
   .08
   .49
   .27
   .99

8
1. 2.4
   3.5

2. 6.2
   9.3
   8.3

3. 4.1
   2.2
   1.1
   9.2

4. 9.2
   3.2
   4.3
   2.1
   7.1

9
1. 7.8
   6.5

2. 4.1
   1.5
   7.4

3. 2.8
   1.4
   8.2
   7.5

4. 8.5
   4.9
   6.3
   7.4
   5.6

10
1. 8.05
   4.08

2. 9.08
   8.03
   1.07

3. 6.01
   8.07
   4.05
   2.09

4. 6.08
   2.03
   9.07
   4.09
   7.03

11
1. 3.26
   2.15

2. 6.27
   2.83
   5.76

3. 2.19
   3.82
   5.27
   1.38

4. 3.86
   2.29
   4.57
   5.18
   1.33

12
1. 6.25
   4.6

2. 6.74
   9.3
   1.87

3. 5.7
   4.3
   9.25
   8.75

4. 4.92
   8.1
   9.6
   2.43
   1.8

13
1. .9
   5.28

3. 1.4
   .7
   .29
   2.45

2. 9.5
   .33
   .19

4. 5.27
   7.4
   .35
   3.82
   .06

14
1. .6
   4.

3. 3.9
   .49
   6.
   .57

2. .94
   7.
   .8

4. .36
   4.68
   .2
   9.
   1.7

15
1. 5.01
   2.999

3. 1.516
   6.24
   .006
   4.518

2. 3.728
   3.517
   9.282

4. 4.783
   1.829
   5.318
   2.175
   9.384

16

1. 20.56
14.27

2. 36.87
8.26
15.84

3. 28.45
42.83
56.19
18.28

4. 32.28
14.85
9.74
98.42
.94

17

1. 138.35
253.42

2. 639.32
182.08
24.19

3. .00532
.13847
.32296
.48325

4. 335.48
618.37
407.54
250.36
195.83

18

1. .08
.57
.13
.42
.83
.66

2. 1.16
2.65
4.27
1.04
3.53
5.21

3. 16.91
28.38
19.43
37.21
4.82
.48

4. 596.74
289.57
193.86
319.29
475.43
628.58

19

1. $.24
.59
.37
.73
.85

2. $.83
.25
.03
.62
.54

3. $.36
.88
.41
.29
.37

4. $.84
.27
.49
.56
.99
.78

20

1. $2.33
.96
4.28
.51
9.16

2. $8.25
3.70
9.64
8.23
4.72

3. $45.63
5.96
.83
14.71
9.43
18.65

4. $355.95
109.82
481.56
247.49
575.37
281.25

21
1. .3 + .6
5. .47 + .1785
2. .05 + .12
6. 18 + .32
3. 3.6 + 5.1
7. .9 + 4
4. .275 + .38
8. .017 + 15

22
1. .17 + .38 + .53
3. .83 + 7 + 4.45 + .049
5. 1.5 + 0.18 + 6.84 + 0.016 + 0.27
7. 4.92 + 1.853 + 9.7 + 60 + 42.6
2. .52 + 1.6 + 8.26
4. .06 + 1.2 + 46 + 3.825 + .075
6. .83 + .37 + 4 + 3.9 + .051
8. 1.4 + 26 + .39 + 5.98 + 9

23
1. $.23 + $.75
3. $1.80 + $2.60 + $4.25
5. $0.60 + $0.38 + $1.50 + $3.25 + $9.70
7. $.74 + $1.60 + $.99 + $4.88 + $.04
2. $.17 + $.49 + $.83
4. $2.75 + $3.35 + $.96 + $1.45
6. $100 + $8.42 + $93.75 + $6.83 + $.14
8. $3.42 + $6.51 + $12.54 + $9.49 + $8.68

Find the sum of:

24
1. 4.23, 6.832, and 4.4
3. .6, 8, and .24
5. 2.05, .156, 4.69, and .08
7. $12.59, $9.47, $1.27, $.56, and $3.46
2. 8.01, .684, and 5.9
4. 1.4, 3.8, and .87
6. $7.26, $.85, $.42, $.94, and $2.35
8. $8.41, $2.25, $2.50, $7.64, and $.85

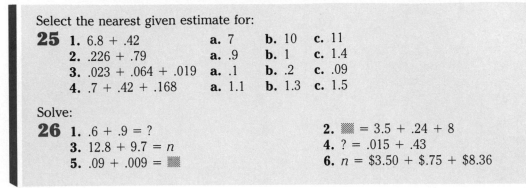

Select the nearest given estimate for:

25
1. 6.8 + .42   **a.** 7   **b.** 10   **c.** 11
2. .226 + .79   **a.** .9   **b.** 1   **c.** 1.4
3. .023 + .064 + .019   **a.** .1   **b.** .2   **c.** .09
4. .7 + .42 + .168   **a.** 1.1   **b.** 1.3   **c.** 1.5

Solve:

26
1. .6 + .9 = ?   2. ▓ = 3.5 + .24 + 8
3. 12.8 + 9.7 = n 4. ? = .015 + .43
5. .09 + .009 = ▓   6. n = $3.50 + $.75 + $8.36

SKILL ACHIEVEMENT TEST

Add:

1. .4	**2.** 2.08	**3.** $.46	**4.** $ 8.51
.9	3.04	.87	49.93
.3	9.06	.54	.89
.5	3.02	.39	256.36
		.68	7.47

5. .07 + 8   **6.** 5.83 + 18.2 + .457   **7.** $8.75 + $14.80 + $5.96 + $.64 + $10.48
8. Find the sum of: .09, 16.6, and 8.45
9. Select the nearest given estimate for 5.6 + .48   **a.** 6   **b.** 5   **c.** .5
10. Solve: .68 + .2 = n

Enrichment

Our Number Symbols

We use number symbols of the Hindu-Arabic System. These symbols were introduced in Europe during the 12th century and are now used generally throughout the world. They were brought to the United States by the colonists. In this system there are 10 symbols: 0, 1, 2, 3, 4, 5, 6, 7, 8, and 9. Each of these number symbols is called a digit or figure.

The number symbols we write to represent numbers are not numbers but names for numbers, which we call numerals. Number is an abstract idea. We cannot see or write a number.

Although it is usually called a number, technically 25 is a numeral or group of number symbols which represent the number named twenty-five. A number may be represented by symbols in many ways; it has many names. The number named twenty-five may be represented by 25, 13 + 12, or 5^2, etc. It should be noted however, that number symbols are generally used to denote both numerals and numbers, as "Add the numbers 35 and 54" and "Write the numeral 796."

BUILDING
PROBLEM-SOLVING
SKILLS

FLASHBACK

1 READ

2 PLAN

3 SOLVE

4 CHECK

PROBLEM: During the five times it rained in August, 1.02 inches, 2 inches, .79 inch, .4 inch, and 1.6 inches fell. What was the total rainfall for the month?

1 **Given facts:** 1.02 inches, 2 inches, .79 inch, .4 inch, and 1.6 inches of rain fell.

Question asked: What was the total rainfall?

2 **Operation needed:** The word *total* with *unequal* quantities indicates addition.

Think or write out your plan: The sum of 1.02 inches, 2 inches, .79 inch, .4 inch, and 1.6 inches of rain is equal to the total rainfall.

Equation: $1.02 + 2 + .79 + .4 + 1.6 = n$

3 **Estimate the answer:**

$1.02 \rightarrow 1$	
$2 \;\;\; \rightarrow 2$	
$.79 \rightarrow 1$	
$.4 \;\; \rightarrow 0$	
$1.6 \;\; \rightarrow \underline{2}$	
6 inches, estimate	

Solution:

```
  1.02
  2.00
   .79
   .40
+ 1.60
  5.81 inches
```

ANSWER: The total rainfall was 5.81 inches.

4 **Check the accuracy:**

```
  1.02
  2.00
   .79      Add.
   .40
  1.60
  5.81 ✔
```

Compare answer to the estimate: The answer 5.81 inches compares reasonably with estimate 6 inches.

PRACTICE PROBLEMS

1. Jim has checks of $21.68, $45.90, and $73.45 to deposit. What is his total deposit?

2. Barbara can run a lap in 41.2 seconds. Susan took 6.9 seconds more to run the same lap. How long did it take Susan?

3. The school orchestra bought a violin for $169.95, a saxophone for $249.75, and a trumpet for $124.50. What is the total cost of the new instruments?

4. A mutual fund reported a dividend of 8.954 shares to a stockholder who already owns 490.573 shares. Find the new total of shares owned by the stockholder.

MICROMETER CALIPER READING

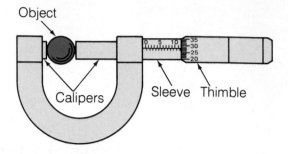

A micrometer caliper is an instrument used to measure very small lengths and thicknesses. The sleeve contains two scales: the top sleeve scale is marked in units of one millimeter (1 mm); the bottom sleeve scale is marked in units of one-half millimeter (.5 mm). The thimble scale contains 50 equal divisions, each representing one hundredth millimeter (.01 mm).

To make a micrometer reading, place an object between the calipers, as shown above, and turn the thimble to tighten the calipers against the object being measured. Add the readings on the three scales.

EXAMPLE: Find the total micrometer reading.

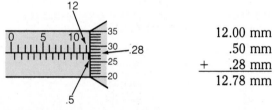

$$
\begin{array}{r}
12.00 \text{ mm} \\
.50 \text{ mm} \\
+ \quad .28 \text{ mm} \\
\hline
12.78 \text{ mm}
\end{array}
$$

The total micrometer reading is 12.78 millimeters.

EXERCISES

Find the total micrometer reading for each object.

Object	Top-Sleeve Reading	Bottom-Sleeve Reading	Thimble Reading	Total Reading
1. A	5 mm	.5 mm	.14 mm	
2. B	8 mm	.5 mm	.29 mm	
3. C	14 mm	.5 mm	.48 mm	
4. D	6 mm	.0 mm	.06 mm	
5. E	11 mm	.5 mm	.37 mm	

2-5 Subtraction of Decimals

PROCEDURE

1 Subtract: .512 − .421

.512 ← **minuend**
− .421 ← **subtrahend**   Check: .421
　　1   + .091
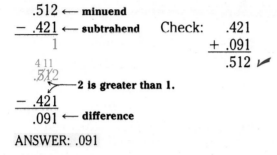

4 11
.5̸1̸2
　　↱ **2 is greater than 1.**
− .421
.091 ← **difference**

ANSWER: .091

Arrange in columns. Line up the decimal points. Subtract, starting from the right.

Regroup when necessary.
5 tenths 1 hundredth = 4 tenths 11 hundredths
Place the decimal point in the difference directly below the decimal points in the given numbers. Check by adding.

2 From 10.6 subtract 5.524.

　5910
10.6̸0̸0̸   Check: 5.524
− 5.524   + 5.076
　5.076   10.600 = 10.6 ✓

ANSWER: 5.076

Write ending zeros so both numbers have the same number of decimal places.
6 tenths = 5 tenths 9 hundredths 10 thousandths
For a whole number, write a decimal point and ending zeros.

3 Find the difference: 5.61 − .9

4 16
5̸.6̸1   Check: .90
− .90   + 4.71
4.71   5.61 ✓

ANSWER: 4.71

4 Subtract: $20 − $6.38

1 9 910
$2̸0̸.0̸0̸   Check: $ 6.38
−   6.38   + 13.62
$13.62 $20.00 ✓

ANSWER: $13.62

---

5 Estimate the difference: 2.892 − .763

2.892 →   2.9
.763 → −   .8
　　　　　 2.1

ANSWER: 2.1

Round the numbers to the same place. Usually, this is the greatest place in the smaller number. Subtract.

---

6 Solve the equation: 18.5 − 9.647 = n

18.500   Check: 9.647
− 9.647   + 8.853
　8.853   18.500 ✓

ANSWER: 8.853

To find the value of n, subtract the given numbers.

DIAGNOSTIC TEST

Subtract:

1 .8
.2

2 .38
.24

3 .67
.48

4 .05
.02

5 .86
.79

6 .57
.37

7 4.6
3.1

8 5.4
2.8

9 6.5
2.5

10 .679
.398

11 3.72
1.95

12 15.8
3.9

13 .6593
.4978

14 83.452
49.596

15 .34617
.14596

16 .4583
.2783

17 .85328
.84793

18 6.531
5.975

19 8.000
1.742

20 3.56
.8

21 .9
.735

22 5.
1.43

23 4.2
.372

24 $176.27
93.48

25 .15 − .08

26 .375 − .2

27 .7 − .625

28 4.2 − .83

29 9 − .05

30 $16 − $1.50

31 Subtract .08 from .3

32 Select the nearest given estimate for:
8.2 − .82 **a.** 0 **b.** 16 **c.** 7

33 Solve: .9 − .06 = *n*

RELATED PRACTICE EXERCISES

Subtract:

1
1. .9
.5
2. .4
.3
3. .6
.1
4. .7
.2

2
1. .29
.03
2. .73
.32
3. .98
.65
4. .64
.21

3
1. .36
.18
2. .44
.27
3. .25
.09
4. .83
.36

4
1. .08
.01
2. .09
.03
3. .07
.05
4. .09
.06

5
1. .85
.76
2. .54
.45
3. .25
.17
4. .13
.08

6
1. .58
.28
2. .46
.36
3. .25
.05
4. .92
.42

7
1. 8.4
4.2
2. 4.7
2.5
3. 5.6
3.2
4. 9.5
4.4

8
1. 7.5
4.7
2. 9.3
1.9
3. 4.6
2.8
4. 6.1
4.3

9
1. 7.6
2.6
2. 9.2
6.2
3. 2.8
1.8
4. 8.4
4.4

10
1. .835
.214
2. .967
.378
3. .076
.043
4. .749
.729

11
1. 7.64
3.53
2. 9.68
5.95
3. 4.29
1.76
4. 9.01
7.84

12
1. 18.5
6.2
2. 17.3
5.8
3. 24.5
3.5
4. 42.7
5.9

13 **1.** .9355
.8492

2. .1327
.1219

3. 60.07
42.38

4. 375.3
190.4

14 **1.** .25683
.14974

2. 34.645
17.859

3. 849.54
258.46

4. 5,986.3
2,894.7

15 **1.** .5844
.2837

2. 6.2937
4.2843

3. .83572
.43564

4. 43.596
35.589

16 **1.** .7856
.3256

2. 5.8362
3.7362

3. .92041
.52041

4. 82.475
17.275

17 **1.** .6848
.6827

2. .04965
.04894

3. .00325
.00255

4. .03741
.03659

18 **1.** 4.6
3.8

2. 5.83
4.96

3. 9.786
8.895

4. 7.0352
6.9446

19 **1.** 4.000
1.753

2. 9.000
4.068

3. 6.0000
5.9325

4. 24.000
13.307

20 **1.** 1.4
.9

2. 2.5
.7

3. 8.65
.8

4. 5.0046
4.307

21 **1.** .36
.034

2. .78
.1561

3. .6
.49

4. .4
.003

22 **1.** 8
7.3

2. 3
1.2

3. 7
2.84

4. 6
4.005

23 **1.** 5
.6

2. 9
.09

3. 3
.753

4. 4.9
.807

24 **1.** $4.85
2.60

2. $1.36
.89

3. $36.80
17.42

4. $25.00
16.68

5. $193.45
108.91

6. $200.00
127.18

7. $476.13
85.75

8. $1,250.00
975.80

Subtract as indicated:

25 **1.** .3 − .2
2. .8 − .3
3. .49 − .25
4. .38 − .09
5. .536 − .008
6. .837 − .325
7. .0045 − .0023
8. .1534 − .0976

26 **1.** .45 − .4
2. .97 − .6
3. .384 − .2
4. .518 − .46
5. .039 − .01
6. .8056 − .74
7. .5842 − .095
8. .0034 − .003

27 **1.** .6 − .45
2. .9 − .83
3. .4 − .176
4. .35 − .285
5. .06 − .043
6. .2 − .1356
7. .685 − .5903
8. .006 − .0005

28
1. 3.6 − .24
2. 4.7 − .83
3. 23.4 − 1.75
4. 5.82 − .004
5. 12.54 − 1.054
6. 81.2 − 4.18
7. 3.4 − .0056
8. 2.875 − .375

29
1. 8 − .5
2. 2 − 1.7
3. 1 − .16
4. 3 − 1.38
5. 6 − .09
6. 7 − 3.0625
7. 10 − .375
8. 12 − 8.2003

30
1. $.50 − $.25
2. $.78 − $.09
3. $2.74 − $.86
4. $10 − $5.60
5. $15.42 − $9
6. $200 − $133.62
7. $446.58 − $279.49
8. $3,500 − $2,938.75

31
1. From .8 subtract .35
2. From 45.9 take 6.38
3. From $2.84 take $1.75
4. Subtract .004 from .08
5. Take .45 from 1.5
6. Take $8.40 from $12
7. Find the difference between .03 and .003
8. Find the difference between 4.81 and .481

32 Select the nearest given estimate for:
1. 17.8 − 3.1   **a.** 13   **b.** 14   **c.** 15
2. .61 − .074   **a.** .08   **b.** .5   **c.** .6
3. $128.50 − $51.97   **a.** $70   **b.** $80   **c.** $90
4. .9 − .09   **a.** .01   **b.** .8   **c.** 0

33 Solve:
1. .8 − .2 = ?
2. 4.5 − .3 = ▨
3. .085 − .08 = n
4. ? = $75.20 − $9.60
5. ▨ = 15 − .625
6. n = $3.64 − $.97

SKILL ACHIEVEMENT TEST

Subtract:

1. $10.61	2. 8	3. .9724	4. 15.3
9.68	1.9	.3632	8.3

5. .865 − .092   6. .071 − .02   7. 24.1 − 5.57
8. Take $3.25 from $20
9. Select the nearest given estimate for $6.19 − $2.88
    a. $3   **b.** $4   **c.** $5
10. Solve .8 − .65 = n

BUILDING PROBLEM-SOLVING SKILLS

FLASHBACK

1 READ

2 PLAN

3 SOLVE

4 CHECK

PROBLEM: The winning automobile speed for the Indianapolis 500 was 162.117 m.p.h. and for the Daytona 500 it was 155.979 m.p.h. How much faster was the Indianapolis winning speed?

1 **Given facts:** Winning automobile speeds: Indianapolis 500: 162.117 m.p.h.
Daytona 500: 155.979 m.p.h.

Question asked: How much faster is the Indianapolis winning speed?

2 **Operation needed:** The words *how much faster* indicate subtraction.

Think or write out your plan: The difference in the two speeds is equal to the number of m.p.h. faster.

Equation: $162.117 - 155.979 = n$

3 **Estimate the answer:**
$162.117 \rightarrow 162$ m.p.h.
$155.979 \rightarrow \underline{156}$ m.p.h.
$ 6$ m.p.h., estimate

Solution:
$ 162.117$ m.p.h.
$\underline{- 155.979}$ m.p.h.
$ 6.138$ m.p.h.

ANSWER: 6.138 m.p.h. faster.

4 **Check for accuracy:**
$ 155.979$ m.p.h.
$\underline{+ 6.138}$ m.p.h.
$ 162.117$ m.p.h. ✔

Compare answer to the estimate: The answer 6.138 compares reasonably with the estimate 6 m.p.h.

PRACTICE PROBLEMS

1. John bought a baseball glove that regularly sold for $27.49 at a reduction of $4.98. How much did he pay?

2. The Dow Jones stock average opened at 1,104.63 and closed at 1,096.45. How many points did it lose?

3. Selma bought a set of weights for $10.79. How much change should she receive from a $20 bill?

4. The World Trade Center in New York is 411.48 meters tall. The Sears Tower in Chicago is 443.18 meters tall. How many meters taller is the Sears Tower?

MIXED PROBLEMS

Solve each of the following problems. For problems 1-3 select the letter corresponding to your answer.

1. The Dow Jones stock averages opened at 885.57 and during the day gained 3.46 points. What was the closing stock average?
 a. 889.03   **b.** 888.93
 c. 898.93   **d.** 988.93

2. A girl ran the 100-meter dash in 15.4 seconds, then later ran the same distance in 13.7 seconds. How many seconds less did she take the second time?
 a. 2.7 seconds   **b.** 1.7 seconds
 c. 2.3 seconds   **d.** 29.1 seconds

3. Marilyn bought her graduation outfit. Her dress cost $49.98; shoes, $21.49; hat, $15.75; and bag, $16.95. How much did she spend?
 a. $105.17   **b.** $106.17
 c. $104.17   **d.** $96.17

5. In measuring a 1-inch block of metal by a precision instrument, a student found the average of all her readings to be 0.9996 inch. Find the amount of error.

7. Charlotte's parents took the family on a motor trip. The expenses included gasoline, $142.90; oil, $6.36; lodging, $175.00; meals, $259.25; amusements, $64.83; miscellaneous, $39.54. How much did the trip cost?

4. Harry bought a new suit costing $94.95; shoes, $21.75; hat, $12.50; shirt, $10.98; and tie, $3.50. If his mother gave him $150, how much money is left over after he pays for these articles?

6. Robert's brother is a salesperson. Last week he received $294.50 salary and $59.68 commission. How much did he earn in all?

8. The passenger ship, *United States*, on its first voyage established a record of 35.59 knots crossing the Atlantic Ocean. If the record speed of the *Queen Mary* is 30.99 knots, how much faster is the *United States*?

Refresh Your Skills

The numerals in boxes indicate Lessons where help may be found.

1. Add: 1-4
149,526
12,997
500,725
36,589
8,763

2. Subtract: 1-5
821,427
291,398

3. Multiply: 1-6
1,760
500

4. Divide: 1-7
407)3,538,865

5. Name the greatest common factor of 15, 40, and 90. 1-8
6. Round 30,246,701 to the nearest million. 1-3
7. Write the numeral 800.08 in words. 2-1
8. Write as a decimal numeral: Three and fifty-nine thousandths. 2-2
9. Round $5.8273 to the nearest cent. 2-3   **10.** Add: 2-4   **11.** Subtract: 2-5
.693 + 4.82 + 37.6   4.37 − .6

2-6 Comparing Decimals

PROCEDURE

1 Which is greater: .03 or .3?

$.03 = .03$ ⎫ **.30 is greater than .03**
$.3 = .30$ ⎭ **So .3 is greater than .03**

ANSWER: .3 is greater.

Write the decimals with the same number of decimal places. Compare.

2 Which is less: .7 or .238?

$.7 \;\; = .700$ ⎫ **.238 is less than .700**
$.238 = .238$ ⎭ **So .238 is less than .7**

ANSWER: .238 is less.

3 Which is greater: 32.72 or .983?

32.72 ⎱ **When comparing a mixed decimal**
.983 ⎰ **or a whole number to a decimal, the mixed decimal or whole number is always greater.**

ANSWER: 32.72 is greater.

---

4 Is the statement .16 < .6 true?
↑
is less than

$.16 = .16$ ⎫ **.16 < .60**
$.6 \;\; = .60$ ⎭ **So .16 < .6**

ANSWER: Yes, true

5 Is the statement .45 > 4.3 true?
↑
is greater than

.45 ⎱ **The mixed number is greater.**
4.3 ⎰

ANSWER: No, false

---

6 Arrange in order from least to greatest: .635, .65, 6.1, .069

$.635 = \;\;.635$ ⎫
$.65 \;\; = \;\;.650$ ⎬ **.069 < .635 < .650 < 6.100**
$6.1 \;\;\; = 6.100$ ⎪ **So .069 < .635 < .65 < 6.1**
$.069 = \;\;.069$ ⎭

Compare.

ANSWER: .069, .635, .65, 6.1

Vocabulary

The < symbol means "is less than."   The > symbol means "is greater than."

DIAGNOSTIC TEST

Which is greater:

1 .5 or .45?   **2** .154 or 1.02?

5 Arrange in order of size (greatest first):
.06, 1.4, .19, and .388

Which is less:

3 .7 or .699?   **4** 5.53 or .553?

6 Arrange in order of size (least first):
4.72, .493, 4.8, and .465

7 Is the statement .725 < .73 true?   **8** Is the statement .9 > .19 true?

9 Which has the same value as .607?   **a.** .067   **b.** .0607   **c.** .6070   **d.** .670

RELATED PRACTICE EXERCISES

Which is greater:

1 **1.** .3 or .29?
   2. .04 or .004?
   3. .91 or .893?
   4. .7 or .074?

2 **1.** 1.47 or .278?
   2. .63 or 4.5?
   3. .28 or 2.8?
   4. 4.504 or 4.5035?

Which is less:

3 **1.** .89 or .9?
   2. .2 or .21?
   3. .50 or .05?
   4. .0051 or .006?

4 **1.** 4.1 or 3.010?
   2. 5.06 or 1.059?
   3. 1.638 or .5376?
   4. 2.0439 or 2.04395?

Arrange in order of size (greatest first):

5 **1.** .01, .001, .1, and .0001
   2. 2.25, .253, .2485, and 2.249
   3. .38, 1.5, .475, and .0506
   4. .006, 5.02, .503, and .1483

Arrange in order of size (least first):

6 **1.** .201, .19, 1.2, and .21
   2. .465, .4053, .47, and 4.5
   3. .51, .583, .60 and .5126
   4. .76, .7, .076, and .0710

Which of the following statements are true?

7 **1.** .036 < 0.36
   2. .3 < .27
   3. .049 < .5
   4. 6.7 < .675

8 **1.** 4.8 > 5
   2. 6.4 > .640
   3. .087 > .1
   4. .0070 > .007

Which has the same value as:

9 **1.** .06?   **a.** .6   **b.** .60   **c.** .060   **d.** .0060
   2. .40?   **a.** .04   **b.** 4.0   **c.** 40.00   **d.** .400
   3. .2?   **a.** 2.0   **b.** .02   **c.** .002   **d.** .200
   4. .087?   **a.** .807   **b.** .0870   **c.** .0807   **d.** .8700

SKILL ACHIEVEMENT TEST

1. Which is less .79 or .8?

2. Which is greater .01 or .009?

3. Which is greater 3.7 or .37?

4. Which is less 4.15 or .4156?

5. Which has the same value as .03?
   a. .30   **b.** 3.0   **c.** .030   **d.** .003

6. Is the statement .69 > .7 true?

7. Is the statement .0023 < .00230 true?

8. Is the statement .04 < .041 true?

Arrange in order of size:

9. Greatest first: .09, .89, .093, and .9

10. Least first: 1.25, 2.3, .04, and .167

2-7 Multiplication of Decimals

PROCEDURE

1 Multiply: 43 × 2.6

factors → 43 ←**0 decimal places**
$\times$ 2.6 ←**1 decimal place**
258 ← **partial products**
86 ←
product → 111.8 ←**1 decimal place**

Check: 2.6
$\times$ 43
78
104
111.8 ✔

ANSWER: 111.8

Arrange in columns. Multiply, starting from the right as you do with whole numbers. Add partial products.

Place a decimal point in the product. The number of decimal places in the product is equal to the sum of the numbers of decimal places in the factors.

Check by interchanging the factors.

2 Find the product of .9 and .7

.9 ← **1 decimal place**   Check: .7
$\times$ .7 ←+ **1 decimal place** $\times$ .9
.63 ← **2 decimal places**   .63 ✔

ANSWER: .63

3 Multiply: .12 × .4

.12 ← **2 decimal places**
$\times$ .4 ←+ **1 decimal place**
Write 1 zero. → .048 ← **3 decimal places**

Check: .4
$\times$ .12
.048 ✔

ANSWER: .048

If necessary, write zeros in the product to produce the proper number of decimal places.

4 Multiply 2.43 by .56

2.43   Check: .56
$\times$ .56 $\times$ 2.43
1458   168
1 215   224
1.3608   112
ANSWER: 1.3608   1.3608 ✔

91

5 Estimate the product: 41.95 × 1.8

41.95 → 40
  1.8 → 2
40 × 2 = 80

ANSWER: 80

Round each factor to its greatest place.

Multiply.

---

6 Solve the equation: 7.62 × 42.9 = n

7.62	Check: 42.9
× 42.9	× 7.62
6858	858
1524	2574
3048	3003
326.898	326.898 ✔

ANSWER: 326.898

To find the value of n, multiply the given numbers.

DIAGNOSTIC TEST

Multiply:

1 .3
8

2 43
.24

3 .351
86

4 .6739
7

5 75
.48

6 37
.05

7 .03
2

8 14
.007

9 .002
4

10 3.14
18

11 .6
.2

12 .3
.3

13 .58
.6

14 .21
.4

15 .56
.37

16 .05
.01

17 16.2
.045

18 34.89
.875

19 .147
.03

20 3.1416
.75

21 .059
.064

22 6 × .005

23 .012 × .07

In examples involving money, find the product to the nearest cent:

24 $.25
8

25 $3.80
24

26 $3.62
.06

27 $4.28
.125

28 Find .46 of 150

29 Select the nearest given estimate for .29 × $51.   **a.** $5   **b.** $10   **c.** $15

30 Solve: .04 × 1.25 = n

RELATED PRACTICE EXERCISES

Multiply:

1 **1.** 4   **2.** 6   **3.** 12   **4.** .9   **2** **1.** .23   **2.** 7   **3.** 84   **4.** 39
        .1   .7   .8   5   9   .56   .42   .61

3 **1.** .247   **2.** .456   **3.** .572   **4.** 28
          5   34   159   .707

4 **1.** .9522   **2.** 5   **3.** 46   **4.** 8725
            8   .4673   .5034   .2839

5 **1.** 25   **2.** .95   **3.** 8   **4.** 246   **6** **1.** .03   **2.** .05   **3.** 10   **4.** 200
        .4   20   .125   .625   8   17   .04   .08

7 **1.** .03   **2.** .02   **3.** 3   **4.** 5   **8** **1.** .006   **2.** .009   **3.** 5   **4.** 21
         3   4   .02   .01   3   10   .013   .004

9 **1.** .001   **2.** .0015   **3.** 13   **4.** 4   **10** **1.** 8.7   **2.** 56.17   **3.** 460   **4.** 3.1416
          9   2   .0006   .0014   6   75   4.8   32

11 **1.** .9   **3.** .5   **3.** 3   **4.** 5.7   **12** **1.** .3   **2.** .1   **3.** .4   **4.** .1
         .8   .6   8.2   2.5   .2   .1   .2   .8

13 **1.** .34   **2.** .95   **3.** .7   **4.** 1.15   **14** **1.** .24   **2.** .3   **3.** .15   **4.** .07
          .3   .4   .66   5.2   .2   .03   .6   1.3

15 **1.** .28   **2.** 5.93   **3.** .45   **4.** 12.52   **16** **1.** .67   **2.** .05   **3.** .02   **4.** 2.14
         .74   .87   4.91   .06   .02   .76   .02   .03

17 **1.** .003   **2.** .7   **3.** .375   **4.** .009
           .4   .002   1.4   36.6

18 **1.** .368   **2.** 70.84   **3.** 95.26   **4.** 453.40
          .26   .034   1.125   .375

19 **1.** .0002   **2.** .057   **3.** 3.14   **4.** .04
            .2   .38   .002   1.225

20 **1.** .268   **2.** 3.1416   **3.** 8.504   **4.** 2.423
         .924   6.25   .015   9.146

21 **1.** .00008   **2.** .003   **3.** .01   **4.** .02167
            .6   .009   .0007   1.8

22 **1.** 5 × .7   **2.** 8 × .04   **23** **1.** .2 × .3   **2.** .3 × .04
        3. 3 × .28   **4.** .06 × 4   **3.** .4 × .35   **4.** .02 × .09
        5. 9 × .0001   **6.** 10 × .14   **5.** .01 × .005   **6.** .025 × .008
        7. 1.2 × 12   **8.** 18 × .05   **7.** .013 × .01   **8.** .04 × .029

In examples involving money, find the products correct to the nearest cent:

24 **1.** $.42
      4

2. $.69
    18

3. $.80
    7

4. $.75
    48

25 **1.** $4.97
    3

2. $10.50
    60

3. $16.31
    96

4. $4.25
    144

26 **1.** $89
    .04

2. $.75
    .06

3. $14.25
    .19

4. $293.28
    .63

27 **1.** $840
    .625

2. $15.61
    .045

3. $2,500
    .4375

4. $675.90
    .002

Find:

28 **1.** .25 of 60   **2.** .04 of 9   **3.** .08 of 1.5   **4.** .75 of $4
    5. .06 of $93   **6.** .39 of $3.40   **7.** .4 of 200   **8.** .13 of $15.64

29 Select the nearest given estimate for:

  1. .21 × $19:   **a.** $4   **b.** $5   **c.** $6
  2. 3.98 × .007:   **a.** 3   **b.** .3   **c.** .03
  3. .06 × $59.85:   **a.** $36   **b.** $3.60   **c.** $.36
  4. 42 × $7.90:   **a.** $500   **b.** $400   **c.** $300

30 Solve:

  1. $6 \times .8 = ?$
  2. $.09 \times .2 = $ ▨
  3. $.004 \times .03 = n$
  4. $? = 1.4 \times 5.8$
  5. ▨ $= .08 \times \$10.50$
  6. $n = 12 \times .3$

SKILL ACHIEVEMENT TEST

In examples involving money, find the product correct to the nearest cent.

  1. .974
    5

2. 3.6
  2.4

3. 5.08
  .009

4. $.95
  12

5. $16.95
  .04

  6. .003 × .012
  7. 10 × $9.67
  8. Find .06 of $4.50
  9. Select the nearest given estimate for: .48 × $69.50
    a. $3 **b.** $35 **c.** $2.40
10. Solve: .05 × $20 = n

BUILDING PROBLEM-SOLVING SKILLS

FLASHBACK

1 READ

2 PLAN

3 SOLVE

4 CHECK

PROBLEM: Richy earns $7.90 per hour and works 38.5 hours each week. How much are his total earnings per week?

1 **Given facts:** Works 38.5 hours at $7.90 per hour.

Question asked: How much are the total earnings?

2 **Operation needed:** The word *total* with *equal* quantities indicates multiplication.

Think or write out your plan: The number of working hours times the pay rate per hour is equal to the total earnings.

Equation: $38.5 \times \$7.90 = n$

3 **Estimate the answer:**

   38.5 →   40 hours
$7.90 → $8 per hour
          $320 weekly wages, estimate

Solution:

    $7.90 rate per hour
$\times$ 38.5 hours
    3950
    6320
    2370
  $304.150

ANSWER: Weekly wages are $304.15.

4 **Check for accuracy:**

    38.5 hours
$\times$ $7.90 rate per hour
  34650
   2695
 $304.150 ✔

Compare answer to the estimate: The answer $304.15 compares reasonably with estimate $320.

PRACTICE PROBLEMS

1. Joan's mother bought 25 shrubs at $7.49 each. What was the total cost?

2. How much lumber is needed to make 12 shelves each 1.3 meters long?

3. Find the cost of 16.4 gallons of gasoline at $1.35 per gallon.

4. A carton contains 48 cans of food each weighing .375 kg. What is the weight of the carton of food?

MIXED PROBLEMS

Solve each of the problems. For problems 1-4 select the letter corresponding to your answer.

1. A certain plane on a flight used 175.2 liters of gasoline per hour. If its flight lasted 4.5 hours, how many liters of gasoline were consumed?
  a. 7,884 liters   **b.** 78.84 liters
  c. 7.8840 liters   **d.** 788.4 liters

2. Find the total expenses in producing the Jacques Cartier School show if renting and making costumes cost $362.55; royalty fee, $75.00; properties, $143.66; tickets, $25.50; lighting, $52.25; and miscellaneous items, $157.19.
  a. $806.15   **b.** $796.15
  c. $826.15   **d.** $816.15

3. In addition to her weekly allowance of $8.50, Wanda earned $15.65 after school. How much money should she have left at the end of the week if her expenses were: bus fare, $3.50; school lunches and supplies, $6.97; movies, $1.85; church, $1.00; and savings, $3.75?
  a. $6.08   **b.** $5.08
  c. $7.08   **d.** Answer not given

4. Find the distance represented by 6.7 centimeters if the scale is 1 centimeter = 50 kilometers.
  a. 325 kilometers   **b.** 415 kilometers
  c. 335 kilometers   **d.** 300.7 kilometers

5. An airplane has a ground speed of 325 knots. What is its ground speed in statute m.p.h. if 1 knot = 1.15 statute m.p.h.?

6. For how much should a dealer sell a CB radio if it cost him $68.75 and he wishes to make a profit of $36.75?

7. The outside diameter of a piece of copper tubing is 2.375 millimeters and its wall thickness is .083 millimeters. What is the inside diameter?

8. A refrigerator costs $625 cash or $80 down and 12 payments of $52.95 each. How much do you save by paying cash?

Refresh Your Skills

The numerals in boxes indicate Lessons where help may be found.

1. Add: 1-4   28,104
              625
         9,302
             29
       75,412
          828

2. Subtract: 1-5
  1,500,020
    690,175

3. Multiply: 1-6
  876
  938

4. Divide: 1-7
  289)285,821

5. Add: 2-4
  .88 + .8 + .888

6. Subtract: 2-5
  9.6 − .45

7. Multiply: 2-7
  3.1416
     48

8. Multiply: 2-7
  .004 × 2.5

2-8 Division of Decimals

PROCEDURE

1 Divide 49.8 by 6

```
      8.3
6)49.8
   48
    1 8
    1 8
      0
```

Check: 8.3
       ×   6
      49.8 ✔

ANSWER: 8.3

When the divisor is a whole number, write the decimal point in the quotient directly above the decimal point in the dividend. Divide as you do with whole numbers.

Check by multiplying the quotient by the divisor.

2 Find the quotient: 35.6 ÷ .4

```
      8 9.
.4)35.6∧    Check:    89
   32                × .4
    3 6              35.6 ✔
    3 6
      0
```

ANSWER: 89

Make the divisor a whole number by moving the decimal point. Then move the decimal point in the dividend the same number of places. The carat (∧) is used to indicate the new position of the decimal point.
Use the *original* divisor to check.

3 Find the quotient: .0015 ÷ .05

```
       0.03
.05∧).00∧15   Check:    .03
       15            × .05
        0            .0015 ✔
```

ANSWER: .03

4 Divide: .625)‾15‾

```
            24.
.625∧)15.000∧←Write 3 zeros.
      12 50
       2 500
       2 500
           0
```
Check: .625
       ×   24
   15.000 = 15 ✔

If necessary, write zeros in the dividend to produce the proper number of places so that the dividend has at least as many places as the divisor.

ANSWER: 24

5 Find the quotient: $8.50 ÷ 6. Round the answer to the nearest cent.

$$
\begin{array}{r}
\underline{\$1.41\underline{6}} \text{ rounds to \$1.42.} \\
6)\overline{\$8.500} \\
\underline{6} \quad \text{⌐Write 1 zero.} \\
2\,5 \\
\underline{2\,4} \\
10 \\
\underline{6} \\
40 \\
\underline{36} \\
4
\end{array}
$$

To round to the nearest cent, carry out the division to the thousandths' place.

ANSWER: $1.42

6 Divide 28.5 by .87. Find the quotient to the nearest tenth.

$$
\begin{array}{r}
\underline{32.7\underline{5}} \text{ rounds to 32.8.} \\
.87_\wedge)\overline{28.50_\wedge00} \leftarrow\text{Write 2 zeros.} \\
\underline{26\,1} \\
2\,40 \\
\underline{1\,74} \\
66\,0 \\
\underline{60\,9} \\
5\,10 \\
\underline{4\,35} \\
75
\end{array}
$$

To round to a given decimal place, carry out the division to the next decimal place, then round to the given place.
Divide to hundredths.

ANSWER: 32.8

7 Estimate the quotient: 119.8 ÷ .42

$$
\begin{array}{l}
.42 \rightarrow .4 \\
11\overline{9}.8 \rightarrow 120 \\
1\overline{2}0 ÷ .4 = 300
\end{array}
$$

Round the divisor to its highest place.
Round the dividend so that it can be divided exactly by the rounded divisor. Divide.

ANSWER: 300

8 Solve the equation: 21 ÷ 56 = n

$$
\begin{array}{r}
\underline{.375} \\
56)\overline{21.000} \\
\underline{16\,8} \\
4\,20 \\
\underline{3\,92} \\
280 \\
\underline{280} \\
0
\end{array}
$$

Write a decimal point and ending zeros.

To find the value of n, divide the given numbers.

Check:
$$
\begin{array}{r}
.375 \\
\times \quad 56 \\
\hline
21.000 = 21 \checkmark
\end{array}
$$

ANSWER: .375

DIAGNOSTIC TEST (divisors—whole numbers)

Divide:

1 4)9.2 **2** 7)8.96 **3** 2)5.328 **4** 3).8226 **5** 6)5.124

6 8).736 **7** 36)91.44 **8** 8)5.000 **9** 200)4 **10** Divide 3.6 by 6

11 Find the quotient correct to nearest thousandth: 29)24

12 Find the answer correct to nearest cent: 12)$2.57

13 2 ÷ 16 (Find the quotient to 3 decimal places.)

14 Select the nearest given estimate for 78.12 ÷ 4 **15** Solve: .96 ÷ 16 = *n*
    a. 1.3 **b.** 20 **c.** 38.12

RELATED PRACTICE EXERCISES

Divide:

1 **1.** 3)6.9 **2.** 7)86.1 **3.** 5)746.5 **4.** 6)6765.6

2 **1.** 2).86 **2.** 4)5.88 **3.** 7)92.96 **4.** 8)971.44

3 **1.** 8).968 **2.** 5).865 **3.** 2)7.942 **4.** 3)73.914

4 **1.** 4).8936 **2.** 6).6432 **3.** 5)6.9185 **4.** 7)8.9789

5 **1.** 3)2.52 **2.** 8)5.216 **3.** 6)2.418 **4.** 9)7.5519

6 **1.** 6).228 **2.** 8).024 **3.** 7).0114 **4.** 5).00375

7 **1.** 12)3.36 **2.** 48)158.4 **3.** 24)11.688 **4.** 144)112.32

8 **1.** 5)2.0 **2.** 4)5.00 **3.** 8)7.000 **4.** 32)13.00000

9 **1.** 60)3 **2.** 200)6 **3.** 48)54 **4.** 7,000)84

10 **1.** 4.2 by 7 **2.** .616 by 4 **3.** .56 by 8 **4.** $10.50 by 10

11 Find the quotient correct to nearest tenth:
    1. 12)7 **2.** 8)43 **3.** 90)375 **4.** 108)20,000

    Find the quotient correct to nearest hundredth:
    5. 9)5 **6.** 6)38 **7.** 46)2.8 **8.** 156)108

    Find the quotient correct to nearest thousandth:
    9. 7)285 **10.** 15)46 **11.** 12)365 **12.** 24)13.59

12 Find the answer correct to nearest cent:
    1. 3)$.72 **2.** 12)$9.60 **3.** 24)$8.16 **4.** 144)$11.52
    5. 4)$.63 **6.** 6)$1.25 **7.** 72)$234 **8.** 144)$1,056

13 **1.** .87 ÷ 3 **2.** 9 ÷ 12 (2 decimal places)
    3. 8 ÷ 7 (nearest thousandth) **4.** $2.16 ÷ 5 (nearest cent)

Select the nearest given estimate for:

14 **1.** $13.92 \div 2$ **a.** .69 **b.** 11.92 **c.** 7   **2.** $.5814 \div 3$ **a.** 175 **b.** 20 **c.** .2
       3. $9.486 \div 9$ **a.** 1 **b.** .125 **c.** 0   **4.** $985.6 \div 49$ **a.** 18.2 **b.** 20 **c.** 2.5

Solve:

15 **1.** $.76 \div 4 = ?$ **2.** $97.2 \div 2 = \blacksquare$ **3.** $125.04 \div 8 = n$
       4. $? = 6.945 \div 5$ **5.** $\blacksquare = 16.5 \div 3$ **6.** $n = 60.3 \div 9$

DIAGNOSTIC TEST   (divisors—tenths)

Divide:

1 $3 \overline{)247.8}$ **2** $.5 \overline{)9.25}$ **3** $.8 \overline{).896}$ **4** $.6 \overline{)2.6898}$

5 $.2 \overline{).0034}$ **6** $1.2 \overline{)108.72}$ **7** $.6 \overline{)12.0}$ **8** $.7 \overline{)42}$

9 $.4 \overline{)2}$ **10** Find the quotient correct to nearest tenth: $2.7 \overline{)18}$

11 Select the nearest given estimate for $101.92 \div 19.6$ **12** Solve: $3.44 \div 8 = n$
      a. 5 **b.** 22.3 **c.** 82

RELATED PRACTICE EXERCISES

Divide.

1 **1.** $.4 \overline{)7.6}$ **2.** $.5 \overline{)89.5}$ **3.** $.3 \overline{)176.7}$ **4.** $1{,}804.2 \div .6$

2 **1.** $.3 \overline{).84}$ **2.** $.2 \overline{)4.76}$ **3.** $.5 \overline{)32.15}$ **4.** $733.08 \div .4$

3 **1.** $.7 \overline{).294}$ **2.** $.4 \overline{)3.024}$ **3.** $.3 \overline{)62.928}$ **4.** $18{,}725 \div .5$

4 **1.** $.5 \overline{).4205}$ **2.** $.3 \overline{).7128}$ **3.** $.9 \overline{)9.3843}$ **4.** $6.4722 \div .7$

5 **1.** $.8 \overline{).0016}$ **2.** $.7 \overline{).0224}$ **3.** $.3 \overline{).0009}$ **4.** $.0552 \div .6$

6 **1.** $1.8 \overline{)43.74}$ **2.** $2.6 \overline{)14.976}$ **3.** $3.5 \overline{)3041.5}$ **4.** $8.8452 \div 24.3$

7 **1.** $.4 \overline{)14.0}$ **2.** $2.8 \overline{)49.00}$ **3.** $6.4 \overline{)104.000}$ **4.** $15.000 \div 1.6$

8 **1.** $.5 \overline{)15}$ **2.** $.4 \overline{)72}$ **3.** $1.8 \overline{)36}$ **4.** $126 \div 4.2$

9 **1.** $.6 \overline{)3}$ **2.** $.8 \overline{)2}$ **3.** $5.6 \overline{)14}$ **4.** $40 \div 12.8$

10 Find the quotient correct to nearest tenth:

      1. $.8 \overline{)9}$ **2.** $7.5 \overline{)456.2}$ **3.** $25.4 \overline{)82.25}$ **4.** $5{,}000 \div 2.2$

      Find the quotient correct to nearest hundredth:

      5. $.3 \overline{)2}$ **6.** $1.4 \overline{)6}$ **7.** $5.4 \overline{).48}$ **8.** $36 \div 13.2$

      Find the quotient correct to nearest thousandth:

      9. $.8 \overline{).45}$ **10.** $2.6 \overline{)740}$ **11.** $3.9 \overline{)85.3}$ **12.** $200 \div 5.7$

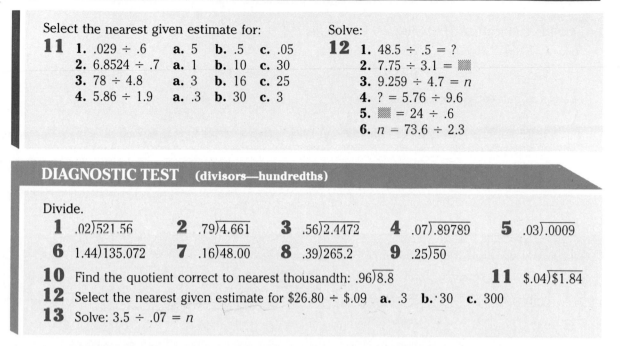

Select the nearest given estimate for:

11 **1.** .029 ÷ .6   **a.** 5   **b.** .5   **c.** .05
    2. 6.8524 ÷ .7   **a.** 1   **b.** 10   **c.** 30
    3. 78 ÷ 4.8   **a.** 3   **b.** 16   **c.** 25
    4. 5.86 ÷ 1.9   **a.** .3   **b.** 30   **c.** 3

Solve:

12 **1.** 48.5 ÷ .5 = ?
    2. 7.75 ÷ 3.1 = ▓
    3. 9.259 ÷ 4.7 = n
    4. ? = 5.76 ÷ 9.6
    5. ▓ = 24 ÷ .6
    6. n = 73.6 ÷ 2.3

DIAGNOSTIC TEST   (divisors—hundredths)

Divide.

1 .02)‾521.56‾   **2** .79)‾4.661‾   **3** .56)‾2.4472‾   **4** .07)‾.89789‾   **5** .03)‾.0009‾

6 1.44)‾135.072‾   **7** .16)‾48.00‾   **8** .39)‾265.2‾   **9** .25)‾50‾

10 Find the quotient correct to nearest thousandth: .96)‾8.8‾   **11** $.04)‾$1.84‾

12 Select the nearest given estimate for $26.80 ÷ $.09   **a.** .3   **b.** 30   **c.** 300

13 Solve: 3.5 ÷ .07 = n

RELATED PRACTICE EXERCISES

Divide:

1 **1.** .04)‾.68‾   **2.** .06)‾15.06‾   **3.** .32)‾16.96‾   **4.** 369.36 ÷ .57

2 **1.** .07)‾.812‾   **2.** .03)‾7.749‾   **3.** .96)‾6.912‾   **4.** 24.175 ÷ .25

3 **1.** .02)‾4.6954‾   **2.** .08)‾.7216‾   **3.** .43)‾1.9694‾   **4.** .4725 ÷ .75

4 **1.** .09)‾.08928‾   **2.** .05)‾.15425‾   **3.** .22)‾.21692‾   **4.** 1.32156 ÷ .36

5 **1.** .06)‾.0018‾   **2.** .01)‾.0005‾   **3.** .03)‾.00012‾   **4.** .04216 ÷ .68

6 **1.** 3.65)‾208.05‾   **2.** 2.27)‾88.303‾   **3.** 4.24)‾7.0808‾   **4.** 404.352 ÷ 8.3

7 **1.** .04)‾76.00‾   **2.** .36)‾18.00‾   **3.** .64)‾56.000‾   **4.** 120.000 ÷ 1.92

8 **1.** .08)‾57.6‾   **2.** .65)‾45.5‾   **3.** 1.47)‾3,719.1‾   **4.** 4,952.5 ÷ 2.83

9 **1.** .09)‾27‾   **2.** .18)‾90‾   **3.** .64)‾16‾   **4.** 119 ÷ 1.36

10 Find the quotient correct to nearest tenth:

    1. 1.15)‾825‾   **2.** .87)‾79.4‾   **3.** 3.14)‾853.76‾   **4.** 432 ÷ 39.37

Find the quotient correct to nearest hundredth:

    5. .06)‾4‾   **6.** .84)‾70‾   **7.** .37)‾2.9‾   **8.** 34 ÷ 1.53

Find the quotient correct to nearest thousandth:

    9. .54)‾98‾   **10.** .69)‾8.45‾   **11.** .26)‾42.7‾   **12.** .162 ÷ 2.43

Find the quotient correct to nearest hundredth:

11 **1.** $.02)\overline{\$3.46}$ **2.** $.05)\overline{\$4}$ **3.** $.38)\overline{\$27.36}$ **4.** $\$30 \div \$1.25$

   5. $.08)\overline{\$1.24}$ **6.** $.25)\overline{\$3.90}$ **7.** $.49)\overline{\$10.43}$ **8.** $\$97.90 \div \$2.67$

Select the nearest given estimate for: Solve:

12 **1.** $.98 \div .04$ **a.** 2.5   **b.** 25   **c.** .25   **13** **1.** $.9 \div .03 = ?$
   2. $15.7 \div 2.15$ **a.** 8   **b.** .08   **c.** .008   **2.** $\$17.55 \div \$.65 = $ ▨
   3. $\$71.86 \div \$.08$ **a.** 9   **b.** 90   **c.** 900   **3.** $\$54 \div \$.06 = n$
   4. $\$74.62 \div \$15.25$ **a.** 15   **b.** 10   **c.** 5   **4.** $? = .0006 \div .01$
                                                                          5. ▨ $ = .00378 \div .42$
                                                                          6. $n = 9.275 \div 1.75$

DIAGNOSTIC TEST   (divisors—thousandths)

Divide:

1 $.006)\overline{74.898}$ **2** $.018)\overline{.4554}$ **3** $.007)\overline{6.53912}$ **4** $.231)\overline{.00924}$

5 $4.375)\overline{11.8125}$ **6** $.048)\overline{60.000}$ **7** $.125)\overline{53.75}$ **8** $.052)\overline{452.4}$

9 $.014)\overline{112}$ **10** Find the quotient correct to nearest hundredth: $.33)\overline{249}$

11 Select the nearest given estimate for $71 \div .069$:   **12** Solve: $.558 \div .018 = n$
   a. .001   **b.** 100   **c.** 1,000

RELATED PRACTICE EXERCISES

Divide:

1 **1.** $.007)\overline{.763}$ **2.** $.003)\overline{17.862}$ **3.** $.175)\overline{49.525}$ **4.** $77.322 \div .526$

2 **1.** $.006)\overline{6.0282}$ **2.** $.035)\overline{.1715}$ **3.** $.216)\overline{8.1864}$ **4.** $1.2144 \div .024$

3 **1.** $.004)\overline{.05964}$ **2.** $.073)\overline{.31828}$ **3.** $.524)\overline{.36156}$ **4.** $.86278 \div .358$

4 **1.** $.009)\overline{.00081}$ **2.** $.018)\overline{.0009}$ **3.** $.105)\overline{.00735}$ **4.** $.04002 \div .667$

5 **1.** $2.548)\overline{8.1536}$ **2.** $4.125)\overline{.94875}$ **3.** $6.875)\overline{39.1875}$ **4.** $4.44296 \div 3.002$

6 **1.** $.008)\overline{16.000}$ **2.** $.056)\overline{140.00}$ **3.** $.144)\overline{108.000}$ **4.** $132.000 \div 1.375$

7 **1.** $.094)\overline{25.38}$ **2.** $.008)\overline{90.16}$ **3.** $.231)\overline{147.84}$ **4.** $893.88 \div .382$

8 **1.** $.005)\overline{438.5}$ **2.** $.072)\overline{691.2}$ **3.** $.337)\overline{1,819.8}$ **4.** $12,799.5 \div .265$

9 **1.** $.059)\overline{177}$ **2.** $.108)\overline{972}$ **3.** $.007)\overline{210}$ **4.** $2,364 \div .591$

10 Find the quotient correct to nearest tenth:
   1. $.542)\overline{158}$ **2.** $.003)\overline{.55}$ **3.** $1.853)\overline{9.647}$ **4.** $365.9 \div .868$

Find the quotient correct to nearest hundredth:
5. .006)25   **6.** .013)56.3   **7.** .314)8.49   **8.** .964 ÷ .592

Find the quotient correct to nearest thousandth:
9. .007)33   **10.** .056).428   **11.** .723)62.5   **12.** 38.25 ÷ 4.007

Select the nearest given estimate for:
11 **1.** 29.7 ÷ .005   **a.** .0006 **b.** 600   **c.** 6,000
        2. 101.25 ÷ 9.814 **a.** 20   **b.** 10   **c.** .5
        3. 12 ÷ .039   **a.** 300   **b.** .4   **c.** .048
        4. 1.50864 ÷ .499 **a.** 3   **b.** 11   **c.** 40

Solve:
12 **1.** 1.2 ÷ .006 = ?
        2. .4 ÷ .016 = ▧
        3. 10 ÷ .125 = n
        4. ? = .21 ÷ 2.625
        5. ▧ = .0036 ÷ .009
        6. n = .3225 ÷ .075

SKILL ACHIEVEMENT TEST

Divide:
1. 6).0474   **2.** 7.2)9   **3.** .87).82563   **4.** .0028 ÷ .4
5. 382 ÷ .043   **6.** 18.7572 ÷ 5.39
7. Find the quotient correct to nearest cent: 1.8)$44.20
8. Find the quotient correct to nearest tenth: 15 ÷ 6.3
9. Select the nearest given estimate for: 278.4 ÷ .39 **a.** .6 **b.** 80 **c.** 700
10. Solve: 1.2 ÷ .03 = n

Enrichment

Order of Operations—Using Parentheses

Parentheses are used to change the standard rule of operations of first multiplying
and/or dividing before adding and/or subtracting from left to right.

EXAMPLES:

$6 + 9 \div 3 = 6 + 3 = 9$
but $(6 + 9) \div 3 = 15 \div 3 = 5$
$20 \times (12 - 7) = 20 \times 5 = 100$
$1.8 \div (.2 + .4) = 1.8 \div .6 = 3$

To evaluate a numerical expression involving
parentheses, first perform the operations within
the parentheses.

Find the value of each of the following expressions:

1. $4 \times 6 + 5$ **2.** $4 \times (6 + 5)$ **3.** $.24 \div .2 \times .6$ **4.** $.24 \div (.2 \times .6)$
5. $(1.2 + .8) \div 4$ **6.** $(.35 - .16) \times .2$ **7.** $(8 \times 6) + 12$ **8.** $8 \times 6 + 12$
9. $16 - (10 + 5)$ **10.** $16 - 10 + 5$ **11.** $80 \div 8 \times 2$ **12.** $80 \div (8 \times 2)$
13. $12 + 8 \div 4$ **14.** $35 - 16 \times 2$ **15.** $50 - 25 \div 5$ **16.** $(50 - 25) \div 5$
17. $(.42 \div 2.1) + (1.2 \times .4)$ **18.** $(2.9 + 3.1) \times (.9 \div .3)$

BUILDING PROBLEM-SOLVING SKILLS

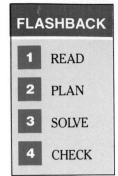

FLASHBACK

1	READ
2	PLAN
3	SOLVE
4	CHECK

PROBLEM: A mutual fund quarterly dividend of $59.40 was reinvested in the fund at the price of $8.64 per share. How many shares of the fund (to three decimal places) should the investor receive?

1 **Given facts:** Dividend of $59.40 reinvested at the price of $8.64 per share.

Question asked: How many shares should the investor receive?

2 **Operation needed:** The words *how many shares* when the *price per share* and *total price* are given indicates division.

Think or write out your plan: The amount of the dividend divided by the price per share is equal to the number of shares.

Equation: $59.40 ÷ $8.64 = n

3 **Estimate** the answer:

$8.64 → $9

$59.40 → $63

$63 ÷ $9 = 7 shares, estimate

Solution:

```
                6.875
$8.64 )$59.40 000
         51 84
          7 56 0
          6 91 2
            64 80
            60 48
             4320
             4320
```

ANSWER: 6.875 shares

4 **Check the accuracy:**

```
     $8.64
   × 6.875
     4320
     6048
     6912
     5184
   $59.40   ✔
```

Compare answer to the estimate: The answer of 6.875 shares compares reasonably with the estimate 7 shares.

PRACTICE PROBLEMS

1. How many times as large is an 18 oz. jar of jelly compared to a 4.5 oz. jar of jelly?

2. Lisa earns $12,680.20 per year. What are her earnings per week?

3. An automobile traveled 247.5 miles on 12.5 gallons of gasoline. Find how many miles per gallon the automobile averaged.

4. A lap around the school track is .25 km long. How many times around the track must you run in a 2 km race?

MIXED PROBLEMS

Solve each of the following problems. For problems 1-4 select the letter corresponding to your answer.

1. The freshmen contributed $139.75 to the Red Cross; the sophomores, $98.40; the juniors, $106.15; and the seniors, $134.90. What was the total amount of their contributions?
   a. $459.20 **b.** $469.20
   c. $479.20 **d.** Answer not given

2. A vessel heads N. 15 W. for 5 hours at 10.4 knots. Find the distance traveled in nautical miles. (1 knot = 1 nautical m.p.h.)
   a. 52 naut. mi **b.** 5.20 naut. mi
   c. 520 naut. mi **d.** 50.2 naut. mi

3. A butcher charged $9.52 for a certain cut of meat at $2.24 a pound. What was the weight of the meat?
   a. 5.32 lb **b.** 4.76 lb
   c. 4.25 lb **d.** Answer not given

4. During the month a merchant made deposits of $439.76, $180.53, $263.98, and $129.49. Checks and cash withdrawals were: $163.20, $248.00, $92.85, $310.94, and $8.52. If his previous balance was $716.91, find his new bank balance.
   a. $897.16 **b.** $887.16
   c. $907.16 **d.** Answer not given

5. Find the amount saved when buying in quantity: 1 dozen cans of peas for $4.50 or $.43 each.

6. A finance company can be repaid on a loan of $100 in 6 monthly payments of $18.15, 12 monthly payments of $9.75, or 18 monthly payments of $6.97. Find the amounts paid back and the interest under each plan.

7. If the postage for sending a certain package by first class mail was $1.56 at the rate of $.20 for the first ounce and $.17 for each additional ounce, how much did the package weigh?

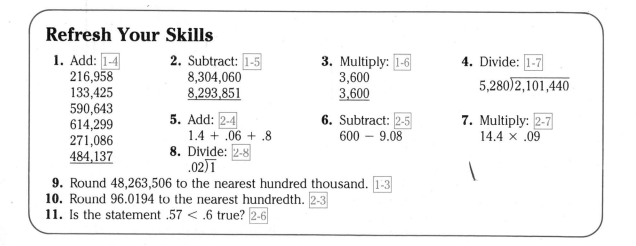

Refresh Your Skills

1. Add: 1-4
   216,958
   133,425
   590,643
   614,299
   271,086
   484,137

2. Subtract: 1-5
   8,304,060
   8,293,851

5. Add: 2-4
   1.4 + .06 + .8

8. Divide: 2-8
   .02)1

3. Multiply: 1-6
   3,600
   3,600

6. Subtract: 2-5
   600 − 9.08

4. Divide: 1-7
   5,280)2,101,440

7. Multiply: 2-7
   14.4 × .09

9. Round 48,263,506 to the nearest hundred thousand. 1-3
10. Round 96.0194 to the nearest hundredth. 2-3
11. Is the statement .57 < .6 true? 2-6

2-9 Finding What Decimal Part One Number Is of Another

PROCEDURE

1 What decimal part of 20 is 9?

$$9 \div 20$$

↑   ↑
part   whole unit

```
        .45 ←—decimal part
   20)9.00
      8 0
      1 00
      1 00
          0
```

Check: .45 × 20 = 9 ✔

ANSWER: .45

Write as a division problem with the given part as the dividend and the number of parts in the whole unit as the divisor.

Divide to find the **decimal part**.*

2 .15 is what decimal part of 2.4?

$$.15 \div 2.4$$

```
         .0625
   2.4)̯.1̯5000
       1 44
         60
         48
        120
        120
          0
```

Check: .0625 × 2.4 = .15 ✔

ANSWER: .0625

Write as a division problem.

Divide.

Remember, use the original divisor to check.

3 What decimal part (rounded to 2 places) of 3 is 2?

```
              .666 rounds to .67
   2 ÷ 3   3)2.000
             1 8
              20
              18
               20
               18
                2
```

ANSWER: .67

*For an alternate method, see Lesson 17-14.

4 What decimal part of 32 is 18?
Round the answer to the nearest thousandth.

$$
\begin{array}{r}
\underline{.5625}\text{ rounds to }.563 \\
18 \div 32 \quad 32)\overline{18.0000} \\
\underline{16\ 0} \\
2\ 00 \\
\underline{1\ 92} \\
80 \\
\underline{64} \\
160 \\
\underline{160} \\
0
\end{array}
$$

ANSWER: .563

DIAGNOSTIC TEST

1 What decimal part of 10 is 3?

2 4 is what decimal part of 5?

3 What decimal part of 100 is 21?

4 What decimal part of 25 is 17?

5 7 is what decimal part of 8?

6 What decimal part (2 places) of 6 is 1?

7 What decimal part (to nearest thousandth) of 14 is 9?

8 27 is what decimal part of 36?

9 24 is what decimal part (2 places) of 72?

10 What decimal part (to nearest hundredth) of 49 is 21?

11 What decimal part of .4 is .02?

12 .65 is what decimal part (2 places) of 7.8?

13 What decimal part of $2 is $.60?

14 $2.25 is what decimal part of $11.25?

RELATED PRACTICE EXERCISES

1
1. What decimal part of 10 is 7?
2. What decimal part of 2 is 1?
3. What decimal part of 5 is 4?
4. What decimal part of 10 is 1?

2
1. 9 is what decimal part of 10?
2. 1 is what decimal part of 5?
3. 3 is what decimal part of 10?
4. 1 is what decimal part of 2?

3
1. What decimal part of 100 is 7?
2. 83 is what decimal part of 100?
3. 31 is what decimal part of 100?
4. What decimal part of 100 is 97?

5
1. 3 is what decimal part of 8?
2. What decimal part of 16 is 11?
3. What decimal part of 8 is 5?
4. 21 is what decimal part of 32?

(Round to nearest thousandth.)

7
1. What decimal part of 9 is 4?
2. 5 is what decimal part of 7?
3. What decimal part of 11 is 8?
4. 2 is what decimal part of 15?

(Round to 2 decimal places.)

9
1. 45 is what decimal part of 54?
2. What decimal part of 84 is 35?
3. 68 is what decimal part of 96?
4. What decimal part of 102 is 85?

11
1. What decimal part of .12 is .3?
2. What decimal part of .96 is .24?
3. .5 is what decimal part of .75?
4. What decimal part of .2 is .18?

13
1. What decimal part of $1 is $.05?
2. $6 is what decimal part of $7.50?
3. What decimal part of $3 is $1.86?
4. $.65 is what decimal part of $13?

4
1. What decimal part of 4 is 3?
2. What decimal part of 20 is 7?
3. 13 is what decimal part of 50?
4. What decimal part of 25 is 22?

(Round to 2 places.)

6
1. What decimal part of 7 is 1?
2. 5 is what decimal part of 6?
3. What decimal part of 12 is 7?
4. 13 is what decimal part of 24?

8
1. 16 is what decimal part of 32?
2. What decimal part of 56 is 21?
3. 40 is what decimal part of 500?
4. 64 is what decimal part of 200?

(Round to nearest thousandth.)

10
1. What decimal part of 84 is 24?
2. 15 is what decimal part of 27?
3. What decimal part of 135 is 36?
4. 99 is what decimal part of 121?

12
1. .4 is what decimal part of 1.6?
2. What decimal part of 3.65 is 1.46?
3. .03 is what decimal part of 300?
4. What decimal part of 9.6 is 6?

14
1. $.18 is what decimal part of $.90?
2. What decimal part of $12.40 is $1.86?
3. What decimal part of $27.30 is $16.38?
4. $1.11 is what decimal part of $1.48?

SKILL ACHIEVEMENT TEST

1. What decimal part of 5 is 3?
2. 11 is what decimal part of 20?
3. What decimal part of 200 is 58?
4. What decimal part of 16 is 9?
5. 24 is what decimal part of 192?
6. What decimal part of $25 is $7.25?
7. What decimal part of 8.5 is .34?
8. .26 is what decimal part of .65?
9. $.86 is what decimal part (2 places) of $1.29?
10. 40 is what decimal part (2 places) of 105?

PROCEDURE

1 .4 of what number is 12?

$$12 \div .4$$

↑ part   ↑ decimal

```
      30. ← whole unit
  .4⟋12.0
     12
      0
```

Check: 30 × .4 = 12.0 = 12 ✔

ANSWER: 30

Write as a division problem with the known part as the dividend and the **decimal** representing this part as the divisor.

Divide to find the number representing the whole unit.*

2 480 is .75 of what number?

$$480 \div .75$$

```
       6 40.
  .75⟋480.00
      450
       30 0
       30 0
          0
```

Check: 640 × .75 = 480.00 = 480 ✔

ANSWER: 640

3 .42 of what number is $225?
Round the answer to the nearest cent.

$$\$225 \div .42$$

```
              $5 35.714  rounds to $535.71
  .42⟋$225.00 000
      210
       15 0
       12 6
        2 40
        2 10
          30 0
          29 4
            60
            42
           180
           168
            12
```

ANSWER: $535.71

*For an alternate method, see Lesson 17-14.

DIAGNOSTIC TEST

1 .2 of what number is 14?
3 .8 of what number is 2?
5 .07 of what number is 56?
7 500 is .625 of what number?

2 20 is .5 of what number?
4 3 is .6 of what number?
6 .18 of what number is 27?
8 .0375 of what number is 150?

9 11 is .8 of what number?   **10** .08 of what number is 100?
11 .24 is .03 of what number?   **12** .25 of what number is $1.50?
13 324 is 1.08 of what number?   **14** 1.15 of what amount is $9,200?

RELATED PRACTICE EXERCISES

1 1. .3 of what number is 12?
   2. .9 of what number is 540?
   3. .8 of what number is 400?
   4. .7 of what number is 63?

2 1. 10 is .2 of what number?
   2. 96 is .4 of what number?
   3. 300 is .6 of what number?
   4. 83 is .1 of what number?

3 1. .8 of what number is 4?
   2. .2 of what number is 1?
   3. .4 of what number is 2?
   4. .5 of what number is 5?

4 1. 3 is .6 of what number?
   2. 6 is .8 of what number?
   3. 4 is .4 of what number?
   4. 1 is .1 of what number?

5 1. .06 of what number is 42?
   2. 8 is .01 of what number?
   3. 72 is .05 of what number?
   4. .09 of what number is 630?

6 1. .12 of what number is 78?
   2. 51 is .75 of what number?
   3. .46 of what number is 1,500?
   4. 123 is .82 of what number?

7 1. 49 is .875 of what number?
   2. .125 of what number is 152?
   3. 180 is .045 of what number?
   4. .098 of what number is 588?

8 1. .3125 of what number is 1,250?
   2. 350 is .4375 of what number?
   3. .1875 of what number is 375?
   4. .0625 of what number is 2,500?

9 1. 15 is .4 of what number?
   2. .5 of what number is 28?
   3. 24 is .08 of what number?
   4. .04 of what number is 16?

10 1. .75 of what number is 150?
   2. .07 of what number is 63?
   3. 24 is .625 of what number?
   4. .18 of what number is 36?

11 1. .92 is .04 of what number?
   2. .15 of what number is .9?
   3. 2.73 is .07 of what number?
   4. .375 of what number is 5.4?

12 1. .06 of what amount is $210?
   2. $1.08 is .75 of what amount?
   3. .12 of what amount is $.84?
   4. $4.75 is .625 of what amount?

13 1. 500 is 1.25 of what number?
   2. 2.5 of what number is $30.75?
   3. 312 is 1.04 of what number?
   4. 1.8 of what number is 9,720?

14 1. 1.75 of what amount is $35?
   2. 1.06 of what amount is $4.77?
   3. $7.80 is 1.2 of what amount?
   4. 2.3 of what amount is $1,978?

SKILL ACHIEVEMENT TEST

1. .5 of what number is 60?
2. 270 is .3 of what number?
3. 9 is .12 of what number?
4. .25 of what number is 775?
5. .09 of what amount is $24.30?
6. 810 is .6 of what number?
7. 267 is .375 of what number?
8. 1.75 of what amount is $2,205?
9. 9 is .18 of what number?
10. $7.02 is 1.08 of what amount?

POWERS OF TEN

The power of a factor is the product determined by multiplying the repeating factor. It usually is expressed as an indicated product using exponents.

The table at right is developed using 10 as the repeated factor.

$$10 = 10 = 10^1$$
$$100 = 10 \times 10 = 10^2$$
$$1,000 = 10 \times 10 \times 10 = 10^3$$
$$10,000 = 10 \times 10 \times 10 \times 10 = 10^4$$
$$100,000 = 10 \times 10 \times 10 \times 10 \times 10 = 10^5$$

Notice in the table that the number of zeros in the product after the digit 1 corresponds to the exponent of 10 in the indicated product.

EXAMPLE: Express 100,000 as a power of ten using exponents.

$$100,000 = 10^5$$

5 zeros   The exponent is 5.

Any numeral having zeros for all digits except the first, such as 7,000, may be expressed as a product of a digit and a power of ten.

EXAMPLE: Express 7,000 as a product of a digit and a power of ten.

$$7,000 = 7 \times 1,000 = 7 \times 10^3$$

3 zeros   3 zeros   The exponent is 3.

Any numeral may be written as the sum of the products of each digit in the numeral and its place value expressed as a power of ten.

EXAMPLES:

Write the numeral 5,963 in expanded form as an indicated sum (polynomial)

5,963
= 5 thousands + 9 hundreds + 6 tens + 3 ones
= $(5 \times 1,000) + (9 \times 100) + (6 \times 10) + (3 \times 1)$
= $(5 \times 10^3) + (9 \times 10^2) + (6 \times 10^1) + (3 \times 10^0)$

Write the numeral .9247 in expanded form.

.9247
= 9 tenths + 2 hundredths + 4 thousandths + 7 ten-thousandths
= $(9 \times .1) + (2 \times .01) + (4 \times .001) + (7 \times .0001)$
= $(9 \times 10^{-1}) + (2 \times 10^{-2}) + (4 \times 10^{-3}) + (7 \times 10^{-4})$

Write the numeral 685.379 in expanded form.
$(6 \times 10^2) + (8 \times 10^1) + (5 \times 10^0) + (3 \times 10^{-1}) + (7 \times 10^{-2}) + (9 \times 10^{-3})$

It can be shown that:
$$.1 = 10^{-1}$$
$$.01 = 10^{-2}$$
$$.001 = 10^{-3}$$
$$.0001 = 10^{-4}$$
and that:
$$1 = 10^0$$

EXERCISES

Express as a power of ten using exponents:
1. 1,000,000,000   2. 10,000,000,000,000,000

Express as a product of a digit and a power of ten:
3. 80,000   4. 500   5. 200,000
6. 3,000,000,000   7. 900,000,000,000   8. 60,000,000

Express each of the following numerals in expanded form as a polynomial:
9. 492   10. 37,026   11. 5,688,247   12. .29
13. .674   14. .58143   15. 215.83   16. 4,153.972

2-11 Multiplying by Powers of Ten

PROCEDURE

1 Multiply 6 by 10; by 100; by 1,000.

$6 \times 10 = 60$

1 zero

$6 \times 100 = 600$

2 zeros

$6 \times 1,000 = 6,000$

3 zeros

Write as many zeros to the right of a whole number as there are in the multiplier.

ANSWERS: 60; 600; 6,000

2 Multiply 95 by 10; by 100; by 1,000.

$95 \times 10 = 950$

1 zero

$95 \times 100 = 9,500$

2 zeros

$95 \times 1,000 = 95,000$

3 zeros

ANSWERS: 950; 9,500; 95,000

3 Multiply .7 by 10; by 100; by 1,000.

$.7 \times 10 = .7 = 7$

1 zero   1 place

$.7 \times 100 = .70 = 70$

2 zeros   2 places

$.7 \times 1,000 = .700 = 700$

3 zeros   3 places

Move the decimal point as many places to the right in a decimal as there are zeros in the multiplier.

ANSWERS: 7; 70; 700

4 Multiply 48.67 by 10; by 100; by 1,000.

$48.67 \times 10 = 48.67 = 486.7$

1 zero   1 place

$48.67 \times 100 = 48.67 = 4,867$

2 zeros   2 places

$48.67 \times 1,000 = 48.670 = 48,670$

3 zeros   3 places

ANSWERS: 486.7; 4,867; 48,670

112

5 Multiply .493 by 1,000,000,000.

.493 × 1,000,000,000 = .493000000 = 493,000,000

                9 zeros   9 places

ANSWER: 493,000,000

DIAGNOSTIC TEST

Multiply each of the following given numbers by 10:

1 8   **2** 60   **3** .4   **4** 78   **5** .06   **6** 25.324

Multiply each of the following given numbers by 100:

7 26   **8** 500   **9** .83   **10** .5   **11** .0987   **12** 67.39

Multiply each of the following given numbers by 1,000:

13 9   **14** 420   **15** .365   **16** .67   **17** .8574   **18** 56.967

Multiply:

19 574.82 by 1,000,000   **20** 125.7 by 1,000,000,000

21 8.273 by 1,000,000,000,000

RELATED PRACTICE EXERCISES

Multiply each of the following given numbers by 10:

1	**2**	**3**	**4**	**5**	**6**
1. 5	**1.** 40	**1.** .3	**1.** .26	**1.** .03	**1.** 5.8
2. 27	**2.** 100	**2.** .1	**2.** .57	**2.** .085	**2.** 96.34
3. 85	**3.** 150	**3.** .9	**3.** .805	**3.** .007	**3.** 49.927
4. 763	**4.** 3,000	**4.** .8	**4.** .4326	**4.** .0625	**4.** 540.653

Multiply each of the following given numbers by 100:

7	**8**	**9**	**10**	**11**	**12**
1. 7	**1.** 20	**1.** .42	**1.** .2	**1.** .721	**1.** 8.54
2. 51	**2.** 300	**2.** .33	**2.** .8	**2.** .039	**2.** 36.46
3. 423	**3.** 590	**3.** .19	**3.** .9	**3.** .5257	**3.** 5.792
4. 9,564	**4.** 7,400	**4.** .95	**4.** .1	**4.** .0416	**4.** 25.875

Multiply each of the following given numbers by 1,000:

13	**14**	**15**	**16**	**17**	**18**
1. 3	**1.** 50	**1.** .657	**1.** .35	**1.** .2653	**1.** 6.582
2. 62	**2.** 200	**2.** .942	**2.** .7	**2.** .0357	**2.** 29.37
3. 597	**3.** 780	**3.** .076	**3.** .09	**3.** .17425	**3.** 81.1
4. 2,055	**4.** 5,000	**4.** .189	**4.** .4	**4.** .00072	**4.** 176.2563

Multiply:

19 By 1,000,000:
1. 952
2. 6.3
3. 47.915
4. 864.26

20 By 1,000,000,000:
1. 83
2. 15.7
3. 203.06
4. 9.658

21 By 1,000,000,000,000:
1. 17
2. 8.9
3. 25.62
4. 1.907

SKILL ACHIEVEMENT TEST

Multiply:

1. 60 by 10
2. 7 by 1,000
3. .83 by 100
4. .5 by 1,000
5. 4.39 by 10
6. 3.092 by 1,000,000
7. 21.6 by 1,000,000,000
8. .9047 by 10
9. .0025 by 1,000
10. 680.4 by 100

Applications

To change kilowatts (kW) to watts, multiply the number of kilowatts by 1,000.

Change to watts:

1. 8 kW
2. 40 kW
3. 6.5 kW
4. 139 kW
5. 27.3 kW

To change centimeters (cm) to millimeters, multiply the number of centimeters by 10.

Change to millimeters:

6. 3 cm
7. 15 cm
8. 4.2 cm
9. 87 cm
10. 18.5 cm

To change meters (m) to centimeters, multiply the number of meters by 100.

Change to centimeters:

11. 5 m
12. 20 m
13. 3.7 m
14. 15.08 m
15. 9.385 m

16. How many dimes are in 28 dollars?
17. Find the number of pennies in 17 dollars.
18. A fuel oil consumer bought 100 gallons of heating oil at $1.19 per gallon. How much did the oil cost?

2-12 Dividing by Powers of Ten

PROCEDURE

1 Divide 85 by 10; by 100; by 1,000.

$$85 \div 10 = 85. = 8.5$$
1 zero   1 place

Move the decimal point as many places to the left as there are zeros in the divisor.

$$85 \div 100 = 85. = .85$$
2 zeros   2 places

Write 1 zero.
$$85 \div 1,000 = 085. = .085$$
3 zeros   3 places

If necessary, write zeros to produce the proper number of decimal places.

ANSWERS: 8.5; .85; .085

2 Divide 7,600 by 10; by 100; by 1,000.

$$7,600 \div 10 = 7,600. = 760$$
1 zero   1 place

Drop any zeros to the right of the decimal point.

$$7,600 \div 100 = 7,600. = 76$$
2 zeros   2 places

$$7,600 \div 1,000 = 7,600. = 7.6$$
3 zeros   3 places

ANSWERS: 760; 76; 7.6

3 Divide 2.46 by 10; by 100; by 1,000.

$$2.46 \div 10 = 2.46 = .246$$
1 zero   1 place

Write 1 zero.
$$2.46 \div 100 = 02.46 = .0246$$
2 zeros   2 places

Write 2 zeros.
$$2.46 \div 1,000 = 002.46 = .00246$$
3 zeros   3 places

ANSWER: .246; .0246; .00246

4 Divide 4,605,000,000 by 1,000,000,000.

$$4,605,000,000 \div 1,000,000,000 = 4,605,000,000. = 4.605$$
9 zeros   9 places

ANSWER: 4.605

DIAGNOSTIC TEST

Divide each of the following given numbers by 10:

1 80 **2** 95 **3** 6 **4** 7 **5** 15.683

Divide each of the following given numbers by 100:

6 400 **7** 875 **8** 92
9 8 **10** .34 **11** 197.2

Divide each of the following given numbers by 1,000:

12 65,000 **13** 2,973 **14** 467
15 72 **16** .675 **17** 527.3

Divide:

18 5,900,000 by 1,000,000 **19** 64,125,000,000 by 1,000,000,000
20 3,720,000,000,000 by 1,000,000,000,000

RELATED PRACTICE EXERCISES

Divide each of the following given numbers by 10:

1
1. 20
2. 600
3. 300
4. 1,000

2
1. 34
2. 276
3. 408
4. 5,426

3
1. 5
2. 3
3. 9
4. 4

4
1. .9
2. .32
3. .08
4. .936

5
1. 3.5
2. 9.82
3. 27.46
4. 39.239

Divide each of the given numbers by 100:

6
1. 200
2. 700
3. 5,000
4. 2,700

7
1. 382
2. 829
3. 4,520
4. 65,726

8
1. 59
2. 32
3. 70
4. 85

9
1. 4
2. 2
3. 3
4. 7

10
1. .21
2. .60
3. .8
4. .045

11
1. 29.5
2. 502.86
3. 68.24
4. 1,500.75

Divide each of the following numbers by 1,000:

12
1. 8,000
2. 10,000
3. 28,000
4. 150,000

13
1. 3,725
2. 2,890
3. 15,925
4. 18,464

14
1. 628
2. 314
3. 200
4. 957

15
1. 85
2. 93
3. 6
4. 8

16
1. .925
2. .56
3. .3
4. .072

17
1. 284.9
2. 500.74
3. 1,526.1
4. 2,963.45

Divide:

18 By 1,000,000:
   1. 617,000,000
   2. 450,000
   3. 9,200,000
   4. 13,670,000

19 By 1,000,000,000:
   1. 39,000,000,000
   2. 6,400,000,000
   3. 775,000,000
   4. 8,350,000,000

20 By 1,000,000,000,000:
   1. 8,000,000,000,000
   2. 12,500,000,000,000
   3. 470,000,000,000
   4. 1,834,000,000,000

SKILL ACHIEVEMENT TEST

Divide:

1. 4,000 by 100

2. 965 by 1,000

3. 16.2 by 10

4. 740.1 by 100

5. 23,000,000 by 1,000,000

6. 5,780 by 100

7. 8,336.5 by 1,000

8. 430.52 by 10

9. 172.914 by 1,000

10. 6,450,000,000 by 1,000,000,000

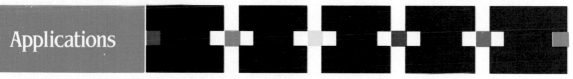

Applications

To change from watt-hours (W·h) to kilowatt-hours (kW·h), divide the number of watt-hours by 1,000.

Change to kilowatt-hours:

1. 20,000 W·h

2. 135,000 W·h

3. 27,500 W·h

4. A bag of sugar, weighing 100 pounds, cost $35.30. What is the cost per pound?

Taxes

5. If the tax rate is $9.80 per hundred dollars, how much must you pay for taxes on a house assessed for $98,900?

6. Find the amount of taxes on properties having the following assessed valuations and tax rates:

Assessed valuation	$3,400	$17,500	$51,800	$45,000	$80,200
Tax rate per $100	$5.70	$6.45	$3.60	$4.25	$5.84

SCIENTIFIC NOTATION

Scientists and mathematicians sometimes work with very large or very small numbers. To simplify their work, they often express such numbers in an abbreviated form known as scientific notation.

To express a number in scientific notation, write the numeral for the given number as the product of a number between 1 and 10 and a power of ten.*

The required power of ten may be determined by counting the number of places the decimal point is moved to get the required whole number or mixed number between 1 and 10.

When you move the decimal point to the left, the exponent is positive.

EXAMPLE: Express 9,800,000 in scientific notation.

$$9,800,000. = 9.8 \times 1,000,000 = 9.8 \times 10^6$$

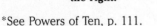

6 places to the left. **The exponent is positive 6.**

When you move the decimal point to the right, the exponent is negative.

EXAMPLE: Express .0000536 in scientific notation.

$$.0000536 = 5.36 \times .00001 = 5.36 \times 10^{-5}$$

5 places to the right. **The exponent is negative 5.**

*See Powers of Ten, p. 111.

EXERCISES

Express each of the following numbers in scientific notation:

1. 60,000
2. 79,000
3. 520,000
4. 4,875,000
5. 2,000,000
6. 308,000,000,000
7. 960,000,000
8. 84,500,000,000,000
9. .0018
10. .00249
11. .06356
12. .0000091
13. .0000365
14. .00000079
15. .000000801
16. .000000000043

Express the numbers in each of the following in scientific notation:

17. The sun at any second develops 500,000,000,000,000,000,000,000 horsepower.

18. The star Alpha Hercules is 3,860,000,000 kilometers in diameter.

19. The weight of the earth is about 6,600,000,000,000,000,000,000 tons.

20. The distance from the sun to the planet Jupiter is about 779,000,000 kilometers.

2-13 Writing Complete Numerals for Shortened Names

PROCEDURE

Write a complete numeral for each of the following:

1 98 thousand

98 thousand = 98 × 1,000

               ↑
            1 thousand

        = 98,000

Multiply by the power of ten equivalent to the value of the period name.

ANSWER: 98,000

2 193 million

193 million = 193 × 1,000,000

                  ↑
              1 million

        = 193,000,000

ANSWER: 193,000,000

3 84 billion

                  1 billion

84 billion = 84 × 1,000,000,000

        = 84,000,000,000

ANSWER: 84,000,000,000

4 7.65 million

7.65 million = 7.65 × 1,000,000 = 7.650000

                    6 zeros   **6 places**

        = 7,650,000

ANSWER: 7,650,000

5 81.9 trillion

81.9 trillion = 81.9 × 1,000,000,000,000 = 81.900000000000

                      12 zeros   **12 places**

ANSWER: 81,900,000,000,000

6 $53.71 billion

53.71 billion = 53.71 × 1,000,000,000 = 53.710000000

                    9 zeros   **9 places**

        = 53,710,000,000

ANSWER: $53,710,000,000

DIAGNOSTIC TEST

Write the complete numeral for each of the following:

1 725 million   **2** 51.9 million   **3** 483.28 million

4 6.071 million   **5** $34.56 million   **6** 49 billion

7 98.4 billion   **8** 2.06 billion   **9** 350.742 billion

10 $691.18 billion   **11** 31 trillion   **12** 85.3 trillion

13 408.57 trillion   **14** 9.529 trillion   **15** $70.04 trillion

16 616 thousand   **17** 18.947 thousand   **18** $5.09 thousand

19 1,826 hundred   **20** 62.3 hundred

RELATED PRACTICE EXERCISES

Write the complete numeral for each of the following:

1
1. 83 million
2. 9 million
3. 460 million
4. 2,805 million

2
1. 6.8 million
2. 84.7 million
3. 358.3 million
4. 1,098.5 million

3
1. 17.45 million
2. 5.07 million
3. 227.93 million
4. 1,540.16 million

4
1. 8.562 million
2. 13.468 million
3. 479.704 million
4. 3,872.835 million

5
1. $4.5 million
2. $27.4 million
3. $75.09 million
4. $408.263 million

6
1. 78 billion
2. 5 billion
3. 609 billion
4. 3,570 billion

7
1. 9.4 billion
2. 72.8 billion
3. 403.7 billion
4. 2,576.3 billion

8
1. 12.06 billion
2. 6.95 billion
3. 820.73 billion
4. 3,505.81 billion

9
1. 7.568 billion
2. 31.047 billion
3. 564.382 billion
4. 1,067.509 billion

10
1. $8.3 billion
2. $30.7 billion
3. $54.61 billion
4. $935.054 billion

11
1. 4 trillion
2. 106 trillion
3. 58 trillion
4. 292 trillion

12
1. 3.9 trillion
2. 207.8 trillion
3. 83.1 trillion
4. 6.7 trillion

13
1. 5.92 trillion
2. 60.07 trillion
3. 139.78 trillion
4. 308.59 trillion

14
1. 8.227 trillion
2. 41.058 trillion
3. 250.829 trillion
4. 11.256 trillion

15
1. $9.6 trillion
2. $3.28 trillion
3. $87.92 trillion
4. $40.071 trillion

16
1. 9 thousand
2. 528 thousand
3. 77 thousand
4. 3,863 thousand

17
1. 8.7 thousand
2. 28.5 thousand
3. 63.09 thousand
4. 592.16 thousand

18
1. $19 thousand
2. $51.6 thousand
3. $200.52 thousand
4. $1,174.49 thousand

19
1. 38 hundred
2. 952 hundred
3. 2,433 hundred
4. $77 hundred

20
1. 8.5 hundred
2. 63.27 hundred
3. $83.6 hundred
4. $4.72 hundred

SKILL ACHIEVEMENT TEST

Write the complete numeral for each of the following:

1. 36.7 million
2. 625 billion
3. 1.8 trillion
4. 849 thousand
5. $9.28 million
6. 70.9 hundred
7. 4.56 billion
8. 14.603 trillion
9. 685.362 million
10. $27.83 trillion
11. $769.05 billion
12. 170.8 thousand

Enrichment

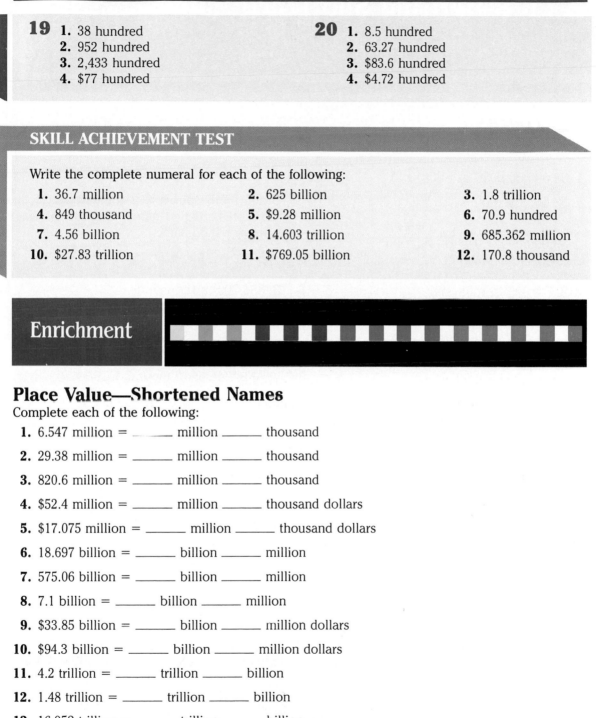

Place Value—Shortened Names

Complete each of the following:

1. 6.547 million = _____ million _____ thousand

2. 29.38 million = _____ million _____ thousand

3. 820.6 million = _____ million _____ thousand

4. $52.4 million = _____ million _____ thousand dollars

5. $17.075 million = _____ million _____ thousand dollars

6. 18.697 billion = _____ billion _____ million

7. 575.06 billion = _____ billion _____ million

8. 7.1 billion = _____ billion _____ million

9. $33.85 billion = _____ billion _____ million dollars

10. $94.3 billion = _____ billion _____ million dollars

11. 4.2 trillion = _____ trillion _____ billion

12. 1.48 trillion = _____ trillion _____ billion

13. 16.952 trillion = _____ trillion _____ billion

14. $8.5 trillion = _____ trillion _____ billion dollars

15. $21.76 trillion = _____ trillion _____ billion dollars

2-14 Writing Shortened Names for Numbers

PROCEDURE

Write the shortened number name for each of the following:

1 58,000,000,000 in billions

$$58,000,000,000 \div 1,000,000,000 = 58$$
↑
1 billion

58,000,000,000 = 58 billion
ANSWER: 58 billion

Divide by the power of ten equivalent to the value of the period name. Write the quotient with the period name.

2 8,910,000 in millions

$$8,910,000 \div 1,000,000 = 8.91$$
↑
1 million

8,910,000 = 8.91 million
ANSWER: 8.91 million

3 556,700 in thousands

$$556,700 \div 1,000 = 556.7$$
↑
1 thousand

556,700 = 556.7 thousand
ANSWER: 556.7 thousand

4 9,832,000,000,000 in trillions

$$9,832,000,000,000 \div 1,000,000,000,000 = 9.832$$
↑
1 trillion

9,832,000,000,000 = 9.832 trillion
ANSWER: 9.832 trillion

5 7,250 in hundreds

$$7,250 \div 100 = 72.50 = 72.5$$
↑
1 hundred

7,250 = 72.5 hundred
ANSWER: 72.5 hundred

6 $68,400,000 in millions of dollars.

$$68,400,000 \div 1,000,000 = 68.4$$
↑
1 million

$68,400,000 = $68.4 million
ANSWER: $68.4 million

DIAGNOSTIC TEST

Write the shortened number name for each of the following:

In hundreds: **1** 21,400   **2** 6,830   **3** 8,745

In thousands: **4** 1,632,000   **5** 750,600   **6** 9,050   **7** 48,409

In millions: **8** 35,000,000   **9** 4,700,000
                   10 264,010,000   **11** 80,557,000

In billions: **12** 167,000,000,000   **13** 4,200,000,000
                   14 379,460,000,000   **15** 50,088,000,000

In trillions: **16** 45,000,000,000,000   **17** 8,100,000,000,000
                   18 66,090,000,000,000   **19** 724,525,000,000,000

20 In thousands of dollars: $27,900
21 In millions of dollars: $838,250,000
22 In billions of dollars: $96,080,000,000
23 In trillions of dollars: $4,202,000,000,000

RELATED PRACTICE EXERCISES

Write the shortened number name for each of the following:

In hundreds: In hundreds: In hundreds:
1 **1.** 900   **2** **1.** 1,340   **3** **1.** 2,568
   2. 3,200   **2.** 870   **2.** 693
   3. 7,100   **3.** 4,950   **3.** 5,756
   4. 12,500   **4.** 20,820   **4.** 33,497

In thousands: In thousands: In thousands:
4 **1.** 6,000   **5** **1.** 8,500   **6** **1.** 7,130
   2. 11,000   **2.** 260,800   **2.** 8,040
   3. 1,437,000   **3.** 53,700   **3.** 37,290
   4. 282,000   **4.** 4,839,100   **4.** 626,750

In thousands: In millions: In millions:
7 **1.** 2,654   **8** **1.** 8,000,000   **9** **1.** 7,300,000
   2. 13,852   **2.** 43,000,000   **2.** 256,800,000
   3. 824,905   **3.** 119,000,000   **3.** 69,200,000
   4. 1,303,056   **4.** 2,640,000,000   **4.** 3,108,500,000

In millions: In millions: In billions:
10 **1.** 11,470,000   **11** **1.** 28,063,000   **12** **1.** 9,000,000,000
    2. 6,080,000   **2.** 4,709,000   **2.** 23,000,000,000
    3. 408,950,000   **3.** 396,677,000   **3.** 102,000,000,000
    4. 1,584,120,000   **4.** 54,096,000   **4.** 2,421,000,000,000

In billions:

13 1. 7,500,000,000
2. 32,700,000,000
3. 184,300,000,000
4. 1,421,900,000,000

In billions:

15 1. 51,006,000,000
2. 4,592,000,000
3. 608,028,000,000
4. 4,312,454,000,000

In trillions:

17 1. 4,700,000,000,000
2. 26,400,000,000,000
3. 870,500,000,000,000
4. 53,600,000,000,000

In trillions:

19 1. 17,526,000,000,000
2. 2,905,000,000,000
3. 56,847,000,000,000
4. 480,061,000,000,000

In millions of dollars:

21 1. $84,000,000
2. $9,360,000
3. $450,844,000
4. $2,655,007,000

In trillions of dollars:

23 1. $6,500,000,000,000
2. $210,680,000,000,000
3. $36,493,000,000,000
4. $9,057,000,000,000

In billions:

14 1. 16,820,000,000
2. 6,370,000,000
3. 309,760,000,000
4. 86,940,000,000

In trillions:

16 1. 6,000,000,000,000
2. 263,000,000,000,000
3. 95,000,000,000,000
4. 107,000,000,000,000

In trillions:

18 1. 32,080,000,000,000
2. 8,450,000,000,000
3. 97,530,000,000,000
4. 208,410,000,000,000

In thousands of dollars:

20 1. $63,000
2. $19,500
3. $420,930
4. $56,078

In billions of dollars:

22 1. $58,000,000,000
2. $15,060,000,000
3. $8,377,000,000
4. $142,522,000,000

SKILL ACHIEVEMENT TEST

Write the shortened number name for each of the following:

1. 47,000,000 in millions
2. 693,000 in thousands
3. 20,300,000,000 in billions
4. 5,460,000,000,000 in trillions
5. 751,400,000 in millions
6. 2,910 in hundreds
7. $807,500 in thousands of dollars
8. $95,620,000 in millions of dollars
9. $6,892,000,000,000 in trillions of dollars
10. $725,314,000,000 in billions of dollars

OPERATIONS WITH NUMBERS
WITH SHORTENED NAMES

> $63.45 *billion* is read:
>
> Sixty-three point four five billion dollars
> and is the shortened name for $63,450,000,000.

We add, subtract, multiply, and divide numbers with shortened names just as we do with whole numbers and decimals.

EXAMPLES:

Add: 82.6 million + 4.7 million

   82.6 million
+   4.7 million
   87.3 million

Subtract: 21.87 trillion − 11.78 trillion

   21.87 trillion
− 11.78 trillion
  10.09 trillion

Multiply: 8 × 4.6 billion

  4.6 billion
×   8
36.8 billion

Divide: 148.54 billion ÷ 2

   74.27 billion
2)148.54 billion

EXERCISES

Read, or write in words, each shortened name in the following:

1. Every day 218 million consumers in the United States spend more than 3 billion dollars.

2. An electric utility requested $190 million increase in rates for the year.

3. Total Canadian newsprint shipments were 2.38 million metric tons.

4. In one recent year the personal income increased to 1.375 trillion dollars.

5. Earnings of a certain corporation rose to $252.3 million from $244.2 million and revenues increased to $3.8 billion from $3.45 billion.

Add:

6. 1.4 trillion + 2.36 trillion + 5.5 trillion
7. $1.8 billion + $675 million

Subtract:

8. 56.2 thousand − 7.8 thousand
9. $2.6 trillion − $73 billion

Multiply:

10. 1.9 × 5.8 billion
11. 25 × $30.64 million

Divide:

12. 3.44 trillion ÷ 8
13. $73.6 million ÷ 2.3 million

14. Earnings of a corporation this year doubled from last year. If the earnings last year were $18.2 million, what are the earnings for this year?

15. The township school budget was $9.79 million last year but is $310,000 more this year. What is the budget this year?

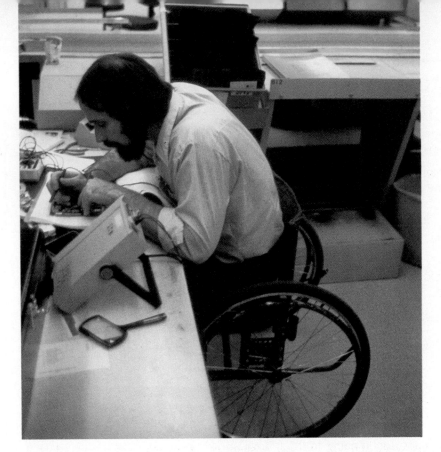

Career Applications

Many products used in industry or by the consumer require factory workers with a wide variety of skills to manufacture them. Pattern makers make very accurate patterns or molds for various articles that are mass-produced. Tool-and-die makers make tools, measuring devices, and other precision parts. Machinists make metal parts for machinery. Welders use heat and/or electricity to join parts together. Assemblers put together articles such as electronic components. Inspectors make sure that parts and finished products meet specifications. Industrial engineers work out ways for people and machines to work most efficiently.

Most of these factory workers are high school graduates. They spend as long as five years as apprentices. Industrial engineers are college graduates.

1. A welder needs three pieces of I-beam with lengths of 5.85 meters, 12.6 meters, and 19.32 meters. What is the total length of I-beam needed?

2. A machinist is using steel rod that weighs .11 kilograms per inch. How much does a piece 27.5 inches long weigh?

3. An inspector measures a part as 2.64 centimeters long. The part should be 2.59 centimeters long. How much shorter should it be?

4. A laminated helicopter rotor blade is 16.31 millimeters thick. It is made of 7 sheets of laminating material. How thick is each sheet?

5. A welder uses 1.138 cubic meters of acetylene gas to make one bracket. How much gas is needed to make 15 brackets?

6. A punch press operator is paid $1.42 per molding. If the operator is paid $306.72, how many moldings were pressed?

7. A machinist makes 73 gears on Monday, 89 on Tuesday, 65 on Wednesday, 91 on Thursday, and 82 on Friday. What is the total number for the week?

8. An industrial engineer finds that a worker can chrome plate 96 trailer hitches per hour. How many hitches can the worker plate in a 40-hour week?

9. A tool-and-die maker measures a part as .0324 centimeter thick. The part should be .0314 centimeter thick. How much thinner should it be?

10. An airplane propeller is made of 5 layers of laminating material, each 4.52 millimeters thick. How thick is the propeller?

11. A welder made 275 parts last week and 302 this week. How many more parts were made this week?

12. A stack of 24 sheets of sheet metal is 9 inches high. How thick is each sheet?

13. An industrial engineer finds that a spray painter can paint 14 panels in an hour. How many workers are needed to spray paint 378 panels in an hour?

14. An assembler is paid $.43 for each flex lead assembled. How much is the assembler paid for assembling 724 flex leads?

15. A pattern maker measures a pattern as 8.225 centimeters wide. The pattern should be 8.243 centimeters wide. How much wider should it be?

16. Copper used in making printed circuits is .4 millimeter thick. What is the thickness of 8 layers of copper?

17. A machinist cuts a steel rod 30 inches long into 8 equal pieces. How long is each piece?

18. A machinist sets a machine to advance a cutting tool 1.125 inches per minute. How long will it take to advance 13.75 inches? (nearest hundredth)

19. A tool-and-die maker drills a hole with its center 8.375 inches from the end of the plate. She drills a second hole 2.125 inches farther from the first. How far from the end of the plate is the second hole?

20. An industrial engineer finds that four welders can weld an average of 41 brackets each per hour. How many brackets can be welded in an 8-hour day by these four welders?

CHAPTER REVIEW

> **VOCABULARY**
> decimal p.66   > symbol p.89   < symbol p.89   mixed decimal p.66

EXERCISES The numerals in boxes indicate Lessons where help may be found.

Write each of the following as a decimal numeral: 2-2

1. Nine hundredths
3. Eight and five tenths
5. Seven thousand eight hundred twenty-two ten-thousandths

2. Five hundred and forty-three thousandths
4. Eleven millionths

Write each of the following numerals in words: 2-1

6. .7   **7.** .006   **8.** 4.09   **9.** .80501   **10.** 62.5483

Round: 2-3

11. .96 correct to the nearest tenth
13. 25.1427 correct to the nearest thousandth

12. 8.2915 correct to the nearest hundredth
14. $4.279 correct to the nearest cent

Add: **15.** 8.2   **16.** 8.05   **17.** 8.2 + .19 + 24
2-4   5.9   2.03
             4.5   5.04
                          1.08

18. Find the sum of: 49.2; .871; and 6.45
19. $6.21 + $58.14 + $.68 + $9.85 + $180.07

Subtract: **20.** 47.512   **21.** .489 − .3   **22.** .8 − .375
2-5   19.835   **23.** 10 − .04   **24.** Subtract $3.25 from $28.

Which is greater: 2-6   **25.** .8 or .62?   **26.** 1.06 or .305?

Which is less: 2-6   **27.** .425 or .94?   **28.** 2.9 or .49?

29. Arrange in order, greatest first: 1.67; .0167; .167; 16.7 2-6

Multiply: **30.** .04   **31.** 90.72   **32.** .018 × .05   **33.** .75   of $5.20
2-7   .2   .125

Divide: 2-8

34. 78)‾.624   **35.** 1.2)‾11.076   **36.** 35 ÷ .875   **37.** $.04)‾$2
38. Find correct to the nearest tenth: 39.37)‾582   **39.** What decimal part of 20 is 11? 2-9
40. .06 of what amount is $450? 2-10

Multiply: 2-11   Divide: 2-12
41. 100 × .08   **42.** 1,000 × 26.9   **43.** 93.2 ÷ 1,000   **44.** 654 ÷ 100

45. Write the complete numeral for $18.38 million. 2-13

46. Write the shortened name for 504,703,000,000 in billions. 2-14

47. Find the cost of 940 liters of fuel oil at $.324 per liter. 2-7

48. A basketball player scored 871 points in 56 games. Find the number of points averaged per game to the nearest tenth of a point. 2-8

COMPETENCY CHECK TEST

The numerals in boxes indicate Lessons where help may be found.

Solve each problem and select the letter corresponding to your answer.

1. Add: 8,416 + 5,267 + 6,329 + 9,755 $\boxed{1\text{-}4}$
   a. 39,776   **b.** 29,667   **c.** 39,767   **d.** 29,767

2. Subtract: 40,805 − 30,945 $\boxed{1\text{-}5}$
   a. 10,860   **b.** 9,860   **c.** 9,850   **d.** 10,850

3. Multiply: 36 × 68 $\boxed{1\text{-}6}$
   a. 504   **b.** 21,888   **c.** 2,448   **d.** Answer not given

4. Divide: 79,550 ÷ 86 $\boxed{1\text{-}7}$
   a. 925   **b.** 840   **c.** 960   **d.** 885

5. Add: 50.8 + .392 $\boxed{2\text{-}4}$
   a. 51.192   **b.** .900   **c.** 900   **d.** 50.8392

6. Subtract: .004 − .0003 $\boxed{2\text{-}5}$
   a. .001   **b.** .0043   **c.** .0037   **d.** .0001

7. Multiply: .007 × 400 $\boxed{2\text{-}7}$
   a. 28   **b.** 2.8   **c.** .028   **d.** .28

8. Divide: 3.5 ÷ .07 $\boxed{2\text{-}8}$
   a. .05   **b.** .5   **c.** 5   **d.** 50

9. Select the smallest number: $\boxed{2\text{-}6}$
   a. .852   **b.** .0825   **c.** .582   **d.** .0852

10. 300 is .25 of: $\boxed{2\text{-}10}$
    a. 75   **b.** 325   **c.** 900   **d.** 1,200

11. Write the complete numeral for 837.4 million. $\boxed{2\text{-}13}$
    a. 8,374,000   **b.** 83,740,000,000
    c. 837,400,000   **d.** .8374

12. Write the shortened name for 25,460,000,000 in billions. $\boxed{2\text{-}14}$
    a. 2.546 billion   **b.** 25.46 billion
    c. 254.6 billion   **d.** Answer not given

13. The leading batter in the American League had a batting average of .349 and in the National League, .355. Who had the higher average and how much higher? $\boxed{2\text{-}5}$ and $\boxed{2\text{-}6}$
    a. National, .060   **b.** National, .014
    c. National, .006   **d.** Answer not given

14. The original funding to build the Northeast Corridor railroad system was $1.75 billion. It appears that funding will be increased to $2.5 billion. Find the amount of increase in funding. $\boxed{2\text{-}5}$ and $\boxed{2\text{-}13}$
    a. $1.25 billion   **b.** $750 million
    c. $75 million   **d.** $1.5 billion

CUMULATIVE ACHIEVEMENT TEST

The numerals in boxes indicate Lessons where help may be found.

1. Add: 2,958
    1-4   47,426
           8,019
             846
          25,625

2. Subtract: 1,600,520
    1-5   549,618

3. Multiply: 6,080
    1-6   320

4. Divide: 456)176,472
    1-7

5. Round 83,499,607,192 to the nearest billion. 1-3

6. Find the greatest common factor (GCF) of 45, 75, and 105. 1-8

7. Add: 8.3 + .75 + 14
    2-4

8. Subtract: $92 − $3.85
    2-5

9. Multiply: 2.5 × .04
    2-7

10. Divide: .02).001
    2-8

11. Write as a decimal numeral: Seven hundred and forty-six thousandths 2-2

12. Round 67.928 to the nearest hundredth. 2-3

13. Multiply 91.04 by 1,000. 2-11

14. Divide 683.1 by 100. 2-12

15. Which is greater: .35 or .279? 2-6

16. What decimal part of 60 is 3.6? 2-9

17. .08 of what amount is $12.64? 2-10

18. Write 6.85 trillion as a complete numeral. 2-13

19. Write the shortened name for 87,420,000 in millions. 2-14

20. One year the national debt ceiling was $1.517 trillion. If Congress increased the ceiling by $53 billion, what was the new debt ceiling? 2-14

21. A package of ground meat weighs 1.85 pounds. At $2.19 per pound, find the cost of the ground meat to the nearest cent. 2-7

22. The average price of a gallon of gasoline in New Jersey is $1.109, but in California it is $1.339. Find the difference in price. 2-5

23. Juan purchased a tennis racket for $37.49 and a can of tennis balls for $2.79. The sales tax is $2.42. How much change should he get from $50? 2-5

24. A student measured a 2-inch metal block by a precision instrument and made the following readings: 1.9962 inches, 2.0008 inches, 2.0012 inches, 1.9993 inches, and 2.0015 inches. Find the average reading. 2-4 and 2-8

25. A color television set costs $595 or $75 down and 12 monthly payments of $51.25 each. How much do you save by paying cash? 2-5 and 2-7

Calculator Skills

Use a calculator to check whether the following answers are correct or incorrect.

Add:

1.	**2.**	**3.**	**4.**	**5.**
29	8,658	49,968	73,893	615,968
496	3,709	38,947	29,529	384,557
5,847	5,266	70,539	83,154	509,664
69	1,975	56,388	40,787	832,375
758	2,763	83,584	78,919	987,976
7,199	22,471	297,426	306,282	3,330,540

Subtract:

6.	**7.**	**8.**	**9.**	**10.**
5,629	36,503	694,000	725,137	1,854,006
4,819	8,307	385,638	269,582	860,849
810	28,196	319,362	455,555	983,157

Multiply:

11. $563 \times 78 = 43,914$
12. $887 \times 645 = 582,115$
13. $609 \times 527 = 320,943$
14. $7,403 \times 9,056 = 67,031,568$

Divide:

15. $36,421 \div 43 = 837$
16. $10,800 \div 16 = 675$
17. $682,864 \div 728 = 948$
18. $409,236 \div 804 = 509$

Add:

19. $.905 + .690 + .347 = 1.942$
20. $.8 + .56 = .64$
21. $.836 + 9 = 9.836$
22. $2.67 + 81.5 = 83.72$

Subtract:

23. $.582 - .496 = .86$
24. $73.8 - 5 = 68.8$
25. $80 - 69.3 = 10.7$
26. $6.834 - 4.62 = 2.214$

Multiply:

27. $36 \times .89 = 32.04$
28. $4.5 \times .008 = .0036$
29. $17.9 \times 82.4 = 1,464.96$
30. $.013 \times .004 = .000052$

Divide:

31. $.84 \div 14 = .06$
32. $60 \div .05 = 120$
33. $9.23 \div .2 = 46.15$
34. $3 \div 40 = .75$

Some calculators can:

(1) Round *off* decimals as explained in Lesson 2-3.
(2) Round *up* decimals. One (1) is always added when the figure dropped is greater than zero (0).
(3) Round *down* decimals. Nothing is added. Figures are just dropped.

Examples:

5.3682 rounded *off* to the nearest hundredth is 5.37

      rounded *up* to hundredths is 5.37
      rounded *down* to hundredths is 5.36

34.2817 rounded *off* to the nearest hundredth is 34.28

      rounded *up* to thousandths is 34.282
      rounded *down* to hundredths is 34.28

Round *up* each of the following:

35. 5.637 to hundredths
38. 3.6192 to hundredths

36. 10.496 to hundredths
39. 24.952 to tenths

37. 8.2527 to thousandths
40. 7.5584 to thousandths

Round *down* each of the following:

41. 18.143 to hundredths
44. 8.2059 to thousandths

42. 9.2528 to hundredths
45. 69.7235 to tenths

43. 7.5148 to tenths
46. 80.1439 to hundredths

Computer Activities

The computer program below is written in the BASIC language. It is designed to help you practice rounding decimals to the nearest *tenth*. Enter the program on your computer keyboard exactly as it appears, one line at a time. Press the RETURN or ENTER key at the end of each line.

10 Clearscreen	Clears the screen. (Use appropriate command for *your* computer.)
20 LET X = RND(1)	Generates a "random" (unpredictable) decimal number and stores it in memory location X.
30 LET Z = INT(10*X+.5)/10	Rounds X to the nearest tenth and stores it in memory location Z. This is the correct answer.
40 PRINT X	Prints the number X (the problem to be solved.)
50 INPUT A	Displays a "?" as a way of asking for your answer.
60 LET D = ABS(A − Z)	Determines D, the difference between your answer and the correct answer.
70 IF D < .0001 THEN PRINT "YES"	Says "YES" if your answer is correct.
80 IF D > = .0001 THEN PRINT "NO"	Says "NO" if your answer is incorrect.
90 GOTO 20	Sends the computer back to line 20 for another problem.

(Note: TRS-80 and IBM users replace RND(1) in line 20 with RND(0)).

If you make a typing mistake:

    (a) Use the ← key to back up and make corrections.

       or

    (b) Simply press RETURN or ENTER and retype the line.

EXERCISES

1. Type RUN and press RETURN or ENTER to start your program. Use the program until you have achieved mastery with rounding to the nearest tenth.

2. Change the program to help you practice rounding decimals to the nearest *hundredth*. Retype your line 30 as follows:

           30 LET Z = INT(100*X+.5)/100

3. Change the program to help you practice rounding decimals to the nearest *thousandth*. Again, only line 30 needs to be changed.

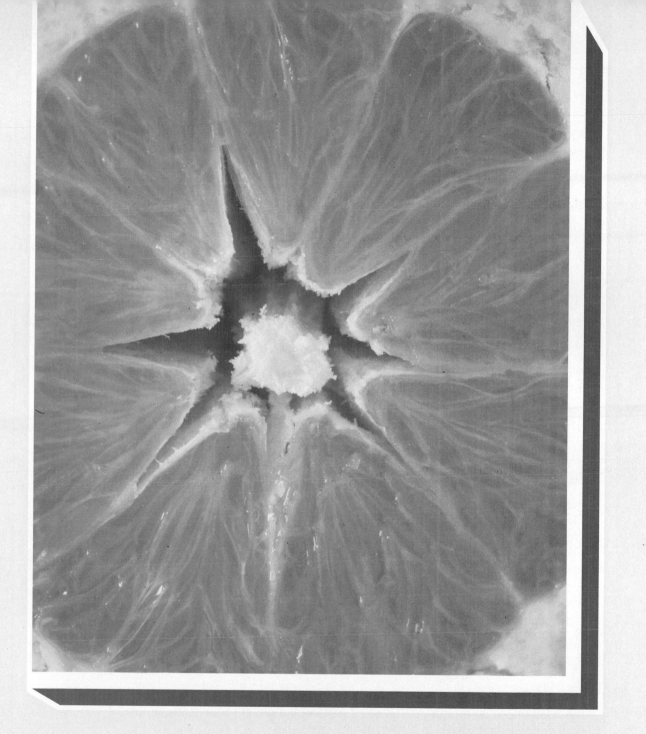

CHAPTER 3 Fractions

INTRODUCTION TO FRACTIONS

When an object or unit is divided into equal parts, the number expressing the relation of one or more equal parts to the total number of equal parts is called a fraction.

If a piece of ribbon is divided into two equal parts, each part is one-half of the whole piece of ribbon and is represented by the fraction symbol $\frac{1}{2}$.

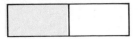

If the piece of ribbon is divided into:

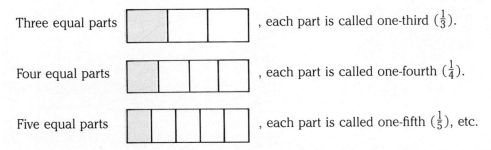

Three equal parts   , each part is called one-third ($\frac{1}{3}$).

Four equal parts   , each part is called one-fourth ($\frac{1}{4}$).

Five equal parts   , each part is called one-fifth ($\frac{1}{5}$), etc.

The symbol for a fraction consists of a pair of numerals, one written above the other with a horizontal bar between them.

In the fraction $\frac{5}{8}$:

The numerals 5 and 8 are the terms of the fraction.
The numeral above the fraction bar, 5, is called the numerator.
The numeral below the fraction bar, 8, is called the denominator

The denominator tells us the number of equal parts into which an object is divided. The numerator tells us how many equal parts are being used. The denominator cannot be zero.

The fraction $\frac{5}{8}$ means 5 parts of 8 equal parts.

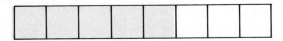

Sometimes a group of objects is divided into equal parts. The number expressing the relation of one or more of the equal parts of the group to the total number of equal parts is considered to be a fraction.

Observe above ($\frac{1}{5}$ compared to $\frac{1}{3}$) that the larger the denominator, the smaller the size of the part.

3-1 Expressing Fractions in Lowest Terms

PROCEDURE

1 Express $\frac{5}{15}$ in lowest terms.

numerator $\longrightarrow$ $\frac{\mathbf{5}}{\mathbf{15}} = \frac{5 \div 5}{15 \div 5}$
denominator $\longrightarrow$

$= \frac{1}{3} \longleftarrow$ **lowest terms, or simplest form**

ANSWER: $\frac{1}{3}$

Divide the numerator and the denominator of the fraction by the greatest number (GCF) that can be divided exactly into both.

5: 1   5
15: 1   3   5   15

2 Express $\frac{63}{72}$ in simplest form.

$\frac{63}{72} = \frac{63 \div 9}{72 \div 9} = \frac{7}{8}$

ANSWER: $\frac{7}{8}$

63: 1 3 7 9 21 63
72: 1 2 3 4 6 8 9 12 18 24 36 72

3 Express $\frac{18}{29}$ in lowest terms.

ANSWER: $\frac{18}{29}$

The GCF of 18 and 29 is 1. So the fraction is already in lowest terms.

Vocabulary

A fraction is in lowest terms, or simplest form, when its numerator and denominator cannot be divided exactly by the same number, except by 1.

DIAGNOSTIC TEST

Express the following fractions in lowest terms:

1 $\frac{3}{27}$ **2** $\frac{18}{36}$ **3** $\frac{6}{9}$ **4** $\frac{48}{64}$ **5** $\frac{26}{39}$ **6** $\frac{648}{852}$ **7** $\frac{250}{1000}$

RELATED PRACTICE EXERCISES

Express the following fractions in lowest terms:

1   1. $\frac{2}{4}$ 2. $\frac{5}{15}$ 3. $\frac{7}{21}$ 4. $\frac{3}{12}$ 5. $\frac{7}{35}$ 6. $\frac{3}{18}$ 7. $\frac{11}{88}$ 8. $\frac{23}{46}$

2   1. $\frac{8}{16}$ 2. $\frac{4}{40}$ 3. $\frac{8}{48}$ 4. $\frac{6}{42}$ 5. $\frac{9}{54}$ 6. $\frac{12}{36}$ 7. $\frac{32}{96}$ 8. $\frac{21}{84}$

3 1. $\frac{8}{10}$ 2. $\frac{4}{6}$ 3. $\frac{6}{20}$ 4. $\frac{10}{25}$ 5. $\frac{9}{12}$ 6. $\frac{6}{15}$ 7. $\frac{10}{16}$ 8. $\frac{18}{32}$

4 1. $\frac{36}{60}$ 2. $\frac{12}{16}$ 3. $\frac{8}{20}$ 4. $\frac{24}{32}$ 5. $\frac{20}{36}$ 6. $\frac{32}{48}$ 7. $\frac{50}{75}$ 8. $\frac{56}{64}$

5 1. $\frac{38}{57}$ 2. $\frac{68}{85}$ 3. $\frac{57}{76}$ 4. $\frac{69}{92}$ 5. $\frac{52}{91}$ 6. $\frac{62}{93}$ 7. $\frac{34}{119}$ 8. $\frac{91}{104}$

6 1. $\frac{70}{112}$ 2. $\frac{28}{400}$ 3. $\frac{84}{192}$ 4. $\frac{135}{144}$ 5. $\frac{180}{216}$ 6. $\frac{420}{756}$ 7. $\frac{405}{567}$ 8. $\frac{680}{765}$

7 1. $\frac{10}{40}$ 2. $\frac{20}{60}$ 3. $\frac{80}{100}$ 4. $\frac{150}{600}$ 5. $\frac{300}{1,000}$ 6. $\frac{1,200}{2,000}$ 7. $\frac{750}{2,500}$ 8. $\frac{2,000}{3,600}$

SKILL ACHIEVEMENT TEST

Express each of the following fractions in lowest terms:

1. $\frac{2}{12}$ 2. $\frac{21}{28}$ 3. $\frac{15}{45}$ 4. $\frac{42}{60}$ 5. $\frac{40}{200}$

6. $\frac{58}{87}$ 7. $\frac{31}{93}$ 8. $\frac{85}{102}$ 9. $\frac{256}{352}$ 10. $\frac{3,500}{4,900}$

Applications

Express the following measurements in lowest terms:

1. $\frac{2}{4}$ inch   2. $\frac{14}{16}$ inch   3. $\frac{6}{8}$ inch   4. $\frac{20}{32}$ inch   5. $\frac{36}{64}$ inch

6. $\frac{8}{12}$ foot   7. $\frac{3}{12}$ foot   8. $\frac{30}{36}$ yard   9. $\frac{220}{1,760}$ mile   10. $\frac{3,960}{5,280}$ mile

11. $\frac{10}{16}$ pounds   12. $\frac{600}{2,000}$ ton   13. $\frac{840}{2,240}$ ton   14. $\frac{24}{32}$ quart   15. $\frac{4}{8}$ peck

16. $\frac{24}{60}$ hour   17. $\frac{45}{60}$ minute   18. $\frac{9}{24}$ day   19. $\frac{10}{12}$ year   20. $\frac{73}{365}$ year

Which is greater: **21.** $\frac{9}{16}$ inch or $\frac{28}{64}$ inch?   **22.** $\frac{12}{16}$ inch or $\frac{7}{8}$ inch?

Which is less: **23.** $\frac{26}{32}$ inch or $\frac{15}{16}$ inch?   **24.** $\frac{3}{8}$ inch or $\frac{40}{64}$ inch?

25. A student read the length of a metal block as $\frac{56}{64}$ of an inch. In what other way could this measurement have been expressed?

Select the fractions which are not in lowest terms:

27. $\frac{7}{10}$ $\frac{16}{21}$ $\frac{18}{81}$ **28.** $\frac{21}{25}$ $\frac{63}{72}$ $\frac{27}{58}$

3-2 Rewriting Improper Fractions and Simplifying Mixed Numbers

PROCEDURE

1 Rewrite $\frac{6}{3}$ as a whole number.

$\frac{6}{3} = 6 \div 3 = 2$

ANSWER: 2

Divide the numerator by the denominator.

2 Rewrite $\frac{8}{5}$ as a mixed number.

$\frac{8}{5} = 8 \div 5$

$$\begin{array}{r} 1 \leftarrow \textbf{whole number} \\ 5\overline{)8} \\ \underline{5} \\ 3 \leftarrow \textbf{numerator of fraction} \end{array}$$

denominator of fraction

$\frac{8}{5} = 1\frac{3}{5}$

ANSWER: $1\frac{3}{5}$

3 Rewrite $\frac{14}{8}$ as a mixed number.

$\frac{14}{8} = 1\frac{6}{8} = 1\frac{3}{4}$

ANSWER: $1\frac{3}{4}$

Simplify the fraction.

4 Simplify $6\frac{14}{9}$.

$6\frac{14}{9} = 6 + \frac{14}{9}$

$\quad = 6 + 1\frac{5}{9}$

$\quad = 7\frac{5}{9}$

ANSWER: $7\frac{5}{9}$

Rewrite mixed numbers as the sum of the whole number and the fraction.
Simplify the fraction.
Add.

5 Simplify $4\frac{8}{8}$.

$4\frac{8}{8} = 4 + \frac{8}{8}$

$\quad = 4 + 1$

$\quad = 5$

ANSWER: 5

6 Simplify $3\frac{6}{4}$.

$3\frac{6}{4} = 3 + \frac{6}{4}$

$\quad = 3 + 1\frac{2}{4}$

$\quad = 3 + 1\frac{1}{2}$

$\quad = 4\frac{1}{2}$

ANSWER: $4\frac{1}{2}$

DIAGNOSTIC TEST

Rewrite each of the following improper fractions as a whole number or a mixed number:

1 $\frac{18}{6}$ **2** $\frac{5}{3}$ **3** $\frac{12}{8}$ **4** $\frac{19}{4}$ **5** $\frac{35}{10}$

Rewrite each of the following mixed numbers in simplest form:

6 $5\frac{6}{6}$ **7** $6\frac{4}{2}$ **8** $3\frac{12}{16}$ **9** $9\frac{13}{10}$ **10** $7\frac{14}{8}$

RELATED PRACTICE EXERCISES

Rewrite each of the following improper fractions as a whole number or a mixed number:

1 1. $\frac{8}{8}$ 2. $\frac{5}{5}$ 3. $\frac{12}{4}$ 4. $\frac{4}{2}$ 5. $\frac{15}{5}$ 6. $\frac{18}{3}$ 7. $\frac{16}{2}$ 8. $\frac{36}{9}$

2 1. $\frac{3}{2}$ 2. $\frac{7}{5}$ 3. $\frac{4}{3}$ 4. $\frac{13}{8}$ 5. $\frac{25}{16}$ 6. $\frac{14}{9}$ 7. $\frac{19}{12}$ 8. $\frac{13}{10}$

3 1. $\frac{6}{4}$ 2. $\frac{10}{6}$ 3. $\frac{12}{9}$ 4. $\frac{14}{8}$ 5. $\frac{28}{16}$ 6. $\frac{18}{10}$ 7. $\frac{38}{24}$ 8. $\frac{52}{32}$

4 1. $\frac{9}{2}$ 2. $\frac{7}{3}$ 3. $\frac{12}{5}$ 4. $\frac{25}{8}$ 5. $\frac{22}{7}$ 6. $\frac{41}{6}$ 7. $\frac{32}{9}$ 8. $\frac{67}{12}$

5 1. $\frac{10}{4}$ 2. $\frac{28}{8}$ 3. $\frac{34}{6}$ 4. $\frac{60}{16}$ 5. $\frac{45}{20}$ 6. $\frac{33}{9}$ 7. $\frac{100}{32}$ 8. $\frac{80}{24}$

Rewrite each of the following numbers in simplest form:

6 1. $2\frac{3}{3}$ 2. $7\frac{2}{2}$ 3. $5\frac{7}{7}$ 4. $6\frac{16}{16}$

7 1. $3\frac{12}{3}$ 2. $9\frac{6}{2}$ 3. $12\frac{32}{4}$ 4. $15\frac{30}{5}$

8 1. $4\frac{6}{9}$ 2. $9\frac{10}{16}$ 3. $3\frac{5}{20}$ 4. $6\frac{21}{24}$
   5. $28\frac{8}{12}$ 6. $11\frac{16}{30}$ 7. $25\frac{24}{32}$ 8. $14\frac{25}{40}$

9 1. $6\frac{8}{5}$ 2. $5\frac{7}{4}$ 3. $9\frac{5}{2}$ 4. $8\frac{8}{3}$
   5. $11\frac{7}{6}$ 6. $23\frac{19}{10}$ 7. $40\frac{16}{9}$ 8. $17\frac{35}{16}$

10 1. $4\frac{10}{4}$ 2. $3\frac{12}{8}$ 3. $7\frac{8}{6}$ 4. $9\frac{32}{12}$
   5. $12\frac{10}{8}$ 6. $16\frac{15}{6}$ 7. $21\frac{26}{16}$ 8. $45\frac{52}{24}$

SKILL ACHIEVEMENT TEST

Rewrite each of the following as a whole number or a mixed number:

1. $\frac{12}{3}$ **2.** $\frac{11}{6}$ **3.** $\frac{18}{12}$ **4.** $\frac{13}{4}$ **5.** $\frac{50}{12}$

6. $8\frac{18}{9}$ **7.** $9\frac{35}{42}$ **8.** $2\frac{13}{8}$ **9.** $6\frac{15}{10}$ **10.** $13\frac{40}{16}$

Applications

Rewrite each of the following measurements as a mixed number:

1. $\frac{10}{8}$ inches   **2.** $\frac{9}{4}$ inches   **3.** $\frac{20}{16}$ inches   **4.** $\frac{7}{2}$ inches

5. $\frac{41}{32}$ inches   **6.** $\frac{27}{12}$ feet   **7.** $\frac{35}{12}$ feet   **8.** $\frac{43}{36}$ yards

9. $\frac{3,080}{1,760}$ miles   **10.** $\frac{7,920}{5,280}$ miles   **11.** $\frac{23}{16}$ pounds   **12.** $\frac{3,500}{2,000}$ tons

13. $\frac{5,040}{2,240}$ tons   **14.** $\frac{17}{8}$ gallons   **15.** $\frac{14}{4}$ bushels   **16.** $\frac{75}{60}$ hours

17. $\frac{108}{60}$ minutes   **18.** $\frac{42}{24}$ days   **19.** $\frac{15}{12}$ years   **20.** $\frac{91}{52}$ years

21. Which is greater: $6\frac{3}{8}$ inches or $\frac{100}{16}$ inches?

22. Which is less: $\frac{72}{32}$ inches or $2\frac{3}{4}$ inches?

23. Leslie wrote $\frac{28}{15}$ as her answer to a multiplication example. How should it have been expressed?

Rewrite each of the following measurements in simplest form:

24. $5\frac{2}{8}$ inches   **25.** $3\frac{6}{16}$ inches   **26.** $1\frac{14}{32}$ inches   **27.** $6\frac{5}{4}$ inches

28. $9\frac{2}{2}$ inches   **29.** $6\frac{4}{12}$ feet   **30.** $1\frac{17}{12}$ feet   **31.** $4\frac{30}{36}$ yards

32. $2\frac{440}{1,760}$ miles   **33.** $3\frac{1,980}{5,280}$ miles   **34.** $4\frac{8}{16}$ pounds   **35.** $5\frac{1,400}{2,000}$ tons

36. $2\frac{1,960}{2,240}$ tons   **37.** $8\frac{11}{4}$ gallons   **38.** $7\frac{6}{32}$ bushels   **39.** $2\frac{18}{60}$ hours

40. $9\frac{72}{60}$ minutes   **41.** $5\frac{16}{24}$ days   **42.** $3\frac{4}{12}$ years   **43.** $1\frac{13}{52}$ years

3-3 Expressing Fractions in Higher Terms

PROCEDURE

1 Express $\frac{1}{3}$ as an equivalent fraction with a denominator of 6.

$$\frac{1}{3} = \frac{?}{6}$$

$$6 \div 3 = 2$$

$$\frac{1}{3} = \frac{1 \times 2}{3 \times 2} = \frac{2}{6}$$

ANSWER: $\frac{2}{6}$

Divide the new denominator by the denominator of the given fraction.

Multiply the numerator and denominator of the given fraction by the quotient.

2 Express $\frac{9}{25}$ in 100ths.

$$\frac{9}{25} = \frac{?}{100}$$

$$100 \div 25 = 4$$

$$\frac{9}{25} = \frac{9 \times 4}{25 \times 4} = \frac{36}{100}$$

ANSWER: $\frac{36}{100}$

3 Express 5 in 20ths.

$$5 = \frac{5}{1}$$

$$\frac{5}{1} = \frac{?}{20}$$

$$20 \div 1 = 20$$

$$\frac{5}{1} = \frac{5 \times 20}{1 \times 20} = \frac{100}{20}$$

ANSWER: $\frac{100}{20}$

Express the given whole number as a fraction with a denominator of 1.

DIAGNOSTIC TEST

Express the following fractions as equivalent fractions having denominators as specified:

1 $\frac{1}{4} = \frac{?}{12}$

2 $\frac{5}{8} = \frac{?}{32}$

3 $\frac{9}{10} = \frac{?}{100}$

4 Express $\frac{3}{5}$ in 100ths.

5 Express 6 in 10ths.

RELATED PRACTICE EXERCISES

Express the following fractions as equivalent fractions having denominators as specified:

1 **1.** $\frac{1}{2} = \frac{?}{8}$ **2.** $\frac{1}{3} = \frac{?}{15}$ **3.** $\frac{1}{5} = \frac{?}{20}$ **4.** $\frac{1}{4} = \frac{?}{28}$

   5. $\frac{1}{8} = \frac{?}{48}$ **6.** $\frac{1}{12} = \frac{?}{72}$ **7.** $\frac{1}{6} = \frac{?}{42}$ **8.** $\frac{1}{16} = \frac{?}{64}$

2 **1.** $\frac{2}{3} = \frac{?}{12}$ **2.** $\frac{3}{4} = \frac{?}{24}$ **3.** $\frac{5}{6} = \frac{?}{12}$ **4.** $\frac{7}{12} = \frac{?}{60}$

   5. $\frac{3}{8} = \frac{?}{40}$ **6.** $\frac{5}{9} = \frac{?}{72}$ **7.** $\frac{9}{32} = \frac{?}{96}$ **8.** $\frac{11}{16} = \frac{?}{80}$

3 **1.** $\frac{7}{10} = \frac{?}{100}$ **2.** $\frac{3}{4} = \frac{?}{100}$ **3.** $\frac{2}{5} = \frac{?}{100}$ **4.** $\frac{9}{20} = \frac{?}{100}$

   5. $\frac{43}{50} = \frac{?}{100}$ **6.** $\frac{21}{25} = \frac{?}{100}$ **7.** $\frac{3}{8} = \frac{?}{100}$ **8.** $\frac{17}{50} = \frac{?}{100}$

4 **1.** $\frac{1}{8}$ in 64ths   **2.** $\frac{1}{4}$ in 32nds   **3.** $\frac{1}{2}$ in 12ths   **4.** $\frac{11}{16}$ in 64ths

   5. $\frac{2}{3}$ in 24ths   **6.** $\frac{4}{5}$ in 100ths   **7.** $\frac{5}{8}$ in 16ths   **8.** $\frac{19}{32}$ in 64ths

   9. $\frac{5}{6}$ in 30ths   **10.** $\frac{7}{9}$ in 36ths   **11.** $\frac{2}{1}$ in 10ths   **12.** 5 in 8ths

5 **1.** $\frac{2}{1}$ in 10ths   **2.** $\frac{4}{1}$ in 12ths   **3.** $\frac{5}{1}$ in 24ths   **4.** $\frac{7}{1}$ in 6ths

   5. 5 in 8ths   **6.** 3 in 5ths   **7.** 7 in 100ths   **8.** 9 in 72nds

SKILL ACHIEVEMENT TEST

Express each as an equivalent fraction with the specified denominator:

1. $\frac{1}{8} = \frac{?}{48}$ **2.** $\frac{17}{32} = \frac{?}{64}$ **3.** $\frac{7}{12} = \frac{?}{36}$

4. $\frac{19}{25} = \frac{?}{100}$ **5.** $\frac{2}{3}$ in 18ths   **6.** $\frac{3}{4}$ in 64ths

7. $\frac{7}{10}$ in 50ths   **8.** $\frac{5}{8}$ in 72nds   **9.** $\frac{5}{8}$ in 32nds

10. 13 in 10ths

Applications

Use a ruler to find the answer to each of the following:

1. $\frac{1}{8}$ in. $= \frac{?}{16}$ in.   **2.** $\frac{5}{8}$ in. $= \frac{?}{16}$ in.

3. $\frac{1}{4}$ in. $= \frac{?}{8}$ in. $= \frac{?}{16}$ in.   **4.** $\frac{3}{4}$ in. $= \frac{?}{8}$ in. $= \frac{?}{16}$ in.

5. $\frac{1}{2}$ in. $= \frac{?}{4}$ in. $= \frac{?}{8}$ in. $= \frac{?}{16}$ in.   **6.** 3 in. $= \frac{?}{2}$ in. $= \frac{?}{4}$ in. $= \frac{?}{8}$ in.

3-4 Equivalent Fractions

PROCEDURE

1 Are $\frac{6}{15}$ and $\frac{4}{10}$ equivalent fractions?

$$\frac{6}{15} = \frac{6 \div 3}{15 \div 3} = \frac{2}{5} \searrow$$
$$\frac{4}{10} = \frac{4 \div 2}{10 \div 2} = \frac{2}{5} \nearrow \text{ same}$$

So, $\frac{6}{15} = \frac{4}{10}$
$\nwarrow \qquad \nearrow$
equivalent
fractions

ANSWER: $\frac{6}{15}$ and $\frac{4}{10}$ are equivalent fractions.

2 Are $\frac{12}{20}$ and $\frac{18}{27}$ equivalent fractions?

$$\frac{12}{20} = \frac{12 \div 4}{20 \div 4} = \frac{3}{5} \searrow$$
$$\frac{18}{27} = \frac{18 \div 9}{27 \div 9} = \frac{2}{3} \nearrow \text{ not the same}$$

So, $\frac{12}{20} \neq \frac{18}{27}$.
$\uparrow$
$\llcorner$ **is not equal to**

ANSWER: $\frac{12}{20}$ and $\frac{18}{27}$ are not equivalent fractions.

Express each fraction in lowest terms.
If the resulting fractions are the same, the given fractions are equivalent.

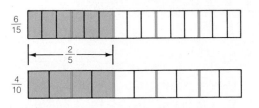

If the resulting fractions are not the same, the given fractions are not equivalent.

---

3 Are $\frac{9}{12}$ and $\frac{12}{16}$ equivalent fractions?

first   **second**
fraction   **fraction**
$\downarrow$ $\downarrow$

$\frac{9}{12} \diagdown\diagup \frac{12}{16}$ $9 \times 16 = 144$
$\qquad\qquad 12 \times 12 = 144$

$144 = 144$ So, $\frac{9}{12} = \frac{12}{16}$

ANSWER: $\frac{9}{12}$ and $\frac{12}{16}$ are equivalent fractions.

ALTERNATE METHOD (cross product):
Find the product of the numerator of the first fraction and the denominator of the second fraction.
Then find the product of the numerator of the second fraction and the denominator of the first fraction.
If the products are equal, the fractions are equivalent.

Vocabulary

Equivalent fractions are fractions that name the same number.
The $\neq$ symbol means "is not equal to."

DIAGNOSTIC TEST

1 Write a group of four fractions equivalent to $\frac{3}{8}$.

2 Write the simplest form and three other fractions equivalent to $\frac{8}{12}$.

3 Test whether $\frac{14}{35}$ and $\frac{6}{15}$ are equivalent fractions using the method of lowest terms.

4 Test whether $\frac{36}{90}$ and $\frac{10}{25}$ are equivalent fractions using the alternate method.

RELATED PRACTICE EXERCISES

1 Write a group of four fractions equivalent to:

1. $\frac{1}{2}$ **2.** $\frac{3}{5}$ **3.** $\frac{7}{10}$ **4.** $\frac{11}{12}$

5. $\frac{3}{4}$ **6.** $\frac{4}{11}$ **7.** $\frac{17}{20}$ **8.** $\frac{13}{16}$

9. $\frac{5}{8}$ **10.** $\frac{19}{50}$

2 Write the simplest form and three other fractions equivalent to:

1. $\frac{7}{42}$ **2.** $\frac{9}{36}$ **3.** $\frac{35}{84}$ **4.** $\frac{48}{54}$

5. $\frac{55}{80}$ **6.** $\frac{24}{30}$ **7.** $\frac{110}{121}$ **8.** $\frac{76}{96}$

9. $\frac{21}{70}$ **10.** $\frac{170}{200}$

Test whether each of the following pairs of fractions are equivalent by using:

3 Lowest-terms method:

1. $\frac{6}{16}$ and $\frac{15}{40}$ **2.** $\frac{15}{18}$ and $\frac{30}{48}$ **3.** $\frac{42}{63}$ and $\frac{16}{24}$

4. $\frac{28}{35}$ and $\frac{35}{42}$ **5.** $\frac{21}{27}$ and $\frac{24}{33}$ **6.** $\frac{39}{52}$ and $\frac{51}{68}$

4 Alternate method:

1. $\frac{16}{28}$ and $\frac{24}{36}$ **2.** $\frac{10}{16}$ and $\frac{15}{24}$ **3.** $\frac{15}{25}$ and $\frac{36}{60}$

4. $\frac{24}{33}$ and $\frac{18}{26}$ **5.** $\frac{42}{54}$ and $\frac{28}{35}$ **6.** $\frac{27}{48}$ and $\frac{45}{80}$

SKILL ACHIEVEMENT TEST

1. Write a group of four fractions equivalent to $\frac{5}{6}$.

2. Write the simplest form and three other fractions equivalent to $\frac{56}{64}$.

Test whether each of the following pairs of fractions are equivalent by using:

3. Lowest-terms method: $\frac{15}{24}$ and $\frac{12}{20}$ **4.** Lowest-terms method: $\frac{24}{32}$ and $\frac{42}{56}$

5. Alternate method: $\frac{48}{75}$ and $\frac{35}{50}$

3-5 Comparing Fractions and Rounding Mixed Numbers

PROCEDURE

1 Which is greater: $\frac{2}{5}$ or $\frac{1}{2}$?

$\frac{2}{5} = \frac{4}{10}$

$\frac{1}{2} = \frac{5}{10}$

$\frac{5}{10} > \frac{4}{10}$, so $\frac{1}{2} > \frac{2}{5}$

ANSWER: $\frac{1}{2}$ is greater.

Express the fractions as equivalent fractions having a common denominator.

The given fraction that is equal to the fraction having the greater numerator is the greater fraction.

2 Which is less: $\frac{3}{4}$ or $\frac{5}{6}$?

$\frac{3}{4} = \frac{9}{12}$

$\frac{5}{6} = \frac{10}{12}$

$\frac{9}{12} < \frac{10}{12}$, so $\frac{3}{4} < \frac{5}{6}$

ANSWER: $\frac{3}{4}$ is less.

---

3 Is the statement $\frac{2}{3} < \frac{3}{5}$ true?

$\frac{2}{3} = \frac{10}{15}$ and $\frac{3}{5} = \frac{9}{15}$

$\overset{\text{is not less than}}{\frac{10}{15} \not< \frac{9}{15}}$, so $\frac{2}{3} \not< \frac{3}{5}$

ANSWER: No, false.

4 Is the statement $\frac{3}{10} > \frac{1}{4}$ true?

$\frac{3}{10} = \frac{6}{20}$ and $\frac{1}{4} = \frac{5}{20}$

$\frac{6}{20} > \frac{5}{20}$, so $\frac{3}{10} > \frac{1}{4}$

ANSWER: Yes, true.

---

5 Is the statement $\frac{3}{5} > \frac{7}{12}$ true?

first **second**
fraction **fraction**

$\frac{3}{5} \diagdown\diagup \frac{7}{12}$ $3 \times 12 = 36$
$5 \times 7 = 35$

$36 > 35$, so the statement is true.

ANSWER: Yes, true.

ALTERNATE METHOD (cross product):
Find the product of the numerator of the first fraction and the denominator of the second fraction.
Then find the product of the numerator of the second fraction and the denominator of the first fraction.
If the first product is greater than the second, then the first fraction is greater.

6 Is the statement $\frac{2}{5} < \frac{3}{8}$ true?

$\frac{2}{5} \diagup\diagdown \frac{3}{8}$ $2 \times 8 = 16$
$5 \times 3 = 15$

$16 \not< 15$, so the statement is false.

ANSWER: No, false.

If the first product is less than the second, then the first fraction is less.

7 Arrange in order from least to greatest: $\frac{3}{4}, \frac{7}{8}, \frac{2}{3}$

$\frac{3}{4} = \frac{18}{24}$

$\frac{7}{8} = \frac{21}{24}$ Express the fractions as equivalent fractions having a common denominator.

$\frac{2}{3} = \frac{16}{24}$

$\frac{16}{24} < \frac{18}{24} < \frac{21}{24}$, so $\frac{2}{3} < \frac{3}{4} < \frac{7}{8}$ Compare.

ANSWER: $\frac{2}{3}, \frac{3}{4}, \frac{7}{8}$

---

8 Round $5\frac{3}{4}$ to the nearest whole number.

$\frac{3}{4}$ is greater than $\frac{1}{2}$ To round a mixed number to the nearest whole number, when the fraction is $\frac{1}{2}$ or more, drop the fraction and add 1 to the whole number.

Therefore, $5\frac{3}{4}$ rounds to 6.

ANSWER: 6

9 Round $3\frac{1}{8}$ to the nearest whole number.

$\frac{1}{8}$ is less than $\frac{1}{2}$ When the fraction is less than $\frac{1}{2}$, drop the fraction and write only the whole number.

Therefore, $3\frac{1}{8}$ rounds to 3.

ANSWER: 3

---

10 Round $8.76\frac{1}{3}$ to the nearest cent.

$\frac{1}{3}$¢ is less than $\frac{1}{2}$¢

Therefore, $8.76\frac{1}{3}$ rounds to $8.76.

ANSWER: $8.76

Vocabulary

The $\nless$ symbol means "is not less than."   The $\ngtr$ symbol means "is not greater than."

DIAGNOSTIC TEST

Which is greater:

1 $\frac{1}{3}$ or $\frac{1}{2}$?   **2** $\frac{5}{6}$ or $\frac{7}{8}$?

Which is less:

3 $\frac{1}{4}$ or $\frac{1}{10}$?   **4** $\frac{5}{8}$ or $\frac{11}{16}$?

5 Is the statement $\frac{4}{5} < \frac{5}{6}$ true?   **6** Is the statement $\frac{1}{3} > \frac{3}{8}$ true?

Use the alternate method to determine which of the following statements are true:

7 $\frac{3}{4} > \frac{5}{8}$ **8** $\frac{2}{3} < \frac{3}{5}$ **9** Arrange in order of size (greatest first): $\frac{7}{32}, \frac{3}{8}$, and $\frac{13}{16}$.

10 Arrange in order of size (least first): $\frac{3}{5}, \frac{7}{12}$, and $\frac{1}{2}$.

Round to the nearest whole number:

11 $2\frac{7}{12}$ **12** $6\frac{2}{5}$

Round to the nearest cent:

13 $6.82\frac{4}{9}$ **14** $21.09\frac{5}{6}$

RELATED PRACTICE EXERCISES

Which is greater in each of the following?

1 1. $\frac{1}{4}$ or $\frac{1}{3}$ 2. $\frac{1}{6}$ or $\frac{1}{10}$ 3. $\frac{1}{2}$ or $\frac{1}{16}$ 4. $\frac{1}{12}$ or $\frac{1}{10}$

2 1. $\frac{2}{3}$ or $\frac{5}{8}$ 2. $\frac{4}{5}$ or $\frac{3}{4}$ 3. $\frac{3}{8}$ or $\frac{1}{4}$ 4. $\frac{13}{16}$ or $\frac{5}{6}$

Which is less in each of the following?

3 1. $\frac{1}{2}$ or $\frac{1}{6}$ 2. $\frac{1}{8}$ or $\frac{1}{10}$ 3. $\frac{1}{16}$ or $\frac{1}{12}$ 4. $\frac{1}{3}$ or $\frac{1}{5}$

4 1. $\frac{5}{16}$ or $\frac{1}{4}$ 2. $\frac{2}{5}$ or $\frac{3}{8}$ 3. $\frac{2}{3}$ or $\frac{11}{16}$ 4. $\frac{7}{10}$ or $\frac{3}{4}$

Which of the following statements are true?

5 1. $\frac{1}{10} < \frac{1}{8}$ 2. $\frac{2}{3} < \frac{7}{12}$ 3. $\frac{5}{6} < \frac{8}{9}$ 4. $\frac{13}{16} < \frac{19}{24}$

6 1. $\frac{3}{5} > \frac{1}{2}$ 2. $\frac{11}{16} > \frac{3}{4}$ 3. $\frac{3}{10} > \frac{2}{7}$ 4. $\frac{3}{8} > \frac{5}{12}$

Use the alternate method to determine which of the following statements are true:

7 1. $\frac{1}{8} > \frac{1}{6}$ 2. $\frac{7}{12} > \frac{9}{16}$ 3. $\frac{15}{24} > \frac{21}{30}$ 4. $\frac{18}{48} > \frac{27}{81}$

8 1. $\frac{7}{8} < \frac{9}{10}$ 2. $\frac{19}{24} < \frac{7}{9}$ 3. $\frac{3}{11} < \frac{5}{18}$ 4. $\frac{49}{56} < \frac{18}{48}$

Arrange in order of size (greatest first):

9 1. $\frac{1}{2}, \frac{1}{5}$, and $\frac{1}{3}$ 2. $\frac{5}{8}, \frac{2}{3}$, and $\frac{3}{5}$ 3. $\frac{3}{4}, \frac{9}{16}$, and $\frac{7}{12}$ 4. $\frac{7}{8}, \frac{5}{6}$, and $\frac{4}{5}$

Arrange in order of size (smallest first):

10 1. $\frac{1}{4}, \frac{1}{2}$, and $\frac{1}{6}$ 2. $\frac{1}{2}, \frac{2}{5}$, and $\frac{3}{10}$ 3. $\frac{11}{16}, \frac{5}{8}$, and $\frac{3}{4}$ 4. $\frac{5}{8}, \frac{3}{5}$, and $\frac{3}{4}$

Round to the nearest whole number:

11 1. $8\frac{9}{16}$ 2. $40\frac{7}{10}$ 3. $11\frac{8}{15}$ 4. $7\frac{1}{2}$

12 1. $2\frac{3}{10}$ 2. $15\frac{3}{7}$ 3. $25\frac{7}{16}$ 4. $5\frac{11}{24}$

Round to the nearest cent:

13 1. $\$3.28\frac{2}{5}$ 2. $\$15.20\frac{1}{6}$ 3. $\$.86\frac{5}{12}$ 4. $\$9.53\frac{7}{15}$

14 1. $\$17.34\frac{5}{8}$ 2. $\$6.55\frac{2}{3}$ 3. $\$9.70\frac{1}{2}$ 4. $\$.69\frac{11}{12}$

SKILL ACHIEVEMENT TEST

1. Which is less: $\frac{1}{8}$ or $\frac{1}{10}$?   **2.** Which is greater: $\frac{3}{5}$ or $\frac{2}{3}$?

3. Arrange in order of size (least first): $\frac{7}{8}, \frac{3}{4}$, and $\frac{5}{6}$

4. Arrange in order of size (greatest first): $\frac{7}{12}, \frac{3}{5}$, and $\frac{9}{16}$

5. Is the statement $\frac{8}{9} > \frac{11}{13}$ true?   **6.** Is the statement $\frac{28}{63} < \frac{36}{84}$ true?

Round to the nearest whole number: **7.** $59\frac{3}{5}$ **8.** $3\frac{5}{12}$

Round to the nearest cent: **9.** $\$1.69\frac{3}{8}$ **10.** $\$7.40\frac{7}{10}$

3-6 Number Multiples

PROCEDURE

1 Find the multiples of 6.

$6 \times 0 = 0$
$6 \times 1 = 6$
$6 \times 2 = 12$ ←— multiples
$6 \times 3 = 18$

The three dots indicate that the numbers continue in the same pattern.

To find the multiples of a number, multiply the number by the whole numbers (0, 1, 2, 3, etc.).

ANSWER: The multiples of 6 are: 0, 6, 12, 18, . . .

---

2 Find the common multiples of 6 and 8.

Zero is a multiple of every number.

Multiples of 6: 0, 6, 12, 18, 24, 30, 36, 42, 48, . . .
Multiples of 8: 0, 8, 16, 24, 32, 40, 48, 56, 64, . . .

common multiples

List all the multiples of each number.

Identify the multiples that are common to both.

ANSWER: The common multiples of 6 and 8 are: 0, 24, 48, . . .

---

3 Find the least common multiple (LCM) of 4 and 6.

Multiples of 4: 0, 4, 8, 12, 16, 20, 24, 28, . . .
Multiples of 6: 0, 6, 12, 18, 24, 30, 36, 42, . . .

least common multiple

The least *nonzero* common multiple is the LCM

ALTERNATE METHOD:

$4 = 2 \cdot 2$ **2 appears twice.**
$6 = 2 \cdot 3$ $2 \cdot 2 \cdot 3 = 12$

3 appears once.

Factor the given numbers as primes.

Then, form a product by using each prime the greatest number of times it appears in the factored form of any one number.

ANSWER: The LCM of 4 and 6 is 12.

Vocabulary

A **multiple** of a given whole number is a product of the number and a whole number. Any multiple of a number is divisible by the number.

A **common multiple** of two or more numbers is any number that is a multiple of all the numbers.

The **least common multiple (LCM)** of two or more numbers is the smallest nonzero number that is a multiple of all of them. It is the smallest possible whole number (excluding zero) that can be divided exactly by all the given numbers.

147

DIAGNOSTIC TEST

1 Which of the following numbers are multiples of 9? **a.** 56 **b.** 27 **c.** 81 **d.** 109

2 Write all the multiples of 7, listing the first five multiples.

3 Name the first four common multiples of 20 and 25.

4 Write all the common multiples of 2, 6, and 8, listing the first six nonzero common multiples.

Name the least common multiple of:

5 2 and 3   **6** 8, 4, and 16   **7** 12 and 18

RELATED PRACTICE EXERCISES

1 Which of the following are multiples:
  1. Of 3? **a.** 19 **b.** 42 **c.** 51 **d.** 73
  2. Of 10? **a.** 25 **b.** 40 **c.** 135 **d.** 200
  3. Of 12? **a.** 36 **b.** 54 **c.** 90 **d.** 132
  4. Of 25? **a.** 75 **b.** 150 **c.** 215 **d.** 340

2 Write all the multiples of each of the following, listing the first six multiples:
  1. 5   **2.** 11   **3.** 2   **4.** 24

3 **1.** Name the first three common multiples of 3 and 5.
  2. Name the first five common multiples of 8 and 4.
  3. Name the first four common multiples of 10 and 12.
  4. Name the first six common multiples of 8, 36, and 18.

4 Write all the common multiples of each of the following, listing the first four nonzero common multiples:
  1. 4 and 5   **2.** 2 and 16   **3.** 8 and 12   **4.** 6, 8, and 16

Name the least common multiple of:

5 **1.** 3 and 8   **2.** 7 and 2   **3.** 4 and 3   **4.** 5 and 7
  5. 2, 3, and 5   **6.** 4, 5, and 7   **7.** 2, 9, and 11

6 **1.** 6 and 12   **2.** 25 and 100   **3.** 20 and 5   **4.** 14 and 56
  5. 4, 8, and 16   **6.** 5, 10, and 2   **7.** 12, 3, and 24

7 **1.** 16 and 24   **2.** 15 and 20   **3.** 12 and 18   **4.** 70 and 42
  5. 8, 10, and 12   **6.** 4, 6, and 15   **7.** 18, 45, and 54

SKILL ACHIEVEMENT TEST

1. Write all the multiples of 20, listing the first six multiples.
2. Name the first four common multiples of 32 and 48.
3. Write all the common multiples of 15, 10, and 75, listing the first five nonzero common multiples.
4. Name the least common multiple of 6 and 16.
5. Name the least common multiple of 10, 15, and 60.

3-7 Least Common Denominator and Equivalent Fractions

PROCEDURE

Find the least common denominator (LCD) of the given fractions. Then express the given fractions as equivalent fractions using the LCD as the new denominator.

1 $\frac{1}{3}$ and $\frac{5}{6}$

3: 0, 3, 6, 9, 12, . . .
6: 0, 6, 12, 18, 24, . . .
The LCD is 6.

$\frac{1}{3} = \frac{?}{6}$ $\frac{5}{6} = \frac{?}{6}$

$\frac{1}{3} = \frac{1 \times 2}{3 \times 2} = \frac{2}{6}$ $\frac{5}{6} = \frac{5 \times 1}{6 \times 1} = \frac{5}{6}$

ANSWER: LCD = 6; $\frac{1}{3} = \frac{2}{6}$ and $\frac{5}{6} = \frac{5}{6}$.

To find the LCD, find the LCM of the denominators.

To express as equivalent fractions, divide the LCD by the denominator of each given fraction. Multiply the numerator and denominator by the quotient.

2 $\frac{1}{2}$ and $\frac{3}{5}$

2: 0, 2, 4, 6, 8, 10, 12, . . .
5: 0, 5, 10, 15, 20, . . .
The LCD is 10.

$\frac{1}{2} = \frac{?}{10}$ $\frac{3}{5} = \frac{?}{10}$

$\frac{1}{2} = \frac{1 \times 5}{2 \times 5} = \frac{5}{10}$ $\frac{3}{5} = \frac{3 \times 2}{5 \times 2} = \frac{6}{10}$

ANSWER: LCD = 10; $\frac{1}{2} = \frac{5}{10}$ and $\frac{3}{5} = \frac{6}{10}$.

3 $\frac{5}{12}$ and $\frac{7}{16}$

12: 0, 12, 24, 36, 48, . . .
16: 0, 16, 32, 48, 64, . . .
The LCD is 48.

$\frac{5}{12} = \frac{?}{48}$ $\frac{7}{16} = \frac{?}{48}$

$\frac{5}{12} = \frac{5 \times 4}{12 \times 4} = \frac{20}{48}$ $\frac{7}{16} = \frac{7 \times 3}{16 \times 3} = \frac{21}{48}$

ANSWER: LCD = 48; $\frac{5}{12} = \frac{20}{48}$ and $\frac{7}{16} = \frac{21}{48}$.

4 $\frac{2}{3}$, $\frac{7}{8}$, and $\frac{1}{6}$

3: 0, 3, 6, 9, 12, 15, 18, 21, 24, 27, 30, . . .
8: 0, 8, 16, 24, 32, 40, 48, . . .
6: 0, 6, 12, 18, 24, 30, 36, 42, . . .
The LCD is 24.

$\frac{2}{3} = \frac{?}{24}$ $\frac{2}{3} = \frac{2 \times 8}{3 \times 8} = \frac{16}{24}$ $\frac{7}{8} = \frac{?}{24}$ $\frac{7}{8} = \frac{7 \times 3}{8 \times 3} = \frac{21}{24}$ $\frac{1}{6} = \frac{?}{24}$ $\frac{1}{6} = \frac{1 \times 4}{6 \times 4} = \frac{4}{24}$

ANSWER: LCD = 24; $\frac{2}{3} = \frac{16}{24}$, $\frac{7}{8} = \frac{21}{24}$, and $\frac{1}{6} = \frac{4}{24}$.

Vocabulary

The **least common denominator (LCD)** is the least possible whole number (excluding zero) that can be divided exactly by the denominators of all the given fractions.

DIAGNOSTIC TEST

Find the least common denominator (LCD) of the given fractions, then express the given fractions as equivalent fractions using the (LCD) as their new denominator:

1 $\frac{1}{2}$ and $\frac{3}{8}$ **2** $\frac{3}{4}$ and $\frac{2}{3}$ **3** $\frac{7}{10}$ and $\frac{9}{16}$ **4** $\frac{5}{6}, \frac{4}{5}$, and $\frac{1}{3}$

RELATED PRACTICE EXERCISES

Find the least common denominator (LCD) of the given fractions, then express the given fractions as equivalent fractions using the (LCD) as their new denominator:

1   **1.** $\frac{1}{2}$ and $\frac{1}{4}$ **2.** $\frac{3}{5}$ and $\frac{9}{10}$ **3.** $\frac{1}{4}$ and $\frac{9}{16}$ **4.** $\frac{17}{20}$ and $\frac{4}{5}$

     5. $\frac{2}{3}$ and $\frac{1}{6}$ **6.** $\frac{7}{8}$ and $\frac{3}{4}$ **7.** $\frac{11}{24}$ and $\frac{5}{8}$ **8.** $\frac{5}{6}$ and $\frac{7}{18}$

2   **1.** $\frac{1}{3}$ and $\frac{1}{2}$ **2.** $\frac{2}{5}$ and $\frac{5}{6}$ **3.** $\frac{2}{3}$ and $\frac{1}{4}$ **4.** $\frac{7}{8}$ and $\frac{3}{5}$

     5. $\frac{4}{5}$ and $\frac{3}{4}$ **6.** $\frac{3}{5}$ and $\frac{5}{12}$ **7.** $\frac{8}{9}$ and $\frac{3}{4}$ **8.** $\frac{1}{6}$ and $\frac{4}{7}$

3   **1.** $\frac{1}{4}$ and $\frac{1}{6}$ **2.** $\frac{7}{8}$ and $\frac{9}{10}$ **3.** $\frac{5}{6}$ and $\frac{3}{8}$ **4.** $\frac{3}{4}$ and $\frac{7}{10}$

     5. $\frac{1}{6}$ and $\frac{15}{16}$ **6.** $\frac{13}{20}$ and $\frac{7}{12}$ **7.** $\frac{19}{24}$ and $\frac{13}{18}$ **8.** $\frac{11}{12}$ and $\frac{27}{32}$

4   **1.** $\frac{1}{2}, \frac{1}{3}$, and $\frac{1}{4}$ **2.** $\frac{2}{3}, \frac{7}{12}$, and $\frac{5}{6}$ **3.** $\frac{4}{5}, \frac{1}{4}$, and $\frac{2}{3}$

     4. $\frac{5}{6}, \frac{3}{4}$, and $\frac{11}{12}$ **5.** $\frac{7}{8}, \frac{5}{6}$, and $\frac{3}{4}$ **6.** $\frac{9}{16}, \frac{3}{8}$, and $\frac{13}{24}$

     7. $\frac{7}{8}, \frac{9}{10}$, and $\frac{1}{12}$ **8.** $\frac{3}{5}, \frac{5}{16}$, and $\frac{9}{20}$

SKILL ACHIEVEMENT TEST

Find the least common denominator (LCD) of the given fractions, then express the given fractions as equivalent fractions having the LCD as their new denominator.

1. $\frac{1}{2}$ and $\frac{13}{16}$ **2.** $\frac{2}{3}$ and $\frac{5}{8}$ **3.** $\frac{7}{10}$ and $\frac{5}{12}$ **4.** $\frac{3}{4}, \frac{1}{2}$, and $\frac{7}{8}$ **5.** $\frac{2}{5}, \frac{5}{6}$, and $\frac{1}{3}$

6. Of which group of fractions is 24 the LCD?   **a.** $\frac{3}{4}$ and $\frac{5}{6}$ **b.** $\frac{5}{8}$ and $\frac{2}{3}$ **c.** $\frac{1}{2}, \frac{11}{12}$, and $\frac{1}{3}$

7. Of which group of fractions is 18 the LCD?   **a.** $\frac{1}{6}$ and $\frac{2}{3}$ **b.** $\frac{7}{9}$ and $\frac{1}{2}$ **c.** $\frac{7}{12}, \frac{5}{6}$, and $\frac{2}{9}$

8. Of which group of fractions is 12 the LCD?   **a.** $\frac{5}{9}$ and $\frac{3}{4}$ **b.** $\frac{1}{2}$ and $\frac{5}{6}$ **c.** $\frac{2}{3}, \frac{3}{4}$, and $\frac{1}{6}$

9. Of which group of fractions is 32 the LCD?   **a.** $\frac{11}{16}$ and $\frac{7}{8}$ **b.** $\frac{3}{4}$ and $\frac{15}{32}$ **c.** $\frac{3}{8}, \frac{1}{2}$, and $\frac{3}{4}$

10. Of which group of fractions is 48 the LCD?   **a.** $\frac{5}{6}$ and $\frac{3}{8}$ **b.** $\frac{7}{12}$ and $\frac{3}{4}$ **c.** $\frac{11}{16}, \frac{1}{6}$, and $\frac{5}{8}$

3-8 Addition of Fractions and Mixed Numbers

PROCEDURE

1 Add: $\frac{1}{5} + \frac{2}{5}$

$$\frac{1}{5} + \frac{2}{5} = \frac{1+2}{5} = \frac{3}{5}$$

ANSWER: $\frac{3}{5}$

To add fractions that have a common denominator, add the numerators. Then write the sum over the common denominator.

2 Add: $\frac{5}{8} + \frac{1}{8}$

$$\begin{array}{r} \frac{5}{8} \\ + \frac{1}{8} \\ \hline \frac{6}{8} = \frac{3}{4} \end{array}$$

ANSWER: $\frac{3}{4}$

Express the answer in lowest terms.

3 Add: $\frac{11}{16} + \frac{8}{16}$

$$\begin{array}{r} \frac{11}{16} \\ + \frac{8}{16} \\ \hline \frac{19}{16} = 1\frac{3}{16} \end{array}$$

ANSWER: $1\frac{3}{16}$

4 Add: $\frac{11}{12} + \frac{5}{12}$

$$\begin{array}{r} \frac{11}{12} \\ + \frac{5}{12} \\ \hline \frac{16}{12} = 1\frac{4}{12} = 1\frac{1}{3} \end{array}$$

ANSWER: $1\frac{1}{3}$

5 Add: $\frac{4}{5} + \frac{7}{10}$

$$\frac{4}{5} = \frac{8}{10}$$

$$\frac{7}{10} = \frac{7}{10}$$

$$\begin{array}{r} \frac{8}{10} \\ + \frac{7}{10} \\ \hline \frac{15}{10} = 1\frac{5}{10} = 1\frac{1}{2} \end{array}$$

ANSWER: $1\frac{1}{2}$

To add fractions that do *not* have a common denominator, find the LCD. Express the fractions as equivalent fractions using the LCD. Add.

Simplify.

6 Add: $2\frac{1}{2} + 5\frac{2}{3}$

$$2\frac{1}{2} = 2\frac{3}{6}$$
$$+ 5\frac{2}{3} = 5\frac{4}{6}$$
$$\overline{\phantom{+ 5\frac{2}{3} =}\ \ \frac{7}{6}}$$

To add mixed numbers, first add the fractions.

$$2\frac{3}{6}$$
$$+ 5\frac{4}{6}$$
$$\overline{7\frac{7}{6} = 8\frac{1}{6}}$$

Then add the whole numbers. Simplify.

ANSWER: $8\frac{1}{6}$

7 Add: $4 + 6\frac{5}{8}$

$$4$$
$$+ 6\frac{5}{8}$$
$$\overline{10\frac{5}{8}}$$

ANSWER: $10\frac{5}{8}$

8 Add: $9\frac{7}{8} + \frac{13}{16}$

$$9\frac{7}{8} = 9\frac{14}{16}$$
$$+ \ \frac{13}{16} = \ \frac{13}{16}$$
$$\overline{\phantom{+ \frac{13}{16} =}\ 9\frac{27}{16} = 9 + 1\frac{11}{16} = 10\frac{11}{16}}$$

ANSWER: $10\frac{11}{16}$

9 Add: $13\frac{1}{6} + 5\frac{1}{3} + 9\frac{1}{2}$

$$13\frac{1}{6} = 13\frac{1}{6}$$
$$5\frac{1}{3} = \ \ 5\frac{2}{6}$$
$$+ \ \ 9\frac{1}{2} = \ \ 9\frac{3}{6}$$
$$\overline{\phantom{+ \ \ 9\frac{1}{2} =}\ 27\frac{6}{6} = 27 + 1 = 28}$$

ANSWER: 28

10 Solve the equation: $7\frac{3}{5} + 8\frac{2}{3} = n$

$$7\frac{3}{5} = \ \ 7\frac{9}{15}$$
$$+ 8\frac{2}{3} = \ \ 8\frac{10}{15}$$
$$\overline{\phantom{+ 8\frac{2}{3} =}\ 15\frac{19}{15} = 16\frac{4}{15}}$$

To find the value of n, add the given numbers.

ANSWER: $16\frac{4}{15}$

DIAGNOSTIC TEST

Add:

1 $\frac{1}{3}$
$\frac{1}{3}$

2 $\frac{5}{16}$
$\frac{7}{16}$

3 $\frac{5}{6}$
$\frac{1}{6}$

4 $\frac{3}{5}$
$\frac{4}{5}$

5 $\frac{3}{4}$
$\frac{3}{4}$

6 $\frac{1}{2}$
$\frac{3}{8}$

7 $\frac{2}{5}$
$\frac{3}{4}$

8 $\frac{7}{8}$
$\frac{5}{6}$

9 $\frac{3}{10}$
$\frac{1}{2}$
$\frac{4}{5}$

10 3
$\frac{1}{8}$

11 $2\frac{3}{4}$
5

12 $3\frac{1}{5}$
$4\frac{3}{5}$

13 $2\frac{3}{10}$
$3\frac{1}{10}$

14 $16\frac{5}{8}$
$23\frac{3}{8}$

15 $7\frac{2}{5}$
$9\frac{4}{5}$

16 $5\frac{7}{12}$
$6\frac{11}{12}$

17 $8\frac{1}{6}$
$\frac{5}{6}$

18 $12\frac{2}{3}$
$5\frac{1}{4}$

19 $6\frac{1}{12}$
$8\frac{1}{6}$

20 $24\frac{9}{10}$
$17\frac{5}{8}$

21 $15\frac{2}{3}$
$9\frac{5}{6}$

22 $6\frac{7}{8}$
$\frac{11}{12}$

23 $\frac{1}{2}$
$3\frac{3}{4}$
$\frac{7}{8}$

24 $4\frac{2}{3}$
$6\frac{1}{2}$
$5\frac{5}{8}$

25 $\frac{7}{8} + \frac{7}{12} + \frac{1}{6}$

26 $2\frac{5}{6} + 3\frac{1}{10} + 4\frac{1}{2}$

27 Find the sum of $6\frac{1}{4}$, $3\frac{13}{16}$, and $7\frac{3}{8}$.

28 Solve: $2\frac{9}{10} + 4\frac{1}{2} = n$

RELATED PRACTICE EXERCISES

Add:

1 1. $\frac{2}{5}$ 2. $\frac{3}{7}$ 3. $\frac{6}{25}$ 4. $\frac{3}{6}$
  $\frac{2}{5}$ $\frac{2}{7}$ $\frac{8}{25}$ $\frac{2}{6}$

2 1. $\frac{3}{8}$ 2. $\frac{7}{12}$ 3. $\frac{13}{32}$ 4. $\frac{27}{64}$
  $\frac{1}{8}$ $\frac{1}{12}$ $\frac{15}{32}$ $\frac{21}{64}$

3 1. $\frac{1}{2}$ 2. $\frac{1}{3}$ 3. $\frac{27}{32}$ 4. $\frac{11}{20}$
  $\frac{1}{2}$ $\frac{2}{3}$ $\frac{5}{32}$ $\frac{9}{20}$

4 1. $\frac{2}{3}$ 2. $\frac{4}{9}$ 3. $\frac{9}{16}$ 4. $\frac{4}{10}$
  $\frac{2}{3}$ $\frac{7}{9}$ $\frac{10}{16}$ $\frac{9}{10}$

5 1. $\frac{5}{6}$ 2. $\frac{3}{8}$ 3. $\frac{17}{24}$ 4. $\frac{15}{32}$
  $\frac{5}{6}$ $\frac{7}{8}$ $\frac{19}{24}$ $\frac{29}{32}$

6 1. $\frac{1}{2}$ 2. $\frac{5}{6}$ 3. $\frac{1}{10}$ 4. $\frac{13}{32}$
  $\frac{1}{4}$ $\frac{5}{12}$ $\frac{2}{5}$ $\frac{3}{4}$

7 1. $\frac{2}{3}$ 2. $\frac{1}{2}$ 3. $\frac{2}{5}$ 4. $\frac{7}{8}$
  $\frac{3}{4}$ $\frac{1}{3}$ $\frac{1}{6}$ $\frac{3}{5}$

8 1. $\frac{5}{12}$ 2. $\frac{3}{4}$ 3. $\frac{1}{8}$ 4. $\frac{13}{24}$
  $\frac{3}{10}$ $\frac{5}{6}$ $\frac{7}{12}$ $\frac{9}{16}$

9 1. $\frac{3}{8}$ 2. $\frac{2}{3}$ 3. $\frac{3}{16}$ 4. $\frac{7}{24}$
  $\frac{4}{8}$ $\frac{3}{4}$ $\frac{1}{8}$ $\frac{3}{16}$
  $\frac{1}{8}$ $\frac{5}{6}$ $\frac{1}{4}$ $\frac{5}{12}$

10 1. 8 2. 12 3. $\frac{3}{16}$ 4. $\frac{4}{7}$
  $\frac{7}{8}$ $\frac{1}{6}$ 7 18

11 1. $6\frac{3}{5}$ 2. $9\frac{5}{8}$ 3. 8 4. 23
  7 4 $7\frac{1}{2}$ $8\frac{9}{32}$

12 1. $4\frac{1}{4}$ 2. $7\frac{2}{9}$ 3. $10\frac{4}{8}$ 4. $32\frac{3}{10}$
  $3\frac{2}{4}$ $2\frac{5}{9}$ $9\frac{3}{8}$ $14\frac{6}{10}$

13 1. $5\frac{1}{8}$ 2. $6\frac{7}{32}$ 3. $10\frac{7}{16}$ 4. $17\frac{1}{2}$
  $8\frac{5}{8}$ $3\frac{9}{32}$ $16\frac{7}{16}$ $11\frac{7}{12}$

14 1. $6\frac{3}{4}$ 2. $7\frac{5}{8}$ 3. $13\frac{1}{6}$ 4. $17\frac{3}{16}$
  $5\frac{1}{4}$ $7\frac{3}{8}$ $12\frac{5}{6}$ $22\frac{13}{16}$

15 1. $1\frac{2}{3}$ 2. $4\frac{3}{5}$ 3. $32\frac{5}{9}$ 4. $4\frac{7}{16}$ **16** 1. $8\frac{3}{4}$ 2. $9\frac{8}{10}$ 3. $5\frac{5}{6}$ 4. $6\frac{3}{8}$
      $\underline{2\frac{2}{3}}$ $\underline{7\frac{4}{5}}$ $\underline{13\frac{8}{9}}$ $\underline{5\frac{12}{16}}$ $\underline{4\frac{3}{4}}$ $\underline{8\frac{7}{10}}$ $\underline{24\frac{5}{6}}$ $\underline{16\frac{7}{8}}$

17 1. $6\frac{1}{5}$ 2. $8\frac{5}{8}$ 3. $\frac{1}{4}$ 4. $\frac{9}{16}$ **18** 1. $8\frac{1}{8}$ 2. $6\frac{1}{6}$ 3. $4\frac{3}{10}$ 4. $15\frac{3}{4}$
      $\underline{\frac{1}{5}}$ $\underline{\frac{7}{8}}$ $\underline{7\frac{3}{4}}$ $\underline{18\frac{5}{16}}$ $\underline{2\frac{1}{3}}$ $\underline{9\frac{5}{8}}$ $\underline{8\frac{5}{12}}$ $\underline{29\frac{3}{16}}$

19 1. $3\frac{1}{3}$ 2. $4\frac{1}{4}$ 3. $8\frac{1}{5}$ 4. $22\frac{7}{12}$ **20** 1. $9\frac{1}{2}$ 2. $2\frac{2}{3}$ 3. $8\frac{3}{4}$ 4. $16\frac{11}{12}$
      $\underline{3\frac{1}{6}}$ $\underline{8\frac{5}{12}}$ $\underline{6\frac{3}{10}}$ $\underline{17\frac{1}{6}}$ $\underline{5\frac{3}{4}}$ $\underline{6\frac{4}{5}}$ $\underline{3\frac{7}{8}}$ $\underline{14\frac{5}{8}}$

21 1. $4\frac{4}{5}$ 2. $3\frac{3}{4}$ 3. $4\frac{5}{6}$ 4. $21\frac{17}{20}$ **22** 1. $7\frac{1}{4}$ 2. $9\frac{2}{3}$ 3. $\frac{3}{8}$ 4. $\frac{5}{6}$
      $\underline{6\frac{7}{10}}$ $\underline{7\frac{5}{12}}$ $\underline{3\frac{1}{2}}$ $\underline{37\frac{2}{5}}$ $\underline{\frac{7}{16}}$ $\underline{\frac{1}{2}}$ $\underline{14\frac{9}{10}}$ $\underline{19\frac{2}{3}}$

23 1. $\frac{5}{6}$ 2. $7\frac{3}{4}$ 3. $\frac{1}{3}$ 4. $6\frac{7}{12}$ **24** 1. $5\frac{1}{8}$ 2. $3\frac{2}{3}$ 3. $4\frac{1}{3}$ 4. $18\frac{5}{8}$
      $\frac{2}{3}$ $\frac{5}{8}$ $8\frac{2}{5}$ $\frac{3}{8}$ $1\frac{1}{4}$ $2\frac{1}{6}$ $8\frac{3}{8}$ $13\frac{11}{12}$
      $\underline{2\frac{1}{2}}$ $\underline{3\frac{13}{16}}$ $\underline{\frac{1}{4}}$ $\underline{5\frac{1}{6}}$ $\underline{4\frac{1}{2}}$ $\underline{2\frac{7}{12}}$ $\underline{2\frac{3}{4}}$ $\underline{42\frac{3}{16}}$

25 1. $\frac{1}{4} + \frac{1}{8}$ 2. $\frac{2}{3} + \frac{1}{6}$ 3. $\frac{5}{16} + \frac{7}{8}$ 4. $\frac{2}{5} + \frac{3}{5}$
     5. $\frac{3}{4} + \frac{7}{8} + \frac{1}{2}$ 6. $\frac{9}{16} + \frac{3}{8} + \frac{3}{4}$ 7. $\frac{1}{2} + \frac{9}{10} + \frac{3}{5}$ 8. $\frac{2}{3} + \frac{5}{6} + \frac{5}{8}$

26 1. $2\frac{3}{8} + 1\frac{1}{4}$ 2. $7 + \frac{5}{16}$ 3. $1\frac{2}{5} + \frac{7}{10} + 9$
     4. $6\frac{1}{4} + 3\frac{3}{8} + 1\frac{7}{16}$ 5. $2\frac{1}{2} + 4\frac{9}{16} + 8\frac{1}{2}$ 6. $7\frac{2}{3} + 5\frac{1}{6} + 3\frac{1}{12}$
     7. $5\frac{1}{10} + 8\frac{2}{5} + 7\frac{1}{2}$ 8. $23\frac{9}{10} + 12\frac{3}{8} + 9\frac{3}{4}$

27 Find the sum of:
     1. $\frac{2}{3}$ and $\frac{7}{12}$ 2. $3\frac{1}{4}$ and $9\frac{3}{4}$ ✓ 3. $9, \frac{4}{5},$ and $\frac{1}{2}$
     4. $9\frac{1}{5}, 6\frac{3}{10},$ and $4\frac{1}{2}$ 5. $3\frac{7}{8}, 4\frac{3}{4},$ and $2\frac{9}{16}$ 6. $8\frac{2}{5}, 14\frac{7}{10},$ and $9\frac{9}{10}$
     7. $3\frac{5}{6}, 5\frac{2}{3},$ and $1\frac{7}{12}$ 8. $4\frac{3}{4}, 5\frac{9}{10},$ and $2\frac{1}{6}$

28 Solve:
     1. $\frac{2}{3} + \frac{5}{6} = ?$ 2. $\frac{11}{16} + \frac{9}{16} = \blacksquare$ 3. $1\frac{2}{5} + 6\frac{1}{2} = n$
     4. $? = 4\frac{3}{8} + 5\frac{13}{32}$ 5. $\blacksquare = 10\frac{1}{2} + 6\frac{2}{3}$ 6. $n = 3\frac{1}{4} + 2\frac{5}{12} + 4\frac{2}{3}$

SKILL ACHIEVEMENT TEST

Add:
1. $\frac{11}{16} + \frac{9}{16}$ 2. $\frac{3}{4} + \frac{4}{5}$ 3. $\frac{7}{10} + \frac{13}{16}$ 4. $9 + 4\frac{2}{3}$
5. $4\frac{1}{6} + 2\frac{5}{6}$ 6. $8\frac{1}{5} + 6\frac{3}{10}$ 7. $5\frac{5}{8} + \frac{3}{4}$ 8. $21\frac{9}{16} + 15\frac{7}{12}$
9. $15\frac{9}{10} + 2\frac{5}{6} + 9\frac{1}{4}$ 10. $4\frac{1}{2} + 5\frac{3}{16} + 2\frac{7}{8}$

BUILDING PROBLEM-SOLVING SKILLS

FLASHBACK

1 READ

2 PLAN

3 SOLVE

4 CHECK

PROBLEM: Joan needed $2\frac{1}{4}$ cups of sifted cake flour to make a plain cake and $1\frac{1}{2}$ cups for a pineapple sponge cake. Find the total amount of flour required.

1 **Given facts:** $2\frac{1}{4}$ cups of flour and $1\frac{1}{2}$ cups of flour needed.
Question asked: What is the total amount of flour needed?

2 **Operation needed:** The word <u>total</u> with unequal quantities indicates addition.
Think or write out your plan: The sum of $2\frac{1}{4}$ cups of flour and $1\frac{1}{2}$ cups of flour is equal to the total number of cups of flour needed.
Equation: $2\frac{1}{4} + 1\frac{1}{2} = n$

3 **Estimate the answer:**

$2\frac{1}{4}$ cups→ 2 cups

$1\frac{1}{2}$ cups→ 2 cups

$\overline{4\text{ cups, estimate}}$

Solution:

$$\begin{array}{r} 2\frac{1}{4} = 2\frac{1}{4} \\ + 1\frac{1}{2} = 1\frac{2}{4} \\ \hline 3\frac{3}{4}\text{ cups of flour} \end{array}$$

ANSWER: $3\frac{3}{4}$ cups of flour needed.

4 **Check the accuracy:**

$$\begin{array}{r} 2\frac{1}{4} = 2\frac{1}{4} \\ + 1\frac{1}{2} = 1\frac{2}{4} \\ \hline 3\frac{3}{4} \end{array}$$

Compare answer to the estimate:
The answer $3\frac{3}{4}$ cups compares reasonably with estimate 4 cups.

PRACTICE PROBLEMS

1. A tailor made a 2-piece dress, requiring $2\frac{7}{8}$ yards of material for one part and $1\frac{3}{4}$ yards for the other. How much material did he use?

2. Carmen works after school. During a certain week, she worked $3\frac{3}{4}$ hours on Monday, $2\frac{1}{2}$ hours on Wednesday, and 4 hours on Friday. How many hours did she work altogether?

3. What is the overall length of a certain machine part consisting of 3 joined pieces measuring $2\frac{9}{16}$ inches, $1\frac{27}{32}$ inches, and $\frac{7}{8}$ inch respectively?

4. In installing an oil burner, Sara finds it necessary to use pieces of pipe measuring $4\frac{11}{16}$ inches, $7\frac{5}{8}$ inches, $3\frac{1}{2}$ inches, and 9 inches. What length of pipe does she need to cut the four pieces, disregarding waste?

3-9 Subtraction of Fractions and Mixed Numbers

PROCEDURE

1 Subtract: $\frac{4}{5} - \frac{1}{5}$

$$\frac{4}{5} - \frac{1}{5} = \frac{4-1}{5} = \frac{3}{5}$$

ANSWER: $\frac{3}{5}$

To subtract fractions that have a common denominator, subtract the numerators. Then write the difference over the common denominator.

2 Subtract: $\frac{11}{16} - \frac{5}{16}$

$$\begin{array}{r} \frac{11}{16} \\ -\ \frac{5}{16} \\ \hline \frac{6}{16} = \frac{3}{8} \end{array}$$

ANSWER: $\frac{3}{8}$

Express the answer in lowest terms.

---

3 Subtract: $\frac{7}{8} - \frac{1}{4}$

$$\begin{array}{l} \frac{7}{8} = \frac{7}{8} \\ \frac{1}{4} = \frac{2}{8} \\ \hline \quad\ \frac{5}{8} \end{array}$$

ANSWER: $\frac{5}{8}$

To subtract fractions that do not have a common denominator, find the LCD. Express the fractions as equivalent fractions using the LCD. Subtract.

---

4 Subtract: $7\frac{5}{6} - 3\frac{1}{6}$

$$\begin{array}{r} 7\frac{5}{6} \\ -\ 3\frac{1}{6} \\ \hline \frac{4}{6} \end{array}$$

To subtract mixed numbers, first subtract the fractions.

$$\begin{array}{r} 7\frac{5}{6} \\ -\ 3\frac{1}{6} \\ \hline 4\frac{4}{6} = 4\frac{2}{3} \end{array}$$

Then subtract the whole numbers.

Simplify.

ANSWER: $4\frac{2}{3}$

5 Subtract: $6\frac{4}{5} - 3\frac{1}{2}$

$6\frac{4}{5} = 6\frac{8}{10}$

$- 3\frac{1}{2} = 3\frac{5}{10}$

$ \phantom{3\frac{1}{2} =} 3\frac{3}{10}$

ANSWER: $3\frac{3}{10}$

Express each fraction as an equivalent fraction using the LCD.

Subtract.

6 Subtract: $8 - 1\frac{7}{16}$

$8 = 7\frac{16}{16}$

$- 1\frac{7}{16} = 1\frac{7}{16}$

$\phantom{- 1\frac{7}{16} =} 6\frac{9}{16}$

ANSWER: $6\frac{9}{16}$

Regroup when necessary.
$8 = 7 + 1 = 7 + \frac{16}{16} = 7\frac{16}{16}$
Subtract.

7 Subtract: $7\frac{2}{3} - 4$

$7\frac{2}{3}$

$- 4$

$\overline{ 3\frac{2}{3}}$

ANSWER: $3\frac{2}{3}$

8 Subtract: $6\frac{1}{8} - \frac{7}{8}$

$6\frac{1}{8} = 5\frac{9}{8}$

$- \frac{7}{8} = \frac{7}{8}$

$\phantom{- 6\frac{1}{8} =} 5\frac{2}{8} = 5\frac{1}{4}$

ANSWER: $5\frac{1}{4}$

9 Subtract: $9\frac{1}{3} - 4\frac{5}{6}$

$9\frac{1}{3} = 9\frac{2}{6} = 8\frac{8}{6}$

$- 4\frac{5}{6} = 4\frac{5}{6} = 4\frac{5}{6}$

$\phantom{- 4\frac{5}{6} = 4\frac{5}{6} =} 4\frac{3}{6} = 4\frac{1}{2}$

ANSWER: $4\frac{1}{2}$

Regroup.
$9\frac{2}{6} = 8 + 1 + \frac{2}{6} = 8 + \frac{6}{6} + \frac{2}{6} = 8\frac{8}{6}$

---

10 Solve the equation: $14\frac{1}{4} - 12\frac{4}{5} = n$

$14\frac{1}{4} = 14\frac{5}{20} = 13\frac{25}{20}$

$- 12\frac{4}{5} = 12\frac{16}{20} = 12\frac{16}{20}$

$\phantom{- 12\frac{4}{5} = 12\frac{16}{20} =} 1\frac{9}{20}$

ANSWER: $1\frac{9}{20}$

To find the value of n, subtract the given numbers.

DIAGNOSTIC TEST

Subtract:

1 $\dfrac{2}{3}$ $\dfrac{1}{3}$ **2** $\dfrac{5}{8}$ $\dfrac{1}{8}$ **3** $\dfrac{5}{6}$ $\dfrac{1}{2}$ **4** $\dfrac{7}{8}$ $\dfrac{1}{5}$ **5** $\dfrac{3}{4}$ $\dfrac{7}{10}$ **6** $4\dfrac{4}{5}$ $3\dfrac{2}{5}$ **7** $8\dfrac{13}{16}$ $5\dfrac{3}{16}$ **8** $45\dfrac{1}{3}$ $32\dfrac{2}{3}$ **9** $12\dfrac{3}{8}$ $7\dfrac{7}{8}$

10 $38\dfrac{9}{16}$ $36\dfrac{9}{16}$ **11** $5\dfrac{15}{32}$ 3 **12** 6 $2\dfrac{3}{4}$ **13** 9 $\dfrac{4}{5}$ **14** $8\dfrac{11}{32}$ $5\dfrac{1}{16}$ **15** $9\dfrac{3}{4}$ $3\dfrac{1}{3}$ **16** $13\dfrac{7}{16}$ $7\dfrac{5}{12}$

17 $14\dfrac{5}{8}$ $5\dfrac{3}{4}$ **18** $8\dfrac{1}{5}$ $2\dfrac{1}{3}$ **19** $6\dfrac{3}{10}$ $3\dfrac{9}{16}$ **20** $14\dfrac{1}{8}$ $13\dfrac{1}{2}$ **21** $1\dfrac{1}{4}$ $\dfrac{5}{6}$ **22** $\dfrac{15}{16} - \dfrac{3}{4}$

23 $14\dfrac{1}{4} - 5\dfrac{2}{3}$ **24** Subtract $1\dfrac{5}{8}$ from 6.   **25** Solve: $7\dfrac{5}{12} - 2\dfrac{1}{6} = n$

RELATED PRACTICE EXERCISES

Subtract:

1 1. $\dfrac{4}{5}$ $\dfrac{3}{5}$ 2. $\dfrac{5}{7}$ $\dfrac{3}{7}$ 3. $\dfrac{3}{4}$ $\dfrac{2}{4}$ 4. $\dfrac{27}{32}$ $\dfrac{18}{32}$ **2** 1. $\dfrac{3}{4}$ $\dfrac{1}{4}$ 2. $\dfrac{9}{16}$ $\dfrac{3}{16}$ 3. $\dfrac{5}{6}$ $\dfrac{1}{6}$ 4. $\dfrac{7}{10}$ $\dfrac{3}{10}$

3 1. $\dfrac{7}{8}$ $\dfrac{3}{16}$ 2. $\dfrac{2}{3}$ $\dfrac{1}{6}$ 3. $\dfrac{3}{4}$ $\dfrac{5}{12}$ 4. $\dfrac{17}{20}$ $\dfrac{3}{5}$ **4** 1. $\dfrac{1}{2}$ $\dfrac{1}{3}$ 2. $\dfrac{3}{4}$ $\dfrac{2}{5}$ 3. $\dfrac{7}{8}$ $\dfrac{2}{3}$ 4. $\dfrac{3}{5}$ $\dfrac{7}{12}$

5 1. $\dfrac{1}{4}$ $\dfrac{1}{6}$ 2. $\dfrac{5}{6}$ $\dfrac{3}{8}$ 3. $\dfrac{7}{12}$ $\dfrac{3}{16}$ 4. $\dfrac{5}{8}$ $\dfrac{1}{12}$ **6** 1. $6\dfrac{3}{5}$ $2\dfrac{1}{5}$ 2. $8\dfrac{8}{9}$ $7\dfrac{3}{9}$ 3. $15\dfrac{2}{4}$ $4\dfrac{1}{4}$ 4. $53\dfrac{7}{8}$ $25\dfrac{2}{8}$

7 1. $3\dfrac{5}{8}$ $1\dfrac{3}{8}$ 2. $6\dfrac{11}{16}$ $3\dfrac{5}{16}$ 3. $9\dfrac{7}{12}$ $7\dfrac{5}{12}$ 4. $11\dfrac{9}{10}$ $6\dfrac{3}{10}$ **8** 1. $9\dfrac{2}{5}$ $4\dfrac{4}{5}$ 2. $5\dfrac{3}{7}$ $1\dfrac{6}{7}$ 3. $16\dfrac{9}{16}$ $7\dfrac{12}{16}$ 4. $18\dfrac{1}{4}$ $14\dfrac{2}{4}$

9 1. $7\dfrac{1}{6}$ $1\dfrac{5}{6}$ 2. $15\dfrac{5}{8}$ $6\dfrac{7}{8}$ 3. $13\dfrac{3}{10}$ $10\dfrac{9}{10}$ 4. $17\dfrac{5}{32}$ $13\dfrac{29}{32}$ **10** 1. $6\dfrac{1}{2}$ $5\dfrac{1}{2}$ 2. $11\dfrac{3}{4}$ $4\dfrac{3}{4}$ 3. $43\dfrac{3}{10}$ $28\dfrac{3}{10}$ 4. $32\dfrac{11}{12}$ $23\dfrac{11}{12}$

11 1. $4\dfrac{1}{2}$ 2 2. $15\dfrac{3}{8}$ 9 3. $43\dfrac{9}{10}$ 27 4. $39\dfrac{25}{32}$ 14 **12** 1. 9 $2\dfrac{1}{4}$ 2. 3 $1\dfrac{3}{5}$ 3. 13 $6\dfrac{5}{8}$ 4. 42 $17\dfrac{7}{20}$

13 1. 5 $\dfrac{3}{8}$ 2. 9 $\dfrac{7}{12}$ 3. 12 $\dfrac{9}{10}$ 4. 10 $\dfrac{2}{3}$ **14** 1. $8\dfrac{5}{6}$ $6\dfrac{1}{3}$ 2. $11\dfrac{11}{12}$ $8\dfrac{1}{4}$ 3. $18\dfrac{9}{10}$ $15\dfrac{2}{5}$ 4. $42\dfrac{7}{8}$ $9\dfrac{3}{16}$

15 1. $7\frac{1}{3}$ 2. $20\frac{3}{5}$ 3. $11\frac{2}{3}$ 4. $40\frac{4}{5}$ **16** 1. $8\frac{3}{4}$ 2. $3\frac{7}{12}$ 3. $7\frac{15}{16}$ 4. $13\frac{9}{10}$
     $\underline{4\frac{1}{5}}$ $\underline{12\frac{7}{16}}$ $\underline{5\frac{1}{2}}$ $\underline{8\frac{3}{4}}$ $\underline{5\frac{1}{6}}$ $\underline{1\frac{3}{8}}$ $\underline{2\frac{5}{6}}$ $\underline{9\frac{3}{4}}$

17 1. $6\frac{1}{2}$ 2. $9\frac{3}{8}$ 3. $16\frac{3}{10}$ 4. $17\frac{5}{32}$ **18** 1. $10\frac{1}{3}$ 2. $13\frac{1}{4}$ 3. $9\frac{1}{6}$ 4. $25\frac{3}{4}$
     $\underline{3\frac{3}{4}}$ $\underline{5\frac{9}{16}}$ $\underline{8\frac{1}{2}}$ $\underline{12\frac{9}{16}}$ $\underline{4\frac{1}{2}}$ $\underline{8\frac{2}{3}}$ $\underline{2\frac{3}{5}}$ $\underline{21\frac{4}{5}}$

19 1. $9\frac{3}{4}$ 2. $7\frac{1}{6}$ 3. $18\frac{5}{6}$ 4. $24\frac{7}{12}$ **20** 1. $8\frac{11}{16}$ 2. $4\frac{7}{8}$ 3. $7\frac{1}{3}$ 4. $13\frac{3}{4}$
     $\underline{1\frac{9}{10}}$ $\underline{5\frac{3}{8}}$ $\underline{12\frac{15}{16}}$ $\underline{14\frac{7}{10}}$ $\underline{8\frac{7}{16}}$ $\underline{4\frac{1}{2}}$ $\underline{6\frac{2}{3}}$ $\underline{12\frac{5}{6}}$

21 1. $5\frac{1}{2}$ 2. $3\frac{9}{16}$ 3. $1\frac{7}{10}$ 4. $1\frac{1}{3}$
     $\underline{\frac{3}{4}}$ $\underline{\frac{7}{8}}$ $\underline{\frac{5}{6}}$ $\underline{\frac{1}{2}}$

22 1. $\frac{3}{5} - \frac{1}{5}$ 2. $\frac{1}{2} - \frac{1}{6}$ 3. $\frac{4}{5} - \frac{3}{4}$ 4. $2\frac{3}{8} - 1$
     5. $2 - 1\frac{3}{8}$ 6. $4 - \frac{9}{16}$ 7. $6\frac{1}{2} - \frac{1}{2}$ 8. $3\frac{5}{6} - \frac{1}{3}$

23 1. $6\frac{7}{8} - 2\frac{3}{8}$ 2. $8\frac{1}{4} - 4\frac{2}{5}$ 3. $11\frac{1}{2} - 2\frac{1}{3}$ 4. $14\frac{3}{8} - 9\frac{1}{10}$
     5. $24\frac{9}{16} - 8\frac{3}{4}$ 6. $21\frac{3}{10} - 13\frac{1}{4}$ 7. $10\frac{5}{6} - 10\frac{7}{12}$ 8. $8\frac{7}{8} - 7\frac{15}{16}$

24 1. From $\frac{7}{8}$ subtract $\frac{5}{16}$.   2. From $1\frac{4}{5}$ take $\frac{9}{10}$.

     3. From 7 subtract $2\frac{11}{16}$.   4. Take 3 from $4\frac{5}{8}$.

     5. Subtract $2\frac{1}{2}$ from $9\frac{1}{2}$.   6. Take $5\frac{3}{4}$ from $7\frac{1}{8}$.

     7. Find the difference between $6\frac{1}{3}$ and $5\frac{5}{6}$.

     8. Find the difference between 9 and $3\frac{3}{4}$.

25 Solve:

     1. $\frac{2}{3} - \frac{5}{8} = ?$ 2. $10 - 9\frac{3}{4} = \blacksquare$ 3. $2\frac{1}{2} - \frac{7}{8} = n$

     4. $? = 5\frac{1}{2} - 1\frac{5}{6}$ 5. $\blacksquare = 3\frac{4}{5} - 1\frac{3}{10}$ 6. $n = 8\frac{1}{6} - 4\frac{2}{3}$

SKILL ACHIEVEMENT TEST

Subtract:

1. $\frac{7}{8} - \frac{3}{8}$ 2. $\frac{15}{16} - \frac{21}{32}$ 3. $\frac{2}{3} - \frac{7}{16}$ 4. $14\frac{3}{4} - 3\frac{1}{8}$

5. $8 - 6\frac{5}{7}$ 6. $7\frac{3}{8} - 4\frac{5}{12}$ 7. $8\frac{7}{10} - 7\frac{1}{2}$ 8. $1\frac{4}{5} - \frac{5}{6}$

9. Subtract $3\frac{2}{3}$ from $4\frac{1}{6}$.

10. Find the difference between $15\frac{3}{10}$ and $8\frac{2}{5}$.

BUILDING PROBLEM-SOLVING SKILLS

FLASHBACK

1 READ

2 PLAN

3 SOLVE

4 CHECK

PROBLEM: Helen bought $4\frac{5}{8}$ yards of material for a dress. If her mother plans to use only $3\frac{3}{4}$ yards, how much extra material did Helen buy?

1 **Given facts:** Bought $4\frac{5}{8}$ yards of material; $3\frac{3}{4}$ yards to be used.
Question asked: How much extra material was bought?

2 **Operation needed:** The words how much extra indicate subtraction.
Think or write out your plan: The difference between the number of yards bought and the number of yards used is equal to the number of extra yards bought.
Equation: $4\frac{5}{8} - 3\frac{3}{4} = n$

3 **Estimate the answer:**

$$4\frac{5}{8} \text{ yd} \rightarrow 5 \text{ yd}$$
$$-\ 3\frac{3}{4} \text{ yd} \rightarrow \underline{4 \text{ yd}}$$
$$1 \text{ yd, estimate}$$

Solution:

$$4\frac{5}{8} \text{ yd} = 4\frac{5}{8} = 3\frac{13}{8} \text{ yd}$$
$$-\ 3\frac{3}{4} \text{ yd} = 3\frac{6}{8} = \underline{3\frac{6}{8} \text{ yd}}$$
$$\frac{7}{8} \text{ yd}$$

ANSWER: $\frac{7}{8}$ yd of extra material was bought.

4 **Check the accuracy:**

$$3\frac{3}{4} \text{ yd} = 3\frac{6}{8} \text{ yd}$$
$$+\ \frac{7}{8} \text{ yd} = \underline{\ \ \frac{7}{8} \text{ yd}}$$
$$3\frac{13}{8} \text{ yd} = 4\frac{5}{8} \text{ yd} ✔$$

Compare answer to the estimate:
The answer $\frac{7}{8}$ yd compares reasonably with the estimate 1 yd.

PRACTICE PROBLEMS

1. Two months ago Marian weighed $123\frac{1}{4}$ pounds. Now she weighs $116\frac{3}{4}$ pounds. How many pounds did she lose?

3. If one mechanic can assemble a motor in $6\frac{1}{2}$ hours while another mechanic can do the same job in $7\frac{5}{6}$ hours, in how much less time can the first mechanic do the job?

2. If normal body temperature is $98\frac{3}{5}$ degrees fahrenheit, how many degrees above normal is a temperature of 101 degrees?

4. A $2\frac{1}{2}$-inch nail is driven through a $1\frac{1}{8}$-inch thick piece of wood supporting a joist. How far into the joist did the nail extend?

MIXED PROBLEMS

Solve each of the following problems. For problems 1-4 select the letter corresponding to your answer.

1. A merchant sold $7\frac{5}{8}$ yards of cloth to a customer. If it was cut from a bolt that contained $18\frac{2}{3}$ yards, what length remained on the bolt?
   a. $10\frac{11}{24}$ yards   b. $11\frac{1}{24}$ yards
   c. $11\frac{1}{6}$ yards   d. $10\frac{5}{6}$ yards

2. Find the net change in a stock if it opened at $63\frac{7}{8}$ and closed at $65\frac{1}{4}$.
   a. $2\frac{3}{8}$ points gain
   b. $1\frac{7}{8}$ points gain
   c. $1\frac{3}{8}$ points gain
   d. Answer not given

3. Find the total thickness of two pieces of wood that Donald glued together if one is $\frac{5}{16}$ inch thick and the other $\frac{7}{8}$ inch thick.
   a. $1\frac{5}{16}$ inches   b. $1\frac{1}{2}$ inches
   c. $1\frac{3}{16}$ inches   d. $1\frac{1}{4}$ inches

4. The running time of a train from Chicago to San Francisco was changed to $49\frac{1}{3}$ hours. If this schedule saves $13\frac{3}{4}$ hours, how long did the trip take before the change was made?
   a. $64\frac{1}{4}$ hours   b. $63\frac{2}{3}$ hours
   c. $63\frac{1}{12}$ hours   d. $63\frac{5}{6}$ hours

5. A pilot finds that winds slow her progress in reaching her destination. Her trip out takes $2\frac{3}{4}$ hours. How long should her return trip take if she must return at the end of $4\frac{1}{2}$ hours flying time?

6. Team B is $4\frac{1}{2}$ games behind team A in the standings. Team C is $2\frac{1}{2}$ games behind team B. How many games behind team A is team C?

7. What is the outside diameter of tubing when the inside diameter is $2\frac{5}{8}$ inches and the wall thickness is $\frac{3}{16}$ inch?

Refresh Your Skills

1. Add: 1-4
   68,322
   87,947
   48,279
   70,976
   91,385

2. Subtract: 1-5
   832,506
   95,708

3. Multiply: 1-6
   869
   798

4. Divide: 1-7
   $2,240\overline{)1,543,360}$

5. Add: 2-4
   $5.6 + .29 + 10$

6. Subtract: 2-5
   $12.69 - \$7$

7. Multiply: 2-7
   $.005 \times .4$

8. Divide: 2-8
   $.6\overline{)54}$

9. What decimal part of 135 is 45? 2-9

10. .8 of what amount is $560? 2-10

11. Add: $4\frac{5}{6}$ 3-8
    $2\frac{3}{4}$
    $5\frac{7}{12}$

12. Subtract: 3-9
    $12\frac{9}{10}$
    $7\frac{2}{5}$

3-10 Writing Mixed Numbers as Improper Fractions

PROCEDURE

1 Write $1\frac{3}{8}$ as an improper fraction.

$$1\frac{3}{8} = 1 + \frac{3}{8} = \frac{8}{8} + \frac{3}{8}$$
$$= \frac{11}{8}$$

ANSWER: $\frac{11}{8}$

Express the whole number as a fraction. Use the same denominator as the fraction.
Add the fractions.

2 Write $4\frac{2}{3}$ as an improper fraction.

$$4\frac{2}{3} = 4 + \frac{2}{3} = \frac{12}{3} + \frac{2}{3}$$
$$= \frac{14}{3}$$

ANSWER: $\frac{14}{3}$

---

3 Write $5\frac{2}{7}$ as an improper fraction.

$$5\frac{2}{7} = \frac{35 + 2}{7} = \frac{37}{7}$$

5 × 7 = 35

ANSWER: $\frac{37}{7}$

Another way is to multiply the whole number by the denominator of the fraction. Add the product to the numerator of the fraction. Write the sum over the denominator of the fraction.

DIAGNOSTIC TEST

Write the following mixed numbers as improper fractions:

1 $1\frac{1}{3}$ **2** $1\frac{3}{4}$ **3** $7\frac{1}{2}$ **4** $5\frac{7}{8}$

RELATED PRACTICE EXERCISES

Write the following mixed numbers as improper fractions:

1 1. $1\frac{1}{5}$ 2. $1\frac{1}{4}$ 3. $1\frac{1}{2}$ 4. $1\frac{1}{16}$

2 1. $1\frac{5}{8}$ 2. $1\frac{2}{3}$ 3. $1\frac{7}{12}$ 4. $1\frac{13}{16}$

3 1. $8\frac{1}{5}$ 2. $3\frac{1}{7}$ 3. $9\frac{1}{6}$ 4. $16\frac{1}{3}$

4 **1.** $4\frac{2}{3}$ **2.** $3\frac{3}{4}$ **3.** $2\frac{5}{8}$ **4.** $3\frac{7}{12}$

    5. $7\frac{4}{5}$ **6.** $6\frac{2}{9}$ **7.** $5\frac{4}{7}$ **8.** $4\frac{13}{32}$

    9. $3\frac{5}{6}$ **10.** $8\frac{7}{10}$ **11.** $10\frac{9}{16}$ **12.** $13\frac{3}{5}$

SKILL ACHIEVEMENT TEST

Write the following mixed numbers as improper fractions:

1. $1\frac{1}{16}$ **2.** $1\frac{3}{5}$ **3.** $10\frac{1}{4}$ **4.** $6\frac{3}{8}$ **5.** $2\frac{11}{12}$

Applications

1. How many half-inches in $5\frac{1}{2}$ inches? In $9\frac{1}{2}$ inches?

2. How many eighths of an Inch in $4\frac{7}{8}$ inches? In $2\frac{3}{8}$ inches?

3. How many sixteenths of an inch in $2\frac{13}{16}$ inches? In $1\frac{9}{16}$ inches?

4. How many quarters of an inch in $6\frac{3}{4}$ inches? In $7\frac{1}{4}$ inches?

5. How many thirty-seconds of an inch in $1\frac{17}{32}$ inches? In $3\frac{5}{32}$ inches?

6. A bag contains $37\frac{1}{2}$ pounds of sugar. How many $\frac{1}{2}$-pound bags can be filled?

7. How many pieces $\frac{1}{4}$ inch long can be cut from a strip of metal $14\frac{3}{4}$ inches long?

8. If it takes $\frac{1}{8}$ yard of material to make a certain necktie, how many similar neckties can be made from a bolt of goods containing $15\frac{3}{8}$ yards?

9. How many pieces of wood $\frac{1}{3}$ foot long can be cut from a board $6\frac{2}{3}$ feet long?

3-11 Multiplication of Fractions and Mixed Numbers

PROCEDURE

1 Multiply: $\frac{7}{8} \times \frac{3}{5}$

$\frac{7}{8} \times \frac{3}{5} = \frac{21}{40}$

ANSWER: $\frac{21}{40}$

Multiply the numerators and multiply the denominators.

2 Multiply: $\frac{4}{5} \times \frac{15}{16}$

$\overset{1}{\underset{1}{\frac{4}{5}}} \times \overset{3}{\underset{4}{\frac{15}{16}}} = \frac{1 \times 3}{1 \times 4} = \frac{3}{4}$

ANSWER: $\frac{3}{4}$

Whenever possible, divide any numerator and denominator by their greatest common factor before multiplying.

Divide 4 and 16 by 4 (GCF).
Divide 5 and 15 by 5 (GCF).

---

3 Multiply: $\frac{2}{3} \times 5$

$\frac{2}{3} \times 5 = \frac{2}{3} \times \frac{5}{1}$

$\qquad = \frac{10}{3}$

$\qquad = 3\frac{1}{3}$

ANSWER: $3\frac{1}{3}$

Rewrite whole numbers as fractions with a denominator of 1.

Multiply.

Simplify.

4 Multiply: $6 \times \frac{3}{8}$

$6 \times \frac{3}{8} = \overset{3}{\frac{6}{1}} \times \underset{4}{\frac{3}{8}}$

$\qquad = \frac{9}{4} = 2\frac{1}{4}$

ANSWER: $2\frac{1}{4}$

Divide 6 and 8 by 2 (GCF).

---

5 Multiply: $2\frac{1}{2} \times 1\frac{1}{5}$

$2\frac{1}{2} \times 1\frac{1}{5} = \overset{1}{\underset{1}{\frac{5}{2}}} \times \overset{3}{\underset{1}{\frac{6}{5}}}$

$\qquad = \frac{3}{1} = 3$

ANSWER: 3

Rewrite mixed numbers as improper fractions.

6 Find the product of 21 and $3\frac{1}{2}$.

$$
\begin{array}{r}
21 \\
\times \ 3\frac{1}{2} \\
\hline
63 \quad \leftarrow \textbf{21} \times \textbf{3} \\
+ \ 10\frac{1}{2} \ \leftarrow \textbf{21} \times \frac{1}{2} \\
\hline
73\frac{1}{2}
\end{array}
$$

ANSWER: $73\frac{1}{2}$

When multiplying a mixed number and a whole number, the vertical form may also be used.

---

7 Multiply: $1\frac{4}{5} \times \frac{2}{3} \times 3\frac{1}{8}$

$$1\frac{4}{5} \times \frac{2}{3} \times 3\frac{1}{8} = \frac{\overset{3}{\cancel{9}}}{\underset{1}{\cancel{5}}} \times \frac{\overset{1}{\cancel{2}}}{\underset{1}{\cancel{3}}} \times \frac{\overset{5}{\cancel{25}}}{\underset{4}{\cancel{8}}}$$

$$= \frac{15}{4} = 3\frac{3}{4}$$

ANSWER: $3\frac{3}{4}$

---

8 Simplify: $\frac{5{,}280 \times 60}{3{,}600}$

$$\frac{5{,}280 \times 60}{3{,}600} = \frac{\overset{88}{5{,}280} \times \overset{1}{60}}{\underset{\underset{1}{60}}{3{,}600}} = \frac{88}{1} \text{ or } 88$$

ANSWER: 88

Division by the GCF may be used to simplify an expression involving both multiplication and division.

---

9 Solve the equation: $2\frac{5}{8} \times 5\frac{1}{7} = n$

$$2\frac{5}{8} \times 5\frac{1}{7} = \frac{\overset{3}{21}}{\underset{2}{8}} \times \frac{\overset{9}{36}}{\underset{1}{7}}$$

$$= \frac{27}{2} = 13\frac{1}{2}$$

ANSWER: $13\frac{1}{2}$

To find the value of n, multiply the given numbers.

---

10 Find $\frac{5}{8}$ of 48.

$$\frac{5}{8} \text{ of } 48 = \frac{5}{8} \times 48$$

$$- \frac{5}{\underset{1}{8}} \times \frac{\overset{6}{48}}{1}$$

$$= \frac{30}{1} = 30$$

ANSWER: 30

To find a fractional part of a number, multiply the given number and the fraction.*

*For an alternate method, see Lesson 17-14.

11 Find $\frac{5}{6}$ of $4.19 correct to the nearest cent.

$$\frac{5}{6} \text{ of } \$4.19 = \frac{5}{6} \times \$4.19$$
$$= \frac{\$20.95}{6}$$
$$= \$3.49\frac{1}{6} \text{ rounds to } \$3.49$$

ANSWER: $3.49

12 Find $.37\frac{1}{2}$ of $28.96.

$$
\begin{array}{r}
\$28.96 \\
\times \quad .37\frac{1}{2} \\
\hline
20272 \\
8688 \\
\hline
107152 \\
1448 \quad \longleftarrow \frac{1}{2} \times 2,896 \\
\hline
\$10.8600
\end{array}
$$

ANSWER: $10.86

DIAGNOSTIC TEST

Multiply:

1 $\frac{1}{5} \times \frac{1}{3}$ **2** $\frac{3}{4} \times \frac{5}{8}$ **3** $\frac{1}{2} \times \frac{2}{3}$ **4** $\frac{3}{8} \times \frac{4}{5}$

5 $\frac{3}{4} \times \frac{8}{15}$ **6** $\frac{9}{16} \times \frac{5}{6}$ **7** $\frac{5}{2} \times \frac{10}{3}$ **8** $\frac{7}{8} \times 8$

9 $\frac{2}{3} \times 6$ **10** $\frac{3}{4} \times 2$ **11** $\frac{5}{6} \times 10$ **12** $\frac{3}{5} \times 7$

13 $10 \times \frac{9}{10}$ **14** $48 \times \frac{7}{12}$ **15** $4 \times \frac{7}{8}$ **16** $12 \times \frac{5}{8}$

17 $5 \times \frac{3}{16}$ **18** $4\frac{1}{2} \times 4$ **19** $1\frac{7}{12} \times 8$ **20** $3\frac{1}{3} \times 5$

21 $12 \times 1\frac{5}{6}$ **22** $10 \times 2\frac{9}{16}$ **23** $7 \times 3\frac{1}{4}$ **24** $2\frac{1}{2} \times \frac{4}{5}$

25 $6\frac{1}{4} \times \frac{3}{8}$ **26** $\frac{5}{16} \times 9\frac{3}{5}$ **27** $\frac{5}{6} \times 1\frac{9}{16}$ **28** $5\frac{1}{3} \times 1\frac{1}{8}$

29 $2\frac{5}{8} \times 2\frac{2}{5}$ **30** $4\frac{1}{2} \times 2\frac{1}{4}$ **31** $\frac{1}{2} \times \frac{8}{15} \times \frac{5}{6}$ **32** $1\frac{3}{4} \times 3\frac{1}{7} \times 1\frac{3}{5}$

33 $\begin{array}{r} 18 \\ 7\frac{1}{3} \end{array}$ **34** $\begin{array}{r} 12\frac{5}{6} \\ 8 \end{array}$

35 Find the product of $2\frac{7}{8}$ and $1\frac{3}{4}$.

36 $\frac{5}{100} \times 900$

37 Simplify: $\frac{132 \times 3,600}{5,280}$

38 Correct to nearest cent:
$$
\begin{array}{r}
\$15.38 \\
\times \quad .62\frac{1}{2}
\end{array}
$$

39 Correct to nearest cent:
$$
\begin{array}{r}
\$23.89 \\
\times \quad .33\frac{1}{3}
\end{array}
$$

40 Solve: $3\frac{3}{4} \times 1\frac{1}{5} = n$

In each of the following, find the indicated fractional part of the given number:

41 Find $\frac{3}{4}$ of 18.   **42** Find $\frac{1}{2}$ of $.84.

43 Find $\frac{7}{8}$ of $3.45 correct to nearest cent.

RELATED PRACTICE EXERCISES

Multiply:

1 1. $\frac{1}{4} \times \frac{1}{2}$ 2. $\frac{1}{8} \times \frac{1}{3}$ 3. $\frac{1}{5} \times \frac{1}{4}$ 4. $\frac{1}{2} \times \frac{1}{10}$

2 1. $\frac{1}{2} \times \frac{3}{5}$ 2. $\frac{5}{6} \times \frac{7}{8}$ 3. $\frac{7}{16} \times \frac{3}{4}$ 4. $\frac{5}{12} \times \frac{1}{3}$

3 1. $\frac{4}{5} \times \frac{3}{4}$ 2. $\frac{3}{10} \times \frac{1}{3}$ 3. $\frac{2}{3} \times \frac{1}{2}$ 4. $\frac{5}{6} \times \frac{6}{7}$

4 1. $\frac{5}{6} \times \frac{3}{8}$ 2. $\frac{6}{7} \times \frac{11}{12}$ 3. $\frac{4}{5} \times \frac{7}{8}$ 4. $\frac{1}{3} \times \frac{9}{10}$

5 1. $\frac{2}{5} \times \frac{5}{12}$ 2. $\frac{9}{10} \times \frac{2}{3}$ 3. $\frac{5}{8} \times \frac{16}{25}$ 4 $\frac{8}{9} \times \frac{3}{4}$.

6 1. $\frac{5}{6} \times \frac{4}{5}$ 2. $\frac{3}{16} \times \frac{6}{7}$ 3. $\frac{7}{8} \times \frac{12}{21}$ 4. $\frac{10}{12} \times \frac{14}{15}$

7 1. $\frac{5}{4} \times \frac{2}{3}$ 2. $\frac{7}{2} \times \frac{4}{5}$ 3. $\frac{4}{3} \times \frac{3}{4}$ 4. $\frac{10}{9} \times \frac{15}{8}$

8 1. $\frac{5}{6} \times 6$ 2. $\frac{11}{8} \times 8$ 3. $\frac{7}{16} \times 16$ 4. $\frac{3}{5} \times 5$

9 1. $\frac{3}{4} \times 8$ 2. $\frac{7}{6} \times 72$ 3. $\frac{5}{12} \times 36$ 4. $\frac{9}{16} \times 64$

10 1. $\frac{3}{8} \times 4$ 2. $\frac{7}{16} \times 2$ 3. $\frac{5}{12} \times 3$ 4. $\frac{11}{24} \times 12$

11 1. $\frac{3}{8} \times 6$ 2. $\frac{9}{16} \times 20$ 3. $\frac{19}{12} \times 8$ 4. $\frac{5}{6} \times 4$

12 1. $\frac{1}{3} \times 7$ 2. $\frac{3}{5} \times 9$ 3. $\frac{9}{8} \times 5$ 4. $\frac{3}{10} \times 21$

13 1. $4 \times \frac{3}{4}$ 2. $8 \times \frac{5}{8}$ 3. $12 \times \frac{13}{12}$ 4. $16 \times \frac{15}{16}$

14 1. $12 \times \frac{5}{6}$ 2. $24 \times \frac{3}{8}$ 3. $32 \times \frac{17}{16}$ 4. $30 \times \frac{2}{5}$

15 1. $2 \times \frac{3}{8}$ 2. $2 \times \frac{7}{4}$ 3. $5 \times \frac{3}{10}$ 4. $6 \times \frac{13}{24}$

16 1. $12 \times \frac{7}{16}$ 2. $18 \times \frac{15}{32}$ 3. $16 \times \frac{19}{12}$ 4. $15 \times \frac{3}{10}$

17 1. $4 \times \frac{2}{3}$ 2. $9 \times \frac{4}{5}$ 3. $3 \times \frac{9}{8}$ 4. $7 \times \frac{5}{12}$

18 1. $2\frac{1}{8} \times 16$ 2. $1\frac{2}{3} \times 9$ 3. $7\frac{1}{2} \times 6$ 4. $4\frac{3}{5} \times 10$

19 1. $2\frac{5}{6} \times 3$ 2. $5\frac{7}{8} \times 6$ 3. $3\frac{11}{12} \times 8$ 4. $1\frac{9}{10} \times 12$

20 1. $2\frac{1}{5} \times 4$ 2. $5\frac{2}{3} \times 2$ 3. $4\frac{5}{8} \times 3$ 4. $3\frac{9}{16} \times 5$

21 1. $8 \times 5\frac{1}{4}$ 2. $16 \times 2\frac{5}{8}$ 3. $48 \times 1\frac{13}{16}$ 4. $24 \times 5\frac{11}{12}$

22 1. $15 \times 1\frac{7}{10}$ 2. $8 \times 2\frac{9}{16}$ 3. $18 \times 5\frac{5}{8}$ 4. $4 \times 3\frac{1}{6}$

23 1. $2 \times 6\frac{1}{3}$ 2. $3 \times 9\frac{1}{2}$ 3. $9 \times 5\frac{7}{16}$ 4. $7 \times 1\frac{3}{4}$

24 1. $3\frac{1}{3} \times \frac{3}{5}$ 2. $6\frac{3}{4} \times \frac{2}{3}$ 3. $3\frac{1}{8} \times \frac{8}{15}$ 4. $4\frac{1}{2} \times \frac{2}{9}$

25 1. $2\frac{1}{2} \times \frac{3}{4}$ 2. $5\frac{2}{3} \times \frac{5}{8}$ 3. $1\frac{3}{8} \times \frac{1}{2}$ 4. $3\frac{7}{16} \times \frac{3}{4}$

26 1. $\frac{1}{2} \times 3\frac{1}{5}$ 2. $\frac{5}{6} \times 2\frac{3}{10}$ 3. $\frac{7}{8} \times 3\frac{1}{7}$ 4. $\frac{9}{16} \times 1\frac{1}{3}$

27 1. $\frac{7}{8} \times 1\frac{1}{4}$ 2. $\frac{1}{2} \times 3\frac{3}{8}$ 3. $\frac{1}{5} \times 4\frac{1}{3}$ 4. $\frac{9}{10} \times 1\frac{1}{2}$

28 1. $1\frac{1}{4} \times 1\frac{3}{5}$ 2. $5\frac{1}{3} \times 4\frac{1}{2}$ 3. $2\frac{5}{8} \times 1\frac{5}{7}$ 4. $2\frac{2}{3} \times 3\frac{3}{8}$

29 **1.** $2\frac{1}{3} \times 1\frac{1}{5}$ **2.** $2\frac{1}{6} \times 2\frac{2}{3}$ **3.** $2\frac{2}{5} \times 1\frac{3}{16}$ **4.** $1\frac{3}{4} \times 3\frac{1}{3}$

30 **1.** $2\frac{1}{8} \times 1\frac{1}{2}$ **2.** $3\frac{3}{4} \times 2\frac{7}{8}$ **3.** $1\frac{9}{16} \times 4\frac{1}{3}$ **4.** $2\frac{5}{6} \times 1\frac{3}{8}$

31 **1.** $\frac{1}{4} \times \frac{5}{6} \times \frac{2}{5}$ **2.** $\frac{2}{3} \times \frac{5}{8} \times \frac{3}{10}$ **3.** $\frac{5}{12} \times \frac{3}{16} \times \frac{4}{5}$ **4.** $\frac{3}{4} \times \frac{3}{5} \times \frac{1}{2}$

32 **1.** $2\frac{3}{4} \times 1\frac{1}{8} \times 3\frac{5}{6}$ **2.** $1\frac{1}{2} \times \frac{4}{5} \times 2\frac{1}{6}$ **3.** $3\frac{1}{5} \times 1\frac{1}{4} \times 1\frac{1}{3}$ **4.** $1\frac{5}{16} \times 2\frac{2}{3} \times 3\frac{1}{7}$

33 **1.** 28 $\quad\quad 4\frac{1}{2}$ **2.** 16 $\quad\quad 5\frac{3}{4}$ **3.** 35 $\quad\quad 7\frac{5}{8}$ **4.** 29 $\quad\quad 2\frac{2}{3}$

34 **1.** $32\frac{1}{4}$ $\quad\quad 8$ **2.** $24\frac{3}{8}$ $\quad\quad 6$ **3.** $17\frac{3}{5}$ $\quad\quad 10$ **4.** $5\frac{5}{6}$ $\quad\quad 9$

35 **1.** Multiply $4\frac{7}{8}$ by 6.

 2. Multiply $2\frac{3}{4}$ by $4\frac{1}{2}$.

 3. Find the product of $1\frac{1}{8}$ and $5\frac{1}{3}$.

 4. Find the product of $2\frac{3}{16}$ and $2\frac{2}{5}$.

36 **1.** $\frac{3}{100} \times 400$ **2.** $\frac{4}{100} \times 500$ **3.** $\frac{2}{100} \times 1,000$ **4.** $\frac{6}{100} \times 2,500$

37 Simplify:

 1. $\frac{1,760 \times 60}{60}$ **2.** $\frac{75 \times 32}{25 \times 24}$

 3. $\frac{22 \times 21 \times 21}{7 \times 2 \times 2}$ **4.** $\frac{33 \times 6 \times 14}{231}$

 5. $\frac{18 \times 12 \times 16}{144}$ **6.** $\frac{5,280 \times 120}{3,600}$

 7. $\frac{3,600 \times 66}{5,280}$ **8.** $\frac{5,280 \times 4}{6,080}$

38 Find the product correct to nearest cent:

 1. $80 $\quad$ .12\frac{1}{2}$ **2.** $3,000 $\quad$ .04\frac{1}{2}$ **3.** $.54 $\quad$ .37\frac{1}{2}$ **4.** $18.48 $\quad$ .05\frac{3}{4}$

39 Find the product correct to nearest cent:

 1. $246 $\quad$ .33\frac{1}{3}$ **2.** $6,000 $\quad$ .08\frac{1}{3}$ **3.** $1.15 $\quad$ .66\frac{2}{3}$ **4.** $45.71 $\quad$ .83\frac{1}{3}$

40 Solve:

 1. $\frac{15}{16} \times \frac{4}{5} = ?$ **2.** $\frac{7}{8} \times \frac{3}{4} = \blacksquare$

 3. $\frac{2}{3} \times \frac{9}{4} = n$ **4.** $? = 1\frac{5}{8} \times 16$

 5. $\blacksquare = 2\frac{1}{2} \times 7\frac{1}{5}$ **6.** $n = 8\frac{3}{4} \times 3\frac{1}{7}$

In each of the following, find the indicated fractional part of the given number:

41 **1.** $\frac{1}{2}$ of 48 **2.** $\frac{1}{2}$ of 32¢ **3.** $\frac{1}{2}$ of $60 **4.** $\frac{1}{2}$ of $2\frac{1}{2}$

 5. $\frac{1}{3}$ of 96 **6.** $\frac{1}{3}$ of 8 **7.** $\frac{1}{4}$ of 52¢ **8.** $\frac{1}{4}$ of $1,000

 9. $\frac{1}{8}$ of $136 **10.** $\frac{1}{8}$ of $\frac{2}{5}$ **11.** $\frac{1}{16}$ of 288 **12.** $\frac{1}{16}$ of 24

 13. $\frac{1}{6}$ of $120 **14.** $\frac{1}{6}$ of 4 **15.** $\frac{1}{12}$ of 84¢ **16.** $\frac{1}{12}$ of $2\frac{1}{4}$

17. $\frac{1}{5}$ of $285   18. $\frac{1}{10}$ of 80¢

19. $\frac{1}{20}$ of 45   20. $\frac{1}{100}$ of 1,400

21. $\frac{3}{4}$ of $16   22. $\frac{3}{4}$ of $\frac{3}{8}$

23. $\frac{2}{3}$ of 87¢   24. $\frac{2}{3}$ of $1\frac{11}{16}$

25. $\frac{2}{5}$ of 45¢   26. $\frac{2}{5}$ of 12

27. $\frac{3}{5}$ of $1\frac{2}{3}$ 28. $\frac{4}{5}$ of $585

29. $\frac{3}{8}$ of 24¢   30. $\frac{7}{8}$ of 6

31. $\frac{15}{16}$ of $1,120   32. $\frac{9}{16}$ of $\frac{2}{3}$

33. $\frac{7}{10}$ of 180   34. $\frac{9}{100}$ of 250

35. $\frac{41}{100}$ of $5,000   36. $\frac{3}{10}$ of $1\frac{1}{2}$

37. $\frac{5}{6}$ of 72¢   38. $\frac{11}{12}$ of 132

39. $\frac{5}{12}$ of $2,400   40. $\frac{7}{12}$ of $2\frac{2}{5}$

42   1. $\frac{1}{2}$ of $.98   2. $\frac{1}{2}$ of $4.64   3. $\frac{1}{4}$ of .56   4. $\frac{3}{4}$ of $5.20

5. $\frac{1}{8}$ of $4.96   6. $\frac{7}{8}$ of 1.44   7. $2\frac{2}{3}$ of $38.16   8. $4\frac{3}{5}$ of $51.40

43   Round to hundredths or nearest cent:

1. $\frac{1}{2}$ of $.39   2. $\frac{3}{4}$ of .58   3. $\frac{1}{6}$ of $1.22   4. $\frac{3}{8}$ of $2.58

5. $\frac{7}{12}$ of $10   6. $1\frac{1}{2}$ of $.75   7. $4\frac{2}{5}$ of 3.64   8. $3\frac{1}{4}$ of $14.20

SKILL ACHIEVEMENT TEST

Multiply:

1. $\frac{9}{10} \times \frac{3}{8}$ 2. $\frac{1}{3} \times \frac{3}{5}$ 3. $\frac{4}{21} \times \frac{7}{8}$ 4. $\frac{5}{8} \times 16$

5. $8 \times \frac{3}{4}$ 6. $\frac{9}{5} \times \frac{7}{12}$ 7. $\frac{9}{10} \times 5$ 8. $13 \times \frac{2}{3}$

9. $3\frac{1}{4} \times 12$ 10. $16 \times 4\frac{3}{10}$ 11. $2\frac{1}{5} \times \frac{5}{6}$ 12. $4\frac{2}{3} \times \frac{11}{16}$

13. $\frac{5}{12} \times 2\frac{1}{4}$ 14. $1\frac{7}{16} \times 1\frac{5}{9}$ 15. $7\frac{1}{2} \times 5\frac{3}{5}$ 16. $3\frac{7}{8}$
                                                                                                             $\times\ 32$

Find the product correct to nearest cent:

17. $\frac{9}{100} \times 4,800$ 18. $2.80
                                                  $.03\frac{1}{4}$

19. $4\frac{1}{6} \times 3\frac{1}{5} \times 1\frac{3}{10}$ 20. $\frac{5,280 \times 100}{3,600}$

Solve:

21. $\frac{3}{4} \times \frac{16}{24} = n$ 22. $n = \frac{5}{6} \times \frac{20}{24}$

In each of the following, find the indicated fractional part of the given number:

23. Find $\frac{7}{12}$ of 96.   24. Find $\frac{5}{6}$ of $12.48.   25. Find $1\frac{2}{3}$ of $27.68 correct to nearest cent.

BUILDING PROBLEM-SOLVING SKILLS

FLASHBACK

1 READ

2 PLAN

3 SOLVE

4 CHECK

PROBLEM: The speed of a certain submarine when submerged is $\frac{3}{5}$ of its surface speed. If its maximum surface speed is $17\frac{1}{2}$ knots, what is its maximum speed when submerged?

1 **Given facts:** Maximum surface speed is $17\frac{1}{2}$ knots.
Speed, when submerged, is $\frac{3}{5}$ of surface speed.
Question asked: What is the maximum speed when submerged?

2 **Operation needed:** Finding a fractional part of a number indicates multiplication.
Think or write out your plan: $\frac{3}{5}$ of the maximum surface speed is equal to maximum speed when submerged.
Equation: $\frac{3}{5} \times 17\frac{1}{2} = n$

3 **Estimate the answer:**
$\frac{3}{5} \to \frac{1}{2}$
$17\frac{1}{2}$ knots $\to 18$ knots
$\frac{1}{2} \times 18 = 9$ knots, estimate

Solution:
$\frac{3}{5} \times 17\frac{1}{2} = \frac{3}{\cancel{5}} \times \frac{\cancel{35}^{7}}{2} = \frac{21}{2} = 10\frac{1}{2}$ knots

4 **Check the accuracy:**

$17\frac{1}{2} \times \frac{3}{5} = \frac{\cancel{35}^{7}}{2} \times \frac{3}{\cancel{5}_{1}} = \frac{21}{2} = 10\frac{1}{2}$ ✔

Compare answer to the estimate:
The answer of $10\frac{1}{2}$ knots compares reasonably with the estimate 9 knots

ANSWER: $10\frac{1}{2}$ knots is maximum speed when submerged.

PRACTICE PROBLEMS

1. How much wood is needed to make 15 shelves each $6\frac{2}{3}$ feet long?

2. Find the cost of $2\frac{1}{4}$ pounds of bananas at 32¢ per pound.

3. The cooking class is divided into six teams. If each team's recipe requires $2\frac{1}{2}$ cups cake flour, $2\frac{1}{4}$ teaspoons baking powder, $\frac{1}{2}$ cup shortening, and $1\frac{1}{4}$ cups sugar, how much of each ingredient does the entire class need?

4. Mr. Rogers earns $9 per hour. If his overtime rate is $1\frac{1}{2}$ times the regular rate, what is his hourly rate for overtime work?

MIXED PROBLEMS

Solve each of the following problems. For problems 1-4 select the letter corresponding to your answer.

1. A sewing class is making costumes for the school play. If each costume requires $3\frac{5}{8}$ yards of goods, how many yards are needed to make 26 costumes?

   a. $85\frac{3}{4}$ yards   **b.** $91\frac{7}{8}$ yards

   c. $94\frac{1}{4}$ yards   **d.** Answer not given

2. The XYZ stock opened Monday morning at the price of $27\frac{5}{8}$. During the 5 business days of the week, the stock gained $\frac{7}{8}$ point, $1\frac{1}{4}$ points, $\frac{5}{8}$ point, $1\frac{1}{2}$ points, and $\frac{3}{4}$ point respectively. What was its closing price on Friday?

   a. $32\frac{7}{8}$ **b.** $32\frac{3}{4}$

   c. $32\frac{5}{8}$ **d.** Answer not given

3. Find the net change in a stock if it opened at $16\frac{1}{8}$ and closed at $15\frac{1}{4}$.

   a. $\frac{3}{8}$ points loss   **b.** $\frac{7}{8}$ points loss

   c. $1\frac{7}{8}$ points loss   **d.** Answer not given

4. A house worth $52,800 is assessed at $\frac{2}{3}$ of its value. What is the assessed value?

   a. $33,600   **b.** $37,500

   c. $35,200   **d.** $30,500

5. What is the perimeter (distance around) of a triangle if its three sides measure $6\frac{3}{8}$ inches, $4\frac{11}{16}$ inches, and $5\frac{3}{4}$ inches respectively?

6. If the scale on a chart is 1 inch = 30 miles, how many miles do $6\frac{1}{2}$ inches represent?

7. A plumber, in installing water pipes, used pieces measuring $5\frac{1}{2}$ feet, $3\frac{3}{4}$ feet, and $1\frac{2}{3}$ feet. If they were cut from a 15-foot length of pipe, how many feet of pipe remained? Disregard waste.

8. A family budgets $\frac{1}{4}$ of its annual income of $18,600 for food, $\frac{3}{10}$ for shelter including operating expenses and furnishings, $\frac{1}{10}$ for transportation, $\frac{1}{20}$ for clothing, $\frac{1}{8}$ for savings, and the remainder for miscellaneous. How much is allowed for each item annually?

Refresh Your Skills

1. Add: 1-4
   627
   4,963
   81,896
   97,988
   8,569

2. Subtract: 1-5
   800,000
   706,094

3. Multiply: 1-6
   5,280
   5,280

4. Divide: 1-7
   $365\overline{)149,285}$

5. Add: 2-4
   .66 + .6 + .666

6. Subtract: 2-5
   .62 − .2

7. Multiply: 2-7
   .08 × $19.75

8. Divide: 2-8
   $.02\overline{).007}$

9. Multiply 81.24 by 100. 2-11

10. Divide .6 by 1,000. 2-12

11. Express $\frac{36}{54}$ in lowest terms. 3-1

12. Add: 3-8
    $3\frac{7}{10}$
    $6\frac{5}{12}$

13. Subtract: 3-9
    $4\frac{1}{3}$
    $3\frac{5}{6}$

14. Multiply: 3-11
    $4\frac{3}{8} \times 3\frac{1}{5}$

MULTIPLICATIVE INVERSE

If the product of two numbers is one (1), then each factor is called the multiplicative inverse or reciprocal of the other.

8 and $\frac{1}{8}$ are multiplicative inverses of each other because $8 \times \frac{1}{8} = 1$.

$\frac{4}{3}$ and $\frac{3}{4}$ are multiplicative inverses of each other because $\frac{4}{3} \times \frac{3}{4} = 1$.

Notice that the numerator of one fraction is the denominator of its reciprocal, and its denominator is the numerator of its reciprocal.

Zero has no inverse for multiplication.

When we divide 12 by 4, the quotient is 3.
$$12 \div 4 = 3$$
When we multiply 12 by $\frac{1}{4}$ the product is 3.
$$12 \times \frac{1}{4} = 3$$

Notice that dividing a number by another number gives the same result as multiplying the first number by the multiplicative inverse of the divisor.

> Thus, to divide a number by another number, we may multiply the first number by the multiplicative inverse of the divisor.
> EXAMPLE: $\frac{2}{3} \div \frac{3}{4} = \frac{2}{3} \times \frac{4}{3} = \frac{8}{9}$

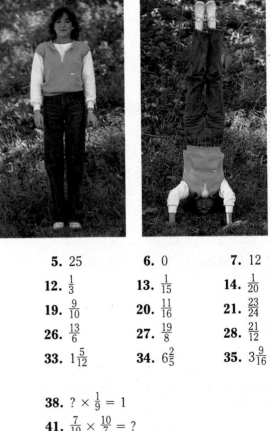

EXERCISES

Write the multiplicative inverse or reciprocal of each of the following:

1. 6	**2.** 3	**3.** 1	**4.** 9	**5.** 25	**6.** 0	**7.** 12
8. $\frac{1}{4}$	**9.** $\frac{1}{10}$	**10.** $\frac{1}{7}$	**11.** $\frac{1}{2}$	**12.** $\frac{1}{3}$	**13.** $\frac{1}{15}$	**14.** $\frac{1}{20}$
15. $\frac{5}{6}$	**16.** $\frac{7}{12}$	**17.** $\frac{3}{8}$	**18.** $\frac{2}{5}$	**19.** $\frac{9}{10}$	**20.** $\frac{11}{16}$	**21.** $\frac{23}{24}$
22. $\frac{8}{5}$	**23.** $\frac{5}{4}$	**24.** $\frac{22}{7}$	**25.** $\frac{9}{2}$	**26.** $\frac{13}{6}$	**27.** $\frac{19}{8}$	**28.** $\frac{21}{12}$
29. $2\frac{1}{2}$	**30.** $6\frac{2}{3}$	**31.** $1\frac{3}{4}$	**32.** $4\frac{3}{8}$	**33.** $1\frac{5}{12}$	**34.** $6\frac{2}{5}$	**35.** $3\frac{9}{16}$

Find the missing numbers:

36. $\frac{3}{8} \times \frac{8}{3} = ?$

37. $12 \times ? = 1$

38. $? \times \frac{1}{9} = 1$

39. $\frac{5}{6} \times ? = 1$

40. $? \times \frac{11}{16} = 1$

41. $\frac{7}{10} \times \frac{10}{7} = ?$

42. $\frac{9}{5} \times ? = 1$

43. $? \times \frac{15}{2} = 1$

3-12 Division of Fractions and Mixed Numbers

PROCEDURE

1 Divide: $\frac{2}{5} \div \frac{5}{8}$

reciprocals

$$\frac{2}{5} \div \frac{5}{8} = \frac{2}{5} \times \frac{8}{5} = \frac{16}{25}$$

ANSWER: $\frac{16}{25}$

To divide by a fraction, invert the divisor and multiply.

---

2 Divide: $\frac{7}{8} \div 3$

$$\frac{7}{8} \div 3 = \frac{7}{8} \div \frac{3}{1} = \frac{7}{8} \times \frac{1}{3} = \frac{7}{24}$$

Express as a fraction.

ANSWER: $\frac{7}{24}$

3 Divide: $15 \div \frac{3}{8}$

$$15 \div \frac{3}{8} = \frac{\overset{5}{15}}{1} \times \frac{8}{\underset{1}{3}} = \frac{40}{1} = 40$$

ANSWER: 40

---

4 Divide: $2\frac{1}{2} \div \frac{3}{4}$

$$2\frac{1}{2} \div \frac{3}{4} = \frac{5}{2} \div \frac{3}{4}$$

$$= \frac{5}{2} \times \frac{\overset{2}{4}}{\underset{1}{3}} = \frac{10}{3} = 3\frac{1}{3}$$

ANSWER: $3\frac{1}{3}$

Rewrite mixed numbers as improper fractions.

5 Divide: $8 \div 2\frac{4}{5}$.

$$8 \div 2\frac{4}{5} = \frac{8}{1} \div \frac{14}{5}$$

$$= \frac{8}{1} \times \frac{5}{\underset{7}{\overset{4}{14}}} = \frac{20}{7} = 2\frac{6}{7}$$

ANSWER: $2\frac{6}{7}$

6 Divide $2\frac{3}{16}$ by $1\frac{1}{4}$.

$$2\frac{3}{16} \div 1\frac{1}{4} = \frac{35}{16} \div \frac{5}{4}$$

$$= \frac{\overset{7}{35}}{\underset{4}{16}} \times \frac{\overset{1}{4}}{\underset{1}{5}} = \frac{7}{4} = 1\frac{3}{4}$$

ANSWER: $1\frac{3}{4}$

7 Simplify $\dfrac{\frac{9}{16}}{\frac{3}{8}}$.

complex fraction

$\dfrac{\frac{9}{16}}{\frac{3}{8}} = \dfrac{9}{16} \div \dfrac{3}{8}$ Divide the numerator by the denominator.

$= \dfrac{\overset{3}{\cancel{9}}}{\underset{2}{\cancel{16}}} \times \dfrac{\overset{1}{\cancel{8}}}{\underset{1}{\cancel{3}}} = \dfrac{3}{2} = 1\dfrac{1}{2}$

ANSWER: $1\dfrac{1}{2}$

Vocabulary

If the product of two numbers is one (1), then each factor is the **reciprocal** of the other. A **complex fraction** is a fraction in which the numerator or denominator or both have a fraction as a term.

DIAGNOSTIC TEST

Divide:

1 $\dfrac{1}{3} \div \dfrac{3}{4}$ **2** $\dfrac{2}{3} \div \dfrac{5}{16}$ **3** $\dfrac{3}{5} \div \dfrac{9}{10}$

4 $\dfrac{3}{4} \div \dfrac{3}{8}$ **5** $\dfrac{5}{6} \div \dfrac{7}{12}$ **6** $\dfrac{3}{4} \div 6$

7 $\dfrac{7}{8} \div 2$ **8** $8 \div \dfrac{1}{2}$ **9** $5 \div \dfrac{15}{16}$

10 $2 \div \dfrac{3}{5}$ **11** $4\dfrac{1}{2} \div 18$ **12** $4\dfrac{2}{3} \div 2$

13 $1\dfrac{2}{3} \div 4$ **14** $15 \div 1\dfrac{7}{8}$ **15** $6 \div 4\dfrac{1}{2}$

16 $4 \div 5\dfrac{1}{3}$ **17** $7 \div 2\dfrac{3}{4}$ **18** $2\dfrac{1}{2} \div \dfrac{5}{6}$

19 $2\dfrac{5}{8} \div \dfrac{3}{5}$ **20** $1\dfrac{7}{16} \div \dfrac{2}{3}$ **21** $\dfrac{7}{8} \div 1\dfrac{3}{4}$

22 $\dfrac{3}{4} \div 1\dfrac{3}{5}$ **23** $11\dfrac{1}{3} \div 2\dfrac{5}{6}$ **24** $3\dfrac{3}{16} \div 2\dfrac{1}{8}$

25 $1\dfrac{1}{6} \div 9\dfrac{1}{3}$ **26** $5\dfrac{3}{8} \div 1\dfrac{2}{5}$ **27** $6\dfrac{3}{5} \div 6\dfrac{3}{5}$

28 Divide $9\dfrac{1}{3}$ by $3\dfrac{1}{7}$.   **29** Simplify: $\dfrac{\frac{2}{3}}{\frac{5}{8}}$ **30** Solve: $1\dfrac{3}{4} \div 8\dfrac{1}{6} = n$

RELATED PRACTICE EXERCISES

Divide:

1 1. $\frac{1}{4} \div \frac{1}{3}$ 2. $\frac{3}{4} \div \frac{4}{5}$ 3. $\frac{1}{2} \div \frac{2}{3}$ 4. $\frac{3}{5} \div \frac{11}{12}$

2 1. $\frac{1}{2} \div \frac{1}{5}$ 2. $\frac{7}{8} \div \frac{2}{3}$ 3. $\frac{9}{10} \div \frac{1}{3}$ 4. $\frac{4}{9} \div \frac{7}{16}$

3 1. $\frac{1}{4} \div \frac{3}{4}$ 2. $\frac{5}{8} \div \frac{5}{6}$ 3. $\frac{3}{16} \div \frac{5}{12}$ 4. $\frac{5}{6} \div \frac{7}{8}$

4 1. $\frac{5}{6} \div \frac{5}{12}$ 2. $\frac{2}{3} \div \frac{2}{9}$ 3. $\frac{5}{8} \div \frac{5}{8}$ 4. $\frac{4}{5} \div \frac{4}{15}$

5 1. $\frac{1}{2} \div \frac{7}{16}$ 2. $\frac{7}{8} \div \frac{5}{12}$ 3. $\frac{4}{5} \div \frac{7}{10}$ 4. $\frac{5}{12} \div \frac{3}{16}$

6 1. $\frac{2}{3} \div 4$ 2. $\frac{7}{8} \div 7$ 3. $\frac{9}{10} \div 6$ 4. $\frac{3}{5} \div 9$

7 1. $\frac{3}{5} \div 8$ 2. $\frac{1}{2} \div 10$ 3. $\frac{11}{12} \div 5$ 1. $\frac{5}{8} \div 4$

8 1. $6 \div \frac{1}{3}$ 2. $7 \div \frac{1}{8}$ 3. $8 \div \frac{4}{5}$ 4. $10 \div \frac{5}{16}$

9 1. $18 \div \frac{9}{10}$ 2. $6 \div \frac{4}{5}$ 3. $10 \div \frac{2}{3}$ 4. $12 \div \frac{15}{16}$

10 1. $5 \div \frac{3}{4}$ 2. $13 \div \frac{2}{3}$ 3. $9 \div \frac{7}{8}$ 4. $7 \div \frac{9}{10}$

11 1. $1\frac{1}{2} \div 3$ 2. $4\frac{2}{3} \div 14$ 3. $5\frac{3}{5} \div 7$ 4. $2\frac{7}{16} \div 13$

12 1. $7\frac{1}{2} \div 5$ 2. $8\frac{2}{5} \div 6$ 3. $18\frac{3}{4} \div 10$ 4. $11\frac{7}{8} \div 5$

13 1. $1\frac{5}{8} \div 2$ 2. $2\frac{1}{16} \div 4$ 3. $5\frac{3}{4} \div 5$ 4. $3\frac{4}{5} \div 8$

14 1. $6 \div 1\frac{1}{2}$ 2. $8 \div 1\frac{1}{3}$ 3. $68 \div 3\frac{2}{5}$ 4. $57 \div 2\frac{3}{8}$

15 1. $8 \div 2\frac{2}{5}$ 2. $10 \div 1\frac{7}{8}$ 3. $14 \div 1\frac{5}{16}$ 4. $40 \div 5\frac{1}{3}$

16 1. $5 \div 6\frac{2}{3}$ 2. $8 \div 9\frac{3}{5}$ 3. $6 \div 7\frac{7}{8}$ 4. $2 \div 4\frac{4}{5}$

17 1. $6 \div 1\frac{2}{3}$ 2. $4 \div 6\frac{3}{5}$ 3. $1 \div 2\frac{1}{3}$ 4. $3 \div 1\frac{5}{8}$

18 1. $1\frac{1}{2} \div \frac{9}{16}$ 2. $1\frac{1}{8} \div \frac{3}{32}$ 3. $8\frac{3}{4} \div \frac{7}{8}$ 4. $9\frac{1}{3} \div \frac{7}{24}$

19 1. $1\frac{3}{5} \div \frac{2}{3}$ 2. $1\frac{5}{6} \div \frac{5}{12}$ 3. $3\frac{1}{4} \div \frac{5}{6}$ 4. $2\frac{1}{8} \div \frac{9}{10}$

20 1. $4\frac{1}{4} \div \frac{3}{5}$ 2. $2\frac{3}{5} \div \frac{5}{8}$ 3. $3\frac{1}{7} \div \frac{3}{4}$ 4. $2\frac{2}{3} \div \frac{9}{16}$

21 1. $\frac{5}{12} \div 8\frac{1}{3}$ 2. $\frac{5}{6} \div 1\frac{1}{9}$ 3. $\frac{3}{5} \div 2\frac{2}{5}$ 4. $\frac{13}{16} \div 1\frac{7}{32}$

22 1. $\frac{2}{3} \div 1\frac{1}{4}$ 2. $\frac{7}{8} \div 3\frac{1}{3}$ 3. $\frac{3}{16} \div 1\frac{3}{5}$ 4. $\frac{1}{3} \div 2\frac{9}{16}$

23 1. $14\frac{3}{8} \div 2\frac{7}{8}$ 2. $7\frac{1}{2} \div 1\frac{1}{4}$ 3. $18\frac{1}{3} \div 1\frac{5}{6}$ 4. $50\frac{1}{4} \div 4\frac{3}{16}$

24 1. $3\frac{1}{5} \div 1\frac{1}{3}$ 2. $1\frac{5}{8} \div 1\frac{7}{32}$ 3. $4\frac{2}{3} \div 1\frac{3}{5}$ 4. $11\frac{1}{4} \div 2\frac{1}{2}$

25 1. $2\frac{1}{4} \div 3\frac{3}{8}$ 2. $1\frac{3}{5} \div 3\frac{1}{5}$ 3. $1\frac{13}{16} \div 2\frac{1}{4}$ 4. $2\frac{1}{12} \div 3\frac{3}{4}$

26 1. $1\frac{2}{5} \div 2\frac{2}{3}$ 2. $3\frac{3}{8} \div 3\frac{1}{5}$ 3. $1\frac{1}{2} \div 1\frac{7}{9}$ 4. $1\frac{2}{3} \div 1\frac{7}{16}$

27 1. $\frac{3}{8} \div \frac{3}{8}$ 2. $1\frac{5}{12} \div 1\frac{5}{12}$ 3. $2\frac{2}{3} \div 2\frac{2}{3}$ 4. $6\frac{13}{16} \div 6\frac{13}{16}$

28 **1.** Divide $4\frac{1}{2}$ by $\frac{3}{5}$.   **2.** Divide $6\frac{1}{4}$ by 8.   **3.** Divide $2\frac{2}{3}$ by $7\frac{1}{2}$.   **4.** Divide $1\frac{7}{8}$ by $1\frac{1}{3}$.

29 Simplify:

1. $\dfrac{\frac{3}{5}}{\frac{3}{8}}$ **2.** $\dfrac{\frac{1}{2}}{\frac{3}{4}}$ **3.** $\dfrac{\frac{7}{16}}{\frac{5}{6}}$ **4.** $\dfrac{\frac{4}{5}}{\frac{7}{8}}$

5. $\dfrac{2\frac{2}{3}}{5\frac{1}{3}}$ **6.** $\dfrac{3\frac{3}{8}}{4\frac{1}{2}}$ **7.** $\dfrac{3\frac{15}{16}}{2\frac{5}{8}}$ **8.** $\dfrac{10\frac{5}{8}}{4\frac{1}{4}}$

9. $\dfrac{\frac{5}{8}}{2}$ **10.** $\dfrac{1\frac{5}{6}}{3}$ **11.** $\dfrac{9\frac{1}{2}}{4}$ **12.** $\dfrac{2\frac{1}{8}}{10}$

13. $\dfrac{87\frac{1}{2}}{100}$ **14.** $\dfrac{33\frac{1}{3}}{100}$ **15.** $\dfrac{62\frac{1}{2}}{100}$ **16.** $\dfrac{16\frac{2}{3}}{100}$

17. $\dfrac{4}{\frac{2}{5}}$ **18.** $\dfrac{6}{1\frac{1}{2}}$ **19.** $\dfrac{5}{3\frac{3}{4}}$ **20.** $\dfrac{\frac{1}{2}}{2\frac{2}{3}}$

21. $\dfrac{\frac{1}{2}+\frac{1}{4}}{\frac{3}{8}}$ **22.** $\dfrac{\frac{5}{6}}{3-1\frac{1}{3}}$ **23.** $\dfrac{\frac{3}{4}+\frac{7}{8}}{\frac{11}{16}-\frac{1}{2}}$ **24.** $\dfrac{4\frac{1}{2}+1\frac{1}{2}}{7\frac{1}{4}+2\frac{5}{8}}$

30 Solve:

1. $\frac{2}{5} \div \frac{9}{16} = ?$ **2.** $\frac{7}{8} \div \frac{7}{16} = $ ▨

3. $3\frac{3}{8} \div 9 = n$ **4.** $? = 15 \div \frac{21}{32}$

5. ▨ $= \frac{7}{12} \div 3\frac{3}{4}$ **6.** $n = 6\frac{1}{8} \div 1\frac{1}{6}$

SKILL ACHIEVEMENT TEST

Divide:

1. $\frac{2}{3} \div \frac{4}{5}$ **2.** $\frac{9}{16} \div \frac{3}{4}$

3. $4 \div \frac{2}{5}$ **4.** $\frac{5}{6} \div 10$

5. $3\frac{3}{4} \div 3$ **6.** $\frac{4}{5} \div 4\frac{5}{8}$

7. $2\frac{1}{3} \div \frac{7}{8}$ **8.** $9\frac{1}{2} \div 9\frac{1}{2}$

9. $2\frac{1}{16} \div 1\frac{3}{8}$ **10.** $1\frac{17}{32} \div 2\frac{5}{8}$

11. $3 \div 6\frac{3}{4}$ **12.** Simplify: $\dfrac{\frac{9}{16}}{\frac{2}{5}}$

BUILDING PROBLEM-SOLVING SKILLS

FLASHBACK

1 READ

2 PLAN

3 SOLVE

4 CHECK

PROBLEM: A bus is scheduled to go a distance of 121 miles in $2\frac{3}{4}$ hours. What average rate of speed must be maintained in order to arrive on schedule?

1 **Given facts:** A distance of 121 miles is to be traveled in $2\frac{3}{4}$ hours.
Question asked: What is the average rate of speed?

2 **Operation needed:** The words <u>what is the average</u> indicate division.
Think or write out your plan: The distance (121 miles) divided by the time of travel ($2\frac{3}{4}$ hours) is equal to the average rate of speed.
Equation: $121 \div 2\frac{3}{4} = n$

3 **Estimate the answer:**
$121 \rightarrow 120$
$2\frac{3}{4} \rightarrow 3$
$120 \div 3 = 40$ m.p.h., estimate
Solution:
$$121 \div 2\frac{3}{4}$$
$$= 121 \div \frac{11}{4}$$
$$= \overset{11}{\cancel{121}} \times \frac{4}{\cancel{11}_1} = 44 \text{ m.p.h.}$$

4 **Check the accuracy:**
$2\frac{3}{4} \times 44$
$= \frac{11}{\cancel{4}_1} \times \overset{11}{\cancel{44}} = 121$ miles ✔

Compare answer to the estimate:
The answer 44 m.p.h. compares reasonably with the estimate 40 m.p.h.

ANSWER: Average rate of speed is 44 m.p.h.

PRACTICE PROBLEMS

1. If an airplane flies 1,440 kilometers in $2\frac{1}{4}$ hours, what is its average ground speed in kilometers per hour?

2. If a board $10\frac{1}{2}$ feet long was cut into 6 pieces of equal length, what would the length of each piece be? Disregard waste.

3. What is the cost of 1 pound of apples if $2\frac{5}{8}$ pounds cost 84¢?

4. How many athletic association membership cards $1\frac{3}{8}$ inches wide can be cut from stock $24\frac{3}{4}$ inches wide?

MIXED PROBLEMS

Solve each of the following problems. For problems 1-4 select the letter corresponding to your answer.

1. If each costume for the school show requires $3\frac{1}{3}$ yards of material, how many costumes can be made from a 30-yard bolt of material?

    a. 10 costumes   **b.** 9 costumes

    c. 8 costumes   **d.** Answer not given

2. What are the actual dimensions of a porch which, drawn to the scale of $\frac{1}{4}$ inch = 1 foot, measures $1\frac{3}{4}$ inches by $2\frac{1}{2}$ inches?

    a. 8 ft. by 10 ft.   **b.** 7 ft. by 10 ft.

    c. 7 ft. by 11 ft.   **d.** Answer not given

3. In arranging an $8\frac{1}{2}$ inch by 11 inch piece of paper in the drawing class, pupils were directed to draw a line $\frac{1}{4}$ inch from each edge. What are the inside dimensions between the lines?

    a. $8\frac{1}{4} \times 10\frac{3}{4}$ inches   **b.** $8 \times 10\frac{3}{4}$ inches

    c. $8 \times 10\frac{1}{2}$ inches   **d.** Answer not given

4. Paul wishes to buy a cassette recorder priced at $60. He pays $\frac{1}{5}$ of the price in cash and the rest in 6 equal monthly installments. How much must he pay each month?

    a. $8   **b.** $9

    c. $10   **d.** Answer not given

5. How many lengths of pipe $3\frac{1}{2}$ feet long can be cut from a pipe 28 feet long? Disregard waste in cutting.

6. If 1 cubic foot of water weighs $62\frac{1}{2}$ pounds, find the weight of a column of water containing 32 cubic feet.

7. A storage bin has a capacity of 1,190 cubic feet. How many bushels of wheat can it hold? (1 bu. = $1\frac{1}{4}$ cu. ft.)

8. If the distance between Chicago and Cleveland, 360 miles, is represented on a map by a length of $4\frac{1}{2}$ inches, what is the distance between Detroit and Memphis represented on the same map by a length of $9\frac{1}{8}$ inches?

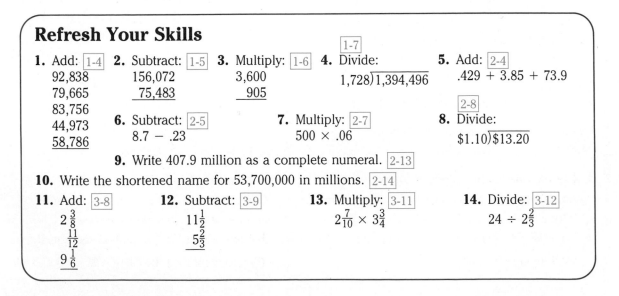

Refresh Your Skills

1. Add: [1-4]
92,838
79,665
83,756
44,973
58,786

2. Subtract: [1-5]
156,072
  75,483

3. Multiply: [1-6]
3,600
  905

4. Divide: [1-7]
$1,728\overline{)1,394,496}$

5. Add: [2-4]
.429 + 3.85 + 73.9

6. Subtract: [2-5]
8.7 − .23

7. Multiply: [2-7]
500 × .06

8. Divide: [2-8]
$\$1.10\overline{)\$13.20}$

9. Write 407.9 million as a complete numeral. [2-13]

10. Write the shortened name for 53,700,000 in millions. [2-14]

11. Add: [3-8]
$2\frac{3}{8}$
$\frac{11}{12}$
$9\frac{1}{6}$

12. Subtract: [3-9]
$11\frac{1}{2}$
$5\frac{2}{3}$

13. Multiply: [3-11]
$2\frac{7}{10} \times 3\frac{3}{4}$

14. Divide: [3-12]
$24 \div 2\frac{2}{3}$

Finding What Fractional Part One Number Is of Another

PROCEDURE

1 What part of 5 is 4?

fractional part $\begin{cases} 4 \longleftarrow \textbf{part} \\ 5 \longleftarrow \textbf{whole unit} \end{cases}$

ANSWER: $\frac{4}{5}$

Write a fraction. Use the given part as the numerator and the number of parts in the whole unit as the denominator.

2 9 is what part of 12?

$\frac{9}{12} = \frac{3}{4}$

Simplify the fraction.

ANSWER: $\frac{3}{4}$

---

3 Compare 10 with 16.

$\frac{10}{16} \longleftarrow$ **number being compared**
$\frac{10}{16} \longleftarrow$ **number it is compared with**
$\frac{10}{16} = \frac{5}{8}$

Write a fraction. Use the number being compared as the numerator and the number it is compared with as the denominator.*

ANSWER: 10 is $\frac{5}{8}$ of 16.

4 Compare 9 with 6.

$\frac{9}{6} = \frac{3}{2} = 1\frac{1}{2}$

ANSWER: 9 is $1\frac{1}{2}$ times 6.

*For an alternate method, see Lesson 17-14.

DIAGNOSTIC TEST

Find the following:

1 What part of 6 is 1?

2 2 is what part of 3?

3 What part of 100 is 40?

4 What part of 56 is 35?

5 Compare 3 with 5.

6 Compare 12 with 16.

7 Compare 11 with 8.

8 Compare 15 with 9.

9 Compare 8 with 1.

10 Compare 24 with 24.

RELATED PRACTICE EXERCISES

Find the following:

1 **1.** What part of 4 is 1?
    3. What part of 8 is 7?

    2. What part of 5 is 1?
    4. What part of 24 is 13?

2 **1.** 3 is what part of 4?
    3. 8 is what part of 15?

    2. 7 is what part of 10?
    4. 29 is what part of 32?

3 **1.** What part of 100 is 28?
    3. $37\frac{1}{2}$ is what part of 100?

    2. 75 is what part of 100?
    4. What part of 100 is $66\frac{2}{3}$?

4 **1.** What part of 12 is 6?
    3. What part of 64 is 16?
    5. 12 is what part of 18?
    7. 27 is what part of 36?
    9. What part of 72 is 30?
   11. What part of 112 is 21?

    2. What part of 18 is 3?
    4. What part of 75 is 25?
    6. 49 is what part of 56?
    8. 45 is what part of 54?
   10. 78 is what part of 90?
   12. 66 is what part of 108?

Compare:

5 **1.** 1 with 8   **2.** 3 with 7   **3.** 5 with 6   **4.** 12 with 25

6 **1.** 6 with 10   **2.** 21 with 28   **3.** 84 with 96   **4.** 25 with 30

7 **1.** 7 with 6   **2.** 13 with 8   **3.** 9 with 4   **4.** 17 with 5

8 **1.** 10 with 8   **2.** 21 with 12   **3.** 24 with 10   **4.** 40 with 16

9 **1.** 2 with 1   **2.** 5 with 1   **3.** 10 with 1   **4.** 28 with 1

10 **1.** 8 with 8   **2.** 10 with 5   **3.** 24 with 6   **4.** 96 with 4

SKILL ACHIEVEMENT TEST

Find the following:

1. What part of 9 is 5?
3. 45 is what part of 75?
5. What part of 100 is 85?

2. 13 is what part of 20?
4. What part of 96 is 36?
6. What part of 144 is 54?

Compare:

7. 2 with 9   **8.** 11 with 3   **9.** 36 with 90   **10.** 18 with 15

Applications

1. There are 16 boys and 20 girls in a class. What part of the class is boys? What part is girls?

2. If 175 out of 200 freshmen passed their physical examination, what part of the freshmen passed?

3. On the final report 6 pupils in the mathematics class received A. The teacher also announced there were 12 Bs, 10 Cs, 8 Ds, and 4 Fs. What part of the class received A? B? C? D? F?

4. Meyer made two errors in 25 fielding chances. What part of his chances did he field the ball cleanly?

5. Janet Miller, a pitcher, won 8 games and lost 4. What part of the games did she win?

6. At bat 36 times, Sue Thompson made 7 singles, 2 doubles, 1 triple, and 2 home runs. What part of the time did she hit safely?

7. If brass contains 3 parts copper and 2 parts zinc, what part of brass is copper? What part zinc?

8. A certain hydrochloric acid solution contains 4 parts acid and 8 parts water. What part of the solution is acid? How many quarts of acid are there in 18 quarts of the solution?

9. **a.** What part of a dollar is a dime?
b. 50 minutes is what part of an hour?
c. 3 is what part of a dozen?
d. 12 ounces is what part of a pound?
e. What part of a bushel is a peck?
f. 6 inches is what part of a foot?

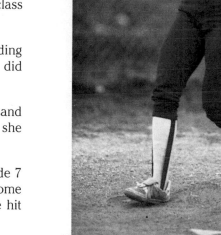

3-14 Finding a Number When a Fractional Part of It Is Known

PROCEDURE

1 $\frac{1}{6}$ of what number is 18?

$18 \div \frac{1}{6}$

part   fraction representing part

$18 \div \frac{1}{6} = \frac{18}{1} \div \frac{1}{6}$

$\quad\quad\quad = \frac{18}{1} \times \frac{6}{1}$

$\quad\quad\quad = 108$

ANSWER: 108

Divide the known part by the given fraction representing this part.*

2 $\frac{3}{4}$ of what number is 21?

$21 \div \frac{3}{4} = \frac{21}{1} \div \frac{3}{4} = \frac{\overset{7}{21}}{1} \times \frac{4}{\underset{1}{3}} = 28$

ANSWER: 28

ALTERNATE METHOD:
Since $\frac{3}{4}$ of the number = 21

$\quad \frac{1}{4}$ of the number = $21 \div 3 = 7$

$\quad \frac{4}{4}$ of the number = $4 \times 7 = 28$

*For the method of solving by equation, see Lesson 17-14.

Therefore, the number = 28.

DIAGNOSTIC TEST

Find the following:

1 $\frac{1}{4}$ of what number is 12?

2 $\frac{5}{8}$ of what number is 45?

3 42 is $\frac{3}{5}$ of what number?

4 560 is $.87\frac{1}{2}$ of what number?

5 $.33\frac{1}{3}$ of what number is 41?

RELATED PRACTICE EXERCISES

Find the following:

1 1. $\frac{1}{2}$ of what number is 5?

2. $\frac{1}{2}$ of what number is 500?

3. $\frac{1}{3}$ of what number is 4?

4. $\frac{1}{3}$ of what number is 18?

5. $\frac{1}{4}$ of what number is 8?

6. $\frac{1}{4}$ of what number is 250?

7. $\frac{1}{8}$ of what number is 2?

8. $\frac{1}{8}$ of what number is 19?

9. $\frac{1}{16}$ of what number is 21?

10. $\frac{1}{6}$ of what number is 7?

11. $\frac{1}{6}$ of what number is 24?

12. $\frac{1}{12}$ of what number is 200?

13. $\frac{1}{5}$ of what number is 6?

14. $\frac{1}{10}$ of what number is 40?

15. $\frac{1}{100}$ of what number is 9?

16. $\frac{1}{20}$ of what number is 150?

2 **1.** $\frac{3}{4}$ of what number is 9?

2. $\frac{3}{4}$ of what number is 26?

3. $\frac{2}{3}$ of what number is 110?

4. $\frac{2}{5}$ of what number is 42?

5. $\frac{3}{5}$ of what number is 96?

6. $\frac{4}{5}$ of what number is 55?

7. $\frac{5}{8}$ of what number is 20?

8. $\frac{7}{8}$ of what number is 35?

9. $\frac{11}{16}$ of what number is 33?

10. $\frac{3}{16}$ of what number is 9?

11. $\frac{7}{10}$ of what number is 105?

12. $\frac{11}{20}$ of what number is 55?

13. $\frac{3}{100}$ of what number is 60?

14. $\frac{47}{100}$ of what number is 94?

15. $\frac{5}{6}$ of what number is 10?

16. $\frac{7}{12}$ of what number is 21?

3 **1.** 16 is $\frac{1}{2}$ of what number?

2. 35 is $\frac{1}{3}$ of what number?

3. 23 is $\frac{1}{4}$ of what number?

4. 9 is $\frac{1}{8}$ of what number?

5. 6 is $\frac{3}{4}$ of what number?

6. 30 is $\frac{5}{8}$ of what number?

7. 27 is $\frac{2}{3}$ of what number?

8. 56 is $\frac{7}{16}$ of what number?

4 **1.** 27 is $.37\frac{1}{2}$ of what number?

2. $.09\frac{3}{4}$ of what number is 780?

3. 1,640 is $.10\frac{1}{4}$ of what number?

4. $.62\frac{1}{2}$ of what number is 900?

5 **1.** $.16\frac{2}{3}$ of what number is 107?

2. 24 is $.66\frac{2}{3}$ of what number?

3. $.83\frac{1}{3}$ of what number is 2,000

4. $.08\frac{1}{3}$ of what number is 960?

SKILL ACHIEVEMENT TEST

Find the following:

1. $\frac{1}{3}$ of what number is 27?

2. 70 is $\frac{5}{6}$ of what number?

3. 25 is $\frac{4}{5}$ of what number?

4. $.66\frac{2}{3}$ of what number is 36?

5. $.04\frac{1}{2}$ of what number is 81?

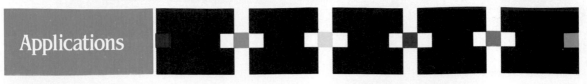

Applications

For problems 1-3, select the letter corresponding to your answer.

1. Charlotte received $\frac{5}{6}$ of all the votes cast in the election for school treasurer. If she received 885 votes, how many students voted?
   a. 1,054 students   **b.** 1,062 students
   c. 1,065 students   **d.** Answer not given

2. Marilyn bought a pair of ice skates at a sale for $18. What was the regular price of the skates if they were reduced one third?
   a. $6   **b.** $12   **c.** $18   **d.** $27

3. If the school baseball team won 16 games or $\frac{2}{3}$ of the games played, how many games were lost?
   a. 6 games   **b.** 7 games
   c. 8 games   **d.** 9 games

4. If 138 students or $\frac{3}{8}$ of the graduating class selected the college preparatory course, how many pupils were in the graduating class?

5. The school athletic association sold 1,295 student membership tickets. If $\frac{7}{8}$ of the school became members, what is the school enrollment?

6. Herbert, being paid at the rate of $\frac{3}{20}$ of his sales, received $51 commission. What was the amount of his sales?

7. Shay weighs 42 kg or $\frac{3}{4}$ as much as her brother Stuart. How much does Stuart weigh?

8. Helen purchased a radio that was reduced $\frac{1}{4}$ of its regular price. If she paid $25.50 for the radio, what was the regular price?

9. How much does a pound of each of the following cost if:
   a. $\frac{1}{2}$ lb of pretzels costs 49¢?
   b. $\frac{3}{4}$ lb of margarine costs 69¢?
   c. $\frac{7}{8}$ lb of chicken costs 84¢?
   d. $\frac{15}{16}$ lb of peppers costs 45¢?

3-15 Writing Fractions as Decimals

PROCEDURE

1 Write $\frac{2}{5}$ as a decimal.

$$5 \overline{)2.0} \quad .4$$
$$\underline{2\,0}$$

Divide the numerator by the denominator.

ANSWER: .4

2 Write $\frac{2}{7}$ as a two-place decimal.

$$7 \overline{)2.00} \quad .28\frac{4}{7}$$
$$\underline{1\,4}$$
$$60$$
$$\underline{56}$$
$$4$$

ANSWER: $.28\frac{4}{7}$

3 Write $\frac{7}{6}$ as a two-place decimal.

$$6 \overline{)7.00} \quad 1.16\frac{2}{3}$$
$$\underline{6}$$
$$1\,0$$
$$\underline{6}$$
$$40$$
$$\underline{36}$$
$$4$$

ANSWER: $1.16\frac{2}{3}$

4 Write $\frac{19}{100}$ as a decimal.

$$\frac{19}{100} = .19$$

2 zeros 2 places

ANSWER: .19

If the given fraction has a denominator that is 10 or 100 or 1,000, and so on, write the numerator and move the decimal point as many places to the left as there are zeros in the denominator.

5 Write $\frac{1,650}{1,000}$ as a decimal.

$$\frac{1,650}{1,000} = 1.650$$

3 zeros 3 places

ANSWER: 1.65

6 Write $\frac{1}{4}$ as a decimal.

$$\frac{1}{4} = \frac{1 \times 25}{4 \times 25} = \frac{25}{100}$$
$$\frac{25}{100} = .25$$

ANSWER: .25

If a fraction has a denominator that can be written as an equivalent fraction with denominator of 10, or 100, or 1,000 and so on, then write as an equivalent fraction.

7 Write $1\frac{7}{8}$ as a decimal.

$1\frac{7}{8} = 1 + \frac{7}{8} = 1 + .87\frac{1}{2} = 1.87\frac{1}{2}$ or 1.875

ANSWER: $1.87\frac{1}{2}$ or 1.875

Write the mixed number as a sum. Then rewrite the fraction as a decimal. Add the whole number and the decimal part.

DIAGNOSTIC TEST

Write the following fractions or mixed numbers as decimals. (Carry out exercises 5, 6, and 9-16 to 2 places.)

1 $\frac{9}{10}$ **2** $\frac{1}{2}$ **3** $\frac{27}{100}$ **4** $\frac{1}{4}$ **5** $\frac{7}{8}$ **6** $\frac{5}{6}$

7 $\frac{125}{100}$ **8** $\frac{37\frac{1}{2}}{100}$ **9** $\frac{8}{9}$ **10** $\frac{20}{25}$ **11** $\frac{49}{56}$ **12** $\frac{24}{28}$

13 $1\frac{3}{4}$ **14** $\frac{12}{8}$ **15** $\frac{18}{14}$ **16** $2\frac{7}{16}$ **17** $\frac{893}{1,000}$ **18** $\frac{7,429}{10,000}$

RELATED PRACTICE EXERCISES

Write the following fractions or mixed numbers as decimals. (Carry out sets 5, 6, and 9-16 to 2 places.)

1 1. $\frac{1}{10}$ 2. $\frac{7}{10}$ 3. $\frac{2}{10}$ 4. $\frac{8}{10}$ **2** 1. $\frac{4}{5}$ 2. $\frac{1}{5}$ 3. $\frac{2}{5}$ 4. $\frac{3}{5}$

3 1. $\frac{39}{100}$ 2. $\frac{3}{100}$ 3. $\frac{6}{100}$ 4. $\frac{91}{100}$ **4** 1. $\frac{3}{4}$ 2. $\frac{17}{20}$ 3. $\frac{14}{25}$ 4. $\frac{41}{50}$

5 1. $\frac{3}{8}$ 2. $\frac{5}{8}$ 3. $\frac{1}{8}$ 4. $\frac{9}{16}$ **6** 1. $\frac{1}{3}$ 2. $\frac{1}{6}$ 3. $\frac{2}{3}$ 4. $\frac{1}{12}$

7 1. $\frac{115}{100}$ 2. $\frac{175}{100}$ 3. $\frac{150}{100}$ 4. $\frac{234}{100}$ **8** 1. $\frac{33\frac{1}{3}}{100}$ 2. $\frac{62\frac{1}{2}}{100}$ 3. $\frac{16\frac{2}{3}}{100}$ 4. $\frac{5\frac{3}{4}}{100}$

9 1. $\frac{4}{7}$ 2. $\frac{3}{11}$ 3. $\frac{7}{9}$ 4. $\frac{13}{15}$ **10** 1. $\frac{18}{36}$ 2. $\frac{21}{28}$ 3. $\frac{30}{75}$ 4. $\frac{56}{80}$

11 1. $\frac{15}{40}$ 2. $\frac{45}{54}$ 3. $\frac{34}{51}$ 4. $\frac{84}{96}$ **12** 1. $\frac{42}{54}$ 2. $\frac{27}{63}$ 3. $\frac{28}{105}$ 4. $\frac{24}{108}$

13 1. $1\frac{1}{2}$ 2. $1\frac{5}{8}$ 3. $1\frac{2}{5}$ 4. $1\frac{7}{16}$ **14** 1. $\frac{8}{5}$ 2. $\frac{14}{8}$ 3. $\frac{12}{9}$ 4. $\frac{57}{48}$

15 1. $\frac{14}{9}$ 2. $\frac{10}{7}$ 3. $\frac{76}{60}$ 4. $\frac{65}{35}$ **16** 1. $2\frac{3}{8}$ 2. $\frac{96}{36}$ 3. $2\frac{5}{6}$ 4. $\frac{66}{21}$

17 1. $\frac{571}{1000}$ 2. $\frac{49}{1000}$ 3. $\frac{8}{1000}$ 4. $\frac{647}{1000}$ **18** 1. $\frac{9514}{10000}$ 2. $\frac{457}{10000}$ 3. $\frac{26}{10000}$ 4. $\frac{5933}{10000}$

SKILL ACHIEVEMENT TEST

Write the following fractions or mixed numbers as a decimal. (Carry out exercises 5-10 to 2 places.)

1. $\frac{4}{10}$ 2. $\frac{3}{5}$ 3. $\frac{54}{100}$ 4. $\frac{11}{16}$ 5. $\frac{5}{12}$

6. $\frac{8}{11}$ 7. $\frac{84}{90}$ 8. $\frac{80}{128}$ 9. $\frac{10}{7}$ 10. $4\frac{2}{3}$

Applications

Which is longer:

1. $\frac{5}{8}$ inch or .619 inch?   2. .5347 inch or $\frac{17}{32}$ inch?

Which is shorter:

3. .742 inch or $\frac{3}{4}$ inch?   4. .8129 inch or $\frac{13}{16}$ inch?

5. To find the batting average first find what fractional part of the times at bat each player hit safely, then change the fraction to a decimal correct to three places:

	At Bat	Hits	Average
Todd	32	12	
Maria	27	9	
José	41	13	

6. To find the team standing average, first find what fractional part of the games played is the games won, then change the fraction to a decimal correct to three places:

	Won	Lost	Average
Blues	8	4	
Yellows	7	5	
Browns	3	9	

7. To find the fielding average, first find what fractional part of the total chances (the sum of the put-outs, assists, and errors) are the chances properly handled (the sum of the put-outs and assists). Then change the fraction to a decimal correct to three places:

	Put-Outs	Assists	Errors	Average
Dave	32	37	6	
Karen	40	15	5	
Felipe	14	49	7	
Frank	17	31	8	
Rosa	29	27	10	

REPEATING DECIMALS

Decimals which have a digit or a group of digits repeating endlessly are called repeating decimals.

Three dots at the end of a decimal indicate that the sequence of digits repeats endlessly. A bar over a digit or a group of digits indicates the sequence of digits that repeats.

$$.333... \text{ or } .\overline{3} \qquad .7272... \text{ or } .\overline{72}$$

When writing the fraction $\frac{1}{3}$ as a decimal, divide the numerator by the denominator. You will notice that the division is not exact. The remainder at each step is the same (1) and the digit 3 repeats in the quotient.

$$
\frac{1}{3} = 3\overline{)1.000} \\
\begin{array}{r}
.333 \\
\underline{9} \\
10 \\
\underline{9} \\
10 \\
\underline{9} \\
1
\end{array}
$$

$\frac{1}{3} = .333...$ or $.\overline{3}$

When writing the fraction $\frac{8}{11}$ as a decimal, divide the numerator by the denominator. Again, the division is not exact, but the remainder is the same (8) each second step and the digits 72 repeat in the quotient.

$$
\frac{8}{11} = 11\overline{)8.000} \\
\begin{array}{r}
.7272 \\
\underline{7\ 7} \\
30 \\
\underline{22} \\
80 \\
\underline{77} \\
30 \\
\underline{22} \\
8
\end{array}
$$

$\frac{8}{11} = .7272...$ or $.\overline{72}$

When $\frac{1}{4}$ is written as a decimal the division process is exact. The quotient is exactly .25 and the remainder is 0. This type of decimal is called a terminating decimal.

Since .25 can be written in repeating form .25$\overline{0}$, terminating decimals may also be called repeating decimals.

$$
\frac{1}{4} = 4\overline{)1.00} \\
\begin{array}{r}
.25 \\
\underline{8} \\
20 \\
\underline{20} \\
0
\end{array}
$$

$\frac{1}{4} = .25$

EXERCISES

Write each of the following fractions as a repeating decimal:

1. $\frac{2}{3}$
2. $\frac{7}{8}$
3. $\frac{1}{6}$
4. $\frac{3}{7}$
5. $\frac{5}{9}$

6. $\frac{11}{16}$
7. $\frac{5}{12}$
8. $\frac{13}{33}$
9. $\frac{7}{11}$
10. $\frac{9}{13}$

11. $\frac{4}{5}$
12. $\frac{11}{18}$
13. $\frac{7}{15}$
14. $\frac{19}{24}$
15. $\frac{8}{27}$

16. $\frac{20}{21}$
17. $\frac{9}{11}$
18. $\frac{3}{4}$
19. $\frac{14}{15}$
20. $\frac{2}{9}$

21. $\frac{11}{12}$
22. $\frac{23}{32}$
23. $\frac{17}{30}$
24. $\frac{28}{33}$
25. $\frac{5}{14}$

3-16 Writing Decimals as Fractions

PROCEDURE

1 Write .15 as a fraction.

$$.15 = \frac{15}{100} = \frac{3}{20}$$

hundredths

ANSWER: $\frac{3}{20}$

Use the digits of the given decimal as the numerator. As the denominator, choose from powers of ten (10, 100, 1,000, and so on) the number that corresponds to the place value of the last digit of the decimal.
Simplify.

2 Write .625 as a fraction.

$$.625 = \frac{625}{1,000} = \frac{5}{8}$$

thousandths

ANSWER: $\frac{5}{8}$

3 Write .0005 as a fraction.

$$.0005 = \frac{5}{10,000} = \frac{1}{2,000}$$

ten thousandths

ANSWER: $\frac{1}{2,000}$

4 Write 5.875 as a mixed number.

$$5.875 = 5 + .875 = 5 + \frac{875}{1,000}$$
$$= 5 + \frac{7}{8} = 5\frac{7}{8}$$

ANSWER: $5\frac{7}{8}$

Write the mixed decimal as a sum. Then write the decimal as a fraction and simplify. Add the whole number and the fraction.

5 Write $.12\frac{1}{2}$ as a fraction.

$$.12\frac{1}{2} = \frac{12\frac{1}{2}}{100} = 12\frac{1}{2} \div 100$$
$$= \frac{25}{2} \div 100$$
$$= \frac{\overset{1}{\cancel{25}}}{2} \times \frac{1}{\underset{4}{\cancel{100}}} = \frac{1}{8}$$

ANSWER: $\frac{1}{8}$

DIAGNOSTIC TEST

Write the following decimals as fractions or mixed numbers:

1 .3 **2** .25 **3** .04 **4** .60 **5** $.66\frac{2}{3}$

6 1.9 **7** 2.85 **8** $1.37\frac{1}{2}$ **9** .672 **10** .028

11 3.125 **12** .4375 **13** .0075 **14** 7.8125

RELATED PRACTICE EXERCISES

Write the following decimals as fractions or mixed numbers:

1 1. .6   **2** 1. .75   **3** 1. .02   **4** 1. .40   **5** 1. $.87\frac{1}{2}$
2. .2   2. .45   2. .07   2. .70   2. $.06\frac{1}{4}$
3. .5   3. .52   3. .01   3. .10   3. $.83\frac{1}{3}$
4. .9   4. .87   4. .08   4. .80   4. $.62\frac{1}{2}$

6 1. 1.2   **7** 1. 1.25   **8** 1. $1.33\frac{1}{3}$ **9** 1. .125   **10** 1. .036
2. 1.5   2. 2.42   2. $1.12\frac{1}{2}$ 2. .875   2. .085
3. 2.8   3. 2.67   3. $2.66\frac{2}{3}$ 3. .946   3. .004
4. 1.7   4. 1.32   4. $4.08\frac{1}{3}$ 4. .384   4. .048

11 1. 1.375   **12** 1. .3125   **13** 1. .0025   **14** 1. 3.5625
2. 1.248   2. .5625   2. .0054   2. 2.0625
3. 2.964   3. .9375   3. .0068   3. 5.0084
4. 3.755   4. .15625   4. .0075   4. 6.6875

SKILL ACHIEVEMENT TEST

Write the following decimals as fractions or mixed numbers:

1. .7   2. .32   3. .62   4. 1.4   5. 2.33
6. .875   7. .008   8. 2.375   9. .8125   10. 6.0025

Enrichment

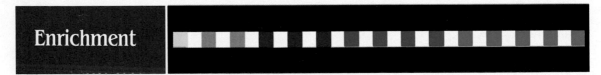

Writing a Repeating Decimal as a Fraction

Let $n = .777....$ $10n = 7.777...$ $10n - n = 7.777... - .777...$ $9n = 7$ $n = \frac{7}{9}$	Let $n = .2424...$ $100n = 24.2424...$ $100n - n = 24.2424... - .2424...$ $99n = 24$ $n = \frac{24}{99} = \frac{8}{33}$	Let $n = .833...$ $100n = 83.333...$ $100n - 10n = 83.333... - 8.333...$ $90n = 75$ $n = \frac{75}{90} = \frac{5}{6}$

Write each of the following repeating decimals as a fraction:

1. .444...   2. $.\overline{5}$ 3. .5454...   4. $.\overline{63}$ 5. $.1\overline{48}$ 6. $.\overline{481}$
7. $.9\overline{4}$ 8. $.58\overline{3}$ 9. $.791\overline{6}$ 10. $.1\overline{5}$ 11. $.1\overline{5}$ 12. $.52\overline{7}$

3-17 Ratio

PROCEDURE

1 Express the ratio of twelve to four.

$$12 \text{ to } 4 = \frac{12}{4}$$

$$= \frac{12 \div 4}{4 \div 4}$$

$$= \frac{3}{1}, \text{ or } 3:1$$

Read "three to one."

ANSWER: $\frac{3}{1}$, or 3:1

Write a fraction. Use the number being compared as the numerator and the number to which it is compared as the denominator. Simplify.

Ratios may be expressed with a fraction bar or with a colon (:).

2 Express the ratio of 4 to 12.

$$4 \text{ to } 12 = \frac{4}{12} = \frac{4 \div 4}{12 \div 4} = \frac{1}{3}$$

ANSWER: $\frac{1}{3}$, or 1:3

3 Express the ratio of 3 meters to 8 meters.

$$3 \text{ meters to } 8 \text{ meters} = \frac{3}{8}$$

ANSWER: $\frac{3}{8}$, or 3:8

If the quantities compared are denominate numbers, they must first be expressed in the same units. The ratio contains no unit of measurement.

---

4 Express 400 kilometers in 5 hours as a ratio.

$$\frac{400}{5} = \frac{400 \div 5}{5 \div 5}$$

$$= \frac{80}{1}$$

ANSWER: $\frac{80}{1}$, or 80 km/h

Sometimes a ratio is used to express a rate.

---

5 Are $\frac{10}{20}$ and $\frac{18}{36}$ equivalent ratios?

$$\frac{10}{20} = \frac{1}{2}$$
$$\frac{18}{36} = \frac{1}{2}$$
same

ANSWER: Yes, $\frac{10}{20}$ and $\frac{18}{36}$ are equivalent ratios.

Write each ratio in simplest form. If the simplest forms of the ratios are the same, the ratios are equivalent.

Vocabulary

The **ratio** of two quantities is the answer we obtain when we compare them by division.
A **rate** is a ratio comparing two different kinds of quantities, such as kilometers and hours.
Equivalent ratios are two ratios that are the same in simplest form.

DIAGNOSTIC TEST

1 Use a fraction bar to express the ratio of six to eleven.

2 Use a colon to express the ratio of nine to five.

Express in simplest form the ratio of:

3 2 to 5

4 6 to 24

5 8 to 3

6 30 to 6

7 27 to 12

8 n to 10

9 25 cm to 75 cm

10 20 kg to 12 kg

11 A nickel to a quarter.

12 A dozen to 8 things.

13 Express the rate 240 miles on 15 gallons of gasoline as a ratio.

14 Are $\frac{18}{42}$ and $\frac{24}{56}$ a pair of equivalent ratios?

15 The pitch of a roof is the ratio of the rise to the span. What is the pitch of a roof if the rise is 2 meters and the span is 8 meters?

RELATED PRACTICE EXERCISES

1 Use the fraction bar to express the ratio of:
  1. Eight to fifteen   **2.** One to seven   **3.** Twelve to three   **4.** Eight to four

2 Use a colon to express the ratio of:
  1. Two to nine   **2.** Fourteen to three   **3.** Six to one   **4.** Five to eleven

Express in simplest form the ratio of:

3 **1.** 5 to 8   **2.** 6 to 17   **3.** 1 to 4   **4.** 2 to 15

4 **1.** 4 to 12   **2.** 10 to 25   **3.** 8 to 14   **4.** 18 to 54

5 **1.** 11 to 2   **2.** 8 to 5   **3.** 20 to 9   **4.** 14 to 3

6 **1.** 24 to 3   **2.** 36 to 12   **3.** 30 to 6   **4.** 42 to 14

7 **1.** 6 to 4   **2.** 15 to 9   **3.** 28 to 10   **4.** 18 to 16

8 **1.** n to 7   **2.** d to 11   **3.** 18 to b **4.** x to y

9 **1.** 8 h to 10 h   **2.** 48 in. to 72 in.   **3.** 2 ft to 5 yd   **4.** 18 min. to 1h

10 **1.** 105 m to 30 m   **2.** 40 min. to 8 min.   **3.** 6 yr to 18 mo   **4.** 3 L to 50 L

11 **1.** A dollar to a nickel.   **2.** A dime to a half-dollar
  3. A quarter to a dime.   **4.** 3 quarters to 4 nickels.

12 **1.** 3 things to 1 dozen.
    2. 2 dozen to 18 things.
    3. 4 dozen to 6 things.
    4. 9 things to 3 dozen.

13 Express each of the following rates as a ratio:
    1. 264 kilometers in 3 hours.
    2. 3,600 liters in 20 minutes.
    3. 9,000 revolutions in 5 minutes
    4. 800 feet in 16 seconds.

14 Which of the following are pairs of equivalent ratios?
    1. $\frac{16}{24}$ and $\frac{10}{15}$ **2.** $\frac{4}{10}$ and $\frac{18}{45}$
    3. $\frac{27}{45}$ and $\frac{45}{72}$ **4.** $\frac{16}{28}$ and $\frac{36}{63}$

15 **1.** There are 18 girls and 12 boys in our class. What is the ratio of:
        (1) boys to girls?   **(2)** girls to boys?
        (3) girls to entire class?   **(4)** boys to entire class?

    2. The aspect ratio is the ratio of the length of an airplane wing to its width. What is the aspect ratio of a wing 32 m long and 4 m wide?

    3. What is the ratio of rising stocks to declining stocks if 750 stocks rose and 625 stocks declined during the day?

    4. The compression ratio is the ratio of the total volume of a cylinder of an engine to its clearance volume. Find the compression ratio of a cylinder of an engine if its total volume is 1,200 cm^3 and its clearance is 25 cm^3?

SKILL ACHIEVEMENT TEST

Express in simplest form the ratio of:
    1. 3 to 7   **2.** 12 to 9
    3. 25 to 40   **4.** n to 15
    5. 18 mm to 4 mm   **6.** 20 min. to 1h
    7. A dime to a dollar
    8. Express the rate 400 meters in 50 seconds as a ratio.
    9. Are $\frac{12}{15}$ and $\frac{28}{35}$ a pair of equivalent ratios?
   10. A bag of mixed grass seed contains 4 pounds of rye grass seed, 3 pounds of fescue seed, 1 pound of clover seed, and 2 pounds of bluegrass seed. What is the ratio of rye grass seed to the total mixture?

3-18 Proportion

PROCEDURE

1 Write as a proportion: 6 compared to 9 is the same as 8 compared to 12.

$\frac{6}{9} = \frac{8}{12}$ or $6:9 = 8:12$

ANSWER: $\frac{6}{9} = \frac{8}{12}$ or $6:9 = 8:12$

Write a fraction bar for "is compared to" and the equal sign for "is the same as." Sometimes colons are used instead of fraction bars.

2 Write as a proportion: Some number compares to 45 as 8 compares to 10.

$\frac{n}{45} = \frac{8}{10}$

ANSWER: $\frac{n}{45} = \frac{8}{10}$

Use n, or some other placeholder, for the unknown number.

---

3 Is $\frac{2}{3} = \frac{10}{15}$ a proportion?

first $\frac{2}{3}$ $\bowtie$ $\frac{10}{15}$ third
second $\qquad$ fourth

first $\qquad$ fourth

$2 \times 15 = 30$
extremes

second $\qquad$ third

$3 \times 10 = 30$
means

$30 = 30$

ANSWER: Yes, $\frac{2}{3} = \frac{10}{15}$ is a proportion.

Find the cross products by multiplying the extremes and means.

If the product of the extremes equals the product of the means, the statement is a proportion.

4 Is $\frac{1}{2} = \frac{9}{12}$ a proportion?

$\frac{1}{2}$ $\bowtie$ $\frac{9}{12}$ $\quad 1 \times 12 = 12$
$\qquad\qquad 2 \times 9 = 18$

$12 \neq 18$

ANSWER: No, $\frac{1}{2} = \frac{9}{12}$ is not a proportion.

---

5 Solve the proportion $\frac{n}{32} = \frac{3}{4}$.

$\frac{n}{32}$ $\bowtie$ $\frac{3}{4}$

$4 \times n = 32 \times 3$

$4n = 96$

$\frac{4n}{4} = \frac{96}{4}$ $\qquad$ Check: $\frac{24}{32} = \frac{3}{4}$?

$n = 24$ $\qquad\qquad\qquad \frac{3}{4} = \frac{3}{4}$ ✔

ANSWER: 24

To find the value of n, divide both sides of the equation by 4.
Check by substituting the value for n. Simplify and compare values.

6 Solve the proportion $\frac{8}{n} = \frac{3}{225}$.

$\frac{8}{n} \bowtie \frac{3}{225}$

$n \times 3 = 8 \times 225$ Check: $\frac{8}{600} = \frac{3}{225}$?

$\qquad 3n = 1,800$ $\qquad\qquad\qquad \frac{1}{75} = \frac{1}{75}$ ✔

$\qquad \frac{3n}{3} = \frac{1,800}{3}$

$\qquad\quad n = 600$

ANSWER: 600

Vocabulary

A **proportion** is a mathematical sentence that states that two ratios are equivalent.
The **terms** of a proportion are:

$$\begin{array}{l}\textbf{first} \longrightarrow \\ \textbf{second} \longrightarrow\end{array} \frac{a}{b} = \frac{c}{d} \begin{array}{l}\longleftarrow \textbf{third} \\ \longleftarrow \textbf{fourth}\end{array}$$

The **extremes** are the first and fourth terms of a proportion.
The **means** are the second and third terms of a proportion.
The **cross products** are the product of the extremes and the product of the means.

DIAGNOSTIC TEST

Write each of the following as a proportion:

1 9 compared to 6 is the same as 24 compared to 16.

2 Some number n is to 8 as 9 is to 12.

3 Is $\frac{9}{15} = \frac{4}{10}$ a true proportion?

Solve and check:

4 $\frac{n}{15} = \frac{4}{5}$ **5** $\frac{12}{21} = \frac{y}{14}$ **6** $\frac{16}{x} = \frac{2}{7}$ **7** $\frac{5}{8} = \frac{15}{a}$ **8** $\frac{2\frac{1}{2}}{x} = \frac{\frac{1}{2}}{10}$

9 Using a proportion, solve: What number compared to 72 is the same as 7 compared to 18?

10 A recipe calls for 4 cups of flour to 6 tablespoons of shortening. How many tablespoons of shortening are needed for 6 cups of flour?

RELATED PRACTICE EXERCISES

Write each of the following as a proportion:

1 **1.** 26 compared to 13 is the same as 6 compared to 3.
2. 4 is to 12 as 5 is to 15.
3. 45 compared to 80 is the same as 18 compared to 32.
4. 56 is to 14 as 76 is to 19.

2 **1.** Some number x compares to 30 as 9 compares to 54.
   2. 24 is to y as 6 is to 11.
   3. 35 compared to 500 is the same as n compared to 60.
   4. 42 is to 63 as 12 is to n.

3 Which of the following are true proportions?
   1. $\frac{2}{3} = \frac{3}{2}$ **2.** $\frac{15}{90} = \frac{2}{12}$ **3.** $18:14 = 27:21$ **4.** $8:20 = 30:100$

Solve and check:

4 **1.** $\frac{t}{32} = \frac{3}{4}$ **2.** $\frac{x}{9} = \frac{49}{63}$ **3.** $\frac{n}{.01} = \frac{.2}{.16}$ **4.** $\frac{b}{5} = \frac{2}{3}$

5 **1.** $\frac{90}{54} = \frac{a}{3}$ **2.** $\frac{5}{6} = \frac{c}{78}$ **3.** $\frac{1}{6} = \frac{s}{5}$ **4.** $\frac{1.5}{.3} = \frac{r}{8}$

6 **1.** $\frac{9}{n} = \frac{3}{38}$ **2.** $\frac{24}{y} = \frac{108}{18}$ **3.** $\frac{.005}{c} = \frac{1.4}{.28}$ **4.** $\frac{3}{b} = \frac{13}{6}$

7 **1.** $\frac{6}{7} = \frac{54}{x}$ **2.** $\frac{105}{35} = \frac{27}{n}$ **3.** $\frac{17}{24} = \frac{17}{a}$ **4.** $\frac{.04}{.12} = \frac{.6}{x}$

8 **1.** $\frac{y}{3\frac{1}{2}} = \frac{8}{7}$ **2.** $\frac{24}{b} = \frac{2\frac{1}{3}}{5\frac{5}{6}}$ **3.** $\frac{n}{2\frac{1}{4}} = \frac{15}{\frac{3}{4}}$ **4.** $\frac{21}{x} = \frac{7\frac{7}{10}}{3\frac{3}{5}}$

9 Using proportions, solve each of the following:
   1. What number compared to 10 is the same as 28 compared to 5?
   2. 12 compared to 3 is the same as what number compared to 18?
   3. 7 compared to what number is the same as 42 compared to 54?
   4. 6 compared to 11 is the same as 84 compared to what number?

10 Solve by using proportions:
   1. At the rate of 3 items for 16¢, how many items can you buy for 80¢?
   2. A motorist traveled 240 kilometers in 3 hours. How long will it take at that rate to travel 400 kilometers?
   3. If the scale distance of 3.5 centimeters on a map represents an actual distance of 175 kilometers, what actual distance does a scale distance of 5.7 centimeters represent?
   4. How many square feet of lawn will 10 pounds of grass seed cover if 3 pounds of seed cover 450 square feet?

SKILL ACHIEVEMENT TEST

Solve and check:

1. $\frac{n}{15} = \frac{2}{3}$ **2.** $\frac{1}{9} = \frac{6}{x}$ **3.** $\frac{25}{8} = \frac{a}{10}$

Solve by using a proportion:

4. 108 compared to 96 is the same as what number compared to 8?

5. Lisa saved $60 in 4 weeks. At that rate how long will it take her to save $300?

Career Applications

In communities all across the country, new homes and office buildings are going up. At the same time, many older structures are being repaired or renovated. The skilled laborers who make all these buildings possible are called construction workers. They include ironworkers, carpenters, plumbers, bricklayers, electricians, drywall workers, glaziers, painters, paperhangers, and floor-covering installers. Generally, workers in the construction trades learn their craft through an apprenticeship that combines on-the-job training with some classroom instruction.

1. A carpenter is given a blueprint for a building. The lengths of all the sections of a wall are marked separately: $1\frac{5}{8}$ inches, $\frac{3}{4}$ inches, $\frac{7}{8}$ inches, and $2\frac{3}{16}$ inches. What is the total length of the wall on the blueprint?

2. A carpenter wants to make a butcher block table from strips of red maple. How many $1\frac{7}{8}$-inch strips of red maple glued together are required to make a butcher block 30 inches wide?

3. An electrician has a wiring job that calls for thirty-two pieces of hollow pipe, each $7\frac{1}{2}$ feet long. What is the total number of feet of pipe needed for the job?

4. A plumber must enlarge a $\frac{1}{8}$-inch hole to $\frac{12}{16}$ inch in order to insert a heating pipe. How much larger than the original hole will this be?

5. A carpenter estimates that 1400 nails are needed for a job. There are about 70 of these nails in a pound. How many pounds are needed?

6. In order to make a basement for a house, 450 cubic yards of earth must be removed. The digging will cost $11.50 per cubic yard. What is the total cost?

7. How much railing is needed for three sides of a deck whose lengths are 12 feet, 20 feet, and 16 feet?

8. One brick, plus mortar, is $8\frac{1}{2}$ inches long. A wall will be 34 feet long. How many bricks are needed for each row?

9. An electrician needs a piece of hollow pipe $6\frac{1}{4}$ feet long. When this is cut from a 10-foot length, how much is left?

10. A plumber uses 226 feet of copper pipe for a job. The pipe costs $.65 per foot. What is the total cost?

11. A bricklayer ordered 12,000 bricks to face a house. When the job was finished, 342 bricks were left over. How many bricks were used?

12. A floor-covering installer plans to cover a floor with boards $2\frac{1}{2}$ inches wide. The room is 15 feet wide. How many boards are needed for the width?

13. A carpenter needs to cut 22 pieces of wood for trim, each $14\frac{3}{8}$ inches long. What is the total length of wood needed?

14. Vinyl floor covering costs $152.28 for 4.5 meters. What is the cost per meter?

15. Paneling a room required 15 sheets of paneling at $7.65 a sheet. What was the total cost?

17. An electrician needs pieces of hollow pipe $1\frac{1}{4}$ feet long. How many pieces can be cut from a 10-foot length of pipe?

16. A plumber cuts pieces of pipe $18\frac{3}{4}$ inches long, 32 inches long, and $12\frac{5}{8}$ inches long. What is the total length?

18. A roofer needs 18 bundles of shingles to roof a house. If each bundle cost $17.98, what is the total cost for shingles?

CHAPTER REVIEW

Vocabulary

common multiple p. 147
complex fraction p. 174
cross products p. 195
denominator p. 134
equivalent fractions p. 142
equivalent ratios p. 191
extremes p. 195

fraction p. 134
lowest terms p. 135
least common
  denominator
  (LCD) p. 149
least common
  multiple
  (LCM) p. 147

means p. 195
multiple p. 147
numerator p. 134
$\neq$ symbol p. 142
$\not>$ symbol p. 145
$\not<$ symbol p. 145
proportion p. 195

rate p. 191
ratio p. 191
reciprocal p. 174
simplest form p. 135
terms (proportion) p. 195

EXERCISES

The numerals in boxes indicate Lessons where help may be found.

Express in lowest terms: 3-1

1. $\frac{16}{36}$ **2.** $\frac{105}{140}$

Rewrite as a whole number or mixed number: 3-2

3. $\frac{21}{3}$ **4.** $\frac{35}{8}$

Rewrite in simplest form: 3-2 **5.** $8\frac{3}{3}$ **6.** $4\frac{11}{5}$

Express each as an equivalent fraction having the denominators as specified: 3-3 and 3-4

7. $\frac{1}{4} = \frac{?}{64}$ **8.** $\frac{17}{20} = \frac{?}{100}$ **9.** Express $\frac{11}{12}$ in 60ths.

10. Arrange in order, least first: 3-5 $\frac{11}{16}$, $\frac{3}{4}$, and $\frac{7}{12}$ **11.** Round $1.37\frac{3}{8}$ to the nearest cent. 3-5

12. Find the least common multiple of 2, 3, and 9. 3-6

Find the least common denominator of each group of fractions: 3-7

13. $\frac{1}{2}$ and $\frac{2}{3}$ **14.** $\frac{5}{6}$ and $\frac{3}{4}$ **15.** $\frac{7}{8}$, $\frac{1}{4}$, and $\frac{5}{16}$

Add: 3-8 Subtract: 3-9

16. $1\frac{7}{8}$ **17.** $4\frac{11}{12} + 3\frac{3}{8} + 1\frac{9}{16}$ **18.** $6\frac{7}{10}$ **19.** Subtract $11\frac{3}{4}$ from 15.
   $2\frac{2}{3}$ $4\frac{1}{2}$

20. Write $1\frac{9}{16}$ as an improper fraction. 3-10 Multiply: 3-11 **21.** $5\frac{1}{4} \times 2\frac{2}{3}$ **22.** 29
Divide: 3-12 $\underline{6\frac{1}{2}}$

23. $4\frac{11}{16} \div 3\frac{3}{4}$ **24.** $14 \div 8\frac{1}{6}$

25. What part of 16 is 9? 3-13 **26.** 45 is what part of 72? 3-13

27. $\frac{1}{3}$ of what number is 20? 3-14 **28.** $27 is $\frac{3}{5}$ of what amount? 3-14

29. Write $\frac{13}{25}$ as a decimal. 3-15 **30.** Write .375 as a fraction. 3-16

31. Find the total weight if a metal can weighs $2\frac{7}{8}$ ounces and its contents $12\frac{3}{4}$ ounces. 3-8

32. The directions tell you to add 3 cans of water to 1 can of orange juice concentrate. What is the ratio of the water to the concentrate? 3-17

33. What number compared to 60 is the same as 7 compared to 15? 3-18

COMPETENCY CHECK TEST

The numerals in boxes indicate Lessons where help may be found.
Solve each problem and select the letter corresponding to your answer.

1. Find the sum of: $897 + 3,568 + 959$ ☐1-4☐
    a. 5,234   **b.** 4,342   **c.** 6,434   **d.** 5,424

2. Subtract: $927 - 696$ ☐1-5☐
    a. 131   **b.** 141   **c.** 231   **d.** Answer not given

3. Multiply: 905×580 ☐1-6☐
    a. 117,650   **b.** 1,485   **c.** 45,400   **d.** 524,900

4. Divide: $79\overline{)63,437}$ ☐1-7☐
    a. 93   **b.** 903   **c.** 803   **d.** 830

5. Add: $.63 + 2.4$ ☐2-4☐
    a. .87   **b.** 8.7   **c.** 8.43   **d.** 3.03

6. Subtract: $.547 - .39$ ☐2-5☐
    a. .508   **b.** .512   **c.** .157   **d.** Answer not given

7. Multiply: $.013 \times .06$ ☐2-7☐
    a. .078   **b.** .00078   **c.** .01306   **d.** .0078

8. Divide: $9 \div .6$ ☐2-8☐
    a. 15   **b.** 1.5   **c.** .15   **d.** 150

9. Add: $\frac{1}{6} + \frac{3}{4}$ ☐3-8☐
    a. $\frac{23}{24}$ **b.** $\frac{11}{12}$ **c.** $\frac{13}{64}$ **d.** $\frac{4}{10}$

10. Subtract: $\frac{7}{10} - \frac{1}{2}$ ☐3-9☐
    a. $\frac{6}{8}$ **b.** $\frac{2}{3}$ **c.** $\frac{1}{5}$ **d.** Answer not given

11. Multiply: $4\frac{2}{3} \times \frac{5}{7}$ ☐3-11☐
    a. $4\frac{7}{10}$ **b.** $3\frac{1}{3}$ **c.** $2\frac{3}{4}$ **d.** $4\frac{10}{21}$

12. Divide: $12 \div 1\frac{3}{5}$ ☐3-12☐
    a. $7\frac{1}{2}$ **b.** $11\frac{2}{5}$ **c.** $12\frac{3}{5}$ **d.** $6\frac{3}{4}$

13. The least common multiple of 10, 16, and 20 is: ☐3-6☐
    a. 320   **b.** 200   **c.** 160   **d.** 80

14. The net change in a stock that opened at $23\frac{7}{8}$ and closed at $24\frac{3}{4}$ is: ☐3-9☐
    a. $\frac{7}{8}$ **b.** $\frac{3}{4}$ **c.** $\frac{5}{8}$ **d.** $1\frac{1}{8}$

CUMULATIVE ACHIEVEMENT TEST

The numerals in boxes indicate Lessons where help may be found.

1. Write 1,978,206,023 in words. $\boxed{\text{1-1}}$

2. Round 3,609,754 to the nearest thousand. $\boxed{\text{1-3}}$

3. Add: $\boxed{\text{1-4}}$
   82,579
   6,429
   51,678
   30,199
   92,389

4. Subtract: $\boxed{\text{1-5}}$
   850,601
   788,963

5. Multiply: $\boxed{\text{1-6}}$
   8,049
   768

6. Divide: $\boxed{\text{1-7}}$
   984)951,528

7. Name the greatest common factor of 64, 80, and 112. $\boxed{\text{1-8}}$

8. Round $9.476 to the nearest cent. $\boxed{\text{2-3}}$

9. Add: $\boxed{\text{2-4}}$
   $8.63 + $.95 + $24.19

10. Subtract: $\boxed{\text{2-5}}$
    $17 − $6.08

11. Multiply: $\boxed{\text{2-7}}$
    .9 × .09

12. Divide: $\boxed{\text{2-8}}$
    1.5)̄.6

13. Express $\frac{54}{60}$ in lowest terms. $\boxed{\text{3-1}}$

14. Express $\frac{5}{6}$ in 48ths. $\boxed{\text{3-3}}$

15. Which is greater: $\frac{5}{8}$ or $\frac{2}{3}$? $\boxed{\text{3-5}}$

16. Find the least common multiple of 4, 6, and 15. $\boxed{\text{3-6}}$

17. Add: $\boxed{\text{3-8}}$
    $7\frac{3}{4}$
    $8\frac{2}{3}$
    $4\frac{1}{6}$

18. Subtract: $\boxed{\text{3-9}}$
    $20\frac{1}{2}$
    $9\frac{7}{8}$

19. Multiply: $\boxed{\text{3-11}}$
    $3\frac{1}{7} \times 8\frac{1}{6}$

20. Divide: $\boxed{\text{3-12}}$
    $\frac{9}{16} \div 6\frac{3}{4}$

21. What part of 80 is 64? $\boxed{\text{3-13}}$

22. $\frac{5}{8}$ of what number is 65? $\boxed{\text{3-14}}$

23. Write $\frac{9}{20}$ as a decimal. $\boxed{\text{3-15}}$

24. Write .36 as a fraction. $\boxed{\text{3-16}}$

25. The Lift-Drag Ratio is the ratio of the lift of the airplane to its drag. Find the Lift-Drag Ratio when the lift is 3,000 lb and the drag is 240 lb. $\boxed{\text{3-17}}$

26. Solve by using a proportion:
    A motorist travels 260 miles in 5 hours. How long will it take at that rate to travel 364 miles? $\boxed{\text{3-18}}$

Computer Activities

The computer program below is written in the BASIC language. It is designed to help you practice writing fractions as decimals. Enter the program on your computer keyboard exactly as it appears, one line at a time. Press the RETURN or ENTER key at the end of each line.

10 Clearscreen	Clears the screen. (Use appropriate command for *your* computer.)
20 LET X = INT(99*RND(1)) + 1	Generates a "random" (unpredictable) whole number X, between 1 and 99 and stores it in memory location X.
30 LET Y = INT(99*RND(1)) + 1	Generates a "random" whole number Y, between 1 and 99 and stores it in memory location Y.
40 LET Z = X/Y	Computes the quotient of X/Y and stores it in Z.
50 LET Z = INT(100*Z+.5)/100	Rounds Z to the nearest hundredth. This is the correct answer.
60 PRINT X; "/"; Y	Prints out the problem for you on your computer.
70 INPUT A	Displays a "?" as a way of asking for your answer.
80 LET D = ABS(A − Z)	Determines D, the difference between your answer and the correct answer.
90 IF D < .001 THEN PRINT "YES"	Says "YES" if your answer is correct.
100 IF D > = .001 THEN PRINT "NO"	Says "NO" if your answer is incorrect.
110 GOTO 20	Sends the computer back to line 20 for another problem.

(Note: TRS-80 and IBM users replace RND(1) in lines 20 and 30 with RND(0)).

If you make a typing mistake:

(a) Use the ← key to back up and make corrections.

or

(b) Simply press RETURN or ENTER and retype the line.

EXERCISES

Type RUN and press RETURN or ENTER to start your program. Use the program until you have achieved mastery with writing fractions as decimals. Round all answers to the nearest *hundredth*.

EXAMPLE: If the problem is: $\frac{1}{7}$
you would answer: .14

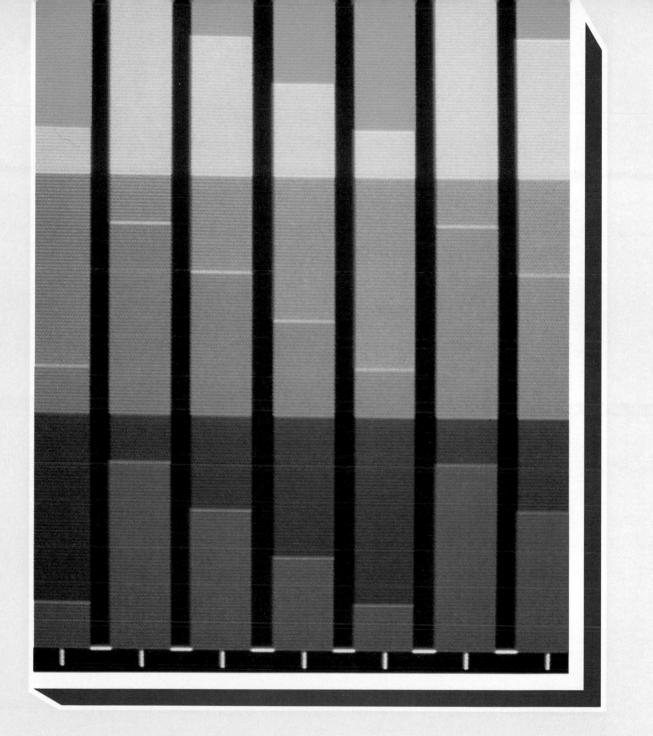

CHAPTER 4 Percent

INTRODUCTION TO PERCENT

Earners, consumers, and business people use percent (symbol %) extensively in their daily affairs.

Earners may find that 14% withholding tax, 7.05% social security tax, and perhaps a 3% state or city income tax are deducted from their paychecks. Salespeople may earn 5% commission on what they sell.

Homeowners may receive a 2% discount for paying their real estate taxes in advance, a 3% discount on their gas and electric bills or a 5% discount on their water bills if they pay before the discount period ends. Consumers may buy merchandise at a department store advertising a 40% reduction on certain sales items. They may pay a 12% carrying charge on a new automobile. They may buy jewelry and cosmetics on which they pay a state sales tax of perhaps 1% to 7%. Labels on clothing may indicate the content of cloth like 30% mohair or 70% wool.

In business, a storekeeper may make a 35% profit on sales. A bank may pay $5\frac{1}{2}$% interest on deposits and charge $14\frac{1}{2}$% interest on loans.

Students and teachers also use percent. A student may receive a mark of 83%. A teacher may find daily class attendance averages 94%.

Percent is very useful in giving a quick comparative picture on a scale from 1 to 100. For example, when a basketball player has a 69% success rate in making foul shots, we immediately understand that the player is successful at the rate of 69 out of every 100 attempts.

4-1 Meaning of Percent

PROCEDURE

1 Express 5 hundredths as a percent.

5 hundredths = 5%

↑
percent sign

ANSWER: 5%

Percent (%) means hundredths. Write % in place of hundredths.

2 Express $3\frac{1}{2}$ hundredths as a percent.

$3\frac{1}{2}$ hundredths = $3\frac{1}{2}$%

ANSWER: $3\frac{1}{2}$%

3 Express 19 out of 100 as a percent.

19 out of 100 = 19 hundredths = 19%

ANSWER: 19%

---

4 How many hundredths are in 4.9%?

4.9% = 4.9 hundredths

ANSWER: 4.9 hundredths

5 Write 37 hundredths as a percent, as a decimal, and as a fraction.

37 hundredths = 37%

37 hundredths = .37

2 decimal places

37 hundredths = $\frac{37}{100}$

ANSWER: 37%; .37; $\frac{37}{100}$

6 Write 100% as a decimal.

100% = 100 hundredths

= 1.00 = 1

2 decimal places

ANSWER: 1

100% of anything is $\frac{100}{100}$ of it, or all of it.

---

7 Write 83% as a ratio.

83% = $\frac{83}{100}$, or 83:100

ANSWER: $\frac{83}{100}$, or 83:100

A percent may be considered as a ratio of a number to 100.

DIAGNOSTIC TEST

1 Express eighteen hundredths as a percent.
2 Express 81 out of 100 as a percent.
3 How many hundredths are in 35%?
4 Write four hundredths as a percent, decimal, and fraction.
5 Write 36% as a ratio.

RELATED PRACTICE EXERCISES

1 Express each of the following as a percent:
   1. 3 hundredths   **2.** 16 hundredths   **3.** 30 hundredths   **4.** 200 hundredths
   5. $5\frac{1}{2}$ hundredths   **6.** $37\frac{1}{2}$ hundredths   **7.** 62.5 hundredths   **8.** $\frac{1}{4}$ hundredth

2 Express each of the following as a percent:
   1. 47 out of 100   **2.** 95 out of 100   **3.** 1 out of 100   **4.** $4\frac{1}{2}$ out of 100
   5. $66\frac{2}{3}$ out of 100   **6.** 12.5 out of 100   **7.** 20 out of 100   **8.** $3\frac{3}{4}$ out of 100

3 How many hundredths are in:
   1. 4%   **2.** 7%   **3.** 45%   **4.** 60%
   5. 140%   **6.** $87\frac{1}{2}$%   **7.** 3.5%   **8.** $\frac{1}{2}$%

4 Write each of the following statements as a percent, decimal, and fraction:
   1. Six hundredths   **2.** Eight hundredths
   3. Twenty-three hundredths   **4.** Seventy hundredths
   5. Twelve and one-half hundredths   **6.** Two and one-fourth hundredths
   7. Three hundred hundredths   **8.** Three-fourths hundredth

5 Write each of the following as a ratio:
   1. 18%   **2.** 45%   **3.** 6%   **4.** 73%
   5. 91%   **6.** 2%   **7.** 64%   **8.** 27%

SKILL ACHIEVEMENT TEST

   1. Express forty-seven hundredths as a percent.
   2. How many hundredths are in 83%?
   3. Express 69 out of 100 as a percent.
   4. Write twenty-one hundredths as a percent, decimal, and fraction.
   5. Write 78% as a ratio.

PROCEDURE

1 Write 84% as a decimal.

$84\% = .84$

2 decimal places

ANSWER: .84

In the numeral naming a whole number, a decimal point is understood after the units' figure. Write the number without the percent sign (%). Move the decimal point 2 places to the left.

2 Write 30% as a decimal.

$30\% = .30$

$= .3$ **Drop ending zeros.**

ANSWER: .3

3 Write 125% as a decimal.

$125\% = 1.25$

ANSWER: 1.25

---

4 Write 5% as a decimal.

Write 1 zero.

$5\% = .05$

ANSWER: .05

5 Write .2% as a decimal.

Write 2 zeros.

$.2\% = .002$

ANSWER: .002

6 Write 8.34% as a decimal.

Write 1 zero.

$8.34\% = .0834$

ANSWER: .0834

---

7 Write $\frac{7}{8}\%$ as a decimal.

$\frac{7}{8}\% = .00\frac{7}{8}$ or $.00875$

ANSWER: $.00\frac{7}{8}$ or $.00875$

8 Write $2\frac{1}{2}\%$ as a decimal.

$2\frac{1}{2}\% = .02\frac{1}{2} = .025$

ANSWER: $.02\frac{1}{2}$ or $.025$

DIAGNOSTIC TEST

Write each of the following percents as a decimal:

1 8%	**2** 53%	**3** 90%	**4** 119%	**5** 160%
6 500%	**7** $12\frac{1}{2}\%$	**8** $5\frac{1}{4}\%$	**9** $40\frac{5}{8}\%$	
10 $105\frac{3}{4}\%$	**11** 27.5%	**12** 12.25%	**13** 8.6%	
14 4.82%	**15** 352.875%	**16** 0.6%	**17** $\frac{3}{4}\%$	

RELATED PRACTICE EXERCISES

Write each of the following percents as a decimal:

1 1. 6%
    2. 9%
    3. 1%
    4. 4%

2 1. 16%
    2. 83%
    3. 45%
    4. 67%

3 1. 40%
    2. 70%
    3. 20%
    4. 10%

4 1. 134%
    2. 157%
    3. 106%
    4. 175%

5 1. 130%
    2. 120%
    3. 180%
    4. 150%

6 1. 100%
    2. 200%
    3. 600%
    4. 400%

7 1. $37\frac{1}{2}$%
    2. $62\frac{1}{2}$%
    3. $83\frac{1}{3}$%
    4. $18\frac{3}{4}$%

8 1. $4\frac{1}{2}$%
    2. $3\frac{3}{4}$%
    3. $2\frac{7}{8}$%
    4. $5\frac{4}{5}$%

9 1. $60\frac{1}{2}$%
    2. $10\frac{3}{8}$%
    3. $50\frac{2}{3}$%
    4. $70\frac{1}{4}$%

10 1. $100\frac{7}{8}$%
    2. $101\frac{1}{4}$%
    3. $106\frac{1}{2}$%
    4. $127\frac{3}{5}$%

11 1. 87.5%
    2. 12.5%
    3. 46.4%
    4. 95.1%

12 1. 17.75%
    2. 56.94%
    3. 99.44%
    4. 82.09%

13 1. 3.5%
    2. 2.8%
    3. 9.3%
    4. 6.4%

14 1. 1.25%
    2. 5.33%
    3. 2.08%
    4. 7.19%

15 1. 26.375%
    2. 31.625%
    3. 128.333%
    4. 895.667%

16 1. .7%
    2. .02%
    3. 0.9%
    4. 0.85%

17 1. $\frac{1}{2}$%
    2. $\frac{1}{4}$%
    3. $\frac{5}{8}$%
    4. $\frac{2}{3}$%

SKILL ACHIEVEMENT TEST

Write each of the following percents as a decimal:

1. 3%	**2.** 39%	**3.** 60%	**4.** 135%	**5.** 140%
6. 800%	**7.** $16\frac{2}{3}$%	**8.** $6\frac{1}{2}$%	**9.** $108\frac{3}{4}$%	**10.** 62.5%
11. 5.18%	**12.** 0.3%	**13.** $\frac{3}{8}$%	**14.** 9.642%	**15.** 12.94%

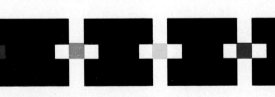

Applications

Write each of the following percents as a decimal:

1. Department store sales increased 8%.
2. The increase in the cost of living was 6.5%.
3. The market price of a corporation bond remained at $100\frac{5}{8}$%.

4-3 Writing Decimals as Percents

PROCEDURE

1 Write .07 as a percent.

$$.07 = .07 = 7\%$$

2 decimal places

ANSWER: 7%

Copy the given numeral. Move the decimal point 2 places to the right and write a percent sign.

2 Write .19 as a percent.

$$.19 = .19 = 19\%$$

ANSWER: 19%

3 Write 1.34 as a percent.

$$1.34 = 1.34 = 134\%$$

ANSWER: 134%

4 Write .5 as a percent.

Write 1 zero.

$$.5 = .50 = 50\%$$

ANSWER: 50%

5 Write 8 as a percent.

Write 2 zeros.

$$8 = 8.00 = 800\%$$

ANSWER: 800%

6 Write 3.687 as a percent.

$$3.687 = 3.687 = 368.7\%$$

ANSWER: 368.7%

7 Write $.06\frac{1}{2}$ as a percent.

$$.06\frac{1}{2} = 6\frac{1}{2}\%$$

ANSWER: $6\frac{1}{2}\%$

8 Write $.00\frac{2}{5}$ as a percent.

$$.00\frac{2}{5} = \frac{2}{5}\%$$

ANSWER: $\frac{2}{5}\%$

9 Write $.4\frac{1}{2}$ as a percent.

$$.4\frac{1}{2} = .45 = 45\%$$

ANSWER: 45%

DIAGNOSTIC TEST

Write each of the following decimals as a percent:

1 .06	**2** .32	**3** .7	**4** 1.12
5 1.4	**6** $.37\frac{1}{2}$	**7** $.04\frac{1}{2}$	**8** $.60\frac{3}{4}$
9 $1.66\frac{2}{3}$	**10** .625	**11** .0467	**12** $.1\frac{1}{4}$
13 2.875	**14** 2	**15** $.00\frac{3}{8}$	**16** .0025

209

RELATED PRACTICE EXERCISES

Write each of the following decimals as a percent:

1 1. .01   2. .08   3. .03   4. .05

2 1. .28   2. .75   3. .16   4. .93

3 1. .6   2. .3   3. .8   4. .1

4 1. 1.39   2. 1.92   3. 1.18   4. 1.50

5 1. 1.2   2. 1.8   3. 1.7   4. 1.3

6 1. $.12\frac{1}{2}$ 2. $.83\frac{5}{6}$ 3. $.42\frac{2}{7}$ 4. $.18\frac{3}{4}$

7 1. $.01\frac{1}{2}$ 2. $.03\frac{3}{4}$ 3. $.04\frac{2}{3}$ 4. $.06\frac{5}{6}$

8 1. $.10\frac{1}{2}$ 2. $.30\frac{2}{3}$ 3. $.70\frac{3}{4}$ 4. $.50\frac{1}{2}$

9 1. $1.37\frac{1}{2}$ 2. $1.00\frac{1}{2}$ 3. $1.16\frac{2}{3}$ 4. $1.05\frac{3}{4}$

10 1. .875   2. .125   3. .078   4. .989

11 1. .2625   2. .0875   3. .0233   4. .65125

12 1. $.2\frac{1}{2}$ 2. $.6\frac{1}{2}$ 3. $.1\frac{3}{4}$ 4. $.8\frac{7}{8}$

13 1. 1.245   2. 1.375   3. 2.667   4. 3.7275

14 1. 1   2. 3   3. 5   4. 6

15 1. $.00\frac{1}{4}$ 2. $.00\frac{1}{2}$ 3. $.00\frac{3}{5}$ 4. $.00\frac{2}{3}$

16 1. .005   2. .0075   3. .008   4. .00875

SKILL ACHIEVEMENT TEST

Write each of the following decimals as a percent:

1. .04   2. .51   3. .9   4. 1.42   5. 1.6

6. $.05\frac{1}{2}$ 7. $.40\frac{7}{8}$ 8. .347   9. .0525   10. $.33\frac{1}{3}$

11. $1.01\frac{1}{4}$ 12. 1.625   13. 7   14. $.00\frac{1}{2}$ 15. 0.0058

Applications

Baseball Averages

1. The pennant-winning team led the league with the standing of .625. What percent of the games played did it win?

2. What percent of the number of times at bat must a batter hit safely to average .300? .375? .400?

3. A pitcher has an average of .800. What percent of the games did she lose?

4. What percent of the games played must a team win to have a standing of .750?

4-4 Writing Percents as Fractions

PROCEDURE

1 Write 25% as a fraction.

$25\% = \frac{1}{4}$

ANSWER: $\frac{1}{4}$

If the fractional equivalent is known, write it directly.

---

2 Write 8% as a fraction.

$8\% = \frac{8}{100}$

$= \frac{2}{25}$

ANSWER: $\frac{2}{25}$

Write a fraction. Use the given numeral as the numerator and 100 as the denominator. Simplify.

3 Write 130% as a mixed number.

$130\% = \frac{130}{100}$

$= 1\frac{30}{100} = 1\frac{3}{10}$

ANSWER: $1\frac{3}{10}$

If the percent is greater than 100%, first write it as an improper fraction then as a mixed number.

DIAGNOSTIC TEST

Write each of the following percents as a fraction or mixed number:

1 75%   **2** $16\frac{2}{3}\%$ **3** 2%   **4** 125%

RELATED PRACTICE EXERCISES

Write each of the following percents as a fraction or mixed number:

1
1. 50%
2. 25%
3. 10%
4. 80%
5. 90%
6. 60%
7. 30%
8. 70%

2
1. $33\frac{1}{3}\%$
2. $87\frac{1}{2}\%$
3. $83\frac{1}{3}\%$
4. $12\frac{1}{2}\%$
5. $37\frac{1}{2}\%$
6. $66\frac{2}{3}\%$
7. $6\frac{1}{4}\%$
8. $62\frac{1}{2}\%$

3
1. 6%
2. 5%
3. 9%
4. 46%
5. 24%
6. 52%
7. 18%
8. 85%

4
1. 110%
2. 150%
3. $133\frac{1}{3}\%$
4. $162\frac{1}{2}\%$
5. 175%
6. 230%
7. 148%
8. 225%

SKILL ACHIEVEMENT TEST

Write each of the following percents as a fraction or mixed number:

1. 4%   **2.** 25%   **3.** 40%   **4.** $37\frac{1}{2}$%   **5.** 150%

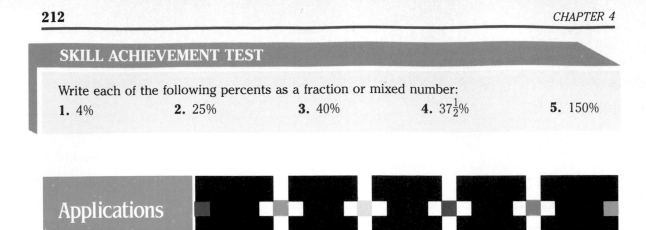

Applications

1. 30% of the pupils at the Southeast High School are in the ninth year. What part of the student body is in the ninth year?

2. If $87\frac{1}{2}$% of a class received a passing mark in a science test, what part of the class failed in the test?

3. What part of a graduating class is planning to go to college if 40% filed applications for entrance?

4. If Mrs. Romano pays 25% down on a house, what part of the purchase price is her down payment?

5. In a certain area $16\frac{2}{3}$% of all the crops were destroyed by floods. What part of the harvest was saved?

6. If a shortstop made 70% of his plays without an error, in what part of the total plays did he make errors?

4-5 Writing Fractions and Mixed Numbers as Percents

PROCEDURE

1 Write $\frac{18}{25}$ as a percent.

$$\frac{18}{25} = 25\overline{)18.00} \quad \begin{array}{r} .72 = 72\% \\ \underline{17\,5} \\ 50 \\ \underline{50} \end{array}$$

ANSWER: 72%

Divide the numerator by the denominator, finding the quotient to two decimal places. Rewrite the quotient, omitting the decimal point and writing a percent sign.

2 Write $\frac{2}{9}$ as a percent.

$$\frac{2}{9} = 9\overline{)2.00} \quad \begin{array}{r} .22\frac{2}{9} = 22\frac{2}{9}\% \\ \underline{1\,8} \\ 20 \\ \underline{18} \\ 2 \end{array}$$

ANSWER: $22\frac{2}{9}\%$

3 Write $\frac{32}{30}$ as a percent.

$$\frac{32}{30} = \frac{16}{15}$$

$$15\overline{)16.00} \quad 1.06\frac{10}{15} = 106\frac{2}{3}\%$$

ANSWER: $106\frac{2}{3}\%$

Simplify the given fraction before dividing.

---

4 Write $\frac{3}{4}$ as a percent.

$$\frac{3}{4} = 75\%$$

ANSWER: 75%

If the percent equivalent is known, write it directly.

---

5 Write $\frac{18}{100}$ as a percent.

$$\frac{18}{100} = 18\%$$

ANSWER: 18%

If the denominator is 100, write the numerator followed by the percent sign.

213

6 Write $\frac{7}{50}$ as a percent.

$$\frac{7}{50} = \frac{7 \times 2}{50 \times 2} = \frac{14}{100}$$
$$= 14\%$$

Write an equivalent fraction with a denominator of 100 by either raising to higher terms or reducing to lower terms.

ANSWER: 14%

7 Write $\frac{24}{40}$ as a percent.

$$\frac{24}{40} = \frac{3}{5} = 60\%$$

Simplify the given fraction.

ANSWER: 60%

---

8 Write $1\frac{1}{4}$ as a percent.

$1\frac{1}{4} = 1.25 = 125\%$ or $1\frac{1}{4} = 1 + \frac{1}{4} = 100\% + 25\% = 125\%$

ANSWER: 125%

DIAGNOSTIC TEST

Write each of the following fractions and mixed numbers as a percent:

1 $\frac{3}{4}$ **2** $\frac{2}{3}$ **3** $\frac{3}{100}$ **4** $\frac{19}{50}$ **5** $\frac{8}{400}$

6 $\frac{9}{36}$ **7** $\frac{7}{9}$ **8** $\frac{24}{56}$ **9** $\frac{18}{18}$ **10** $1\frac{1}{2}$

11 $\frac{7}{4}$ **12** $\frac{72}{64}$ **13** $2\frac{2}{3}$ **14** $\frac{12}{5}$

RELATED PRACTICE EXERCISES

Write each of the following fractions and mixed numbers as a percent:

1
1. $\frac{1}{4}$
2. $\frac{2}{5}$
3. $\frac{1}{2}$
4. $\frac{7}{10}$

2
1. $\frac{5}{6}$
2. $\frac{3}{8}$
3. $\frac{1}{3}$
4. $\frac{1}{6}$

3
1. $\frac{7}{100}$
2. $\frac{39}{100}$
3. $\frac{145}{100}$
4. $\frac{87\frac{1}{2}}{100}$

4
1. $\frac{27}{50}$
2. $\frac{4}{25}$
3. $\frac{13}{20}$
4. $\frac{16}{25}$

5
1. $\frac{18}{200}$
2. $\frac{35}{700}$
3. $\frac{12}{300}$
4. $\frac{60}{500}$

6
1. $\frac{16}{24}$
2. $\frac{39}{65}$
3. $\frac{40}{48}$
4. $\frac{63}{72}$

7
1. $\frac{5}{7}$
2. $\frac{8}{11}$
3. $\frac{4}{9}$
4. $\frac{7}{15}$

8
1. $\frac{20}{36}$
2. $\frac{45}{99}$
3. $\frac{42}{49}$
4. $\frac{65}{75}$

9
1. $\frac{45}{45}$
2. $\frac{7}{7}$
3. $\frac{84}{84}$
4. $\frac{156}{156}$

10
1. $1\frac{3}{4}$
2. $1\frac{2}{5}$
3. $1\frac{1}{3}$
4. $1\frac{5}{8}$

11 1. $\frac{5}{3}$

2. $\frac{13}{8}$

3. $\frac{8}{7}$

4. $\frac{9}{5}$

12 1. $\frac{57}{38}$

2. $\frac{65}{52}$

3. $\frac{70}{49}$

4. $\frac{90}{48}$

13 1. $2\frac{5}{8}$

2. $3\frac{4}{5}$

3. $4\frac{2}{3}$

4. $2\frac{3}{4}$

14 1. $\frac{8}{3}$

2. $\frac{14}{4}$

3. $\frac{17}{6}$

4. $\frac{23}{10}$

SKILL ACHIEVEMENT TEST

Express each of the following fractions and mixed numbers as a percent:

1. $\frac{4}{5}$

2. $\frac{7}{8}$

3. $\frac{81}{100}$

4. $\frac{31}{50}$

5. $\frac{33}{88}$

6. $\frac{5}{18}$

7. $\frac{30}{105}$

8. $\frac{39}{39}$

9. $1\frac{1}{4}$

10. $\frac{11}{3}$

Applications

1. If Maureen received $\frac{3}{5}$ of all votes cast in a homeroom election, what percent of the votes did she get?

2. What percent of the community contributed to the welfare fund if $\frac{5}{6}$ of the community contributed?

3. If $\frac{7}{10}$ of the student body participated in the interclass athletic games, what percent of the students did not participate?

4. Salespeople receive a commission of $\frac{3}{20}$ of the amount of their sales. What percent commission do they receive?

5. A department store advertises a $\frac{1}{4}$ reduction on all merchandise. What is the percent mark-down in this sale?

Table of Equivalents
Percents, Decimals, and Fractions

Percent	Decimal	Fraction
5%	.05	$\frac{1}{20}$
$6\frac{1}{4}$%	$.06\frac{1}{4}$	$\frac{1}{16}$
$8\frac{1}{3}$%	$.08\frac{1}{3}$	$\frac{1}{12}$
10%	.10 or .1	$\frac{1}{10}$
$12\frac{1}{2}$%	$.12\frac{1}{2}$ or .125	$\frac{1}{8}$
$16\frac{2}{3}$%	$.16\frac{2}{3}$	$\frac{1}{6}$
20%	.20 or .2	$\frac{1}{5}$
25%	.25	$\frac{1}{4}$
30%	.30 or .3	$\frac{3}{10}$
$33\frac{1}{3}$%	$.33\frac{1}{3}$	$\frac{1}{3}$
$37\frac{1}{2}$%	$.37\frac{1}{2}$ or .375	$\frac{3}{8}$
40%	.40 or .4	$\frac{2}{5}$
50%	.50 or .5	$\frac{1}{2}$
60%	.60 or .6	$\frac{3}{5}$
$62\frac{1}{2}$%	$.62\frac{1}{2}$ or .625	$\frac{5}{8}$
$66\frac{2}{3}$%	$.66\frac{2}{3}$	$\frac{2}{3}$
70%	.70 or .7	$\frac{7}{10}$
75%	.75	$\frac{3}{4}$
80%	.80 or .8	$\frac{4}{5}$
$83\frac{1}{3}$%	$.83\frac{1}{3}$	$\frac{5}{6}$
$87\frac{1}{2}$%	$.87\frac{1}{2}$ or .875	$\frac{7}{8}$
90%	.90 or .9	$\frac{9}{10}$
100%	1.00 or 1	$\frac{1}{1}$ or $\frac{100}{100}$

4-6 Finding a Percent of a Number

PROCEDURE

1 Find 23% of 64.

23% = .23

$$\begin{array}{r} 64 \\ \times\ .23 \\ \hline 1\ 92 \\ 12\ 8 \\ \hline 14.72 \end{array}$$

ANSWER: 14.72

Rewrite the percent as a decimal.
Multiply the given number by the decimal.

2 Find 3% of 18.

3% = .03

$$\begin{array}{r} 18 \\ \times\ .03 \\ \hline .54 \end{array}$$

ANSWER: .54

3 Find 114% of 240.

114% = 1.14

$$\begin{array}{r} 240 \\ \times\ 1.14 \\ \hline 96\ 0 \\ 240 \\ 240 \\ \hline 273.60 \end{array}$$

ANSWER: 273.6

---

4 Find 4% of $200.

$4\% = \frac{4}{100}$

$\frac{4}{100} \times 200 = \frac{4}{\underset{1}{100}} \times \frac{\overset{2}{200}}{1} = 8$

ANSWER: $8

Sometimes it is easier to rewrite the percent as a fraction.

---

5 Find $3\frac{1}{2}$% of 40.

$3\frac{1}{2}\% = .03\frac{1}{2} = .035$

$$\begin{array}{r} 40 \\ \times\ .035 \\ \hline 200 \\ 1\ 20 \\ \hline 1.400 = 1.4 \end{array}$$ or $$\begin{array}{r} 40 \\ \times\ .03\frac{1}{2} \\ \hline 1\ 20 \\ 20 \\ \hline 1.40 \end{array} \leftarrow 40 \times \frac{1}{2}$$

ANSWER: 1.4

6 Find $\frac{3}{4}$% of 650.

$\frac{3}{4}\% = .00\frac{3}{4} = .0075$

$$\begin{array}{r} 650 \\ \times\ .0075 \\ \hline 3250 \\ 4\ 550 \\ \hline 4.8750 = 4.875 \end{array}$$

ANSWER: 4.875

7 Find .6% of 52.

.6% = .006

$$\begin{array}{r} 52 \\ \times\ .006 \\ \hline .312 \end{array}$$

ANSWER: .312

8 Find 7% of $12.96 to the nearest cent.

7% = .07

$$\begin{array}{r} \$12.96 \\ \times\ \ \ .07 \\ \hline \$.9072 = \$.91 \end{array}$$

ANSWER: $.91

9 Solve the equation: 15% of 35 = n

.15 × 35 = 5.25

ANSWER: 5.25

To find the value of n, multiply the given numbers.

DIAGNOSTIC TEST

Find the following:

1 18% of 46   **2** 6% of 24   **3** 39% of 6.75

Write the percents as fractions in exercises 4, 5, and 6.

4 3% of 5,000   **5** 25% of 36   **6** $83\frac{1}{3}$% of 582   **7** 127% of 743

8 140% of 295   **9** 200% of 75   **10** $4\frac{1}{2}$% of 624   **11** $50\frac{3}{4}$% of 840

12 $\frac{1}{4}$% of 300   **13** 7.8% of 45   **14** .3% of 160   **15** 5% of $29

16 2% of $4.68 (to nearest cent)   **17** Solve: 30% of $75 = n

RELATED PRACTICE EXERCISES

Find the following:

1 1. 24% of 52   2. 63% of 75   3. 14% of 80   4. 92% of 48
     5. 16% of 240   6. 45% of 819   7. 32% of 2,907   8. 21% of 4,384

2 1. 2% of 18   2. 5% of 84   3. 9% of 50   4. 1% of 85
     5. 3% of 200   6. 6% of 325   7. 8% of 1,540   8. 4% of 8,462

3 1. 5% of 8.24   2. 3% of 9.62   3. 18% of 4.7   4. 27% of 8.73
     5. 1% of 15.40   6. 6% of 27.75   7. 5% of 295.08   8. 38% of 251.69

Write the percents as fractions in sets 4, 5, and 6.

4 1. 5% of 200   2. 4% of 4,000   3. 8% of 9,000   4. 6% of 650
     5. 18% of 500   6. 29% of 10,000   7. 95% of 5,300   8. 83% of 7,250

5 1. 10% of 30   2. 50% of 62   3. 40% of 65   4. 75% of 92
     5. 60% of 200   6. 25% of 328   7. 20% of 735   8. 80% of 1,225

6 1. $12\frac{1}{2}$% of 800   2. $37\frac{1}{2}$% of 312   3. $62\frac{1}{2}$% of 672   4. $33\frac{1}{3}$% of 768
     5. $16\frac{2}{3}$% of 804   6. $66\frac{2}{3}$% of 987   7. $87\frac{1}{2}$% of 4,368   8. $6\frac{1}{4}$% of 9,328

7 1. 116% of 28   2. 105% of 57   3. 138% of 120   4. 235% of 1,500

8 1. 120% of 45   2. 125% of 848   3. $133\frac{1}{3}$% of 246   4. $137\frac{1}{2}$% of 5,640

9 1. 100% of 72   2. 500% of 400   3. 200% of 325   4. 300% of 3,854

10 1. $2\frac{1}{2}$% of 30   2. $1\frac{1}{2}$% of 2,500   3. $4\frac{3}{4}$% of 3,000   4. $5\frac{1}{4}$% of 824

11 1. $24\frac{1}{2}$% of 20   2. $142\frac{3}{4}$% of 64   3. $130\frac{1}{2}$% of 245   4. $98\frac{7}{8}$% of 5,000

12 1. $\frac{1}{2}$% of 48   2. $\frac{1}{3}$% of 930   3. $\frac{3}{4}$% of 219   4. $\frac{5}{8}$% of 600

13 1. 2.5% of 36   2. 4.7% of 840   3. 6.75% of 725   4. 5.625% of 192

14 1. .2% of 8   2. .1% of 95   3. .25% of 500   4. .375% of 848

15 **1.** 4% of $5 **2.** 6% of $273 **3.** 1% of $3,000 **4.** 10% of $50

  5. 40% of $240 **6.** 23% of $19 **7.** $33\frac{1}{3}$% of $6,900 **8.** 41% of $5,643

  9. 113% of $42 **10.** 150% of $264 **11.** $2\frac{1}{2}$% of $300 **12.** $133\frac{1}{3}$% of $600

  13. 300% of $493 **14.** .5% of $28 **15.** 6.13% of $4,000 **16.** $\frac{3}{4}$% of $2,000

16 Find to nearest cent:

  1. 3% of $.90 **2.** 5% of $4.60 **3.** 6% of $3.29 **4.** 4% of $10.55

  5. 15% of $8.75 **6.** 79% of $25.38 **7.** 50% of $94.81 **8.** 25% of $285.96

  9. $87\frac{1}{2}$% of $.56 **10.** 124% of $6.73 **11.** 160% of $2.25 **12.** $3\frac{1}{2}$% of $9.40

  13. 200% of $35.86 **14.** $137\frac{1}{2}$% of $11.12 **15.** 6.05% of $58 **16.** .4% of $18.54

17 Solve:

  1. 10% of $98 = ? **2.** 8% of $6.50 − ▨ **3.** 75% of 300 = n

  4. ? = 5.4% of 296 **5.** ▨ = $102\frac{1}{2}$% of 3,000 **6.** n = 12.63% of $10,000

SKILL ACHIEVEMENT TEST

Find:

1. 94% of 36 **2.** 7.9% of 92.8 **3.** 25% of 6,000 **4.** 130% of 400

5. $66\frac{2}{3}$% of 819 **6.** 143% of 750 **7.** 6.85% of 9,000 **8.** .3% of $2,500

9. $4\frac{1}{2}$% of 8,500 **10.** 16% of $36.48 to nearest cent

Refresh Your Skills

1. Add: 29 **2.** Subtract: 1,000,000 **3.** Multiply: 1-6 **4.** Divide: 1-7
  1-4 6,842 1-5 908,147 846 $1,728\overline{)1,655,424}$
  98,386 259
  2,594
  73,967 **5.** Add: 6.03 + 2.39 + 3.26 + 7.32 2-4 **6.** Subtract: $9 − $1.32 2-5

7. Multiply: $70.29 (nearest cent) **8.** Divide: 2-8
  2-7 .01$\frac{1}{2}$ $144\overline{)\$12.96}$

9. Add: 3-8 **10.** Subtract: 3-9 **11.** Multiply: 3-11 **12.** Divide: 3-12
  $4\frac{3}{8} + 5\frac{9}{16}$ $8\frac{1}{4} − 2\frac{1}{3}$ $3\frac{1}{7} \times 4\frac{2}{3}$ $7\frac{1}{2} \div 6\frac{3}{4}$

13. Rewrite $\frac{33}{7}$ as a mixed number. 3-2 **14.** Express $\frac{7}{25}$ in 100ths. 3-3

15. Are $\frac{6}{14}$ and $\frac{18}{42}$ equivalent fractions? 3-4 **16.** Which is greater: $\frac{3}{8}$ or $\frac{5}{16}$? 3-5

17. Write .18 as a percent. 4-3 **18.** Write 60% as a fraction. 4-4 4-1

19. Write 56% as a decimal. 4-2 **20.** Express 26.7 hundredths as a percent.

21. Write $\frac{13}{20}$ as a decimal. 3-15 **22.** Find 3% of $56.25 (nearest cent). 4-6

23. Find $133\frac{1}{3}$% of 924. 4-6

BUILDING PROBLEM-SOLVING SKILLS

FLASHBACK

1 READ

2 PLAN

3 SOLVE

4 CHECK

PROBLEM: How much do you save when you buy at a 20%-reduction sale a camera that regularly sells for $58?

1 **Given facts:** Regular price of camera is $58; Purchased at 20% reduction.

Question asked: How much do you save?

2 **Operation needed:** *Finding a percent of a number* indicates multiplication.
Think or write out your plan: The rate of reduction times the regular price is equal to the amount saved.

Equation: 20% × $58 = n

3 **Estimate the answer:**
20% = .20 = .2
$58 → $60

.2 × $60 = $12, estimate

Solution:
$58
× .20
————
$11.60 saved

ANSWER: $11.60 saved

4 **Check the accuracy:**
.20
× $58
————
1 60
10 0
————
$11.60 saved ✔

Compare answer to the estimate:
The answer $11.60 compares reasonably with estimate $12.

PRACTICE PROBLEMS

1. Richard received a grade of 60% in a spelling test of 25 words. How many words did he spell correctly?

2. If the sales tax is 6%, what would be the tax on a T.V. set costing $495?

3. Susan stored her fur coat for the summer and was charged 2% of its value. If the coat is valued at $1,275, how much did she pay?

4. Betty's mother bought a house for $69,500 and made a down payment of 25%. What was the amount of the down payment?

Finding What Percent One Number Is of Another

PROCEDURE

1 27 is what percent of 36?

$\frac{27}{36} = \frac{27 \div 9}{36 \div 9}$

$\quad = \frac{3}{4}$

$\frac{3}{4} = 75\%$

Write a fraction indicating what fractional part one number is of the other.
Simplify.
Rewrite the fraction as a percent when the percent equivalent is known.*

ANSWER: 75%

---

2 What percent of 48 is 30?

$\frac{30}{48} = \frac{30 \div 6}{48 \div 6} = \frac{5}{8}$

$\begin{array}{r} .62\frac{1}{2} = 62\frac{1}{2}\% \\ 8\overline{)5.00} \\ \underline{4\ 8} \\ 20 \\ \underline{16} \\ 4 \end{array}$

Divide the numerator by the denominator to 2 decimal places.

Write the decimal quotient as a percent.

ANSWER: $62\frac{1}{2}\%$

3 18 is what percent of 16?

$\frac{18}{16} = \frac{18 \div 2}{16 \div 2} = \frac{9}{8}$

$\begin{array}{r} 1.12\frac{1}{2} = 112\frac{1}{2}\% \\ 8\overline{)9.00} \\ \underline{8} \\ 1\ 0 \\ \underline{8} \\ 20 \\ \underline{16} \\ 4 \end{array}$

ANSWER: $112\frac{1}{2}\%$

---

4 9 is ?% of 20?

$\frac{9}{20} = \frac{45}{100} = 45\%$

ANSWER: 45%

5 What percent of 26 is 26?

$\frac{26}{26} = 1 = 100\%$

ANSWER: 100%

*For an alternate method, see Lesson 17-14.

6 $1.50 is what percent of $7.50?

$\frac{\$1.50}{\$7.50} = \frac{1.50}{7.50} = \frac{1}{5} = 20\%$

ANSWER: 20%

DIAGNOSTIC TEST

Find the following:

1 4 is what percent of 5?

2 What percent of 12 is 6?

3 What percent of 8 is 7?

4 45 is __% of 54?

5 2 is what percent of 7?

6 What percent of 18 is 10?

7 37 is what percent of 37?

8 8 is __% of 4?

9 25 is what percent of 20?

10 What percent of 36 is 48?

11 561 is what percent of 935?

12 What percent of $17 is $3.40?

13 $3\frac{1}{2}$ is what percent of $10\frac{1}{2}$?

14 9 is __% of 100?

15 13 is what percent of 25?

16 What percent of 400 is 16?

RELATED PRACTICE EXERCISES

Find the following:

1 1. 3 is what percent of 5?
   3. 3 is __% of 4?
   5. What percent of 5 is 1?
   7. What percent of 2 is 1?

   2. 1 is what percent of 4?
   4. 9 is __% of 10?
   6. What percent of 5 is 2?
   8. What percent of 10 is 7?

2 1. What percent of 36 is 9?
   3. 39 is __% of 52?
   5. 5 is __% of 50?
   7. What percent of 80 is 56?

   2. What percent of 50 is 15?
   4. 42 is what percent of 60?
   6. 72 is what percent of 96?
   8. What percent of 90 is 81?

3 1. What percent of 6 is 5?
   3. 3 is __% of 8?

   2. What percent of 3 is 2?
   4. 5 is what percent of 8?

4 1. 9 is __% of 27?
   3. 35 is what percent of 42?
   5. 9 is what percent of 54?
   7. What percent of 16 is 10?

   2. 16 is __% of 24?
   4. What percent of 72 is 63?
   6. 28 is what percent of 32?
   8. 5 is __% of 80?

5 1. 3 is what percent of 7?
   3. What percent of 13 is 8?

   2. What percent of 9 is 7?
   4. 17 is __% of 18?

6 1. What percent of 14 is 4?
   3. 15 is __% of 84?
   5. 32 is __% of 72?
   7. What percent of 96 is 27?

   2. What percent of 54 is 30?
   4. 42 is what percent of 49?
   6. 70 is what percent of 75?
   8. What percent of 48 is 2?

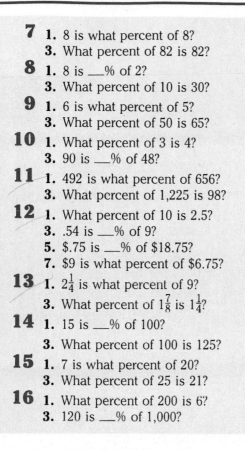

7 **1.** 8 is what percent of 8?
3. What percent of 82 is 82?

2. What percent of 53 is 53?
4. 250 is ___% of 250?

8 **1.** 8 is ___% of 2?
3. What percent of 10 is 30?

2. 60 is what percent of 12?
4. What percent of 12 is 72?

9 **1.** 6 is what percent of 5?
3. What percent of 50 is 65?

2. What percent of 24 is 54?
4. 63 is ___% of 35?

10 **1.** What percent of 3 is 4?
3. 90 is ___% of 48?

2. What percent of 12 is 32?
4. 85 is what percent of 75?

11 **1.** 492 is what percent of 656?
3. What percent of 1,225 is 98?

2. 645 is ___% of 1,032?
4. 1,014 is ___% of 2,535?

12 **1.** What percent of 10 is 2.5?
3. .54 is ___% of 9?
5. $.75 is ___% of $18.75?
7. $9 is what percent of $6.75?

2. What percent of 7.5 is 3?
4. 4.3 is what percent of 6.45?
6. What percent of $6.50 is $1.30?
8. What percent of $15 is $3.75?

13 **1.** $2\frac{1}{4}$ is what percent of 9?
3. What percent of $1\frac{7}{8}$ is $1\frac{1}{4}$?

2. What percent of $4\frac{2}{3}$ is $3\frac{1}{2}$?
4. $4\frac{1}{8}$ is ___% of $6\frac{7}{8}$?

14 **1.** 15 is ___% of 100?
3. What percent of 100 is 125?

2. 4.5 is what percent of 100?
4. What percent of 100 is $33\frac{1}{3}$?

15 **1.** 7 is what percent of 20?
3. What percent of 25 is 21?

2. What percent of 50 is 34?
4. 2 is ___% of 25?

16 **1.** What percent of 200 is 6?
3. 120 is ___% of 1,000?

2. What percent of 900 is 60?
4. 32 is what percent of 800?

SKILL ACHIEVEMENT TEST

Find:

1. 3 is what percent of 10?
3. What percent of 40 is 15?
5. 14 is what percent of 14?
7. $.06 is what percent of $1.80?
9. What percent of $12.50 is $5?

2. 36 is ___% of 45?
4. What percent of 24 is 3?
6. 7 is what percent of 4?
8. $1\frac{2}{3}$ is what percent of $8\frac{1}{3}$?
10. What percent of $2,400 is $480?

Applications

Find each of the following correct to nearest tenth of a percent:

1. 5 is what % of 9?
4. What % of $12 is $7?
7. $.60 is ___% of $1.44?

2. What % of 21 is 13?
5. What % of 63 is 27?
8. What % of $171 is $135?

3. $.11 is ___% of $.15?
6. 30 is what % of 78?

BUILDING
PROBLEM-SOLVING
SKILLS

FLASHBACK

1 READ

2 PLAN

3 SOLVE

4 CHECK

PROBLEM: Juan answered 31 questions correctly on a test of 40 questions. What percent of the questions did he answer correctly?

1 **Given Facts:** Test contained 40 questions; 31 answered correctly

Question asked: What percent of the questions were answered correctly?

2 **Operation needed:** *Finding what percent one number is of another* indicates division and changing decimal quotients to a percent.

Think or write out your plan: The number of questions answered correctly divided by the total number of questions is equal to the decimal or fractional part of the test answered correctly. Express this quotient as a percent.

Equation: $31 \div 40 = n$, or $n = \frac{31}{40}$

3 **Estimate the answer:**

$31 \rightarrow 30$ $\frac{30}{40} = \frac{3}{4} = 75\%$, estimate
$40 \rightarrow 40$

Solution:

$$\frac{31}{40} = 40\overline{)31.00} \quad .77\tfrac{1}{2} = 77\tfrac{1}{2}\% \text{ or } 77.5\%$$

4 **Check the accuracy:**

$40 \times .775 = 31$ ✔

Compare answer to the estimate: The answer 77.5% compares reasonably with the estimate 75%.

ALTERNATE METHOD: Solution by Equation (Also see Lesson 17-14.)

Think or write out your plan: What percent of 40 questions are 31 questions?

Equation: $n \times 40 = 31$
                 or $40n = 31$

Solution: $40n = 31$

$\frac{40n}{40} = \frac{31}{40}$

$n = .775$ or 77.5%

ANSWER: 77.5%

PRACTICE PROBLEMS

1. Mr. Huang saves $150 each month. If his annual income is $18,000, what percent of his income does he save annually?

2. Find the rate of reduction when a basketball that regularly sells for $25 was purchased for $20.

3. The local soccer team won 9 games and lost 6. What percent of the games did the team lose?

4. What is the rate of increase when the school enrollment increased from 250 students to 300 students?

MIXED PROBLEMS

Solve each of the following problems. For problems 1-4, select the letter corresponding to your answer.

1. There are 40 pupils in a class. On a certain day 38 pupils were present. What percent of the class attended school?
a. 38%   **b.** 2%
c. 95%   **d.** Answer not given

2. How many problems did Joan have right if she received a grade of 85% in a mathematics test of 20 problems?
a. 15 problems   **b.** 17 problems
c. 3 problems   **d.** Answer not given

3. There are 18 women and 27 men in a class. What percent of the students are men?
a. $66\frac{2}{3}$%   **b.** 60%
c. 50%   **d.** 40%

4. The enrollment in the Weston H.S. is 850. If the attendance for June was 92%, how many absences were there?
a. 88 absences   **b.** 782 absences
c. 68 absences   **d.** Answer not given

5. Charles answered 19 questions correctly and missed 6 questions. What percent of the questions did he answer correctly?

6. A certain manganese bronze contains 59% copper. How many pounds of copper are in 300 pounds of manganese bronze?

7. Find the rate of decrease if the price of an item is changed from 40¢ to 30¢.

Refresh Your Skills

1. Add: 7,124   **2.** Subtract: 1-5   **3.** Multiply: 1-6   **4.** Divide: 1-7   **5.** Add: 2-4
1-4   83,259   815,008   403   3.8 + .47 + 25
       61,435   747,598   807   794)235,818
       816
       9,527   **6.** Subtract: 2-5   **7.** Multiply: 2-7   **8.** Divide: 2-8
       84,349   6.825 − .97   3.14 × 80   1.15)16

9. Add: $7\frac{2}{3}$ **10.** Subtract: $8\frac{3}{4}$ **11.** Multiply: 3-11   **12.** Divide: 3-12
3-8 $4\frac{5}{6}$ 3-9 $2\frac{4}{5}$ $4\frac{3}{8} \times 2\frac{3}{10}$ $\frac{13}{16} \div 1\frac{1}{12}$

13. Name the least common multiple (LCM) of 12, 24, and 36. 3-6
14. Find the least common denominator (LCD) and write equivalent fractions for: $\frac{7}{16}, \frac{13}{24}$, and $\frac{15}{32}$.
3-7

15. Write $6\frac{3}{7}$ as an improper fraction. 3-10   **16.** What part of 52 is 39? 3-13
17. Find 8% of 60.9. 4-6   **18.** Find $87\frac{1}{2}$% of $526. 4-6
19. What percent of 144 is 64? 4-7   **20.** What percent of $1.25 is $.50? 4-7

4-8 Finding a Number When a Percent of It Is Known

PROCEDURE

1 16% of what number is 48?

$$48 \div \mathbf{16\%}$$

↑   ↖
part   **percent representing part**

$16\% = .16$

$$.16\overline{)48.00}_{\wedge} \quad \frac{300}{48}$$

ANSWER: 300

Divide the given part by the **percent** representing this part.

Rewrite the percent as a decimal before dividing.*

2 31 is 4% of what number?

$$31 \div \mathbf{4\%} = 31 \div .04$$

$$.04\overline{)31.00}_{\wedge}$$
$$\underline{28}$$
$$3\,0$$
$$\underline{2\,8}$$
$$20$$
$$\underline{20}$$
(775)

ANSWER: 775

3 .2% of what number is 6.4?

$$6.4 \div \mathbf{.2\%} = 6.4 \div .002$$

$$.002\overline{)6.400}_{\wedge}$$
$$\underline{6}$$
$$4$$
$$\underline{4}$$
(3 200)

ANSWER: 3,200

---

4 50% of what number is 40?

$$40 \div \mathbf{50\%}$$

$$40 \div \tfrac{1}{2} = 40 \times \tfrac{2}{1}$$
$$= 80$$

ANSWER: 80

Sometimes it is easier to write the percent as a fraction.

5 125% of what number is 15?

$$15 \div \mathbf{125\%} = 15 \div \tfrac{125}{100}$$

$$= 15 \div \tfrac{5}{4}$$

$$= \frac{\overset{3}{\cancel{15}}}{1} \times \frac{4}{\underset{1}{\cancel{5}}}$$

$$= 12$$

ANSWER: 12

*For an alternate method, see Lesson 17-14.

6 $\frac{3}{8}$% of what number is 15?

$15 \div \frac{3}{8}\% = 15 \div .00\frac{3}{8}$

$\qquad\quad = 15 \div .00375$

$$.00375\,\overline{)15.00000}_{\wedge} \quad \overset{4000}{}$$

$\qquad\qquad \underline{15\ 00}$

ANSWER: 4,000

DIAGNOSTIC TEST

Find the missing numbers:

1 12% of what number is 24?

3 25% of what number is 6?

5 6% of what number is 12?

7 100% of what number is 70?

9 40% of what number is 12.6?

11 $4\frac{1}{2}$% of what number is 90?

13 .5% of what number is 4?

15 8% of what amount is $180?

2 18 is 36% of what number?

4 $66\frac{2}{3}$% of what number is 14?

6 20 is 20% of what number?

8 120% of what number is 108?

10 2.5% of what number is 2?

12 $187\frac{1}{2}$% of what number is 105?

14 $\frac{3}{4}$% of what number is 27?

RELATED PRACTICE EXERCISES

Find the missing numbers:

1 1. 45% of what number is 90?
3. 74% of what number is 370?

2. 65% of what number is 260?
4. 15% of what number is 18?

2 1. 12 is 24% of what number?
3. 9 is 15% of what number?

2. 44 is 55% of what number?
4. 17 is 85% of what number?

3 1. 20% of what number is 3?
3. 18 is 60% of what number?

2. 40% of __ = 48
4. 91 = 70% of __

4 1. $33\frac{1}{3}$% of what number is 78?
3. 4 is $16\frac{2}{3}$% of what number?

2. $62\frac{1}{2}$% of __ = 200
4. $462 = 66\frac{2}{3}$% of __

5 **1.** 2% of what number is 10?   **2.** 4% of ___ = 26
   3. 15 = 1% of ___   **4.** 5% of what number is 45?

6 **1.** 50 is 50% of what number?   **2.** 30 = 30% of ___
   3. 79% of what number is 79?   **4.** 43% of ___ = 43

7 **1.** 100% of what number is 59?   **2.** 300% of ___ = 240
   3. 5 is 100% of what number?   **4.** 36 = 200% of ___

8 **1.** 160% of what number is 72?   **2.** 175% of ___ = 42
   3. 513 is 114% of what number?   **4.** 78 = 156% of ___

9 **1.** 60% of ___ = 46.8   **2.** 102.7 is 65% of what number?
   3. 371.2 = 58% of ___   **4.** 70% of what number is 667.8?

10 **1.** 4.75% of what number is 38?   **2.** 8.1 is 8.1% of what number?
   3. 16.5 = 4.125% of ___   **4.** 26.3% of ___ = 18.41

11 **1.** $4\frac{3}{4}$% of what number is 19?   **2.** $6\frac{1}{8}$% of ___ = 12.25
   3. 18 is $2\frac{1}{2}$% of what number?   **4.** 72 = $28\frac{4}{5}$% of ___

12 **1.** $116\frac{2}{3}$% of what number is 21?   **2.** $287\frac{1}{2}$% of ___ = 230
   3. 39 is $162\frac{1}{2}$% of what number?   **4.** 644 = $233\frac{1}{3}$% of ___

13 **1.** .4% of what number is 2?   **2.** .875% of ___ = 70
   3. 2 is .1% of what number?   **4.** 12 = $.33\frac{1}{3}$% of ___

14 **1.** $\frac{1}{2}$% of what number is 10?   **2.** $\frac{5}{12}$% of ___ = 100
   3. 9 is $\frac{3}{20}$% of what number?   **4.** 60 is $\frac{2}{3}$% of ___.

15 **1.** 75% of what amount is $150?   **2.** 6% of what amount is $78?
   3. $.12 is 10% of what amount?   **4.** $5,075 is 101.5% of what amount?

SKILL ACHIEVEMENT TEST

Find the missing numbers:

1. 31% of what number is 279?   **2.** 108 is 75% of what number?

3. 96 is 6% of what number?   **4.** $83\frac{1}{3}$% of what number is 290?

5. 100% of what number is 280?   **6.** 15.4% of what number is 130.9?

7. 125% of what number is 32.5?   **8.** 4.92 is $10\frac{1}{4}$% of what number?

9. $76 is .8% of what amount?   **10.** $37\frac{1}{2}$% of what amount is $8.13?

BUILDING PROBLEM-SOLVING SKILLS

FLASHBACK

1 READ

2 PLAN

3 SOLVE

4 CHECK

PROBLEM: The school baseball team won 39 games or 52% of the games it played. How many games did it play?

1 **Given facts:** Team won 52% or 39 games of the games it played.

Question asked: How many games did it play?

2 **Operation needed:** *Finding a number when a percent of it is known* indicates division.

Think or write out your plan: The number of games won divided by the fractional or decimal equivalent of the given percent is equal to the number of games played.

Equation: $39 \div .52 = n$, or $n = \frac{39}{.52}$

3 **Estimate the answer:**

$39 \rightarrow 40$ $\underline{80}$ games played,
$.52 \rightarrow .5$ $.5\overline{)40.0}$ estimate

Solution:
$$.52\overline{)39.00} \begin{array}{l} 75 \text{ games} \\ \text{played} \end{array}$$
$$\underline{364}$$
$$260$$
$$\underline{260}$$

ANSWER: 75 games played

4 **Check the accuracy:**

$$\begin{array}{r} 75 \text{ games played} \\ \times\ .52 \\ \hline 1\ 50 \\ \underline{37\ 5} \\ 39.00 \text{ games won } \checkmark \end{array}$$

Compare answer to the estimate: The answer 75 games compares reasonably with estimate 80 games.

ALTERNATE METHOD: Solution by Equation (Also see Lesson 17-14.)
Think or write out your plan: 52% of how many games played is 39 games?
Equation: $.52n = 39$ **Solution:** $.52n = 39$
$$\frac{.52n}{.52} = \frac{39}{.52}$$
$$n = 75 \text{ games played}$$

PRACTICE PROBLEMS

1. How much money must be invested at 9% to earn $4,500 per year?

2. If an ore contains 16% copper, how many metric tons of ore are needed to get 20 metric tons of copper?

3. Find the regular price of a computer that sold for $1,025 at an 18%-reduction sale.

4. A real estate agent received $2,700 commission, which is 6% of the selling price of a house. At what price did she sell the house?

MIXED PROBLEMS

Solve each of the following problems. For problems 1-4, select the letter corresponding to your answer.

1. A house worth $47,900 is insured for 80% of its value. How much would the owner receive if the house were destroyed by the fire?
   a. $47,900   **b.** $9,580
   c. $35,000   **d.** $38,320

2. Betty bought a camera at a 20%-reduction sale. If she paid $36 for it, what was the regular price?
   a. $18.20   **b.** $42.50
   c. $45.00   **d.** Answer not given

3. School lateness dropped from 42 cases per month to 35 cases. The rate of decrease is:
   a. 20%   **b.** 10%
   c. $16\frac{2}{3}$%   **d.** 7%

4. If 45% of the students are boys, and the girls number 858, how many boys are enrolled?
   a. 386 boys   **b.** 702 boys
   c. 1,560 boys   **d.** Answer not given

5. Leon saved $7.50 by buying a shirt at a reduction of 25%. What was the regular price?

6. At what price should a dealer sell a T.V. set which costs $420 to make a profit of 30% on the selling price?

Refresh Your Skills

1. Add: 23,845
   1-4 74,583
        96,148
        31,599
        83,172

2. Subtract: 1-5
   403,070
   392,989

3. Multiply: 1-6
   8,004
   9,050

4. Divide: 1-7
   365)79,205

5. Add: $8.29 + $.58 + $46.75 2-4

6. Subtract: $49 − $.89 2-5

7. Multiply: 2.8 × .003 2-7

8. Divide: 20).1 2-8

9. Add: $4\frac{5}{8}$
   3-8 $\frac{3}{4}$
        $8\frac{11}{16}$

10. Subtract: $1\frac{1}{6}$
    3-9 $\frac{1}{2}$

11. Multiply: 3-11
    $6\frac{2}{3} \times 7\frac{7}{8}$

12. Divide: 3-12
    $4\frac{1}{6} \div 2\frac{13}{16}$

13. Write $3.83\frac{1}{3}$ as a fraction. 3-16

14. Express in simplest form the ratio of 30 min to 2 h. 3-17

15. Solve the proportion: $\frac{1\frac{1}{3}}{12} = \frac{1\frac{1}{3}}{x}$ 3-18

16. Write $\frac{19}{20}$ as a percent. 4-5

17. Find $2\frac{1}{2}$% of $6,000. 4-6

18. What percent of 120 is 75? 4-7

19. 2% of what number is 5? 4-8

20. 96% of what number is 288? 4-8

PERCENT:

Solving by Proportion

The three basic types of percentage problems may be treated as one through the use of the *proportion*. Also see Lesson 3-18.

To find 8% of 25 means to determine *the number (n) which compared to 25 is the same as 8 compared to 100.*

Solve the proportion.

$$\frac{n}{25} = \frac{8}{100}$$
$$100n = 200$$
$$n = 2$$

ANSWER: 2

To find what percent of 25 is 2 means to find *the number (n) per 100 or the ratio of a number to 100 which has the same ratio as 2 to 25.*

Solve the proportion.

$$\frac{n}{100} = \frac{2}{25}$$
$$25n = 200$$
$$n = 8$$

ANSWER: 8%

To find the number of which 8% of it is 2 means to determine *the number (n) such that 2 compared to this number is the same as 8 compared to 100.*

Solve the proportion.

$$\frac{2}{n} = \frac{8}{100}$$
$$8n = 200$$
$$n = 25$$

ANSWER: 25

EXERCISES

Write each of the following as a ratio:

1. 6%　　　　　**2.** 18%　　　　　**3.** 70%　　　　　**4.** $62\frac{1}{2}$%　　　　　**5.** 2.4%

Find each of the following:

6. 37% of 16　　　　　　**7.** 60% of $1,200　　　　　　**8.** 3% of $940

9. 180% of 685　　　　　**10.** $16\frac{2}{3}$% of 732　　　　　**11.** 5.9% of 28

12. 2 is what percent of 5?　　　　　**13.** What percent of 84 is 63?

14. 49 is what percent of 56?　　　　**15.** What percent of 8 is 10?

16. What percent of $.45 is $.27?　　　**17.** $1.50 is what percent of $9?

18. 9% of what number is 72?　　　　**19.** 24 is 75% of what number?

20. 7.8 is 4% of what number?

When you shop at a large retail store, you benefit from the work of people in many different jobs. Buyers decide what merchandise the store will sell. Salespeople assist customers with their purchases. Salespeople may be paid a commission, which is a percentage of the purchase price. Cashiers wrap customers' purchases and collect the money. Credit-department clerks use computers to prepare bills for customers who have charge accounts.

Many buyers have college degrees in marketing. Salespeople, cashiers, and credit-department clerks are usually high-school graduates who have had some special training.

1. A cashier rings up a $48.80 purchase. The sales tax rate is 5%. What is the sales tax?

2. A salesperson receives a $560 commission on sales of $16,000. What is the rate of commission?

3. A buyer estimates that the store will sell 10,000 books next year. This year 9,150 books were sold. What will be the percent increase to the nearest tenth of a percent?

4. A bowling ball costs a store $22.50. It is marked up 50%. What is the selling price?

5. A credit-department clerk records a finance charge of $3.50 on a previous balance of $250. What is the finance-charge rate?

6. A customer has a previous balance of $92, a finance charge of $1.38, and new purchases of $46.95. What is the total bill?

7. A buyer buys 450 tape recorders at $21.73 each. What is the total cost?

8. A buyer buys 60 cameras for $7,345.20. What is the cost of one camera?

9. A cashier is paid $1\frac{1}{2}$ times the regular rate for overtime work. The regular rate is $4.20 per hour. What is the overtime rate?

10. A credit-department clerk charges a customer a finance charge of $1.20. The finance-charge rate is 1.5% of the previous balance. What is the previous balance?

11. A salesperson receives a commission of $52 for selling a refrigerator. The commission rate is 8%. What is the selling price of the refrigerator?

12. A certain style of sweater is available in red, green, and blue. This month, $\frac{5}{12}$ of those sold were red and $\frac{1}{4}$ were green. What fraction were blue?

13. A customer buys three pairs of shoes for $38.95, $16.50, and $26.98. What is the total price?

14. A salesperson receives a commission of $648 on sales of $5400. What is the rate of commission?

15. A cashier receives a $50 bill for a purchase of $38.92. What is the change?

16. A cashier rings up a $62.50 purchase. The sales tax rate is 6%. What is the tax?

17. A buyer buys shirts at a trade discount of 40% off the list price of $13.50. How much is the trade discount on each shirt?

18. A customer buys four books for $11.95, $16.50, $3.75, and $7.29. What is the total price?

19. A salesperson receives a 7% commission on sales of $8,650. What is the amount of commission?

20. A cashier receives a $20 bill for the purchase of a $7.95 record. What is the change?

CHAPTER REVIEW

Vocabulary

percent p. 204

EXERCISES

The numerals in boxes indicate Lessons where help may be found.

Write each of the following as a percent, decimal, and fraction: 4-1

1. Seven hundredths
3. 53 out of 100
5. Thirty-nine and one-third hundredths

2. Fifty-six hundredths
4. 91 out of 100

Express each of the following percents as a decimal: 4-2

6. 4%   **7.** 59%   **8.** 150%   **9.** $100\frac{3}{4}$%   **10.** 6.9%

Express each of the following decimals as a percent: 4-3

11. .03   **12.** .3   **13.** 1.18   **14.** .942   **15.** $.66\frac{2}{3}$

Express each of the following percents as a fraction or mixed number: 4-4

16. 25%   **17.** $83\frac{1}{3}$%   **18.** 6%   **19.** $162\frac{1}{2}$%   **20.** 250%

Express each of the following fractions or mixed numbers as a percent: 4-5

21. $\frac{3}{5}$ **22.** $\frac{11}{25}$ **23.** $\frac{5}{4}$ **24.** $\frac{71}{71}$ **25.** $3\frac{2}{3}$

Find the following: 4-6

26. 82% of 350
28. 75% of 2,000
30. 180% of 560

27. 6% of $39
29. $4\frac{1}{2}$% of $92.75 (to nearest cent)

Find the following: 4-7

31. 16 is what percent of 24?
33. $.48 is what percent of $3.20?
35. What percent of 9 is 4? (correct to nearest tenth)

32. What percent of 50 is 29?
34. 5 is what percent of 2?

Find the missing numbers: 4-8

36. 28% of what number is 7?
38. 54 is 3.6% of what number?
40. $137\frac{1}{2}$% of what number is 55?

37. $72 is 9% of what amount?
39. $2\frac{1}{4}$% of what number is 81?

41. Mrs. Chow bought a new car for $8,460. She paid 20% down. If the balance is to be paid in 24 equal monthly installments, how much must she pay each month? 4-6
42. What percent was a television set reduced if it was marked $450 and sold for $390? 4-7

COMPETENCY CHECK TEST

The numerals in boxes indicate Lessons where help may be found.
Solve each problem and select the letter corresponding to your answer.

1. Add: 7,798 + 82,596 + 39,495 + 9,389 ⌑1-4⌑
    a. 129,378   **b.** 139,278   **c.** 138,168   **d.** 137,298

2. Subtract: 1,650,000 − 938,074 ⌑1-5⌑
    a. 712,916   **b.** 728,074   **c.** 711,926   **d.** 712,926

3. Multiply: 398 × 207 ⌑1-6⌑
    a. 82,386   **b.** 10,746   **c.** 8,286   **d.** 605

4. Divide: 506)$\overline{15,180}$ ⌑1-7⌑
    a. 3   **b.** 30   **c.** 300   **d.** 3,000

5. Find the sum: .8 + .88 + .888 ⌑2-4⌑
    a. 2.568   **b.** .2568   **c.** .888888   **d.** Answer not given

6. Subtract: $57 − $9.43 ⌑2-5⌑
    a. $48.67   **b.** $47.57   **c.** $47.67   **d.** $48.57

7. Multiply: .45 × .004 ⌑2-7⌑
    a. .00018   **b.** .0180   **c.** 1.8   **d.** .0018

8. Divide: $.09)$\overline{\$63}$ ⌑2-8⌑
    a. 7   **b.** 70   **c.** 700   **d.** .07

9. Add: $6\frac{3}{4} + 2\frac{5}{8} + 4\frac{1}{2}$ ⌑3-8⌑
    a. $12\frac{1}{2}$ **b.** $13\frac{3}{8}$ **c.** $13\frac{11}{16}$ **d.** $13\frac{7}{8}$

10. Subtract: $12 − 8\frac{7}{10}$ ⌑3-9⌑
    a. $4\frac{3}{10}$ **b.** $4\frac{7}{10}$ **c.** $3\frac{3}{10}$ **d.** $3\frac{7}{10}$

11. Multiply: $4\frac{1}{2} \times \frac{5}{6}$ ⌑3-11⌑
    a. $3\frac{3}{4}$ **b.** $5\frac{1}{4}$ **c.** $4\frac{5}{12}$ **d.** $2\frac{7}{8}$

12. Divide: $\frac{2}{3} \div 6$ ⌑3-12⌑
    a. 4   **b.** $\frac{1}{4}$ **c.** $\frac{1}{9}$ **d.** 9

13. Find 4% of $245. ⌑4-6⌑
    a. $.98   **b.** $9.80   **c.** $98   **d.** $980

14. What percent of 96 is 36? ⌑4-7⌑
    a. 30%   **b.** $33\frac{1}{3}$%   **c.** $37\frac{1}{2}$%   **d.** 40%

15. 8% of what number is 74? ⌑4-8⌑
    a. 592   **b.** 59.20   **c.** 5.92   **d.** 925

16. The school soccer team won 60% of the 25 games played.
How many games did it lose? ⌑4-6⌑
    a. 5 games   **b.** 10 games   **c.** 15 games   **d.** 20 games

CUMULATIVE ACHIEVEMENT TEST

The numerals in boxes indicate Lessons where help may be found.

1. Round 93,482,576 to the nearest thousand. 1-3

2. Write as a decimal: Sixty and five thousandths. 2-2

3. Add: 82,197
   1-4   4,825
            399
          47,286
           8,678

4. Subtract: 250,105
   1-5   176,298

5. Multiply: 1-6
   6,043 × 7,008

6. Divide: 1-7
   809)247,554

7. Name the greatest common factor of 27, 90, and 108. 1-8

8. Add: 2-4
   4.29 + 97.2 + .687

9. Subtract: 2-5
   $6.50 − $3

10. Multiply: 2-7
    .24 × .002

11. Divide: 2-8
    .06).3

12. Find the least common multiple of 4, 6, and 10. 3-6

13. Add: 3-8
    $6\frac{1}{3} + 4\frac{3}{4} + 3\frac{7}{12}$

14. Subtract: 3-9
    $8\frac{1}{2} − 2\frac{2}{3}$

15. Multiply: 3-11
    $6\frac{1}{4} × 3\frac{3}{5}$

16. Divide: 3-12
    $2\frac{7}{16} ÷ \frac{3}{8}$

17. Find the ratio of 21 to 15. 3-17

18. Write $\frac{7}{8}$ as a decimal. 3-15

19. Write .8 as a fraction. 3-16

20. Write .069 as a percent. 4-3

21. Find 5% of $32.68 (to nearest cent). 4-6

22. 45 is what percent of 72? 4-7

23. 6% of what amount is $270? 4-8

24. Solve by proportion: 3-18
    What number compared to 16 is the same as 27 compared to 18?

Is the quotient <u>greater than</u>, <u>equal to</u>, or <u>less than</u> a given number when the given number is divided by: 3-12

25. 1?

26. A number greater than 1?

27. A number less than 1
    (zero excluded)?

Calculator Skills

Use a calculator to check whether the following answers are correct or incorrect.

Add:

1. 693 + 4,856 + 96 + 829 + 9,069 = 15,543
2. 5,826 + 9,658 + 4,178 + 3,907 + 6,455 = 31,024
3. 82,476 + 985 + 7,593 + 6,847 + 91,328 = 188,229
4. 79,327 + 87,419 + 76,678 + 50,724 + 23,586 = 317,734

Subtract:

5. 8,315 − 5,672 = 2,643
7. 626,000 − 498,608 = 207,392

6. 40,681 − 31,584 = 9,197
8. 8,039,547 − 6,929,759 = 1,109,888

Multiply:

9. 69 × 82 = 5,658
11. 5,276 × 416 = 2,294,816

10. 705 × 980 = 690,900
12. 8,509 × 6,053 = 51,404,977

Divide:

13. 41,993 ÷ 49 = 847
15. 248,472 ÷ 609 = 408

14. 671,550 ÷ 726 = 925
16. 618,576 ÷ 1,578 = 392

Add:

17. 1.46 + 8.91 + 6.87 = 17.24
19. .367 + .659 + .348 = 1.374

18. .914 + 5 = .915
20. 6.2 + .276 = .896

Subtract:

21. .74 − .69 = .05
23. 60.7 − 3.54 = 57.16

22. 9.435 − 3.15 = 6.385
24. 8 − 5.6 = 3.6

Multiply:

25. .4 × .19 = .76
27. 3,000 × .0002 = .6

26. .008 × .025 = .002
28. 3.1416 × 28.03 = 88.05948

Divide:

29. 60.8 ÷ 16 = 3.8
31. 65.86 ÷ 8.9 = .74

30. .002 ÷ .04 = .5
32. .00054 ÷ .009 = .006

Find:

33. 5% of 73 = 3.65
35. 8% of 17.26 = 138.08

34. 175% of 20 = 350
36. 4.9% of 52.4 = 2.5676

Find each of the following:

37. What percent of 96 is 72? $62\frac{1}{2}$
39. 30% of what number is 180? 600

38. 51 is what percent of 150? 34%
40. 6 is 4% of what number? 15

Computer Activities

The computer program below is written in the BASIC language. It is designed to help you practice writing percents as decimals. Enter the program on your computer keyboard exactly as it appears, one line at a time. Press the RETURN or ENTER key at the end of each line.

10 Clearscreen	Clears the screen. (Use appropriate command for *your* computer.)
20 LET Q = RND(1)	Generates a "random" (unpredictable) decimal Q which will act as a "coin flip."
30 LET X = INT(1000*RND(1))	Generates a "random" whole number X, between 1 and 999 and stores it in memory location X.
40 IF Q < .5 THEN LET X = X/10	Says "if Q is heads" then divide X by 10, to get a number between .1 and 99.9. This will happen 50% of the time.
50 PRINT X; "%"	Prints out the problem on your computer: X%
60 LET Z = X/100	Computes the answer Z, which equals X/100.
70 INPUT A	Displays a "?" as a way of asking for your answer.
80 LET D = ABS(A − Z)	Determines D, the difference between your answer and the correct answer.
90 IF D < .0001 THEN PRINT "YES"	Says "YES" if your answer is correct.
100 IF D > = .0001 THEN PRINT "NO"	Says "NO" if your answer is incorrect.
110 GOTO 20	Sends the computer back to line 20 for another problem.

(Note: TRS-80 and IBM users replace RND(1) in lines 20 and 30 with RND(0)).

If you make a typing mistake:

(a) Use the ← key to back up and make corrections.

or

(b) Simply press RETURN or ENTER and retype the line.

EXERCISES

Type RUN and press RETURN or ENTER to start your program. Use the program until you have achieved mastery with writing percents as decimals.

EXAMPLES:

If the problem is: 35% you would answer: .35

If the problem is: 125% you would answer: 1.25

If the problem is: 50.9% you would answer: .509

Achievement Test I (BASIC SKILLS)

1-1   **1.** Read the numeral 85,267,428 or write it in words.

1-2   **2.** Write the numeral naming: Ninety-two million, four hundred six thousand, fifty.

1-3   **3.** Round 581,560 to the nearest thousand.

1-4   **4.** Add: 472 + 5,973 + 15,682 + 33 + 280

1-5   **5.** Take 638 from 4,000.

1-6   **6.** Multiply: 14 × 1,728

1-7   **7.** Divide: 48)$\overline{33,744}$

1-8   **8.** Name the greatest common factor of 42 and 70.

2-1   **9.** Read the numeral .0626 or write it in words.

2-2   **10.** Write as a decimal numeral: Fifteen and seven tenths.

2-3   **11.** Round to nearest hundredth: 26.8539

2-4   **12.** Add: $6.57
.36
12.40
.82

2-5   **13.** Subtract: 54.3 − 8.28

2-6   **14.** Which is greater: .46 or .462?

2-7   **15.** Multiply: 3.14 × 15

2-8   **16.** Divide: $.06)$\overline{\$7.44}$

2-9   **17.** What decimal part of 20 is 13?

2-10   **18.** .06 of what number is 18?

2-11   **19.** Multiply: 10 × 8.69

2-12   **20.** Divide: 1.95 ÷ 100

2-13   **21.** Write 14.6 million as a complete numeral.

2-14   **22.** Write the shortened name for 63,700,000,000 in billions.

3-1   **23.** Express $\frac{28}{48}$ in lowest terms.

3-2   **24.** Rewrite $\frac{9}{2}$ as a mixed number.

3-3   **25.** Express $\frac{9}{25}$ as a fraction with the denominator 100.

3-4   **26.** Are $\frac{24}{36}$ and $\frac{18}{24}$ equivalent fractions?

3-5   **27.** Which is less: $\frac{5}{8}$ or $\frac{5}{6}$?

3-6   **28.** Find the least common multiple of 8 and 18.

3-7   **29.** Find the least common denominator of the fractions: $\frac{3}{8}$ and $\frac{11}{16}$

3-8   **30.** Add: $9\frac{1}{2} + 3\frac{2}{5}$

3-9   **31.** Subtract: $4\frac{1}{4}$
$2\frac{1}{3}$

3-10   **32.** Write $6\frac{2}{3}$ as an improper fraction.

3-11   **33.** Multiply: $3\frac{7}{8} \times 2$

3-12   **34.** Divide: $\frac{3}{4} \div 6$

3-13   **35.** What part of 32 is 20?

3-14   **36.** $\frac{9}{20}$ of what number is 36?

3-15   **37.** Write $\frac{11}{16}$ as a decimal (2 places).

3-16   **38.** Write .65 as a fraction.

3-17   **39.** Find the ratio of 15 to 35.

3-18   **40.** Solve and check: $\frac{n}{56} = \frac{9}{14}$

4-1   **41.** Write sixty hundredths as: a percent, a decimal, and a fraction.

4-2   **42.** Express 7% as a decimal.

4-3   **43.** Express .49 as a percent.

4-4   **44.** Express 20% as a fraction.

4-5   **45.** Express $\frac{3}{25}$ as a percent.

4-6   **46.** Find 35% of 560.

4-7   **47.** 52 is what percent of 64?

4-8   **48.** 75% of what number is 60?

Achievement Test II (BASIC SKILLS)

1-1 **1.** Read the numeral 7,015,673 or write it in words.

1-2 **2.** Write the complete numeral naming: 816 million

1-3 **3.** Round 496,280,074 to the nearest million.

1-4 **4.** Add: 15,976
2,734
38,592
21,085
6,588

1-5 **5.** Subtract: 178,346
77,482

1-6 **6.** Multiply 3,600 by 807.

1-7 **7.** Divide: 4,843,630 ÷ 605

1-8 **8.** Name the greatest common factor of 36, 54, and 144.

2-1 **9.** Read the numeral 740.835 or write it in words.

2-2 **10.** Write as a decimal numeral: Five hundred and six thousandths.

2-3 **11.** Round to nearest cent: $4.2046

2-4 **12.** Add: $1.93 + .541 + 43.8$

2-5 **13.** Subtract: $452.46
189.57

2-6 **14.** Which is greater: .92 or .919?

2-7 **15.** Multiply 6.42 by .25.

2-8 **16.** Divide: $.04\overline{)\,.00012}$

2-9 **17.** What decimal part of 64 is 40?

2-10 **18.** .375 of what number is 900?

2-11 **19.** Multiply: 100×250

2-12 **20.** Divide: $18.53 \div 1{,}000$

2-13 **21.** Write 532.7 billion as a complete numeral.

2-14 **22.** Write the shortened name for 594,200,000 in millions.

3-1 **23.** Express $\frac{63}{105}$ in lowest terms.

3-2 **24.** Rewrite $\frac{21}{12}$ as a mixed number.

3-3 **25.** Raise to higher terms: $\frac{2}{3} = \frac{?}{27}$

3-4 **26.** Are $\frac{38}{57}$ and $\frac{32}{48}$ equivalent fractions?

3-5 **27.** Which is less: $\frac{9}{10}$ or $\frac{11}{12}$?

3-6 **28.** Find the least common multiple of 2, 5, and 15.

3-7 **29.** Find the least common denominator of the fractions; $\frac{2}{3}, \frac{3}{4}$, and $\frac{11}{12}$

3-8 **30.** Add: $\frac{3}{4}$
$2\frac{5}{16}$
$4\frac{19}{32}$

3-9 **31.** Subtract $3\frac{3}{8}$ from 4.

3-10 **32.** Write $4\frac{5}{16}$ as an improper fraction.

3-11 **33.** Multiply: $2\frac{1}{3} \times 3\frac{1}{7}$

3-12 **34.** Divide: $\frac{15}{16} \div 1\frac{1}{2}$

3-13 **35.** 60 is what part of 100?

3-14 **36.** $\frac{1}{6}$ of what number is 17?

3-15 **37.** Write $\frac{3}{7}$ as a decimal (2 places).

3-16 **38.** Write $.37\frac{1}{2}$ as a fraction.

3-17 **39.** Find the ratio of 72 to 48.

3-18 **40.** Solve and check: $\frac{21}{45} = \frac{n}{30}$

4-1 **41.** Write ninety-four hundredths as: a percent, a decimal, and a fraction.

4-2 **42.** Express 4.5% as a decimal.

4-3 **43.** Express 1.05 as a percent.

4-4 **44.** Express 175% as a mixed number.

4-5 **45.** Express $\frac{24}{300}$ as a percent.

4-6 **46.** Find 140% of $30.

4-7 **47.** What percent of $7.00 is $1.75?

4-8 **48.** $6\frac{1}{2}$% of what number is 26?

Achievement Test III (BASIC SKILLS)

1-1   **1.** Read the numeral 5,965,408,000 or write it in words.

1-2   **2.** Write the numeral naming: Six billion, four hundred million, ten thousand

1-3   **3.** Round 147,492,350 to the nearest hundred thousand.

1-4   **4.** Add: 93,574
69,898
74,379
58,486
96,798

1-5   **5.** From 823,849 subtract 812,958.

1-6   **6.** Find the product of 6,009 and 7,908.

1-7   **7.** Divide 72,576 by 1,728.

1-8   **8.** Name the greatest common factor of 48, 80, and 112.

2-1   **9.** Read the numeral 583.29 or write it in words.

2-2   **10.** Write as a decimal: Four hundred seventy-one thousandths.

2-3   **11.** Round to nearest ten-thousandth: .00385

2-4   **12.** Add: $8.24 + $5.62 + $.76 + $10.08

2-5   **13.** Subtract .02 from .2.

2-6   **14.** Is the statement .201 > .21 true?

2-7   **15.** Find $1\frac{3}{4}$ of $6.25 (to nearest cent).

2-8   **16.** Divide 27 by 1.5.

2-9   **17.** $3.50 is what decimal part of $5?

2-10   **18.** 78 is .12 of what number?

2-11   **19.** Multiply: 1,000 × 48.21

2-12   **20.** Divide: 37.2 ÷ 10

2-13   **21.** Write 17.594 billion completely as a numeral.

2-14   **22.** Write the shortened name for 4,129,000,000,000 in trillions.

3-1   **23.** Express $\frac{1400}{3600}$ in lowest terms.

3-2   **24.** Rewrite $\frac{51}{16}$ as a mixed number.

3-3   **25.** Raise to higher terms: $\frac{13}{16} = \frac{?}{64}$

3-4   **26.** Are $\frac{65}{78}$ and $\frac{30}{36}$ equivalent fractions?

3-5   **27.** Is the statement $\frac{9}{32} < \frac{5}{16}$ true?

3-6   **28.** Find the least common multiple of 6, 9, and 24.

3-7   **29.** Find the least common denominator of the fractions: $\frac{3}{5}$, $\frac{1}{4}$, and $\frac{2}{3}$

3-8   **30.** Add: $4\frac{1}{8} + \frac{7}{10}$

3-9   **31.** Subtract: $8\frac{3}{16} - \frac{7}{8}$

3-10   **32.** Write $3\frac{7}{12}$ as an improper fraction.

3-11   **33.** Multiply: $1\frac{1}{2} \times 2\frac{1}{4} \times 1\frac{1}{3}$

3-12   **34.** Divide: $(3\frac{3}{4} \div 2\frac{1}{2}) \div \frac{3}{4}$

3-13   **35.** What part of 184 is 115?

3-14   **36.** $\frac{9}{16}$ of what number is 45?

3-15   **37.** Write $\frac{25}{30}$ as a decimal (2 places).

3-16   **38.** Write .016 as a fraction.

3-17   **39.** Find the ratio of 20 to 45.

3-18   **40.** Solve and check: $\frac{80}{48} = \frac{15}{n}$

4-1   **41.** Write eighteen and one-half hundredths as: a percent, a decimal, and a fraction.

4-2   **42.** Express 150% as a decimal.

4-3   **43.** Express .046 as a percent.

4-4   **44.** Express 8% as a fraction.

4-5   **45.** Express $\frac{8}{18}$ as a percent.

4-6   **46.** Find 8.3% of $15 (nearest cent).

4-7   **47.** 12 is what percent of 6?

4-8   **48.** 125% of what number is 70?

UNIT 2

Problem Solving and Consumer Applications

Inventory Test

5-1 **1.** What is the question in the following problem?

Kim earns $19,750 per year and her brother earns $24,500 per year. Find out how much less Kim earns per year.

2. Are enough facts given to solve the following problem? If not, which fact is missing?

George cut a piece of wood into 4 equal lengths. How long is each piece?

3. Which fact is not needed to solve the following problem?

Ruth had $1,219.45 in her bank account. If she withdrew $100 from her account and bought a dress for $49.98, how much does she now have in her bank account?

5-2 **4.** What operation is needed to solve the above problem 3?

5. Express your plan in words as a verbal sentence using the facts given and the question asked.

Given facts: XB10 computer costs $957.75; RC5 computer costs $1,094.50. Question Asked: How much more does the RC5 computer cost?

6. Express the following verbal mathematical sentence as an equation:

Twelve percent of some amount is equal to $720.

7. Write an equation that represents your plan of solution in the following:

An apple orchard has 24 rows of trees with 15 trees in each row. Find the total number of trees.

5-3 **8.** First estimate the answer, then perform the indicated operation: $239,000 \div 59$

5-4 **9.** Select the estimate that makes the answer reasonable.

The answer 4,470 is reasonable if the estimate is:
a. 10,000   **b.** 1,000   **c.** 5,000

5-5 **10.** If a piece of metal tubing $8\frac{3}{4}$ feet long is cut into 5 equal pieces, what is the length of each piece? Disregard waste.

5-6 **11.** Write the question which when answered will provide the missing fact that is needed to solve the following problem:

The enrollment at the local school is 520. If 60% of the enrollment are girls, how many boys are enrolled?

5-7 **12.** If a parking garage charges $1.50 for the first hour and $.75 for each additional hour or fractional part of an hour, what are the charges for parking 5 hours?

6-1 **13.** How much does Ricardo earn for a 42-hour week at $6.60 per hour and time and a half for overtime (over 40 hours)?

6-2 **14.** Ling receives 2% commission on sales. What is her commission if she sold $9,283 worth of merchandise?

6-3 **15.** A merchant buys a Z10 computer for $840 and sells it at a mark-up of 30% on cost. How much profit does he make?

6-4 **16.** Use the table of withholding tax (p. 287) to compute the amount of income tax to be withheld when a married person's weekly wages are $402.50 and 4 allowances are claimed.

6-5 **17.** Ann Williams earns $345 a week. How much social security tax does she pay if the social security tax rate is 7.05% of earnings?

6-6 **18.** If the city wage rate is $2\frac{1}{2}$%, how much wage tax is paid by a person earning $269 weekly?

6-7 **19.** Use the table of withholding tax (p. 287) and the social security tax rate of 7.05% to find the take-home pay when a married person's weekly wages are $375 and 2 allowances are claimed.

6-8 **20.** Use the Federal tax table (p. 294) to find how much Federal income tax you still owe if you are single, your taxable income is $14,225 and your withholding tax payments are $1,580.

7-1 **21.** What is the cost of 4 cans of tuna fish at $1.29 per can.

7-2 **22.** Find the unit price (cost per quart) of a pint of vinegar costing 49¢.

7-3 **23.** Find the sale price of a bicycle that regularly sells for $129.95 and is now reduced 20%.

7-4 **24.** Julia bought a pair of shoes for $40.98. How much change should she get from $50?

7-5 **25.** The meal check for Tim's party at the restaurant amounted to $146.50. If a 6% sales tax is added on, how much is the total bill?

7-6 **26.** The electric meter reading at the end of the previous month was 3569 and this month 3724. At 10.1¢ per kilowatt hour, what is the cost of the electricity consumed?

7-7 **27.** Julio's father traveled 1,158 miles by automobile using 60 gallons of gasoline. How many miles per gallon did the automobile average?

7-8 **28.** Laura's mother could purchase a refrigerator for $499 cash or pay $49 down and 12 equal payments of $42.75. How much can she save by paying cash?

7-9 **29.** If the monthly finance charge rate is $1\frac{1}{4}\%$, what is the annual percentage rate?

7-10 **30.** Find the cost per year of $60,000 fire insurance if the annual rate is $.26 per $100.

7-11 **31.** How much will $50,000 of 20-payment life insurance cost at the rate of $28.06 per $1,000?

7-12 **32.** Fred's house was purchased for $73,500 and is assessed for $56,500. If the tax rate is $6.25 per $100, what are the property taxes?

7-13 **33.** Betty's sister bought her house costing $81,000. She paid 20% down and gave a mortgage for the balance. What is the amount of the mortgage?

8-1 **34.** Write a check payable to the Town Center Store in the amount of $418.45 for the purchase of a television set. Use today's date.

8-2 **35.** What is the amount due on $5,000 borrowed at 12% annual interest and repaid at the end of 3 years?

8-3 **36.** What is the actual amount of money the borrower receives and how much is the monthly payment if the amount of the note of a discount loan is $2,500, the interest is 12%, and the period of the loan is 12 months?

8-4 **37.** Using the table on p. 343, find the total finance charge if you borrow $3,500 for 30 months to finance your car.

8-5 **38.** Find the total interest that is paid on a loan of $25,000 at 13% annual interest and amortized in 15 years. Use the table on p. 344.

8-6 **39.** Using the table on p. 346, find the amount and the interest earned on $5,000 deposited for 4 years at 8% compounded semi-annually.

8-7 **40.** Excluding broker's fees, how much profit do you make if you buy 200 shares of stock at $13\frac{3}{4}$ and sell them at $16\frac{1}{2}$?

41. What is the cost of ten bonds, face value $1,000 each, if they are quoted to sell at $106\frac{1}{2}$?

42. How much interest does the Federal Treasury send semi-annually to a person who owns $20,000 in T-notes bearing 11.25% annual interest?

CHAPTER 5 Problem Solving

INTRODUCTION TO PROBLEM SOLVING

Below is an easy step-by-step approach to help you master the techniques of problem solving.

1 READ the problem carefully to find:

    a. The given facts.

    b. The question asked.

2 PLAN how to solve the problem.

    a. Choose the operation needed.

    b. Think or write out your plan.

    c. Express as an equation.

3 SOLVE the problem.

    a. Estimate the answer.

    b. Find the solution.

4 CHECK.

    a. Check the accuracy.

    b. Compare answer to the estimate.

In this chapter, you will learn specific strategies that will help in applying the steps above.

In the algebra and geometry units, you will find techniques for solving algebraic and geometric problems.

Reading to Understand the Problem

STRATEGY

To help identify the given facts and the question asked, rewrite the problem in your own words.

PROBLEM: The highest point in Asia is Mount Everest, with an elevation of 8,848 meters. Mount McKinley, in Alaska, with an elevation of 6,194 meters, is the highest point in North America. What is the difference in their elevations?

REWRITE:
Mt. Everest is 8,848 m high.
Mt. McKinley is 6,194 m high.
How much higher is Mt. Everest?

GIVEN FACTS:
Elevation of Mt. Everest is 8,848 m.
Elevation of Mt. McKinley is 6,194 m.

QUESTION ASKED:
What is the difference in elevation?

STRATEGY

To find if all the given facts are needed or if a needed fact is missing, ask yourself: Do I have all the information to answer the question?

PROBLEM: An airplane took off from Boston at 8:45 A.M. and arrived in Atlanta in 2 hours. What is the average rate of speed in miles per hour?

MISSING FACT:
The distance in miles
from Boston to Atlanta.

FACT NOT NEEDED:
Took off at 8:45 A.M.

DIAGNOSTIC TEST

1 Rewrite the following problem in your own words:
A heating engineer, when installing an oil burner, finds it necessary to use pieces of pipe measuring $4\frac{11}{16}$ inches, $7\frac{5}{8}$ inches, $3\frac{1}{2}$ inches, and 9 inches. What length of pipe does he need to be able to cut the four pieces, disregarding waste?

2 Find the given facts in the following:
How long was a telephone call if the charge between the two cities was $2.54 at the rate of $.38 for the first minute and $.27 for each additional minute?

3 Write a question for the following:
During spring planting, a farmer planted 135 rows of corn with 24 plants in each row.

4 What is the question in the following?
Find the sale price of a computer which regularly sells for $849.95 but can now be purchased at a 20% reduction.

5 Which fact is not needed to solve the following problem?
The Bells live 16.5 miles from the airport. It takes them 25 minutes to get there. They live 26.4 miles from the nearest railroad station. How much closer do they live to the airport?

6 Are enough facts given to solve the problem? If not, which fact is missing?
Lisa's piano lesson lasted $1\frac{1}{2}$ hours this week. How much longer was her piano lesson last week?

RELATED PRACTICE PROBLEMS

1 Rewrite each of the following problems in your own words:

  1. The Lawrence family drove 424 miles to their seashore home. If it took 8 hours to drive this distance, what was the average rate of speed?

  2. At the 3-game home series, the local baseball team played before 14,526 spectators for the first game, 9,277 spectators for the second game, and 12,965 spectators for the third game. Find the total attendance.

  3. Ten years ago the census showed that the population of a nearby town was 3,684 people. A recent census reveals that the town population is 10,069 people. Find the amount of increase in population.

2 Find the given facts in each of the following:

  1. A ribbon 42 inches long is cut into 3 equal pieces. How long is each piece?

  2. The previous closing price of the BGX was $28\frac{1}{2}$. If it gained $1\frac{1}{4}$ points for the day, what was the new closing price?

  3. How much will a taxi ride cost if you travel 2 miles at the cost of $.90 for the first $\frac{1}{4}$ mile and $.50 for each additional $\frac{1}{4}$ mile?

3 Write a question for each of the following:

1. A train travels for 6 hours at the average speed of 50 m.p.h.

2. Tom saved $5 of his $15 weekly allowance.

3. Bob purchased 20 rugs, each measuring 9 ft by 12 ft, for $139 each.

4 What is the question in each of the following?

1. Juan earns $6.80 an hour. If he works 35 hours each week, how much is his weekly salary?

2. Teresa bought a bicycle that regularly sells for $125 at a 20% reduction sale. How much did she save?

3. The Metroliner travels 68.4 miles from Wilmington to Baltimore at an average speed of 94.5 m.p.h. How long does it take the Metroliner to travel this distance?

5 Which fact is not needed to solve each of the following problems?

1. Gary plans to see a movie at the 8:15 P.M. showing. The movie lasts $1\frac{1}{2}$ hours. It is now 6:30 P.M. How much time does he have to get to the movie theater?

2. Charlotte's station wagon holds 18 gallons of gasoline. She drove 126 miles in one day. How many gallons are in the tank when the gauge shows $\frac{1}{2}$?

3. Of the 823 students enrolled, 155 students are going on a bus trip. Each bus holds only 30 students. How many buses will be needed?

6 Are enough facts given to solve the problem? If not, which fact is missing?

1. At the Todd family's garage sale, they sold $\frac{2}{3}$ of the items that were available. How many items were sold?

2. John and David are running for student council president at the Evan school. John received 923 votes. How many more votes did David receive?

3. The Jordans bought 8 flowering bushes at $6.95 per bush and 12 tomato plants. What was the total amount that the Jordans spent?

SKILL ACHIEVEMENT TEST

1. Are enough facts given to solve the problem? If not, which fact is missing?
The utility bill for May was $76.83. Find the amount of increase during June.

2. Write a question for the following: At the end of the model year, an automobile sold for $8,590. The reduction was 20% off the regular price.

3. Find the given facts in the following: Mike needs 138 flooring tiles for the kitchen. There are 6 tiles in each box. How many boxes should he buy?

4. What is the question in the following? Mrs. Watson cut $6\frac{3}{8}$ yards from a piece of fabric $18\frac{2}{3}$ yards long. How much uncut fabric remains?

5. Which fact is not needed to solve the following problem?
A school athletic field has 32 sections each seating 40 persons. There are 2,540 students enrolled at the school. What is the total seating capacity at the field?

FLASHBACK

2 PLAN how to solve the problem.
  a. Choose the operation needed.
  b. Think or write out your plan.
  c. Express as an equation.

STRATEGY

Look for operation indicators to help choose the operation needed.

PROBLEM: At the supermarket Selma bought ground meat in packages weighing 1.65 lb, 2.14 lb, and 1.85 lb. What is the total weight of the ground meat?

OPERATION: Addition

When combining two or more quantities of *unequal* (or equal) size to find a *sum* or *total,* use addition.

PROBLEM: On Friday the Dow Jones Average increased from 1,191.45 to 1,200.93. What is the amount of increase?

OPERATION: Subtraction

When finding the *difference* between two quantities or how many more or less or the amount of *increase* or decrease, use subtraction.

PROBLEM: Julio bought 3 shirts, each costing $11.49. What is the total cost?

OPERATION: Multiplication

When combining two or more quantities of *equal* size to find a *total,* use multiplication.

PROBLEM: A 15-foot board is cut into 6 equal pieces. What is the length of each piece?

OPERATION: Division

When separating a given quantity into a number of groups (or parts) of equal size, use division.
—Divide the given quantity by the *number of groups* to find the *size of the group.*
—Divide the given quantity by the size of the group to find the *number of groups.*

STRATEGY

Use simpler numbers in place of the given numbers to help find the operation needed.

PROBLEM: Just off the coast of California is a submarine mountain called San Juan Seamount. How far below the surface of the water is the top of the mountain if the ocean floor depth is 3,659 meters and the mountain rises 3,105 meters?

Use 100 meters instead of 3,659 meters, and 90 meters instead of 3,105 meters.

$$100 - 90 = ?$$

OPERATION: Subtraction

STRATEGY

In formalizing your plan, express in words the relationship between the given facts and the question asked.

PROBLEM: Given facts:
The distance between two cities is 204 miles. The car travels at an average speed of 48 m.p.h.

Question asked:
How long does it take the car to travel between cities?

Operation needed:
Division.

ANSWER: The distance (204 miles) *divided by* the average speed (48 m.p.h.) is equal to the time of travel.

STRATEGY

After you have expressed the word relationship between the given facts and the question asked, translate it into symbols.

EXAMPLES: The sum of eight and five | is equal to | thirteen.

$$8 + 5 \qquad\qquad = \qquad 13$$

ANSWER: $8 + 5 = 13$

The difference between ten and four | is equal to | some number n.

$$10 - 4 \qquad\qquad = \qquad\qquad n$$

ANSWER: $10 - 4 = n$

Write a word relationship and then an equation that represents your plan of solution to the following problem.

PROBLEM: At $.29 per pound, what is the cost of 3 pounds of onions?

ANSWER: Word relationship: Number of pounds (3) of onions times the cost per pound ($.29) is equal to the total cost (n).

Equation: $3 \times \$.29 = n$

DIAGNOSTIC TEST

Determine the operation needed to solve each of the problems 1-4:

1 Tony repaid his loan by paying $97.50 per month for 12 months. Find the total amount paid.

2 Ben practiced playing the piano $2\frac{1}{2}$ hours, $1\frac{3}{4}$ hours, 2 hours, and $1\frac{1}{2}$ hours last week. How many hours did he practice in all?

3 Denise's father purchased a house costing $63,500. He paid $12,700 as a down payment. What is the balance owed?

4 A Jet helicopter flew 420 miles in 6 hours. What was its average rate of speed?

5 Use simpler numbers to determine the operation needed to solve the following problem:
During a recent year 666,081,544 nickels, 1,062,188,584 dimes, and 16,725,504,368 pennies were manufactured by the mints. What was the total number of coins manufactured?

6 Express your plan in words as a verbal mathematical sentence, using the facts given and the question asked. Choose the needed operation first.
Given facts: skirt costs $24.95; blouse costs $19.90
Question asked: What is the total cost?

7 Write the number sentence symbolically:
The product of nine and eight is equal to seventy-two.

Write each of the following as an equation:

8 Fourteen increased by nine is equal to some number *n*.

9 Six hundredths of some amount (*a*) is equal to $30.

10 Express the following verbal mathematical sentence as an equation:
The distance (4,840 mi) a Boeing 727 flies divided by its average rate of speed
(605 m.p.h.) is equal to the time (number of hours) of flight *n*.

11 Write an equation that represents your plan of solution in the following problem:
Sara's sister purchased a new car selling for $8,750. If she received $1,990 for her
old car as a trade-in, what is the balance owed?

RELATED PRACTICE PROBLEMS

Determine the operation needed to solve each problem of sets 1-4:

1 **1.** There are 38 rows of 24 orange trees in the grove. How many trees are there
altogether?

  2. Andy's recipe makes $6\frac{1}{2}$ dozen cookies. If he doubles the recipe, how many
dozen cookies will he have?

  3. An airline has 12 flights between Atlanta and Boston each day. If each flight
has the capacity to hold 226 passengers, what is the total number of
passengers that the airline could fly between the two cities daily?

2 **1.** Joe saved $8.95 when he paid $24.80 for a pair of slacks at a clearance sale.
What was the regular price?

  2. The interest rate on insured money market funds is 9.54% at the commercial
banks and .11% higher at the thrift banks. What is the rate at the thrift banks?

  3. How many miles in all did Florence drive if on Friday she drove 288 miles,
Saturday, 198 miles, and Sunday, 269 miles?

3 **1.** The Nile River is 4,145 miles long. The length of the Mississippi River is 2,348
miles. How much longer is the Nile River?

  2. The barometric pressure at New York City is 29.59 inches and at Montreal is
29.94 inches. Find the difference in pressures.

  3. How much profit does a merchant make when she sells for $449.95 a TV set
that cost her $351.29?

4 **1.** Ray cut a piece of wood 20 feet long into 8 equal pieces. How long is each piece?

  2. Five pounds of lamb chops cost $16.45. What was the cost per pound?

  3. Maria's mother drove her car a distance of 429 miles on 26 gallons of gasoline.
How many miles per gallon did her car average?

5 Use simpler numbers to determine the operation needed to solve each of the following problems:

1. There were 23,595,102 half-dollar coins minted one year. What is the total value of these coins?

2. The Three Rivers Stadium in Pittsburgh seats 50,350 people while the seating capacity of Candlestick Park in San Francisco is 61,185 people. How many more people can be seated in Candlestick Park?

3. A corporation earned $9,600,000 last year. On the basis of 1,500,000 shares of stock, what are the earnings per share of stock?

4. The Pacific Ocean has an area of 64,186,300 sq. mi, Atlantic Ocean, 33,420,000 sq mi, Indian Ocean, 28,350,500 sq. mi and Arctic Ocean, 5,105,700 sq. mi. What is the total area of these four major bodies of water?

6 Express your plan in words as a verbal mathematical sentence using the given facts and the question asked. Choose the needed operation first.

1. Given facts: 979 girls and 896 boys are enrolled at school.
   Question asked: How many more girls than boys are enrolled?

2. Given facts: 6 reams of paper were purchased; each ream costs $8.95.
   Question asked: What is the total cost of the paper?

3. Given facts: Distance is 338 miles; average speed of car is 52 m.p.h.
   Question asked: How long does it take to drive this distance?

4. Given facts: Five test scores are: 87, 74, 93, 81, and 70
   Question asked: What is the total score?

7 Write each of the following number sentences symbolically:

1. The difference between twelve and seven is equal to five.

2. The quotient of twenty-seven divided by nine is equal to three.

3. The sum of fifteen and forty-two is equal to fifty-seven.

4. Twenty times eight is equal to one hundred sixty.

8 Write each of the following as an equation:

1. The product of twenty-five and six is equal to some number *n*.

2. Sixty divided by fifteen is equal to some number *n*.

3. Eighty-six decreased by forty-nine is equal to some number *x*.

4. Twelve more than thirty is equal to some number *y*.

9 Write each of the following as an equation:

1. Nine percent of some amount is equal to $225.

2. Fifteen percent of some amount is equal to $12.

3. Thirty-five hundredths of some number is equal to 105.

4. One-third of some number is equal to 37.

10 Express each of the following verbal mathematical sentences as an equation:

1. The length of a pipe (60 in.) divided by the number of equal parts (8) is equal to the length of each piece (n).

2. The gain for the day ($1\frac{3}{4}$ points) added to the previous closing price (27) is equal to the new closing price (p).

3. The average rate of speed (45 m.p.h.) times the travel time (5 hours) is equal to the distance traveled by the train (d).

4. The regular price of a dishwasher ($395) less the reduction ($72.50) is equal to the sale price (s).

11 Write an equation that represents your plan of solution in each problem:

1. The school auditorium has 24 rows of seats. In each row there are 18 seats. Find the total number of seats.

2. Cory Ann bought a baseball glove for $19.95 at a sale. If she saved $8.50, what was the regular price?

3. A bus traveled 325 miles in $7\frac{1}{2}$ hours. What was its average rate of speed?

4. Diane's brother purchased a house for $74,250. He paid $19,250 as a down payment. What is the balance owed?

SKILL ACHIEVEMENT TEST

Determine the operation needed to solve each of the problems 1-4.

1. A Beechcraft 99 flew 1,960 miles in 7 hours. What is its average rate of speed?

2. The Dow Jones Industrial average closed yesterday at 1,198.14. It gained 10.16 points today. What is the new closing price?

3. The mean annual snowfall record is 105.8 inches in Juneau, Alaska, and 113 inches in Sault Ste. Marie, Michigan. How many inches less fell in Sault Ste. Marie?

4. Tim works 38 hours per week. If his hourly wage is $7.50 per hour, how much does he earn each week?

5. Write symbolically: Forty divided by eight is equal to five.

6. Write as an equation: Four less than sixty is equal to some number n.

7. Express your plan in words as a verbal mathematical sentence, using the facts given and the question asked. Choose the needed operation first.
Given facts: 16-foot board cut into 2 pieces. One piece measures $9\frac{1}{2}$ feet.
Question asked: What is the length of the second piece?

8. Express the following verbal mathematical sentence as an equation:
The number of hours worked (35 hours) per week multiplied by the hourly rate ($7.40) is equal to the weekly wage (w).

9. Write an equation that represents your plan of solution to the following problem:
How much change should you get if you bought a shirt for $14.90 and offered a $20 bill in payment?

5-3 Solving the Problem

FLASHBACK	STRATEGY

FLASHBACK

3 SOLVE the Problem
a. Estimate the answer.
b. Find the solution.

STRATEGY

Round the numbers to be operated on in order to estimate the answer before finding the solution.

In each of the following models, *first* estimate, then find the solution.

PROBLEM: 683 + 2,907

$$683 \rightarrow 700$$
$$+ 2,907 \rightarrow + 2,900$$
3,590, solution   3,600, estimate

ANSWER: 3,590

To estimate the sum of two or more whole numbers or decimals:

Round the numbers to the same place. Usually this is the greatest place in the smallest number.
Then add.

PROBLEM: .245 + .587

$$.245 \rightarrow .2$$
$$+ .587 \rightarrow + .6$$
.832, solution   .8, estimate

ANSWER: .832

PROBLEM: 48,305 − 29,219

$$48,305 \rightarrow 50,000$$
$$- 29,219 \rightarrow - 30,000$$
19,086, solution   20,000, estimate

ANSWER: 19,086

To estimate the difference of two whole numbers or decimals:

Round the numbers to the same place. Usually this is the greatest place in the smaller number.
Then subtract.

PROBLEM: .94 − .637

$$.940 \rightarrow .9$$
$$- .637 \rightarrow - .6$$
.303, solution   .3, estimate

ANSWER: .303

PROBLEM: 39 × 615

$$39 \rightarrow 40$$
$$\times 615 \rightarrow \times 600$$
23,985, solution   24,000, estimate

ANSWER: 23,985

To estimate a product of two whole numbers or decimals:

Round each factor to its greatest place.
Then multiply.

PROBLEM: 2.1 × .682

$$
\begin{array}{ccc}
2.1 & \rightarrow & 2 \\
\times\ .682 & \rightarrow & \times\ .7 \\
\hline
1.4322,\ \text{solution} & & 1.4,\ \text{estimate}
\end{array}
$$

ANSWER: 1.4322

PROBLEM: 1,760 ÷ 280

$$1,760 \div 280 = 6\tfrac{2}{7},\ \text{solution}$$
$$\downarrow \qquad \downarrow$$
$$1,800 \div 300 = 6,\ \text{estimate}$$

ANSWER: $6\tfrac{2}{7}$

PROBLEM: 24.19 ÷ .59

$$24.19 \div .59 = 41,\ \text{solution}$$
$$\downarrow \qquad \downarrow$$
$$24 \div\ .6 = 40,\ \text{estimate}$$

ANSWER: 41

To estimate a quotient of two whole numbers or two decimals:

Round the divisor to its greatest place.
Round the dividend so that it can be divided exactly by the rounded divisor.
Then divide.

DIAGNOSTIC TEST

In each of the following, first estimate the answer, then perform the indicated operation.

1 72 × 28   **2** 639 + 795 + 389   **3** 810 ÷ 59   **4** 8,960 − 3,012
5 .98 + .492   **6** 17 ÷ .3   **7** 62 × .98   **8** 8.51 − .293

RELATED PRACTICE PROBLEMS

In each of the following, first estimate the answer, then perform the indicated operation.

1 **1.** 605 × 98   **2.** 529 × 386   **3.** 406 × 9,117   **4.** 8,523 × 6,973
2 **1.** 916 + 789   **2.** 5,824 + 683   **3.** 945 + 279 + 382   **4.** 1,321 + 582 + 907
3 **1.** 2,094 ÷ 7   **2.** 4,131 ÷ 51   **3.** 39,758 ÷ 198   **4.** 544,849 ÷ 891
4 **1.** 8,529 − 3,702   **2.** 7,811 − 968   **3.** 21,000 − 12,063   **4.** 56,302 − 9,294
5 **1.** .61 + .58   **2.** 9.2 + .893   **3.** .17 + .6 + .502   **4.** 8.5 + 21.4 + 1.29
6 **1.** .99 ÷ 4   **2.** 59.84 ÷ .5   **3.** .79 ÷ .2   **4.** 3.2 ÷ .6
7 **1.** .37 × .42   **2.** 6.4 × .018   **3.** 61 × $8.85   **4.** .09 × $59.75
8 **1.** .9 − .637   **2.** 94 − 59.9   **3.** $81.05 − $48.99   **4.** 1.4 − .77

SKILL ACHIEVEMENT TEST

In each of the following, first estimate the answer, then perform the indicated operation.

1. 7,192 ÷ 89   **2.** 4.6 + .853   **3.** 79,698 − 47,715   **4.** 7.9 × 3.16
5. .82 − .517   **6.** 1.493 × 704   **7.** 953 + 826 + 1,294   **8.** 15.7 ÷ .82

5-4 Checking the Answer

STRATEGY

Check the reasonableness of your answer. An answer is reasonable if it is within the range of closeness to the estimate.

PROBLEM: Is the answer 95 reasonable if the estimate is 100?

ANSWER: Yes

The answer 95 is reasonable if the estimate is 100.

PROBLEM: Is the answer 95 reasonable if the estimate is 1,500?

ANSWER: No

The answer 95 makes no sense if the estimate is 1,500.

DIAGNOSTIC TEST

Select the letter corresponding to the estimate that makes the answer reasonable.

1 The answer 189 is reasonable if the estimate is: **a.** 50   **b.** 200   **c.** 1,000

2 The answer 13.29 is reasonable if the estimate is: **a.** 135   **b.** 1.5   **c.** 15

RELATED PRACTICE PROBLEMS

Select the letter corresponding to the estimate that makes the answer reasonable.

1   **1.** The answer 67 is reasonable if the estimate is: **a.** 25   **b.** 110   **c.** 70

     2. The answer 346 is reasonable if the estimate is: **a.** 400   **b.** 40   **c.** 4,000

     3. The answer 1,529 is reasonable if the estimate is: **a.** 1,000   **b.** 2,500   **c.** 1,600

2   **1.** The answer 2.25 is reasonable if the estimate is: **a.** 2.5   **b.** 20   **c.** 25

     2. The answer 44.9 is reasonable if the estimate is: **a.** 400   **b.** 500   **c.** 50

     3. The answer $219.67 is reasonable if the estimate is: **a.** $300   **b.** $200   **c.** $2,000

SKILL ACHIEVEMENT TEST

Select the letter corresponding to the estimate that makes the answer reasonable.

1. The answer 928 is reasonable if the estimate is: **a.** 90   **b.** 1,000   **c.** 100

2. The answer 39.8 is reasonable if the estimate is: **a.** 40   **b.** 400   **c.** 300

STRATEGY—GUESS AND TEST

In using this strategy, the important part is *and test.* Make an educated guess at the answer. Then, test it with the facts of the problem to determine how to improve your guess. Keep guessing and testing until the correct answer is found.

PROBLEMS

Use this strategy to find the answer in each of the following:
1. $\frac{2}{3}$ of what number is 24?
2. There are 7 more girls than boys in the science class of 31. How many boys are there?
3. If you add 10 years to twice my age, you get 40 years. What's my age?
4. What score do you need in a second test to average 80 if you scored 70 in the first test?
5. Your scores in the first two tests are 75 and 90. What score do you need in the third test to average 85?
6. What two consecutive numbers add up to 35?
7. First-class mail costs $.22 for the first ounce and $.17 for each additional ounce. If you were charged $1.58, what is the weight of the package?
8. How long is a telephone call if it cost $1.70 at the rate of $.30 for the first minute and $.20 for each additional minute?
9. How many prints were made if you paid $3.25 at the rate of $1.00 for processing the roll of film and $.15 per print?
10. What distance did you travel by taxi if the rate is $.90 for the first $\frac{1}{4}$ mile and $.50 for each additional $\frac{1}{4}$ mile and you paid $3.40?

11. How fast must you drive the second hour to average 50 m.p.h. for the two hours if your speed for the first hour averaged 47 m.p.h.?
12. Our baseball team won 82 games. What part of the 24 remaining games must the team win in order to wind up with 100 games won?

5-5 Solving Problems— One-Step Problems

PROBLEM: A merchant purchased 48 radios at $79 each. What was the total cost?

READ the problem carefully.
a. Find the given facts.
    48 radios were purchased; each radio costs $79.
b. Find the question asked.
    What was the total cost?

PLAN how to solve the problem
a. Choose the operation needed.
    The word *total* with *equal* quantities indicates *multiplication*.
b. Think or write out your plan relating the given facts to the question asked.
    The number of radios (48) times the cost of a radio ($79) is equal to the total cost.
c. Express as an equation.
    $48 \times \$79 = n$

SOLVE the problem.
a. Estimate the answer.
    $48 \rightarrow 50 \quad \$79 \rightarrow \$80$
    $50 \times \$80 = \$4,000$, estimate
b. Solution.
$$\begin{array}{r} \$79 \text{ cost per radio} \\ \times \ \underline{48} \text{ radios} \\ 632 \\ \underline{316 \ \ } \\ \$3,792, \text{ total cost} \end{array}$$

ANSWER: $3,792 total cost

CHECK.
a. Check the accuracy of your arithmetic.
$$\begin{array}{r} 48 \text{ radios} \\ \times \ \underline{\$79} \text{ cost per radio} \\ 432 \\ \underline{336 \ \ } \\ \$3,792 \text{ total cost } \checkmark \end{array}$$
b. Compare answer to the estimate.
    The answer $3,792 compares reasonably with the estimate, $4,000.

DIAGNOSTIC TEST

1 Using the following given facts, write a question that could be asked.
Kwon is 8 years old. His mother is 32 years old.

2 Solve: Mrs. Milligan planned a trip for her 23 students. The bus fare is $391. What was the cost per pupil?

RELATED PRACTICE PROBLEMS

1 Using the given facts in each of the following, write a question that could be asked.

1. Teresa weighs 75 pounds. Her brother weighs 150 pounds.

2. Joan's father bought a computer for $895. The sales tax is 4%.

3. A plane flew a distance of 1,250 miles. Its average speed was 375 m.p.h.

4. Jose works 40 hours per week. He receives $6.25 per hour.

2 Solve each of the following problems:

1. Eric bought $6\frac{1}{2}$ yards of fabric at $2.84 a yard. How much did the fabric cost?

2. The school auditorium seats 1,150 people. If there are 179 students in the graduating class, how many seats are available for guests at the graduation exercises?

3. The Dow Jones Industrial average gained 3.58 points today, closing at 1,207.63. What was the closing average the previous day?

4. A balloon flight covered 648 kilometers. If the balloon completed the trip in 36 hours, what was its average speed?

5. Susan's father bought a house for $58,000, paying 20% down. What is the amount of the down payment?

6. Donna had a balance of $468.85 in her checking account. If she deposits a check in the amount of $196.07 in her account, what is her new balance?

7. When buying a new automobile, Terry agreed to pay $165.95 per month for 48 months. What is the total amount of his payments?

8. If sound travels at a speed of 332 meters per second, how long does it take to hear an explosion 1,660 meters away?

SKILL ACHIEVEMENT TEST

Using the given facts in each of the following, write a question that could be asked.

1. There are 634 girls and 589 boys enrolled at the local school.

2. The Chicago Cubs won their divisional title with an average of .596 and the San Diego Padres won their title with an average of .568.

Solve each of the following problems:

3. Leon bought a VCR at a sale for $467.75. If the amount he saved was $131.25, what was the regular price?

4. On a science test Sharon had 7 times as many correct answers as incorrect answers. If she had 91 correct answers, how many incorrect answers did she have?

5. If a seafood diet caused Tim to lose 1.2 pounds each day, how much weight did he lose during the month of June?

PROBLEM: For lunch Harry bought a sandwich for $2.95, a beverage for $.45, and a dessert for $.75. How much change should he get back from $10?

READ the problem carefully.
a. Find the given facts.   Sandwich costs $2.95; beverage, $.45; dessert, $.75
b. Find the question.   How much change should he get from $10?
   Think: To find amount of change, you must first find total cost of lunch.
c. Hidden question: What is the total cost of the lunch?

PLAN how to solve the problem.
a. Choose the operations needed.
      Step 1 The word *total* with *unequal* quantities indicates *addition*.
            Use the given facts to find the missing fact (answer the hidden question needed to solve the problem.)
      Step 2 The words *how much . . . from* indicates *subtraction*.
b. Think or write out your plan.
      The amount of change is equal to the amount of money offered in payment (step 2) minus the total cost of the lunch (step 1).
      The two steps may be translated into a single number sentence.
c. Express as an equation. $n = \$10 - (\$2.95 + \$.45 + \$.75)$

SOLVE the problem.
a. Estimate the answer.
      Step 1 $2.95 + $.45 + $.75
                 ↓   ↓   ↓
              $3   + $0   + $1   = $4
      Step 2 $10 − $4 = $6, estimate

b. Solution.
      Step 1 $2.95 + $.45 + $.75 = $4.15
      Step 2 $10.00 offered
              −   4.15 cost of lunch
                  $5.85 change

ANSWER: $5.85 change

CHECK.
a. Check the accuracy of your arithmetic.
      $2.95 + $.45 + $.75 = $4.15 ✔
            $4.15 cost of lunch
         + 5.85 change
            $10.00 offered ✔

b. Compare answer to the estimate.
      The answer $5.85 compares reasonably with the estimate, $6.

DIAGNOSTIC TEST

1 Write the question which when answered will provide the missing fact that is needed to solve the following problem:

If a table sells for $575 and each of 6 chairs sells for $125, what is the total cost of the furniture?

2 Use the given facts to find the missing fact that is needed to solve the following problem:

Debbie works 20 hours per week at $5.50 per hour. How much did she earn in all during a week when she received a bonus of $48?

3 Solve:

The Silhouette Stationery Store had 1,764 loose stickers. They sold 524. The remaining stickers were placed in boxes each holding 10. How many boxes of stickers are there?

RELATED PRACTICE PROBLEMS

1 For each of the following problems, write the question which when answered will provide the missing fact that is needed to solve the problem:

1. Jill spent $17.95 for a blouse and $29.55 for a skirt. How much change should she receive from $50?

2. Bill needs to rent a tuxedo and shirt for his school's prom. Tuxedo Unlimited's rental price for these items is $45.50 and $6.00, respectively. Frank's Formal Wear offers everything at $54.95. Which rental is lower? How much lower?

3. A sofa costing $550 is marked to sell at a profit of 40% on the cost. What is the selling price?

2 For each of the following problems use the given facts to find the missing fact that is needed to solve the problem:

1. Harry wears basketball sneakers all year round. If he wears out a pair of sneakers in 13 weeks and they cost $39 a pair, how much does it cost to keep Harry in sneakers for the year?

2. A watch that regularly sells for $59.95 can be bought at a reduction of 25%. What is the sale price?

3. Tanya has scores of 75, 91, and 89 in three tests. What is her average score?

3 Solve each of the following problems:

1. Mr. Perkins has $1,248.73 in his checking account. During the week he deposited $897.67 to his account and wrote a check in the amount of $346.58. What was the new balance in his checking account?

2. Ronald loses an average of 4 balls per round of golf. If he plays twice a week and each golf ball costs $1.95, what is the cost of the golf balls lost each week?

3. A computer selling regularly for $1,298 is reduced 10%. What is its sale price?

4. Ralph's Pizza Shop had a total revenue of $1,398 this week. If his expenses were: salaries, $485; rent, $125; supplies, $292; and utilities, $140; how much profit did Ralph make?

5. If a car rental charges $20 per day and $.05 for each mile traveled, how much would it cost to keep the car one day and drive 175 miles?

6. A merchant wishes to sell a camera that cost him $69 at a 30% profit on the cost. What should the selling price be?

SKILL ACHIEVEMENT TEST

1. Use the given facts to find the missing fact that is needed to solve the following problem:
   Irma works 40 hours each week at $5.25 per hour. If she has weekly expenses of $60, how much does Irma clear each week?

2. Ed spent $35 in all at a rock concert. If he paid $6.25 for food and $9.95 for a concert T-shirt, how much did he spend for his ticket?

3. Sara's mother bought a house costing $52,000, paying 25% down and taking a mortgage for the rest. What is the amount of the mortgage?

4. Which job offers a higher salary, one that pays $19,450 per year or one that pays $1,650 each month?

5. You can buy a tire that costs regularly $58 at a 25% reduction. How much do you save on 4 tires?

STRATEGY—SOLVING ANOTHER WAY

Many problems can be approached and solved in different ways. Each approach can sometimes give you a new or better understanding of the problem situation.

By solving the problem below in different ways, you gain a better understanding of the relationship between the games won, games lost, and games played.

PROBLEM: The local baseball team won 75% of the 48 games it played. How many games did it lose?

<table>
<tr><td colspan="2" align="center">Solution 1</td><td colspan="2" align="center">Solution 2</td></tr>
</table>

Solution 1

75% = .75

```
   48 games played        48 games played
 ×  .75               −   36 games won
  240                     12 games lost
  336
36.00 games won
```

ANSWER: 12 games lost

Solution 2

100% − 75% = 25%
played − won = lost

```
25% = .25          48 games played
                 ×   .25
                   240
                    96
                 12.00 games lost
```

ANSWER: 12 games lost

PROBLEMS

Solve each of the following problems in two ways:

1. The regular price of a steel-belted radial tire is $65. What is its sale price if it is reduced 20%?

2. $\frac{2}{3}$ of the home economics class of 18 students are girls. How many boys are enrolled in the class?

3. A computer sells for $849. If the sales tax is 5%, what is the total cost?

4. Find the total amount owed if $1,200 is borrowed for 1 year at 15% annual interest.

STRATEGY—WORKING BACKWARDS

Some problems are stated in such a way that working backwards becomes an easier and faster way than starting from the beginning.

In the problem below, instead of working with an unknown value, working backwards turns the problem into a simple arithmetic calculation.

PROBLEM: If 3 is added to 4 times the number, the answer is 51. What is the number?
Reverse the operations: First subtract 3 from 51 to get 48. Then, divide 48 by 4.

ANSWER: 12

PROBLEMS

1. If you add 8 years to 3 times my age, you get 50 years. What is my age?

2. If you subtract 4 from 5 times the number, you get 36. Find the number.

3. If you divide the sum of 6 and a number by 2, the quotient is 5. What is the number?

4. If you multiply a number less 8 by 7, the product is 28. Find the number.

PROBLEM: Mrs. Romero can buy an electric sweeper for the cash price of $149.95 or pay the installment price of $15 down and $16.09 a month for 10 months. How much can she save by paying cash?

READ the problem carefully.
a. Find the given facts.
   Cash price is $149.95;
   Installment price is $15 down and $16.09 per month for 10 months.
b. Find the question.
   How much can be saved by paying cash?
   Think: To find amount saved by paying cash, you first must find installment price. To find installment price you first must find total amount of the monthly payments and add to it the amount of the down payment.
c. Hidden questions:
   What is the total amount of the monthly payments?
   What is the total installment price?

PLAN how to solve the problem.
a. Choose the operations needed.
   Step 1 The words *total amount* with *equal* quantities indicates *multiplication*.
   Step 2 The words *total installment price* indicates *addition*.
   Step 3 The words *how much . . . save* indicates *subtraction*.
b. Think or write out your plan.
   The number of payments times the monthly payment (step 1) plus the down payment (step 2) is equal to the installment price. The installment price (step 2) minus the cash price (step 3) is equal to the savings.
c. Express as an equation. $(10 \times \$16.09 + \$15) - (\$149.95) = n$
   installment price   cash price   savings

SOLVE the problem.
a. Estimate the answer.
   $(10 \times \$16.09 + \$15) - (\$149.95)$
   $(10 \times \ \ \$16 \ \ + \$15) - \ \ (\$150)$
   $= 160 + 15 - 150 = \$25$, estimate

b. Solution.
Step 1 10 × $16.09 = $160.90
Step 2 $160.90 + $15 = $175.90 installment price
Step 3 $175.90 installment price
         − 149.95 cash price
            $25.95 savings

CHECK
a. Check the accuracy of your arithmetic.

$16.09 × 10 = $160.90 ✔ $149.95 cash price
$15 + $160.90 = $175.90 ✔   +   25.95 savings
                                            $175.90 installment price ✔

b. Compare answer to the estimate.
The answer $25.95 compares reasonably with the estimate, $25.

ANSWER: $25.95 savings

DIAGNOSTIC TEST

1 What are the hidden questions which when answered will provide the necessary facts to solve the following problem?

A $35 vest is on sale at a 40% reduction. If the sales tax is 5%, what is the total cost of the vest?

2 Solve:
A dealer bought 24 softballs for $48. At what price must she sell each one to realize a profit of 35% on the cost?

RELATED PRACTICE PROBLEMS

1 What are the hidden questions which when answered will provide the necessary facts to solve each of the following problems?

1. Helen scored 82, 91, and 76 in three tests while Jules scored 75, 88, and 89 in the same tests. Who has the higher average and how much higher?

2. Bart found a $48.95 briefcase on sale at 20% off. He can buy the same briefcase at another store for $39.95. Which is the better buy?

3. What is the cost of a telephone call lasting 7 minutes if the rate is $.35 for the first minute and $.24 for each additional minute?

2 Solve each of the following problems:

1. David's team scored 348 points in the relay races and 589 points in the field events. Gloria's team scored 476 points and 469 points in the same events. Which team had the higher score and how many points higher?

2. A parking lot charges $1.25 for the first hour and $.75 for each additional hour. How much will it cost to park for 4 hours?

3. A VCR was bought by a merchant for $275 and marked to sell for $495. It was sold at a discount of 20% on the marked price. What was the amount of profit?

4. Ann's father can purchase an air conditioner for the cash price of $1,550 or $350 down and 12 equal monthly payments of $115 each. How much can be saved by paying cash?

5. Mrs. Jordan bought a suitcase priced at $79.95 and a tote bag for $15.85. She had to pay a 4% sales tax. How much change did she get back from $100?

SKILL ACHIEVEMENT TEST

1. What are the hidden questions which when answered will provide the necessary facts to solve the following problem?

The golf scores of 4 rounds ended as:

| Charles Boland | 71 | 72 | 69 | 71 |
| Harry Yates | 68 | 73 | 70 | 73 |

Who won and by how many strokes?

Solve each of the following problems:

2. How much does it cost to send by first-class mail a package weighing 12 ounces if the rate is $.22 for the first ounce and $.17 for each additional ounce?

3. Mr. Barnes purchased a new car for $7,488. His old car had a trade-in value of $3,988. If he pays $111.80 for 36 months for the balance, how much is he paying for finance charges?

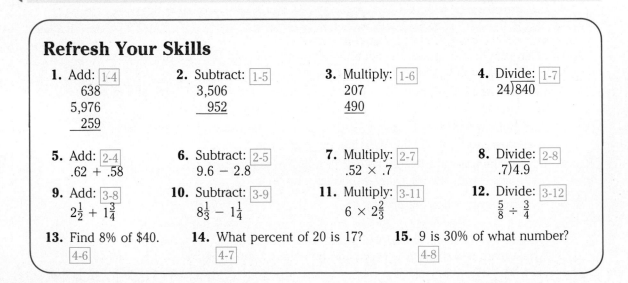

Refresh Your Skills

1. Add: 1-4
638
5,976
259

2. Subtract: 1-5
3,506
952

3. Multiply: 1-6
207
490

4. Divide: 1-7
24)840

5. Add: 2-4
.62 + .58

6. Subtract: 2-5
9.6 − 2.8

7. Multiply: 2-7
.52 × .7

8. Divide: 2-8
.7)4.9

9. Add: 3-8
$2\frac{1}{2} + 1\frac{3}{4}$

10. Subtract: 3-9
$8\frac{1}{3} - 1\frac{1}{4}$

11. Multiply: 3-11
$6 \times 2\frac{2}{3}$

12. Divide: 3-12
$\frac{5}{8} \div \frac{3}{4}$

13. Find 8% of $40.
4-6

14. What percent of 20 is 17?
4-7

15. 9 is 30% of what number?
4-8

STRATEGY—PATTERNS

Recognizing a pattern is sometimes easily determined by first constructing a table or making a diagram and then examining it. After you have recognized the pattern, you can extend the table or diagram to obtain the missing information.

PROBLEM: Find the number of diagonals that can be drawn from any one vertex of a polygon with 7 sides; 8 sides; 10 sides; and 12 sides.

Construct a table:

No. of sides	3	4	5	6	7	8	9	10	11	12
No. of diagonals	0	1	2	3	4	5	6	7	8	9

Pattern: number of sides − 3 = number of diagonals

ANSWER: 7 sides, 4 diagonals; 8 sides, 5 diagonals; 10 sides, 7 diagonals; 12 sides, 9 diagonals

PROBLEMS

1. The sum of the first two consecutive odd numbers is 4, of the first three consecutive odd numbers is 9, the first four is 16, the first five is 25.

$$1 + 3 = 4$$
$$1 + 3 + 5 = 9$$
$$1 + 3 + 5 + 7 = 16$$
$$1 + 3 + 5 + 7 + 9 = 25$$

What is the sum of the first six consecutive odd numbers? First 9? First 20? First 100?

2. The *Fibonacci Sequence* is a sequence of natural numbers.

$$1, 1, 2, 3, 5, 8, 13, 21, \ldots$$

Find the next five numbers in the sequence after 21.

3. *Pascal's Triangle* is a triangular arrangement of rows of numbers, each row increasing by one number. Each row, except the first, begins and ends in a 1 written diagonally as shown.

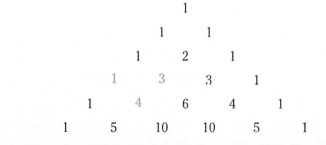

Find the numbers that belong in the next four rows. Hint: look for an addition pattern.

Career Applications

"I'm traveling to Zurich on business and my schedule is tight. Can you get me good flight connections and a hotel with rooms for meetings?"

"Our family can't afford to spend a lot of money on our vacation. We want a place where we can swim, climb mountains, and go to concerts. Can you find a good place and make all the arrangements?"

Travel agents solve problems like these every day. They have to know about airlines, hotels, and many other things travelers need. They use computers to get up-to-the-minute information on schedules and prices. A travel agent asks clients what they need, and helps them make decisions. Travel agents get reduced prices when they travel, but spend most of their time at their desks.

People who want to be travel agents must be patient and helpful. They should have taken high school courses in social studies and mathematics. Many travel agencies have training programs for agents.

1. The adult round-trip fare to Minneapolis is $270. A child can fly for $\frac{2}{3}$ the adult fare. What is the total fare for two adults and one child?

2. The regular round-trip fare to Tampa is $180. A traveler can save 25% by buying a ticket two weeks in advance. How much money can the traveler save?

3. A travel agent makes arrangements for four executives going to a conference. The airplane tickets cost $1,929.44, the hotel rooms cost $1,138.60, and car rental is $206.58. What is the total cost?

4. The base price of a 4-day vacation in New Orleans is $400. A tax and service charge of $60 is added. What percent of the base price is this?

CHAPTER REVIEW

The numerals in the boxes indicate the Lessons where help may be found.

1. Find the given facts in the following: 5-1

    The average speed of winds at Philadelphia were clocked at 9.6 m.p.h. and at San Francisco at 10.5 m.p.h. Find the difference in wind speeds.

2. What is the question in the following? 5-1

    Denise's father purchased food at the supermarket costing $27.34. How much change should he get from $30?

3. Are enough facts given to solve the problem? If not, which fact is missing? 5-1

    Carmela's mother bought a refrigerator that regularly sells for $950. She paid $100 down and $35 each month. How much does she save by paying cash?

4. Which fact is not needed to solve the following problem? 5-1

    A jogger ran 3.2 kilometers on Thursday, 4.6 kilometers on Friday, and 3.7 kilometers on Saturday. He slept 8.5 hours on Sunday. Find the total distance he ran.

Determine the operation needed to solve each of the following problems.

5. An aircraft carrier cruised for 36 hours at the average speed of 29 knots (nautical miles per hour). What distance did it travel? 5-2

6. Mrs. Johnson bought a veal roast that weighed 11.36 pounds and a chicken that weighed 6.29 pounds. How much more did the veal weigh? 5-2

7. Express your plan in words as a verbal mathematical sentence using the facts given and the question asked. Choose the needed operation first. 5-2

    Given facts: Sam works 35 hours each week. He is paid $8.40 per hour.
    Question asked: How much does he earn each week?

8. Express the following verbal mathematical sentence as an equation: 5-2

    The sum of the cost of the dress ($39) and the sales tax ($2.34) is equal to the total cost (t).

9. Write an equation that represents your plan of solution in the following problem: 5-2

    A roll of ribbon 72 inches long is cut into pieces each 4 inches long. How many pieces of ribbon are there?

10. First estimate the answer, then perform the indicated operation: 5-3 98×105

11. Select the estimate that makes the answer reasonable. 5-4

    The answer 25 is reasonable if the estimate is: **a.** 1,000 **b.** 30 **c.** 100

12. At a clearance sale Betty purchased a pair of jeans for $24.49. If she saved $4.50, what was the regular price of the jeans? 5-5

13. Write the question which when answered will provide the missing fact that is needed to solve the problem: 5-6

    Pedro's father purchased 10 gallons of gasoline at $1.089 per gallon. How much change should he get from $15?

14. An agent sold a house for $59,900. If her commission is 6%, how much money does the seller get? 5-6

15. If the telephone rate is $.28 for the first minute and $.16 for each additional minute, what would an 11 minute call cost? 5-7

COMPETENCY CHECK TEST

The numerals in the boxes indicate the Lessons where help may be found.

Solve each problem and select the letter corresponding to your answer.

Problem: It takes $2\frac{1}{2}$ hours for an airplane with the seating capacity of 205 passengers to fly from New York to Miami, a distance of 1,350 miles. What is the average rate of speed of the airplane?

1. The operation needed to solve the above problem is: 5-2

   a. addition   **b.** subtraction   **c.** multiplication   **d.** division

2. The fact not needed in the above problem is: 5-1

   a. distance: 1,350 miles   **b.** the time of flight: $2\frac{1}{2}$ hours
   c. seating capacity: 205   **d.** Answer not given
       passengers

Problem: A merchant sold a computer for $949. Find the amount of profit he made.

3. The question in the above problem is: 5-1

   a. How much did the merchant pay   **b.** What was the selling price?
       for the computer?
   c. How much profit did he make?   **d.** Answer not given.

4. The needed fact that is missing in the above problem is: 5-1

   a. amount of profit   **b.** age of merchant
   c. selling price of computer   **d.** cost of computer

5. The nearest estimate to the product of 31×48 is: 5-3

   a. 2,000   **b.** 1,500   **c.** 1,000   **d.** 500

6. The answer 383 is reasonable when the estimate is: 5-4

   a. 1,000   **b.** 750   **c.** 400   **d.** 100

7. There are 1,086 girls and 993 boys at the Midtown High School. The total enrollment is: 5-5

   a. 2,079   **b.** 1,999   **c.** 73   **d.** Answer not given

8. The missing fact needed to solve the problem: 5-6

A tennis racquet that regularly sells for $39.88 is now reduced 25%. What is the sale price?

   a. the regular price   **b.** amount of reduction
   c. the sale price   **d.** the percent of reduction

9. Harry has two pieces of wood, one piece measuring $5\frac{3}{4}$ feet and the other twice as long. The total length of the two pieces of wood is: 5-6

   a. $5\frac{3}{4}$ feet   **b.** $11\frac{1}{2}$ feet   **c.** $17\frac{1}{4}$ feet   **d.** 23 feet

10. A taxicab charges $1 for the first $\frac{1}{8}$ mile and $.15 for each additional $\frac{1}{8}$ mile. A taxicab ride of 4 miles would cost: 5-7

   a. $4.00   **b.** $4.15   **c.** $3.15   **d.** $5.65

CHAPTER 6   Income, Take-Home
Pay, Income Tax

Computing Income

HOURLY WAGES

In some occupations salaries or wages are based on hourly rates of pay which vary depending on the job.

PROCEDURE

To find the weekly wages when the hourly rate and the number of working hours are known:

(1) Multiply the number of hours of work by the hourly rate. (If the number of hours is more than 40, multiply only the first 40 hours by this rate.)
(2) Multiply the number of hours more than 40 by the overtime rate.
(3) Add steps (1) and (2) above.

EXAMPLE: Find the weekly wage for a person who works $45\frac{1}{2}$ hours at $4.80 per hour.

$40 \times \$4.80 = \$192.00 \leftarrow$ first 40 hours

$5\frac{1}{2} \times \$7.20 = \$39.60 \leftarrow$ overtime

$(1\frac{1}{2} \times \$4.80)$

$\$192.00 \leftarrow$ wage/40h
$+ \quad 39.60 \leftarrow$ overtime wage
$\$231.60 \leftarrow$ total wage

ANSWER: $231.60 weekly wage

PRACTICE PROBLEMS

1 List several occupations in which wages are paid on an hourly basis.

2 What is the weekly wage of a person who works 40 hours per week at each of the following rates?

1. $4.10   **2.** $7.95   **3.** $12.80   **4.** $5.65   **5.** $10.75

Find the weekly earnings of a person who works:

3 **1.** 28 hours at $6.90 per hour   **2.** 35 hours at $4.30 per hour
3. 31 hours at $11.65 per hour   **4.** 39 hours at $5.45 per hour
5. $33\frac{1}{2}$ hours at $6.84 per hour   **6.** $37\frac{1}{2}$ hours at $10.00 per hour
7. $25\frac{1}{2}$ hours at $9.20 per hour   **8.** $18\frac{1}{2}$ hours at $8.25 per hour

4 **1.** 44 hours at $5.80 per hour

2. 41 hours at $8.50 per hour

3. 47 hours at $12.85 per hour

4. $42\frac{1}{2}$ hours at $6.20 per hour

5. 50 hours at $7.75 per hour

6. $46\frac{1}{2}$ hours at $9.18 per hour

7. $49\frac{1}{2}$ hours at $10.45 per hour

8. 53 hours at $7.05 per hour

5 Find the weekly wages for each of the following persons:

		Hours Worked					Total Hours	Hourly Rate	Wages
		M	T	W	T	F			
1.	Mr. A	8	7	8	8	8		$6.35	
2.	Mr. B	7	7	8	7	7		$8.15	
3.	Mrs. C	8	$6\frac{1}{2}$	8	4	7		$9.10	
4.	Mr. D	5	5	7	$6\frac{1}{2}$	8		$5.75	
5.	Ms. E	8	$4\frac{1}{2}$	$5\frac{1}{2}$	7	$7\frac{1}{2}$		$7.60	

6 For each time card below, compute the number of hours worked each day and the total hours for the week. Then find the amount of wages due each employee.

1.

Time Card

No. 25   NAME: LISA ROSS

WEEK ENDING _____

DAY	IN	OUT	IN	OUT	HOURS
M.	8:00	12:00	1:00	5:00	
T.	8:30	12:00	1:00	5:00	
W.	8:00	11:00	12:30	5:00	
T.	9:00	12:30	1:00	5:30	
F.	8:00	1:00	2:00	5:00	
S.					

RATE $8.35   TOTAL HOURS ___

WAGES _____

2.

Time Card

No. 57   NAME: PETER CHANG

WEEK ENDING _____

DAY	IN	OUT	IN	OUT	HOURS
M.	9:00	12:00	12:30	5:00	
T.	8:30	11:30	12:00	4:00	
W.	8:30	12:00	1:30	6:00	
T.	8:00	12:30	1:00	4:00	
F.	8:30	12:00	1:00	5:30	
S.					

RATE $5.40   TOTAL HOURS ___

WAGES _____

WEEKLY SALARIES, MONTHLY SALARIES, ANNUAL SALARIES

In other occupations the employed persons have fixed salaries on a weekly, semi-monthly, monthly, or annual basis.

PRACTICE PROBLEMS

1 List several occupations in which salaries are paid on a:

1. weekly basis   **2.** annual basis   **3.** monthly basis

2 Find the hourly rate of pay when a person's weekly earnings are:

1. $190 for 40 hours of work   **2.** $234 for 40 hours of work

 3. $150.10 for 38 hours of work   **4.** $247.50 for 33 hours of work

 5. $193.45 for $36\frac{1}{2}$ hours of work

3 Find the annual salary of a person earning each week:

 1. $85   **2.** $425   **3.** $104.50   **4.** $257.50   **5.** $362.25

4 Find the annual salary of a person earning each month:

 1. $700   **2.** $535   **3.** $1,000   **4.** $1,915   **5.** $2,450

5 Find the annual salary of a person earning:

 1. $172.50 per week   **2.** $965 per month
 3. $592.75 semi-monthly   **4.** $5.60 per hour, 40-hour week

6 Find your weekly salary if your annual salary is:

 1. $9,360   **2.** $18,200   **3.** $14,560   **4.** $26,942   **5.** $33,000

7 Find your monthly salary if your annual salary is:

 1. $8,100   **2.** $23,180   **3.** $13,320   **4.** $6,246   **5.** $30,500

8 Find your weekly salary if your monthly salary is ($4\frac{1}{3}$ weeks per month):

 1. $816   **2.** $936   **3.** $1,383   **4.** $1,750   **5.** $1,100

9 Find your semi-monthly salary if your annual salary is:

 1. $14,400   **2.** $12,000   **3.** $25,200   **4.** $8,280   **5.** $20,436

10 Find the weekly earnings of a person who works during the week:

 1. 5 days at $59 per day   **2.** $2\frac{1}{2}$ days at $66 per day
 3. 4 days at $54.75 per day   **4.** $3\frac{1}{2}$ days at $27.50 per day

11 Find the rate of pay per day when a person earns:

 1. $177 for 3 days' work   **2.** $87.20 for 2 days' work
 3. $81 for $1\frac{1}{2}$ days' work   **4.** $216 for $4\frac{1}{2}$ days' work

12 Which is a better wage:

 1. $180 per week or $750 per month?   **2.** $15,400 per year or $1,250 per month?
 3. $7,700 per year or $145 per week?   **4.** $510 semi-monthly or $240 per week?
 5. $225 per week or $11,500 per year?   **6.** $1,650 per month or $400 per week?

PIECE WORK

Some people work where their earnings are based on the number of pieces
or units of work that a person completes at a given pay rate per piece or unit.

PRACTICE PROBLEMS

1 A farm worker picked 49 bushels of apples per day at $.68 a bushel. How much did
he earn for the day?

2 A typist charges $.99 per page and types on an average 40 pages per day. How
much does the typist earn during a week of 5 workdays?

3 For each of the following persons, find the earnings for the week when the number of units and the rate per unit are as follows:

		Number of Units					Total Units	Unit Rate	Total Wages
		M	T	W	T	F			
1.	Charlotte	28	25	33	29	36		$.70	
2.	Marilyn	32	41	37	45	39		$.65	
3.	Leon	147	159	162	154	167		$.30	
4.	Ronald	119	107	99	98	125		$.38	
5.	Elaine	206	195	189	214	179		$.13	

FEES, TIPS, PENSIONS, INTEREST, DIVIDENDS

Other ways people earn income are by charging a fee for their work, by receiving tips, by collecting pensions from work and social security, and by receiving interest and dividends from their investments.

PRACTICE PROBLEMS

1 A TV repair person charges a $20 service fee to come to a home to repair a television. How much money in service fees were made on a day when she had 7 service calls?

2 If a doctor sees 4 patients an hour and charges $25 an office visit, how much does she earn during a day when her office hours are from 2 P.M. to 5 P.M. and 7 P.M. to 9 P.M. and her appointment schedule is full?

3 If a bellhop averages $39 per day in tips, how much does he earn in tips for a week when he works 6 days?

4 If a retired worker receives a social security pension of $250 per month and a pension of $310 per month:

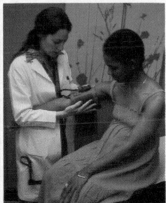

  1. How much is her total monthly pension?
  2. How much is her total annual pension?
  3. How much is her total weekly pension?

5 How much interest should a person receive for the year if his savings account in the bank averages over the year:

  1. $2,000 and the bank pays $5\frac{1}{2}$% annual interest?
  2. $1,840 and the bank pays 6% annual interest?
  3. $900 and the bank pays $5\frac{3}{4}$% annual interest?
  4. $10,000 and the bank pays 5.5% annual interest?

6 Sal owns 100 shares of XYZ stock that pays a $37\frac{1}{2}$¢ dividend per share every three months. Find the total quarterly dividend and the total dividend for the year.

7 Becky's mother has $25,000 invested at the rate of $10\frac{3}{4}$% per year. What is her annual income from this investment?

Commission

> ## Vocabulary
>
> **Commission** is money paid to a person who sells or buys goods for another person.
> **Rate of commission** is the percent that commission is of amount of sale or of purchase.
> **Net proceeds** is amount remaining after commission is deducted from total selling price.

When a buyer purchases goods for another person, the buyer is usually paid
a commission for the services based on the cost of the goods.
Salespeople and buyers are sometimes called agents.

PROCEDURE

To find the commission and net proceeds when the amount of sales and the rate of
commission are given:
(1) Multiply the amount of sales by the rate of commission to find the commission.
(2) Subtract the commission from the amount of sales to find the net proceeds.

EXAMPLE: Find the commission and net proceeds when the sales are $364.89
and the rate of commission is 4%.

$$\begin{array}{rl} \$364.89 & \leftarrow \text{ sales} \\ \times \quad\ .04 & \leftarrow \text{ rate of commission} \\ \hline \$14.5956 \text{ or } \$14.60 & \leftarrow \text{ commission} \end{array} \qquad \begin{array}{rl} \$364.89 & \leftarrow \text{ sales} \\ - \quad \$14.60 & \leftarrow \text{ commission} \\ \hline \$350.29 & \leftarrow \text{ net proceeds} \end{array}$$

ANSWER: $14.60 commission; $350.29 net proceeds

To find a buyer's commission and total cost when the cost of goods purchased and
the rate of commission are given:
(1) Multiply the cost of goods purchased by the rate of commission to find the
commission.
(2) Add the commission to the cost of goods to find the total cost.

To find the rate of commission:
Find what percent the commission is of the amount of sales (or cost of
goods purchased).

EXAMPLE: Find the rate of commission if the commission is $18.40 on sales of $230.

$$\text{commission} \rightarrow \frac{\$18.40}{\$230} = \begin{array}{r} .08 = 8\% \leftarrow \text{ rate of commission} \\ 230\overline{)\$18.40} \\ \underline{18\ 40} \end{array}$$
$$\text{sales} \longrightarrow$$

ANSWER: 8% commission

To find the sales when the commission and rate of commission are given:
Divide the commission by the rate of commission.

PRACTICE PROBLEMS

1 A real estate agent sold a house for $83,500 and was paid a 7% commission. How much was the agent's commission? What were the net proceeds the owner received?

2 Find the commission and net proceeds of each of the following:

	1.	2.	3.	4.	5.
Sales	$426	$93.40	$6,570	$39.15	$229.48
Rate of commission	14%	5%	6%	20%	9%

3 Find the rate of commission if the commission is $78 on sales amounting to $650.

4 Find the rate of commission of each of the following:

	1.	2.	3.	4.	5.
Sales	$860	$965	$824.60	$748.50	$3,267
Commission	$129	$173.70	$164.92	$29.94	$295.83

5 What is the rate of commission when sales amount to $6,300 and net proceeds are $6,111?

6 What must the sales be for a 6% rate of commission to bring a commission of $150?

7 Find the amount of sales of each of the following:

	1.	2.	3.	4.	5.
Commission	$253	$65	$6.83	$54.60	$9.87
Rate of commission	11%	$12\frac{1}{2}$%	5%	14%	15%

8 Find the commission and the total cost of goods purchased for each:

	1.	2.	3.	4.	5.
Cost of goods	$628	$879	$329.75	$956.40	$1,297.50
Rate of commission	9%	$5\frac{1}{2}$%	7%	$8\frac{1}{4}$%	10%

9 Alice works in a clothing store. She receives $105 per week and 1% commission on sales. How much did she earn when she sold $12,028 worth of merchandise?

10 A salesperson sold 9 window fans at $77.99 each and 7 air conditioners at $224.95 each. At 4% commission, how much did he earn?

11 A commission merchant, charging $7\frac{1}{2}$% commission, sold for a grower 290 boxes of melons at $5.65 a box and 475 boxes of plums at $6.40 a box. What net proceeds should the grower receive?

12 A lawyer collected a debt of $824 for a client, charging $123.60 for her services. What rate of commission did she charge?

13 Abe's brother receives $90 per week as salary and an additional 2% commission on the amount of his sales. If he earned $261.34 as his total income in a week, find the amount of his sales.

Vocabulary

The **cost** is the amount merchants pay for goods.

The **selling price** is the amount merchants receive for selling goods.

The **operating expenses** or **overhead** are expenses of running the business.

The **margin**, sometimes referred to as gross profit, spread, or markup, is the difference between the selling price and the cost.

The **net profit** is the amount that remains after both the cost and the operating expenses are deducted from the selling price.

If the selling price is less than the sum of the cost and operating expenses, the goods are sold at a **loss**.

The **rate of profit** is the profit expressed as a percent of either the cost or the selling price.

The **percent markup** is the percent their margin is of the cost or of the selling price.

Merchants are engaged in business to earn a profit. To determine the profit, the merchant must consider the cost of the goods, the selling price of the same goods, and the operating expenses. Some common operating expenses include wages, rent, heat, light, telephone, taxes, insurance, advertising, repairs, supplies, and delivery costs.

To earn a profit the merchant must sell goods at a price which is greater than the sum of the cost of the goods and the operating expenses, otherwise it will be a loss.

To determine the selling price to earn a profit on the articles they sell, some merchants estimate a percent markup based on the cost or on the selling price.

PROCEDURE

To find the selling price:
(1) Multiply the cost by the percent markup on the cost.
(2) Add this markup to the cost.

EXAMPLE: If the rate of markup on the cost is 38%, what is the selling price of an article that costs $65?

```
  $65  ← cost
× .38  ← rate              $65.00  ← cost
  520                    +  24.70  ← markup
  195                      $89.70  ← selling price
$24.70 ← markup
```

ANSWER: $89.70 selling price

To find the selling price when the cost and percent markup or gross profit on the selling price are given:
(1) Subtract the percent markup from 100%.
(2) Divide the cost by the difference in percent.

EXAMPLE: Find the selling price of an article that costs $54 if the percent markup on the selling price is 28%.

$$
\begin{array}{rl}
100\% & \leftarrow \text{selling price} \\
-\ \ 28\% & \leftarrow \text{markup} \\
\hline
72\% & \leftarrow \text{cost}
\end{array}
$$

$72\% = .72$

$$
\begin{array}{r}
\$75 \\
.72\overline{)\$54.00} \\
\underline{50\ 4} \\
3\ 60 \\
\underline{3\ 60}
\end{array}
$$

ANSWER: $75 selling price

According to the information given, margin, or gross profit or spread or markup, can be found in three ways:
• Subtract the cost from the selling price.
• Multiply the cost by the percent markup on the cost.
• Add the operating expenses and the net profit.

To find the percent markup (or gross profit) on cost or on selling price:
Find what percent the margin is of the cost (selling price).

EXAMPLE: Find the rate of markup on cost if an article costs $60 and sells for $80.

$$
\begin{array}{rl}
\$80 & \leftarrow \text{selling price} \\
-\ \ 60 & \leftarrow \text{cost} \\
\hline
\$20 & \leftarrow \text{markup}
\end{array}
$$

$$
\frac{\$20}{\$60} = \frac{1}{3} = 33\frac{1}{3}\%
$$

ANSWER: $33\frac{1}{3}\%$ markup on cost

Net profit can be found either by subtracting the operating expenses from the margin or adding the cost and the operating expenses and subtracting this sum from the selling price. Loss can be found by adding the cost and the operating expenses and then subtracting the selling price from this sum.

To find the rate of net profit or loss on the selling price or on the cost:
Find what percent the net profit (or loss) is of the selling price (cost).

EXAMPLE: Find the net profit and rate of net profit on the selling price if an article costs $25.50, with operating expenses of $2.50 and the selling price of $32.

$25.50 ← cost
$+$ 2.50 ← operating expenses
$28.00 ← total cost

$32 ← selling price
$-$ 28
$4 ← net profit

$$\frac{\$4}{\$32} = \frac{1}{8} = 12\frac{1}{2}\%$$

ANSWER: $4 net profit; $12\frac{1}{2}\%$ rate of net profit on selling price

PRACTICE PROBLEMS

1 A video cassette recorder costs $590. If the rate of markup on cost is 35%, what is the selling price?

2 Find the selling price of each of the following:

	1.	**2.**	**3.**	**4.**	**5.**
Cost	$67	$809	$133.50	$492.75	$52.80
Markup on cost	34%	30%	40%	25%	$37\frac{1}{2}\%$

3 Find the net profit of each of the following:

	1.	**2.**	**3.**	**4.**	**5.**
Cost	$72	$61.90	$3.98	$203	$429.75
Operating expenses	$9	$6.75	$.45	$8.59	$31.40
Selling price	$109	$98.50	$6.25	$266.99	$589.50

4 Find the loss if the cost is $96.50, operating expenses are $3.60, and the selling price is $89.95.

5 What is the selling price of a stereo digital clock radio if it costs $63 and the rate of markup on the selling price is 30%?

6 Find the selling price of each of the following:

	1.	**2.**	**3.**	**4.**	**5.**
Cost	$36	$7.50	$.28	$162.90	$87.50
Markup on selling price	25%	40%	50%	$37\frac{1}{2}\%$	$33\frac{1}{3}\%$

7 Find the rate of markup (margin or gross profit) on cost of each of the following:

	1.	**2.**	**3.**	**4.**	**5.**
Cost	$150	$316	$24.90	$496.50	$59.80
Selling price	$240	$417.12	$37.35	$695.10	$74.75

8 Find the rate of gross profit on selling price of each of the following:

	1.	2.	3.	4.	5.
Cost	$280	$.48	$6.65	$508.69	$86.50
Selling price	$350	$.75	$9.50	$782.60	$129.75

9 What is the rate of net profit on the selling price if the net profit is $3.50 and the selling price is $28?

10 Find the rate of net profit on the selling price of each of the following:

	1.	2.	3.	4.	5.
Cost	$53	$93.40	$2.98	$69.89	$415.36
Operating expenses	$7	$14.60	$.24	$4.61	$52.44
Selling price	$75	$120	$3.50	$89.40	$584.75

11 What is the rate of loss on the selling price if the loss is $10.50 and the selling price is $50?

12 What is the rate of net profit on the cost of each of the following:

	1.	2.	3.	4.	5.
Cost	$60	$425	$19.40	$.80	$1,000
Operating expenses	$10	$37.50	$2.50	$.15	$60.40
Selling price	$85	$550	$27.75	$1.35	$1,149.50

13 Find the selling price of each of the following articles:
  1. Watch, costing $86.50 and 40% markup on cost.
  2. Refrigerator, costing $270 and 25% markup on cost.
  3. Set of golf clubs, costing $140 and 30% markup on selling price.
  4. Shoes, costing $24 and $33\frac{1}{3}$% markup on selling price.

14 Find the rate of gross profit on the selling price of each of the following:
  1. Desk which costs $160 and sells for $240.
  2. Tire which costs $37.50 and sells for $45.
  3. Mirror which costs $38.10 and sells for $57.15.
  4. Dress which costs $18.60 and sells for $31.89.

15 At what price should a dealer sell a hand calculator costing $63 to realize a profit of 40% of the selling price?

16 Shirts cost a merchant $54 a dozen. At what price must each shirt be sold to make a profit of $33\frac{1}{3}$% on the total cost?

17 A dealer bought a table for $76.80 and marked it to sell for a profit of 35% on the cost. If it was finally sold at a reduction of 10% on the marked price, what was the selling price and the amount of profit?

18 Barry's aunt bought a car for $4,160. She sold it for $3,640. What was her rate of loss on the cost of the car?

Vocabulary

Withholding tax is the amount of income tax that employers are required by the federal government to withhold from their employees' earnings.

The withholding tax money is forwarded to the office of the Collector of Internal Revenues, where it is entered and recorded in each employee's account. There are two ways to compute the amount of tax withheld, the *percentage method* and the *income tax withholding table method*.

PERCENTAGE METHOD

PROCEDURE

To compute the amount of weekly withholding tax for a married person:
(1) List the total wage payment.
(2) Subtract the withholding allowance ($19.23 per allowance).
(3) Use the table to calculate the tax.

If the amount of wages is:		The amount of income tax to be withheld shall be:	
Not over $46 0			
Over—	But not over—		of excess over—
$46	—$185	12%	—$46
$185	—$369	.$16.68 plus 17%	—$185
$369	—$454	.$47.96 plus 22%	—$369
$454	—$556	.$66.66 plus 25%	—$454
$556	—$658	.$92.16 plus 28%	—$556
$658	—$862	.$120.72 plus 33%	—$658
$862	. . . .	.$188.04 plus 37%	—$862

EXAMPLE: Compute the withholding tax when a married person's weekly wages are $345 and 4 allowances are claimed.

(1) Total wage payment $345.00
(2) Subtract the withholding allowance ($19.23 × 4) . . . . −   76.92
                                                                           $268.08

(3) Tax from the table: tax on first $185 . . . . $16.68
                       tax on remainder $83.08 @ 17% . . . . . + $14.12
                                                         $30.80

                                  $268.08 − $185

ANSWER: $30.80

INCOME TAX WITHHOLDING TABLE METHOD

PROCEDURE

To compute the amount of weekly withholding tax for a married person:

(1) Locate the row indicating the total wage payment.

(2) Select amount of tax from column corresponding to number of allowances.

Weekly Payroll Period—Married Persons

the wages are—		And the number of withholding allowances claimed is—					
At least	But less than	0	1	2	3	4	5
		The amount of income tax to be withheld shall be—					
92	94	5.60	3.30	1.00			
94	96	5.90	3.60	1.20			
96	98	6.10	3.80	1.50			
98	100	6.30	4.00	1.70			
100	105	6.80	4.50	2.10			
105	110	7.40	5.10	2.70	.40		
110	115	8.00	5.70	3.30	1.00		
115	120	8.60	6.30	3.90	1.60		
120	125	9.20	6.90	4.50	2.20		
125	130	9.80	7.50	5.10	2.80	.50	
130	135	10.40	8.10	5.70	3.40	1.10	
135	140	11.00	8.70	6.30	4.00	1.70	
140	145	11.60	9.30	6.90	4.60	2.30	
145	150	12.20	9.90	7.50	5.20	2.90	.60
150	160	13.10	10.80	8.40	6.10	3.80	1.50
160	170	14.30	12.00	9.60	7.30	5.00	2.70
170	180	15.50	13.20	10.80	8.50	6.20	3.90
180	190	16.70	14.40	12.00	9.70	7.40	5.10
190	200	18.40	15.60	13.20	10.90	8.60	6.30
200	210	20.10	16.80	14.40	12.10	9.80	7.50
210	220	21.80	18.50	15.60	13.30	11.00	8.70
220	230	23.50	20.20	16.90	14.50	12.20	9.90
230	240	25.20	21.90	18.60	15.70	13.40	11.10
240	250	26.90	23.60	20.30	17.10	14.60	12.30
250	260	28.60	25.30	22.00	18.80	15.80	13.50
260	270	30.30	27.00	23.70	20.50	17.20	14.70
270	280	32.00	28.70	25.40	22.20	18.90	15.90
280	290	33.70	30.40	27.10	23.90	20.60	17.30
290	300	35.40	32.10	28.80	25.60	22.30	19.00
300	310	37.10	33.80	30.50	27.30	24.00	20.70
310	320	38.80	35.50	32.20	29.00	25.70	22.40
320	330	40.50	37.20	33.90	30.70	27.40	24.10
330	340	42.20	38.90	35.60	32.40	29.10	25.80
340	350	43.90	40.60	37.30	34.10	30.80	27.50
350	360	45.60	42.30	39.00	35.80	32.50	29.20
360	370	47.30	44.00	40.70	37.50	34.20	30.90
370	380	49.30	45.70	42.40	39.20	35.90	32.60
380	390	51.50	47.40	44.10	40.90	37.60	34.30
390	400	53.70	49.50	45.80	42.60	39.30	36.00
400	410	55.90	51.70	47.50	44.30	41.00	37.70
410	420	58.10	53.90	49.60	46.00	42.70	39.40
420	430	60.30	56.10	51.80	47.70	44.40	41.10
430	440	62.50	58.30	54.00	49.80	46.10	42.80
440	450	64.70	60.50	56.20	52.00	47.80	44.50
450	460	66.90	62.70	58.40	54.20	50.00	46.20
460	470	69.40	64.90	60.60	56.40	52.20	47.90
470	480	71.90	67.10	62.80	58.60	54.40	50.10
480	490	74.40	69.60	65.00	60.80	56.60	52.30
490	500	76.90	72.10	67.30	63.00	58.80	54.50
500	510	79.40	74.60	69.80	65.20	61.00	56.70

EXAMPLE: Compute the withholding tax when a married person's weekly wages are $278 and 2 allowances are claimed.

    (1) Locate the total wage payment.

        Row reads: "At least 270 but less than 280."

    (2) Select the amount of tax from the allowances column.

        Read down column headed "2" until you get to $25.40.

ANSWER: $25.40

PRACTICE PROBLEMS

1 Use the percentage method to compute the amount of income tax to be withheld each week if a married person's weekly wages are:

   1. $119 and 2 withholding allowances are claimed
   2. $97 and 1 withholding allowance is claimed
   3. $342 and 3 withholding allowances are claimed
   4. $260 and 4 withholding allowances are claimed
   5. $495 and 2 withholding allowances are claimed
   6. $183.50 and 1 withholding allowance is claimed
   7. $502.75 and 3 withholding allowances are claimed
   8. $439.50 and 5 withholding allowances are claimed
   9. $247.50 and 3 withholding allowances are claimed
  10. $385.75 and 2 withholding allowances are claimed

2 Use the table of withholding tax to compute the amount of income tax to be withheld each week if a married person's weekly wages are:

   1. $93 and 1 withholding allowance is claimed
   2. $410 and 3 withholding allowances are claimed
   3. $226 and 0 withholding allowance is claimed
   4. $314 and 2 withholding allowances are claimed
   5. $191 and 4 withholding allowances are claimed
   6. $242.50 and 5 withholding allowances are claimed
   7. $508.75 and 1 withholding allowance is claimed
   8. $381.25 and 5 withholding allowances are claimed
   9. $270.50 and 3 withholding allowances are claimed
  10. $162.25 and 2 withholding allowances are claimed

3 Use the table of withholding tax to compute the amount of income tax withheld for the year if a married person's weekly wages are:

   1. $98 and 1 withholding allowance is claimed
   2. $175 and 2 withholding allowances are claimed
   3. $312.50 and 5 withholding allowances are claimed
   4. $430.75 and 2 withholding allowances are claimed
   5. $274.25 and 3 withholding allowances are claimed

4 Use the percentage method to compute the amount of income tax withheld for the year if a married person's weekly wages are: (Round weekly withholding before calculating for a year).

   1. $165 and 2 withholding allowances are claimed
   2. $220 and 3 withholding allowances are claimed
   3. $445 and 1 withholding allowance is claimed
   4. $287.75 and 4 withholding allowances are claimed
   5. $362.50 and 2 withholding allowances are claimed

6-5 Social Security Tax

Vocabulary

Social security tax is the money deducted from employees' earnings to help pay old-age pensions, disability etc.

Employees pay a tax at the rate of 7.05% of the first $39,600 of their wages. Employers pay the same amount, matching the amount paid by the employees.

PROCEDURE

To find the social security tax deducted from an employee's wages:
Multiply the employee's wages by 7.05% to find the amount deducted from the employee's earnings.

EXAMPLE: Find the amount deducted from an employee's weekly wage of $330 for social security tax.

$330 ← weekly wage
× .0705 ← tax rate
1650
2310
$23.2650 or $23.27 ← social security tax

ANSWER: $23.27 social security tax

PRACTICE PROBLEMS

1 Find the amount deducted from the employee's weekly wages for social security tax.

1. $60	**2.** $115	**3.** $89	**4.** $227
5. $405	**6.** $615	**7.** $584	**8.** $770
9. $800	**10.** $725	**11.** $72.50	**12.** $101.75
13. $377.50	**14.** $532.45	**15.** $218.40	**16.** $539.25
17. $445.80	**18.** $209.75	**19.** $315.50	**20.** $608.25

2 Find the amount deducted from the employee's wages for social security tax for the given work period.

1. $395 semi-monthly	**2.** $2,000 monthly	**3.** $32,400 annually
4. $687.50 monthly	**5.** $14,925 annually	**6.** $581.75 semi-monthly

Wage Tax—State or Local

Vocabulary

Wage tax is an additional income tax that a number of states and cities in the United States levy on a person's earnings.

The wage tax rate is usually a fixed percent of the earnings.

PROCEDURE

To find the wage tax:
Multiply a person's wages for the given pay period by the wage tax rate.

EXAMPLE: If the wage tax rate is 3% of the wages, find the tax a person owes for the week on weekly earnings of $429.

$429 ← wage
× .03 ← wage tax rate
$12.87 ← wage tax

ANSWER: $12.87 wage tax

PRACTICE PROBLEMS

1 If the wage tax rate is 2% of the wages, find the tax a person owes for the week if he earns weekly:

1. $85   **2.** $116   **3.** $220   **4.** $174.25   **5.** $369.50

2 If the wage tax rate is 3% of the wages, find the tax a person owes for the week if she earns weekly:

1. $68   **2.** $245   **3.** $407   **4.** $191.75   **5.** $382.25

3 If the wage tax rate is $2\frac{1}{2}$% of the wages, find the tax a person owes for the semi-monthly pay period if he earns semi-monthly:

1. $182   **2.** $634   **3.** $570   **4.** $245.50   **5.** $406.75

4 If the wage tax rate is $1\frac{3}{4}$% of the wages, find the tax a person owes for the month if she earns monthly:

1. $900   **2.** $1,385   **3.** $1,025   **4.** $652.75   **5.** $1,774.50

5 If the wage tax rate is $2\frac{1}{4}$% of the wages, find the tax a person owes for the year if his annual salary is:

1. $6,200   **2.** $8,300   **3.** $11,000   **4.** $14,850   **5.** $12,695

Take-Home Pay

Vocabulary

Take-home pay is the amount of a person's earnings remaining after federal withholding tax, social security tax, and wage tax, if any, are deducted from a person's wages.

In addition to the deductions mentioned above, union dues and health insurance costs may also be deducted.

People sometimes refer to take-home pay as net pay or net wage.

PROCEDURE

To determine the take-home pay:
(1) Find the federal withholding tax.
(2) Find the social security tax.
(3) Find the wage tax, if any.
(4) Subtract the sum of the taxes from the amount of the person's wage.

EXAMPLE: Determine the take-home pay when a married person's weekly wages are $215, there is a wage tax rate of 2%, and 2 allowances are claimed.

(1) Federal withholding tax (by table) $15.60
(2) Social security tax ($215 × 7.05%). $15.16
(3) Wage tax ($215 × 2%) $4.30
$\qquad\qquad\qquad\qquad\qquad\qquad\qquad\qquad$ sum of taxes $35.06
(4) Take-home pay (earnings − sum of taxes) $179.94

ANSWER: $179.94 take-home pay   **$215 − $35.06**

PRACTICE PROBLEMS

In the following problems, use the table of withholding tax (p. 287) for all wages between $92 and $510; for all other wages use the rates given on p. 286. The social security tax rate is 7.05%.

1 Mrs. Brown earns $182 each week and claims 3 withholding allowances. How much federal income tax is withheld each week? How much is deducted for social security? What is her take-home pay?

2 Find the take-home pay for each of the following weekly wages:

	Wage	Withholding Allowances			Wage	Withholding Allowances
1.	$94	4		**6.**	$99.75	5
2.	$105	2		**7.**	$308.50	4
3.	$320	1		**8.**	$191.60	3
4.	$480	3		**9.**	$420.25	1
5.	$214	2		**10.**	$145.50	2

3 Find the take-home pay for each of the following weekly wages when a wage tax is also levied:

	Wage	Withholding Allowances	Wage Tax			Wage	Withholding Allowances	Wage Tax
1.	$96	1	2%		**6.**	$93.50	1	$1\frac{5}{8}$%
2.	$207	2	1%		**7.**	$409.25	2	2%
3.	$193	4	$1\frac{1}{2}$%		**8.**	$191.60	4	$1\frac{1}{2}$%
4.	$500	2	$2\frac{1}{4}$%		**9.**	$279.00	2	2%
5.	$318	3	$1\frac{3}{4}$%		**10.**	$327.50	3	$2\frac{1}{2}$%

4 Which take-home pay is greater, a weekly wage of $140.25 with 1 withholding allowance claimed or $129.75 with 5 withholding allowances claimed?

5 Mr. Wilson earns $217.50 each week and claims 2 withholding allowances. He works in a city where a tax of $1\frac{1}{2}$% is levied on wages. How much federal income tax is withheld each week? How much is deducted for social security? What is his weekly city wage tax? What is his take-home pay?

Refresh Your Skills

1. Add: [1-4]
7,399
586
92,157

2. Subtract: [1-5]
63,514
21,735

3. Multiply: [1-6]
697
38

4. Divide: [1-7]
$205\overline{)7,380}$

5. Add: [2-4]
$4.96 + 8.9$

6. Subtract: [2-5]
$.325 - .29$

7. Multiply: [2-7]
$5.4 \times .25$

8. Divide: [2-8]
$.2\overline{).09}$

9. Add: [3-8]
$3\frac{2}{3} + 1\frac{5}{6}$

10. Subtract: [3-9]
$1\frac{2}{5} - \frac{1}{2}$

11. Multiply: [3-11]
$\frac{5}{8} \times \frac{3}{10}$

12. Divide: [3-12]
$1\frac{3}{4} \div 6$

13. Find 20% of 300. [4-6] **14.** 6 is what percent of 9? [4-7] **15.** $5 is 25% of what amount? [4-8]

Vocabulary

The **total income** includes wages, salaries, tips, interest income, business income, pensions, etc.

The **adjusted gross income** is the total income less adjustments, such as IRA payments, moving expenses, etc.

The **taxable income** is the adjusted gross income decreased by the sum of exemptions claimed credit and itemized deductions.

The federal income tax forms describe different types of income.

Page 294 contains a part of the Federal tax table used to determine the amount of tax based on taxable income. For the rest of this tax table use the Internal Revenue Service publications. Income tax rates are subject to change.

PROCEDURE

(1) Use the Federal tax form 1040, 1040A, or 1040EZ to determine the amount of taxable income.
(2) Turn to the Federal tax table.
(3) Read down the income column of the tax table until you find the line covering this taxable income.
(4) Then read across that income line until you find the column describing this taxpayer to find the amount of tax.

EXAMPLE: Hilda Perry is the head of a household whose taxable income is $10,382. Determine the amount of her Federal income tax.

Read down the income column of the table on page 294 until you reach:

At least	But less than
10,350	10,400

Read across the column headed "Head of a Household."

ANSWER: $1,076, amount of tax

EXAMPLE: If Hilda's withholding tax payments for the year amounted to $898.50, how much income tax does she still owe?

$1,076.00 ← amount of tax
−   898.50 ← withholding tax payments
$177.50 ← owed

ANSWER: $177.50 owed

Taxable Income At least	But less than	Single	Married filing jointly *	Married filing separately	Head of a household
			Your tax is—		
8,000					
8,000	8,050	764	543	875	697
8,050	8,100	771	550	884	704
8,100	8,150	779	557	893	711
8,150	8,200	786	564	902	718
8,200	8,250	794	571	911	725
8,250	8,300	801	578	920	732
8,300	8,350	809	585	929	739
8,350	8,400	816	592	938	746
8,400	8,450	824	599	947	753
8,450	8,500	831	606	956	760
8,500	8,550	839	613	965	767
8,550	8,600	847	620	974	774
8,600	8,650	855	627	983	781
8,650	8,700	863	634	992	788
8,700	8,750	871	641	1,001	795
8,750	8,800	879	648	1,010	804
8,800	8,850	887	655	1,019	812
8,850	8,900	895	662	1,028	821
8,900	8,950	903	669	1,037	829
8,950	9,000	911	676	1,046	838
9,000					
9,000	9,050	919	683	1,055	846
9,050	9,100	927	690	1,064	855
9,100	9,150	935	697	1,073	863
9,150	9,200	943	704	1,082	872
9,200	9,250	951	711	1,091	880
9,250	9,300	959	718	1,100	889
9,300	9,350	967	725	1,109	897
9,350	9,400	975	732	1,118	906
9,400	9,450	983	739	1,127	914
9,450	9,500	991	746	1,136	923
9,500	9,550	999	753	1,145	931
9,550	9,600	1,007	760	1,154	940
9,600	9,650	1,015	767	1,163	948
9,650	9,700	1,023	774	1,172	957
9,700	9,750	1,031	781	1,181	965
9,750	9,800	1,039	788	1,190	974
9,800	9,850	1,047	795	1,199	982
9,850	9,900	1,055	802	1,208	991
9,900	9,950	1,063	809	1,217	999
9,950	10,000	1,071	816	1,226	1,008
10,000					
10,000	10,050	1,079	823	1,235	1,016
10,050	10,100	1,087	830	1,244	1,025
10,100	10,150	1,095	837	1,254	1,033
10,150	10,200	1,103	844	1,265	1,042
10,200	10,250	1,111	851	1,276	1,050
10,250	10,300	1,119	858	1,287	1,059
10,300	10,350	1,127	865	1,298	1,067
10,350	10,400	1,135	872	1,309	1,076
10,400	10,450	1,143	879	1,320	1,084
10,450	10,500	1,151	886	1,331	1,093
10,500	10,550	1,159	893	1,342	1,101
10,550	10,600	1,167	900	1,353	1,110
10,600	10,650	1,175	907	1,364	1,118
10,650	10,700	1,183	914	1,375	1,127
10,700	10,750	1,191	921	1,386	1,135
10,750	10,800	1,199	928	1,397	1,144
10,800	10,850	1,208	935	1,408	1,152
10,850	10,900	1,217	942	1,419	1,161
10,900	10,950	1,226	949	1,430	1,169
10,950	11,000	1,235	956	1,441	1,178

Taxable Income At least	But less than	Single	Married filing jointly *	Married filing separately	Head of a household
			Your tax is—		
11,000					
11,000	11,050	1,244	963	1,452	1,186
11,050	11,100	1,253	970	1,463	1,195
11,100	11,150	1,262	977	1,474	1,203
11,150	11,200	1,271	984	1,485	1,212
11,200	11,250	1,280	991	1,496	1,220
11,250	11,300	1,289	998	1,507	1,229
11,300	11,350	1,298	1,005	1,518	1,237
11,350	11,400	1,307	1,012	1,529	1,246
11,400	11,450	1,316	1,019	1,540	1,254
11,450	11,500	1,325	1,026	1,551	1,263
11,500	11,550	1,334	1,033	1,562	1,271
11,550	11,600	1,343	1,040	1,573	1,280
11,600	11,650	1,352	1,047	1,584	1,288
11,650	11,700	1,361	1,054	1,595	1,297
11,700	11,750	1,370	1,061	1,606	1,305
11,750	11,800	1,379	1,068	1,617	1,314
11,800	11,850	1,388	1,075	1,628	1,323
11,850	11,900	1,397	1,082	1,639	1,332
11,900	11,950	1,406	1,089	1,650	1.341
11,950	12,000	1,415	1,097	1,661	1,350
12,000					
12,000	12,050	1,424	1,105	1,672	1,359
12,050	12,100	1,433	1,113	1,683	1,368
12,100	12,150	1,442	1,121	1,694	1,377
12,150	12,200	1,451	1,129	1,705	1,386
12,200	12,250	1,460	1,137	1,716	1,395
12,250	12,300	1,469	1,145	1,727	1,404
12,300	12,350	1,478	1,153	1,739	1,413
12,350	12,400	1,487	1,161	1,751	1,422
12,400	12,450	1,496	1,169	1,764	1,431
12,450	12,500	1,505	1,177	1,776	1,440
12,500	12,550	1,514	1,185	1,789	1,449
12,550	12,600	1,523	1,193	1,801	1,458
12,600	12,650	1,532	1,201	1,814	1,467
12,650	12,700	1,541	1,209	1,826	1,476
12,700	12,750	1,550	1,217	1,839	1,485
12,750	12,800	1,559	1,225	1,851	1,494
12,800	12,850	1,568	1,233	1,864	1,503
12,850	12,900	1,577	1,241	1,876	1,512
12,900	12,950	1,586	1,249	1,889	1,521
12,950	13,000	1,596	1,257	1,901	1,530
13,000					
13,000	13,050	1,606	1,265	1,914	1,539
13,050	13,100	1,616	1,273	1,926	1,548
13,100	13,150	1,626	1,281	1,939	1,557
13,150	13,200	1,636	1,289	1,951	1,566
13,200	13,250	1,646	1,297	1,964	1,575
13,250	13,300	1,656	1,305	1,976	1,584
13,300	13,350	1,666	1,313	1,989	1,593
13,350	13,400	1,676	1,321	2,001	1,602
13,400	13,450	1,686	1,329	2,014	1,611
13,450	13,500	1,696	1,337	2,026	1,620
13,500	13,550	1,706	1,345	2,039	1,629
13,550	13,600	1,716	1,353	2,051	1,638
13,600	13,650	1,726	1,361	2,064	1,647
13,650	13,700	1,736	1,369	2,076	1,656
13,700	13,750	1,746	1,377	2,089	1,665
13,750	13,800	1,756	1,385	2,101	1,674
13,800	13,850	1,766	1,393	2,114	1,683
13,850	13,900	1,776	1,401	2,126	1,692
13,900	13,950	1,786	1,409	2,139	1,701
13,950	14,000	1,796	1,417	2,151	1,710

Taxable Income At least	But less than	Single	Married filing jointly *	Married filing separately	Head of a household
			Your tax is—		
14,000					
14,000	14,050	1,806	1,425	2,164	1,719
14,050	14,100	1,816	1,433	2,176	1,728
14,100	14,150	1,826	1,441	2,189	1,737
14,150	14,200	1,836	1,449	2,201	1,746
14,200	14,250	1,846	1,457	2,214	1,755
14,250	14,300	1,856	1,465	2,226	1,764
14,300	14,350	1,866	1,473	2,239	1,773
14,350	14,400	1,876	1,481	2,251	1,782
14,400	14,450	1,886	1,489	2,264	1,791
14,450	14,500	1,896	1,497	2,276	1,800
14,500	14,550	1,906	1,505	2,289	1,809
14,550	14,600	1,916	1,513	2,301	1,818
14,600	14,650	1,926	1,521	2,314	1,827
14,650	14,700	1,936	1,529	2,326	1,836
14,700	14,750	1,946	1,537	2,339	1,845
14,750	14,800	1,956	1,545	2,351	1,854
14,800	14,850	1,966	1,553	2,364	1,863
14,850	14,900	1,976	1,561	2,376	1,872
14,900	14,950	1,986	1,569	2,389	1,881
14,950	15,000	1,996	1,577	2,402	1,890
15,000					
15,000	15,050	2,007	1,585	2,416	1,899
15,050	15,100	2,018	1,593	2,430	1,909
15,100	15,150	2,030	1,601	2,444	1,919
15,150	15,200	2,041	1,609	2,458	1,929
15,200	15,250	2,053	1,617	2,472	1,939
15,250	15,300	2,064	1,625	2,486	1,949
15,300	15,350	2,076	1,633	2,500	1,959
15,350	15,400	2,087	1,641	2,514	1,969
15,400	15,450	2,099	1,649	2,528	1,979
15,450	15,500	2,110	1,657	2,542	1,989
15,500	15,550	2,122	1,665	2,556	1,999
15,550	15,600	2,133	1,673	2,570	2,009
15,600	15,650	2,145	1,681	2,584	2,019
15,650	15,700	2,156	1,689	2,598	2,029
15,700	15,750	2,168	1,697	2,612	2,039
15,750	15,800	2,179	1,705	2,626	2,049
15,800	15,850	2,191	1,713	2,640	2,059
15,850	15,900	2,202	1,721	2,654	2,069
15,900	15,950	2,214	1,729	2,668	2,079
15,950	16,000	2,225	1,737	2,682	2,089
16,000					
16,000	16,050	2,237	1,746	2,696	2,099
16,050	16,100	2,248	1,755	2,710	2,109
16,100	16,150	2,260	1,764	2,724	2,119
16,150	16,200	2,271	1,773	2,738	2,129
16,200	16,250	2,283	1,782	2,752	2,139
16,250	16,300	2,294	1,791	2,766	2,149
16,300	16,350	2,306	1,800	2,780	2,159
16,350	16,400	2,317	1,809	2,794	2,169
16,400	16,450	2,329	1,818	2,808	2,179
16,450	16,500	2,340	1,827	2,822	2,189
16,500	16,550	2,352	1,836	2,836	2,199
16,550	16,600	2,363	1,845	2,850	2,209
16,600	16,650	2,375	1,854	2,864	2,219
16,650	16,700	2,386	1,863	2,878	2,229
16,700	16,750	2,398	1,872	2,892	2,239
16,750	16,800	2,409	1,881	2,906	2,249
16,800	16,850	2,421	1,890	2,920	2,259
16,850	16,900	2,432	1,899	2,934	2,269
16,900	16,950	2,444	1,908	2,948	2,279
16,950	17,000	2,455	1,917	2,962	2,289

PRACTICE PROBLEMS

Use the Tax Table to find each of the following:

1 What is your Federal income tax?

1. You are single and your taxable income is $11,800.
2. You are married and filing jointly with a taxable income of $14,745.
3. You arc head of a household and your taxable income is $12,375.
4. You are married and filing separately with a taxable income of $10,264.

2 How much Federal income tax do you still owc?

1. You are married and filing jointly. Your taxable income is $13,930 and your withholding tax payments are $1,200.
2. You are single. Your taxable income is $8,725 and your withholding tax payments are $790.
3. You are head of a household. Your taxable income is $15,600 and your withholding tax payments are $1,750.
4. You are married and filing separately. Your taxable income is $16,500 and your withholding tax payments are $2,500.

3 How much tax do you still owe or what is the amount of your refund?

1. You are single. Your taxable income is $13,425 and your withholding tax payments are $1,750.
2. You are married and filing jointly. Your taxable income is $16,600 and your withholding tax payments are $2,000.
3. You are married and filing separately. Your taxable income is $10,840 and your withholding tax payments are $1,124.
4. You are head of a household. Your taxable income is $12,309 and your withholding tax payments are $1,510.

Refresh Your Skills

1. Add: 4,396 [1-4] 5,287 8,962 6,435

2. Subtract: 80,000 [1-5] 9,607

3. Multiply: 409 [1-6] 506

4. Divide: [1-7] 48⟌46,080

5. Add: [2-4] $12.46 + $9.07 + $.59

6. Subtract: [2-5] $10 − $5.26

7. Multiply: [2-7] 24 × $.69

8. Divide: [2-8] $1.25⟌$20

9. Add: [3-8] $8\frac{3}{4} + 4\frac{5}{12}$

10. Subtract: [3-9] $6 − 5\frac{3}{8}$

11. Multiply: [3-11] $4\frac{1}{2} × 2\frac{2}{3}$

12. Divide: [3-12] $9 ÷ 1\frac{4}{5}$

13. Find $16\frac{2}{3}$% of $24 [4-6]

14. $25 is what percent of $400? [4-7]

15. 3% of what amount is $12 [4-8]

Form 1040 EZ—FEDERAL INCOME TAX RETURN

This form is only for single filers, with no dependents, earning less than $50,000 with interest income $400 or less.

To prepare this tax return, enter the following information:

Name and Address
1. Use the IRS mailing label, or
   print your name, address, and social security number.

Figure Your Tax
2. Use lines 1 thru 3 to find your *adjusted gross income*.

3. Use lines 4 thru 7 to find your *taxable income*. On line 4, enter whichever is *lower*—$75 or the allowable part of your charitable contributions.

4. Use line 8 to enter the amount of your Federal income tax withheld.

5. Use line 9 to enter the amount of *tax* after finding it in the *single* column of the tax table.

Compare Tax with Withholding Payments
6. Use line 10 to enter amount of your refund, or

   Use line 11 to enter amount of tax you still owe.

7. Sign and date your return.
Attach both copy B of your W2 form and your tax payment.

EXERCISES

Use copies of the 1040 EZ form to prepare income tax returns for the following:

1. Rosa Gomez of 2518 Green St., Chicago, IL 36941 earned $10,548 in wages and $279.93 interest. Her Federal withholding tax payments for the year were $1,125. Rosa's social security number is 555-12-1200.

2. Tom Wong of 25 So. Pine St., Los Angeles, CA 94216 has the social security number 808-08-0808. His wages for the year were $11,629 and the interest on his savings account was $316.85. His Federal withholding tax payments for the year were $1,430. The allowable part of his charitable contribution was $105.

3. Hal Turner's wages for the year were $14,500. His Federal withholding tax payments were $1,775. He resides at 8900 West Ave., Dallas, TX 46952. His social security number is 444-22-6666.

4. Mary Williams of 125 Main St., New York, NY 10000 has the social security number 999-99-0000. Last year she earned $15,200 in salary and $389 in interest. She paid $1,925 Federal withholding tax for the year. The allowable part of her charitable contribution amounted to $93.

Department of the Treasury · Internal Revenue Service

1984

Form 1040EZ Income Tax Return for Single filers with no dependents (0)

OMB No. 1545-0675

Name & address

Use the IRS mailing label. If you don't have one, please print:

Please print your numbers like this.

1234567890

Print your name above (first, initial, last)

Present home address (number and street)

City, town, or post office, State, and ZIP code

Social security number

Presidential Election Campaign Fund
Check box if you want $1 of your tax to go to this fund. ▶

Dollars   Cents

Figure your tax

1 Total wages, salaries, and tips. This should be shown in Box 10 of your W-2 form(s). (Attach your W-2 form(s).) **1**

2 Interest income of $400 or less. If the total is more than $400, you cannot use Form 1040EZ. **2**

Attach Copy B of Form(s) W-2 here

3 Add line 1 and line 2. This is your **adjusted gross income.** **3**

4 Allowable part of your charitable contributions. Complete the worksheet on page 21 of the instruction booklet. Do not enter more than $75. **4**

5 Subtract line 4 from line 3. **5**

6 Amount of your personal exemption. **6**   1,000 00

7 Subtract line 6 from line 5. This is your **taxable income.** **7**

8 Enter your Federal income tax withheld. This should be shown in Box 9 of your W-2 form(s). **8**

9 Use the **single** column in the tax table on pages 31-36 of the instruction booklet to find the **tax** on your taxable income on line 7. Enter the amount of tax. **9**

Refund or amount you owe

10 If line 8 is larger than line 9, subtract line 9 from line 8. Enter the **amount of your refund.** **10**

11 If line 9 is larger than line 8, subtract line 8 from line 9. Enter the **amount you owe.** Attach check or money order for the full amount, payable to "Internal Revenue Service." **11**

Attach tax payment here

Sign your return

I have read this return. Under penalties of perjury, I declare that to the best of my knowledge and belief, the return is true, correct, and complete.

Your signature   Date

For IRS Use Only—Please do not write in boxes below.

1 2 3 4 5

Career Applications

Many people cannot prepare their own income tax forms. Others worry that they will make a mistake or pay more tax than is necessary. *Tax preparers* help people fill out income tax forms. A tax preparer must have a good understanding of tax laws. Tax preparers who prepare tax returns incorrectly may be subject to penalties from the federal government.

Tax preparers may work for tax preparation companies or on their own. The federal government gives examinations on tax laws. Some tax preparation companies have training courses.

1. Taxpayers who use Form 1040 can itemize deductions. A taxpayer has the following deductions: $138.44 in medical expenses, $2,954.86 in state income taxes and real estate taxes, $3,129.57 in interest on mortgages and charge accounts, $827.41 in contributions, and $50 for the tax preparer's fee. What is the total deduction?

3. Taxpayers using Form 1040 are allowed an energy credit of 15% on items that save energy in their homes. A taxpayer spent $1,482.37 on insulation and storm windows. What is the energy credit to the nearest cent?

2. After calculating the total deduction, a tax preparer subtracts the zero bracket amount to find the net deduction. For a married couple the zero bracket amount is $3,400. A couple's total deduction is $8,265.17. What is their net deduction?

4. A taxpayer paid $4,236.81 in withholding tax during the year. Her total tax is $4,189.52. Does she owe more tax, or will she get a refund? How much?

CHAPTER REVIEW

Vocabulary

adjusted gross income p. 293	net profit p. 282	selling price p. 282
commission p. 280	operating expenses p. 282	social security tax p. 289
cost p. 282	overhead p. 282	take-home pay p. 291
double time p. 276	percent markup p. 282	taxable income p. 293
loss p. 282	rate of commission p. 280	time and a half p. 276
margin p. 282	rate of profit p. 282	total income p. 293
net proceeds p. 280	salary p. 276	wage tax p. 290
		withholding tax p. 286

The numerals in the boxes indicate Lessons where help may be found.

1. Find the weekly salary for a person who works 40 hours per week at $6.25 per hour. [6-1]
2. Which is the better wage: $15,400 per year or $1,300 per month? [6-1]
3. At the rate of 3% commission on sales, how much will a salesperson receive when she sells $4,583 worth of merchandise? [6-2]
4. If an agent charges 2% commission for buying 125 bushels of apples at $9.35 per bushel, what is the total cost of the apples? [6-2]
5. A dealer bought video cassettes at $216 a dozen. At what price each must she sell them to make a profit of 40% on the cost? [6-3]
6. At what price should a merchant sell a computer costing $520 to realize a profit of 35% on the selling price? [6-3]
7. Use the table of withholding tax (p. 287) to compute the amount of income tax to be withheld when a married person's weekly wages are $278.50 and 3 allowances are claimed. [6-4]
8. How much social security tax is deducted weekly from a wage of $675? The social security tax rate is 7.05% [6-5]
9. If the wage tax rate is $2\frac{1}{2}$% of the wages, find the tax a person owes for the week when she earns $305.80. [6-6]

Use the table of withholding tax (p. 287) and the social security tax rate of 7.05% in each of the following problems? [6-7]

10. Find the take-home pay when a married person's weekly wages are $410 with 3 allowances claimed.
11. Find the take-home pay when a married person's weekly wages are $295 with 4 allowances claimed and a levy of $1\frac{3}{4}$% wage tax.

Use the Federal tax table (p. 294) to find each of the following: [6-8]

12. What is your Federal income tax if you are single and your taxable income is $13,500?
13. How much Federal income tax do you still owe if you are married and filing jointly? Your taxable income is $15,225 and your withholding tax payments are $1,450.
14. How much tax do you still owe or what is the amount of your refund if you are head of a household? Your taxable income is $12,875 and your withholding tax payments are $1,700.

COMPETENCY CHECK TEST

The number in the boxes indicate Lessons where help may be found.

Solve each problem and select the letter corresponding to your answer.

1. Betty earns $302 per week. She works 40 hours. Her wage per hour is: 6-1
    a. $3.42   **b.** $6.75   **c.** $7.55   **d.** $8.40

2. Larry earns $245 per week. His salary for the year (52 weeks) is: 6-1
    a. $10,000   **b.** $12,740   **c.** $16,450   **d.** $24,552

3. Shirley's father receives a salary of $875 twice a month and a bonus of $2,500. His total income for the year is: 6-1
    a. $23,500   **b.** $26,500   **c.** $32,500   **d.** $21,000

4. A real estate agent sold a house for $67,500. At 6%, her commission is: 6-2
    a. $4,500   **b.** $3,950   **c.** $675   **d.** $4,050

5. A merchant buys a rug costing $600 and sells it for $840. The rate of profit on the cost is: 6-3
    a. 20%   **b.** 25%   **c.** $33\frac{1}{3}$%   **d.** 40%

6. If you use the withholding income tax rate: 6-4

Wages over $369 but not over $454—amount of tax is $47.96 plus 22% of excess over $369 the withholding income tax of a person earning $400 per week with no exemptions is:
    a. $69.96   **b.** $54.78   **c.** $47.96   **d.** Answer not given

7. When the tax rate is 7.05%, the social security tax on a weekly salary of $416.80 is: 6-5
    a. $29.38   **b.** $41.68   **c.** $35.44   **d.** $70.50

8. At the tax rate of $2\frac{1}{2}$%, the wage tax on a weekly salary of $363.50 is: 6-6
    a. $7.27   **b.** $36.35   **c.** $9.09   **d.** Answer not given

9. If the withholding tax is $23.48 and the social security tax is $15.86 on a weekly salary of $225, the take-home pay is: 6-7
    a. $264.34   **b.** $39.34   **c.** $201.52   **d.** $185.66

10. If the withholding tax is $80.66, the social security tax is $35.96, and the wage tax is 2% on a weekly salary of $510, the take-home pay is: 6-7
    a. $393.38   **b.** $383.18   **c.** $499.80   **d.** $510

11. If your Federal income tax for the year is $2,363 and your total withholding tax payments are $1,975, the amount of tax you still owe is: 6-8
    a. $4,338   **b.** $412   **c.** $388   **d.** $2,363

12. If your Federal income tax for the year is $1,449 and your total withholding tax payments are $1,525, the amount of your refund is: 6-8
    a. $175   **b.** $124   **c.** $24   **d.** $76

CHAPTER 7 Consumer Spending

7-1 Buying Food

When the purchase price of a food item is computed and the result contains any fractional part of a cent, the fraction is dropped and one cent is added to the price.

EXAMPLE: If 3 bars of soap sell for 67¢, what is the purchase price of 1 bar?

Soap, 3 bars for 67¢

$$1 \text{ bar} \quad 3\overline{)67¢}^{\,22\frac{1}{3}¢}$$

ANSWER: purchase price of 1 bar is 23¢ or $.23

PRACTICE PROBLEMS

GROCERY PRICE LIST

Beverages, 2 bottles for 95¢
Flour, 10 lb for $1.85
Peas, 6 cans for $2.16
Sugar, 5 lb for $1.89

Coffee, $2.69 a pound
Detergent, $1.45 a pkg.
Soap, 4 bars for 95¢
Tomato juice, 3 cans for $1.49

Corn, 5 cans for $1.79
Peaches, 2 cans for $1.29
Soup, 2 cans for 79¢
Tuna, 2 cans for $2.25

1 Find the cost of:

1. 1 can of corn
2. 1 can of peaches
3. 1 can of soup
4. 2 pkg. of detergent
5. 8 bars of soap
6. 6 cans of tuna
7. 10 cans of corn
8. 3 cans of peas
9. 2 bars of soap
10. 1 bottle of beverage
11. 20 lb of flour
12. 12 cans of soup
13. 2 cans of tomato juice
14. 1 can of tuna
15. 5 bottles of beverage

BAKERY PRICE LIST

Assorted cakes, $1.92 a dozen
Bread, 2 loaves for $1.49

Cookies, $3.20 a pound
Rolls, 6 for 78¢

2 Find the cost of:

1. 1 loaf of bread
2. 1 cake
3. 1 roll
4. $\frac{1}{2}$ doz. cakes
5. $\frac{3}{4}$ lb cookies
6. $1\frac{1}{2}$ doz. rolls
7. 3 loaves of bread
8. 16 cakes
9. 19 oz cookies

MEATS, FISH, AND DAIRY PRICE LIST

Bacon, $1.69 a pound
Chicken, 95¢ a pound
Bluefish, $1.79 a pound
Rib roast, $2.96 a pound
Margarine, 96¢ a pound
Eggs, 89¢ a dozen
Lamb, $2.24 a pound
Sirloin, $2.64 a pound
Cheese, $2.56 a pound
Flounder, $1.92 a pound
Milk, 2 qt for $1.25
Smoked ham, $1.76 a pound

3 Find the cost of:

1. 1 qt milk
2. $2\frac{3}{4}$ lb flounder
3. $3\frac{1}{4}$ lb rib roast
4. $\frac{3}{4}$ lb smoked ham
5. $3\frac{1}{2}$ lb chicken
6. $2\frac{1}{4}$ lb bluefish
7. 6 oz cheese
8. 11 oz margarine
9. 3 lb 5 oz sirloin
10. 4 lb 14 oz rib roast
11. 2 lb 10 oz lamb
12. $1\frac{1}{2}$ doz. eggs

FRUIT AND PRODUCE PRICE LIST

Apples, 49¢ a pound
Grapes, 89¢ a pound
Peaches, 39¢ a pound
Celery, 59¢ a stalk
Peppers, 3 for 35¢
Bananas, 32¢ a pound
Lemons, 6 for 84¢
Pears, 48¢ a pound
Lettuce, 2 heads for 95¢
Potatoes, 10 lb for $1.49
Grapefruit, 3 for 69¢
Oranges, 98¢ a dozen
Carrots, 2 bunches for 67¢
Onions, 3 lb for 81¢
String beans, 40¢ a pound

4 Find the cost of:

1. 2 lb string beans
2. 1 lb onions
3. 1 head lettuce
4. 1 bunch carrots
5. $\frac{3}{4}$ lb pears
6. 5 lb potatoes
7. 1 pepper
8. $2\frac{1}{2}$ doz. oranges
9. $1\frac{1}{4}$ lb grapes
10. 3 lemons
11. 6 grapefruit
12. 2 lb 5 oz bananas
13. 1 lb 11 oz pears
14. 4 lb 6 oz apples
15. 3 oranges

Use the price lists given in problem sets 1 through 4 to find the cost of each of the following orders:

5   1. 1 loaf bread, 1 doz. eggs, $\frac{1}{2}$ lb margarine, 10 lb flour, 1 lb coffee

2. $1\frac{1}{2}$ doz. oranges, 8 oz margarine, 2 lb 3 oz flounder, $4\frac{1}{2}$ lb chicken, 3 peppers, 1 lemon, 1 lb string beans, 1 lb onions

3. 4 cans tuna, $1\frac{1}{2}$ doz. rolls, 1 can tomato juice, 1 bar soap, 1 pkg. detergent, 1 can peaches, 2 grapefruit, 2 lb 4 oz smoked ham

4. 2 doz. eggs, 4 loaves bread, 3 qt milk, 6 lb onions, 9 oranges, $1\frac{3}{4}$ lb bananas, 4 lb 5 oz sirloin, 10 oz cheese

5. 5 lb sugar, 1 can peaches, 3 cans corn, 8 rolls, 2 lemons, $1\frac{1}{2}$ doz. eggs, 12 cans soup, 4 bottles beverage, $4\frac{3}{4}$ lb rib roast, 2 lb 3 oz flounder, 6 peppers

6. 2 bunches carrots, 3 heads of lettuce, $4\frac{1}{2}$ lb chicken, $1\frac{1}{2}$ lb string beans, 4 cans corn, 10 rolls, 1 loaf bread, $1\frac{1}{4}$ lb peaches, 4 grapefruit, 12 oz margarine, 1 lb 8 oz lamb, 2 lb 15 oz bananas

7. 1 qt milk, $3\frac{1}{2}$ lb bluefish, 4 oz cheese, 1 lb 4 oz margarine, 3 lb 9 oz sirloin, 2 lb 14 oz apples, 1 can peaches, 2 pkgs. detergent, 12 bars soap, 3 bunches carrots, 12 peppers, 3 doz. eggs, 12 cans corn, 18 cans soup, 1 lb 3 oz cookies, 20 rolls

6   1. When 1 kg of beef costs $4.16, find the cost of:
   a. 1.6 kg of beef
   b. 900 g of beef
   c. 2.74 kg of beef

2. When 2 kg of apples cost $1.50, find the cost of:
   a. 3.4 kg of apples
   b. 750 g of apples
   c. 1.25 kg of apples

3. When 500 g of cake cost $2.48, find the cost of:
   a. 1.4 kg of cake
   b. 800 g of cake
   c. 450 g of cake

4. When 100 g of cheese cost 50¢, find the cost of:
   a. 600 g of cheese
   b. 1.7 kg of cheese
   c. 380 g of cheese

5. When 250 g of bologna cost $1.25, find the cost of:
   a. 540 g of bologna
   b. 1.1 kg of bologna
   c. 1.75 kg of bologna

6. When gasoline costs 39.9¢ per liter, find the cost of:
   a. 46 L of gasoline
   b. 31.5 L of gasoline
   c. 53.8 L of gasoline

Unit Pricing—Determining the Better Buy

When buying food or other commodities packaged in two different sizes or quantities, one would like to know which is a better buy. Although the larger size or larger quantity is generally marked at the lower unit price, sometimes the smaller size or smaller quantity can be the better buy.

The unit price may be the cost per item, per ounce, per pound, per quart, per liter, per kilogram, or per any other unit of measure. The symbol @ means "at" and indicates the unit price.

> 3 shirts @ $14.95 is read "three shirts at $14.95 each."

Sometimes in working with unit prices, we use parts of a cent represented by a numeral like $.467, which means 46.7 cents or simply 46.7¢.

PROCEDURE

To determine the better buy when two different quantities of the same item are priced at two rates:
(1) Find the unit price (cost per unit) of the item at each rate.
(2) Select the lower unit price as the better buy.

EXAMPLE:

Which is the better buy: 10 oranges for 85¢ or 12 oranges for 99¢?

10 for 85¢ cost 8.5¢ each.   12 for 99¢ cost 8.25¢ each.

ANSWER: 12 oranges for 99¢ is the better buy.

PRACTICE PROBLEMS

1 Rewrite each of the following, using a dollar sign ($) and a decimal point:

1. 37¢	**2.** 95¢	**3.** 4¢	**4.** 8¢	**5.** 5.2¢
6. 23.9¢	**7.** $6\frac{1}{2}$¢	**8.** $30\frac{3}{4}$¢	**9.** 29.1¢	**10.** 43.57¢

2 Rewrite each of the following, using the ¢ symbol:

1. $.86	**2.** $.03	**3.** $.09	**4.** $.63	**5.** $.017
6. $.744	**7.** $2.81	**8.** $.029	**9.** $.9402	**10.** $.0075

3 Read or write, each of the following in terms of cents:

1. 34.6¢	**2.** 5.4¢	**3.** $.713	**4.** $.082	**5.** $.6925

4 Read or write, each of the following in terms of dollars and cents:

1. $2.374	**2.** $1.329	**3.** $5.067	**4.** $3.108	**5.** 1.43\frac{9}{10}$

5 Find the unit price (cost per pound) to the nearest tenth of a cent:

    1. 2-lb package bacon costing $2.98   **2.** 8-oz package sliced cheese costing $1.69

    3. 10-oz jar jelly costing 72¢   **4.** 1-lb 6-oz pie costing $1.49

6 Find the unit price (cost per quart) to the nearest tenth of a cent:

    1. 1 pt vinegar costing 43¢   **2.** 1 gal bleach costing 96¢

    3. 48-oz bottle cranberry juice cocktail costing $1.28

    4. 1-pt 6-oz bottle liquid detergent costing $1.10

    5. 1-qt 14-oz can juice costing 83¢

7 Find the unit price (cost per ounce) to the nearest tenth of a cent:

    1. 3-oz jar olives costing 89¢   **2.** 12-oz box cereal costing $1.29

    3. 6.5-oz can tuna costing $1.09   **4.** $7\frac{3}{4}$-oz can salmon costing $1.95

8 Find the unit price (cost per 50 pieces) to the nearest tenth of a cent:

    1. Package of 100 tea bags costing $1.80

    2. Package of 10 soap pads costing 63¢

    3. Package of 75 storage bags costing $1.50

    4. Package of 160 paper napkins costing 81¢

9 Find the unit price (cost per quart) for each of the bottles of vegetable oil and determine the difference between the highest and lowest unit prices:

    1. 1 gal at $5.49   **2.** 1 pt at 81¢   **3.** 24 oz at $1.13

10 Find the unit price (cost per pound) for each of the packages of cereal and determine the difference between the highest and lowest unit prices:

    1. 12 oz at $1.39   **2.** 8 oz at 97¢   **3.** 15 oz at $1.68

11 Find the unit price (cost per quart) for each of the bottles of syrup and determine the difference between the highest and lowest prices:

    1. 12 oz at $1.12   **2.** 1 pt 8 oz at $1.79   **3.** 1 qt 4 oz at $2.25

12 Find the unit price (cost per single item) in each of the following:

    1. 5 jars of baby food costing $1.45   **2.** 10 oranges costing 95¢

    3. 2 shirts costing $37   **4.** 4 tires costing $165

13 In each, the same item is priced at two rates. Which is a better buy?

    1. Pears: 5 for 49¢ or 2 for 25¢

    2. Apples: 4 for 39¢ or 10 for 89¢

    3. Doughnuts: 12 for $1.89 or 3 for 49¢

    4. Grass seed: 5 lb for $6.50 or 25 lb for $27.75

    5. Soup: 7 cans for $2 or 3 cans for 95¢

    6. Rolls: 6 for 59¢ or 99¢ per dozen

14 Find how much you save by buying the larger size of each of the following items instead of an equivalent quantity in the smaller size:

    1. Pretzels: 4-oz bag at 33¢; 12-oz bag at 89¢

2. Bleach: 1-qt bottle at 45¢; 1-pt bottle at 27¢

3. Paint: 1-qt can at $7.49; 1-gal can at $12.95

4. Spaghetti: 8-oz box at 37¢; 1-lb box at 69¢

15 How much do you save per ounce by buying the larger size?

    1. Peanut butter: 6-oz jar at 85¢; 12-oz jar at $1.59

    2. Mayonnaise: 1-qt jar at $1.69; 1-pt jar at 95¢

    3. Mouthwash: 14-oz bottle at $1.09; 20-oz bottle at $1.39

    4. Tomato juice: 1-pt 2-oz. can at 40¢; 1-qt 14-oz can at 83¢

16 Which is the better buy per unit price (cost per quart or cost per pound)?

    1. Milk: 1 qt at 62¢; a half-gallon at $1.12

    2. Vinegar: 1 qt at 73¢; 1 pt at 43¢

    3. Salad dressing: 8-oz jar at 67¢; 1-pt jar at $1.08

    4. Laundry detergent: 1-lb 4-oz pkg at 99¢; 5-lb 4-oz pkg. at $3.34

    5. Grape jelly: 10-oz jar at 79¢; 20-oz jar at $1.19

    6. Floor wax: 27-oz can at $1.39; 46-oz can at $2.29

    7. Baked beans: 1-lb can at 41¢; 1-lb 10-oz can at 59¢

    8. Beverage: 6 16-oz bottles at $2.39; 8 10-oz bottles at $2.49

    9. Toothpaste: 5-oz tube at 99¢; 7-oz tube at $1.25

17 Find how much you save on each can when buying in quantity:

    1. 1 dozen cans of peas for $4.20 or 41¢ each

    2. 7 cans of soup for $1.75 or 31¢ each

    3. 10 cans of tomato paste for $2.49 or 29¢ each

    4. 5 cans of beans for $1.75 or 37¢ each

    5. 6 cans of tomato juice for $2.24 or 41¢ each

    6. 12 cans of beverage for $2.99 or 29¢ each

18 Find the unit price (cost per kilogram) to the nearest tenth of a cent:

    1. 5 kg of potatoes costing $1.19   **2.** 2.27-kg bag of sugar costing $1.65

    3. 283-g jar of jelly costing 69¢   **4.** 964-g jar of applesauce costing 73¢

19 Find the unit price (cost per liter) to the nearest tenth of a cent:

    1. 2-L bottle of beverage costing $1.25   **2.** 1.36-L can of fruit punch costing 64¢

    3. 532-mL can of tomato juice costing 45¢   **4.** 946-mL carton of milk costing 68¢

20 Which is the better buy per unit price (cost per kilogram or cost per liter)?

    1. Flour: 2.26-kg bag at 95¢ or 907-g bag at 51¢

    2. Bleach: 950-mL bottle at 45¢ or 1.89-L bottle at 67¢

    3. Crackers: 454-g box at $1.05 or 340-g box at 93¢

    4. Pickles: 946-mL jar at $1.09 or 473-mL jar at 65¢

21 Find how much you save buying the larger size of each of the following
items instead of an equivalent quantity in the smaller size:

    1. Fabric softener: .94-L bottle at 33¢; 3.76-L bottle at 87¢

    2. Grapefruit sections: 227-g can at 31¢; 1.362-kg can at $1.29

Vocabulary

Discount is the amount an article is reduced in price.
The **list price** or **marked price** is the regular or full price of an article.
The **net price** or **sale price** is the reduced price.
The **rate of reduction** or **rate of discount** is the percent taken off the list price.

PROCEDURE

To find the discount and net price when the list price and rate of discount are given:
(1) Multiply the list price by the rate of discount to find the discount.
(2) Subtract the discount from the list price to find the net price.

EXAMPLE: Find the discount and the net price of a portable TV when the list price is $290 and the rate of discount is 15%.

$$\begin{array}{r} \$290 \\ \times\ .15 \\ \hline 1450 \\ 290\ \ \\ \hline \$43.50 \text{ discount} \end{array} \qquad \begin{array}{r} \$290.00 \text{ list price} \\ -\ \ 43.50 \text{ discount} \\ \hline \$246.50 \text{ net price} \end{array}$$

ANSWER: $43.50 discount; $246.50 net price

To find the rate of discount:
   Determine what percent the discount is of the regular price by dividing the discount by the list price.

EXAMPLE: What is the rate of discount when the list price of a calculator is $48 and the discount is $12?

$$\frac{\$12}{\$48} = \frac{1}{4} = 25\%$$

ANSWER: 25%

To find the list price when the sale price and the rate of discount are given:
(1) Subtract the given rate from 100%.
(2) Divide the sale price by the answer found in Step 1.

EXAMPLE: Find the list price of a dress that sold for $60 at a 20% discount sale.

$$100\% - 20\% = 80\% = .8$$
$$\$60 \div .8 = \$75$$

ANSWER: $75 list price

CORY'S DEPARTMENT STORE

FURNITURE—Reduced 30%   CLOTHING—Reduced $33\frac{1}{3}$%   APPLIANCES—Reduced 20%
  Tables, $99   Dresses, $69   Refrigerators, $595
  Lamps, $54.50   Coats, $90   Toaster-ovens, $39.95
  Sofas, $425   Shoes, $49.95   Hair dryers, $36.75
  Bedroom sets, $899   Suits, $117.75   Color T.V.s, $445
  Bookcases, $72.75   Sweaters, $18.98   Blenders, $29.95
  Mirrors, $37.98   Skirts, $21.45   Can openers, $14.69

PRACTICE PROBLEMS

1 Use the advertisement above to find the sale price of each listed:

   1. furniture items   **2.** items of clothing   **3.** appliances.

2 How much is a football reduced in price at a 25% reduction sale if its regular price is $15? What is its sale price?

3 At a year-end clearance sale an automobile, regularly selling for $6,850, can now be purchased at a reduction of 17%. What is its sale price?

4 Find the sale price of each of the following articles:

   1. Baseball; regular price, $3.50; reduced 40%
   2. Camera; regular price, $79.95; reduced 20%
   3. Stereo system; regular price, $225; $\frac{1}{3}$ off.
   4. Sewing machine; regular price, $204.50; reduced 15%.
   5. Automatic washer; regular price, $349; reduced 23%

5 Find the rate of reduction allowed when a lawn mower that regularly sells for $185 was purchased at $148.

6 Find the rate of discount allowed on each of the following articles:

   1. Typewriter; regular price, $240; sale price, $180
   2. Kitchen clock; regular price, $18; sale price, $15
   3. Tire; regular price, $48; sale price, $42
   4. Vacuum sweeper; regular price $87.50; sale price, $75
   5. Blanket; regular price, $21.75; sale price $14.50

7 Find the list price of a desk that sold for $149.85 at a 25% reduction sale.

8 Find the regular price of each of the following:

   1. Battery, $28 sale price when reduced 30%
   2. Tablecloth, $20.40 sale price when reduced 15%
   3. Sprinkler, $9.60 sale price when reduced $37\frac{1}{2}$%
   4. Freezer, $246.66 sale price when reduced $33\frac{1}{3}$%
   5. Toaster, $17.55 sale price when reduced 10%

9 A calculator which regularly sells for $49.88 is now on sale at $\frac{1}{4}$ off. What is its sale price?

10 What is the marked price of a pair of slacks that sold for $23.60 at a 20% reduction sale?

11 Mr. Warner bought a camera at a 15% discount. If the regular price was $56, how much discount was he allowed? What net price did he pay?

12 How much trade discount is allowed if the catalog lists a metric tool set at $129.95 and the discount sheet shows a 16% discount? What is the net price?

13 Find the trade discount and the net price of each:

	1.	2.	3.	4.	5.
Catalog list price	$183	$13.49	$93.75	$514	$262.50
Rate of trade discount	27%	12%	20%	$16\frac{2}{3}$%	9%

14 What is the net price of a drill press listed at $342.95 with a trade discount of 12% and an additional cash discount of 5%?

15 Find the net price of each:

	1.	2.	3.	4.	5.
List price	$160	$539.80	$316.50	$875	$920.25
Rates of discount	25%, 4%	30%, 2%	15%, 10%	20%, 8%	10%, 3%

16 Find rate of discount when a radio, listed for $125, is sold for $95.

17 What is the rate of trade discount allowed when the catalog list price of a water heater is $190 and the net price is $161.50?

18 On what list price is the discount $63 when rate of discount is 7%?

19 Find the list price of a stereo when the net price is $280 and the rate of discount is $12\frac{1}{2}$%.

20 Find the list price of each:

	1.	2.	3.	4.	5.
Net price	$42	$368	$73.80	$31.32	$208.95
Rate of discount	25%	8%	10%	$16\frac{2}{3}$%	30%

21 The net price of a bicycle is $102 when a 15% discount is allowed. What is its list price?

A FAMILY BUDGET

A budget is a plan for spending income.

Housing	25%
Food	30%
Clothing	9%
Transportation	15%
Health	5%
Recreation	6%
Savings	8%
Miscellaneous	2%

The Gomez family estimates it will earn a total take-home income of $20,000 for the year.

EXERCISES

Using the above budget, find how much money is allotted per year for:

1. Food   **2.** Housing   **3.** Transportation

4. What is the amount of expected savings for the year?

Find the average amount of money allotted per month for:

5. Housing   **6.** Clothing   **7.** Health

Find the average amount of money allotted per week for:

8. Food   **9.** Recreation   **10.** Transportation

11. During the year the total food costs amounted to $5,809.69. How much over or under budget was this amount?

12. Health costs for the year amounted to $1,075. Is this over or under the budget? How much?

13. Car payments for the year amounted to $1,344.96, gasoline and oil costs were $585.75, repairs and maintenance costs were $239.50, insurance cost $360, and public transportation for the year cost $695. How much over or under budget were the total costs for transportation.

The following year the Gomez family's take-home income will increase to $23,000.

Using the same budget, find how much more money per year will be allotted for:

14. Housing   **15.** Recreation   **16.** Clothing

17. How much more money per month will be allotted for transportation than last year?

18. How much more money per week will be allotted for food than last year?

7-4 Making Change

Today, more and more stores are using computerized cash registers that automatically calculate the correct amount of change for a purchase. However, many purchases are made in situations that require you to figure the amount of change without the aid of a machine.

PROCEDURE

To calculate the correct amount of change:
(1) Add the value of coins, starting with pennies, to the cent portion of the purchase until the sum equals a dollar amount.
(2) Then, if necessary, add the value of bills, starting with ones, to the dollar portion of the purchase until the sum equals the amount offered in payment.

EXAMPLE: Find how much change you should receive from $10 if your purchase is $7.19.

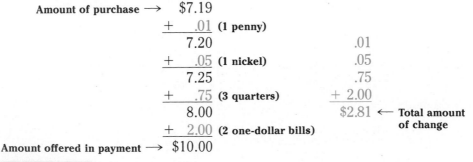

ANSWER: $2.81

PRACTICE PROBLEMS

Find how much change you should receive:

1 From $1 if your purchases cost:
  1. $.79   **2.** $.48   **3.** $.16   **4.** $.09   **5.** $.83

2 From $5 if your purchases cost:
  1. $1.34   **2.** $2.07   **3.** $4.58   **4.** $.72   **5.** $3.96

3 From $10 if your purchases cost:
  1. $6.41   **2.** $3.15   **3.** $9.69   **4.** $7.87   **5.** $.92

4 From $20 if your purchases cost:
  1. $8.08   **2.** $17.94   **3.** $5.16   **4.** $11.29   **5.** $15.47

5 For each of the following purchases, determine how much change should be given when the specified amount of money is offered in payment:

Articles Purchased	Amount Offered in Payment
1. 3 candy bars @ 25¢, 2 ice cream cones @ 65¢	$5.00
2. 1 shirt @ $17.95, 2 ties @ $5.50	$30.00
3. 2 dresses @ $39.98, 3 sweaters @ 19.99	$150.00
4. 48 L gasoline @ 33.9¢, 1 L oil @ $2.25	$20.00
5. 9 cans soup @ 33¢, 3 cans beans @ 37¢	$10.00
6. 8 grapefruit @ 19¢, 15 apples @ 9¢, 10 pears @ 11¢	$5.00
7. 5 gal paint @ $15.49, 2 brushes @ $2.69	$100.00
8. 1 suit @ $109.50, 2 pr. slacks @ $19.95, 4 shirts @ $10.75, 1 tie @ $4.50	$200.00

Refresh Your Skills

1. Add: [1-4]
9,246
3,851
8,678
5,962

2. Subtract: [1-5]
146,592 − 96,478

3. Multiply: [1-6]
246 × 725

4. Divide: [1-7]
$144\overline{)5,616}$

5. Add: [2-4]
8.32 + .527 + 26.6

6. Subtract: [2-5]
9.4 − .51

7. Multiply: [2-7]
.005 × .04

8. Divide: [2-8]
$.6\overline{).036}$

9. Add: [3-8]
$6\frac{3}{5} + 7\frac{3}{4}$

10. Subtract: [3-9]
$3\frac{1}{3} - 2\frac{5}{6}$

11. Multiply: [3-11]
$12\frac{1}{2} \times 2\frac{2}{5}$

12. Divide: [3-12]
$\frac{5}{16} \div 10$

13. Find $12\frac{1}{2}\%$ of 96.
[4-6]

14. What percent of 75 is 30?
[4-7]

15. 18% of what number is 27?
[4-8]

Sales tax is a tax that many states and cities have on the selling price of articles which consumers purchase.

When figuring sales tax, any fractional part of a cent is rounded to the next higher cent.

EXAMPLE: Find the tax on a purchase of $15.50 when the tax rate is 7%.

$$.07 \text{ of } \$15.50 = .07 \times \$15.50 = \$1.085 \text{ or } \$1.09$$

ANSWER: $1.09 sales tax

PRACTICE PROBLEMS

Find the tax on each of the following sales when the tax rate is:

1 2%:   **1.** $5.00   **2.** $10.40   **3.** $84.25   **4.** $134.80

2 6%:   **1.** $16.00   **2.** $.48   **3.** $8.97   **4.** $107.55

3 5%:   **1.** $3.00   **2.** $9.50   **3.** $57.30   **4.** $212.45

4 8%:   **1.** $91.00   **2.** $6.89   **3.** $731.75   **4.** $2,451.49

5 4%:   **1.** $175.00   **2.** $5.12   **3.** $27.64   **4.** $1,783.50

6 $1\frac{1}{2}$%:   **1.** $2.00   **2.** $.79   **3.** $45.80   **4.** $362.75

7 $3\frac{1}{2}$%:   **1.** $12.00   **2.** $9.60   **3.** $78.35   **4.** $825.88

8 Find the selling price, including sales tax, of each of the following articles:

1. Bag, $18.98, tax 8%   **2.** Calculator, $24.99, tax 3%

3. Slacks, $29.50, tax 4%   **4.** Bedspread, $31.75, tax 6%

5. Crib, $87.49, tax $4\frac{1}{2}$%   **6.** Basketball, $14.89, tax 2%

7. Wading pool, $39.95, tax 5%   **8.** Heating pad, $10.88, tax $3\frac{1}{2}$%

9. Pajamas, $9.79, tax $2\frac{1}{2}$%   **10.** Electric dryer, $269.95, tax $1\frac{1}{2}$%

Sales Tax Table—4%					
Purchase	**Tax**	**Purchase**	**Tax**	**Purchase**	**Tax**
$.01–$ .10	$.00	$1.11–$1.25	$.05	$2.11–$2.25	$.09
.11– .25	.01	1.26– 1.50	.06	2.26– 2.50	.10
.26– .50	.02	1.51– 1.75	.07	2.51– 2.75	.11
.51– .75	.03	1.76– 2.10	.08	2.76– 3.10	.12
.76– 1.10	.04				

9 Use the table above to find the tax on each of the following amounts of purchase:

1. $.28   **2.** $.75   **3.** $1.47   **4.** $1.09   **5.** $2.61

6. $1.24   **7.** $2.98   **8.** $.61   **9.** $2.51   **10.** $3.05

7-6 Utility Services

In order to check bills for the gas, electricity, and water services, one must be able to read a meter measuring the consumption.

PROCEDURE

To read a gas, electric, or water meter:
(1) Select on each dial the numeral that was just passed by the pointer. Note: Some pointers move clockwise and others move counterclockwise.
(2) Then put the dial readings together and read from left to right.

EXAMPLE: Read

ANSWER: 4627

We read the meter in the same way for each of the services, but different units are used to express their consumption.

> Gas is expressed in hundred cubic feet.
> The reading 5486 represents 5,486 × 100 cu. ft or 548,600 cu. ft

> Electricity is expressed in kilowatt hours (kWh).
> The reading 8291 represents 8,291 kilowatt hours.

> Water is expressed in tens of gallons.
> The reading 25736 represents 25,736 × 10 gallons or 257,360 gal.

PROCEDURE

To find the consumption of a service during a given period:
(1) Subtract the reading at the beginning of the period from the reading at the end of the period.
(2) Then express the difference in the appropriate units.

EXAMPLE: What is the gas consumption for the period when the beginning reading is 4276 and the end reading is 4737?

$$4737 - 4276 = 461 \qquad 461 = 461 \times 100 \text{ cu. ft} = 46,100 \text{ cu. ft}$$

ANSWER: 46,100 cu. ft

PRACTICE PROBLEMS

1 **1.** What is the reading of the gas meter on April 1? On May 1?

Cubic Feet   Cubic Feet

April 1   May 1

2. How many cubic feet of gas were consumed during the month of April?

3. Find the cost of the gas consumed if the rates were:

First 100 cu. ft $3.47
Next 900 cu. ft $.90 per 100 cu. ft
Next 1,500 cu. ft $.79 per 100 cu. ft
Next 5,000 cu. ft $.73 per 100 cu. ft
Next 12,500 cu. ft $.71 per 100 cu. ft
Over 20,000 cu. ft $.70 per 100 cu. ft

2 Given the reading at the beginning and at the end of the period, use the above rates in each of the following to find the cost of the gas consumed during the period.

	READING				READING	
	At Beginning	At End			At Beginning	At End
1.	2418	2487		**6.**	5523	5541
2.	8594	8691		**7.**	9905	0298
3.	6956	7143		**8.**	6182	6425
4.	1009	1216		**9.**	3527	3982
5.	7231	7495		**10.**	0064	0639

3 **1.** What is the reading of the electric meter on November 1? on December 1?

Kilowatt Hours   Kilowatt Hours

November 1   December 1

2. How many kilowatt hours of electricity were used during the month of November?

3. Find the cost of the electricity used if the rates were:

> First 12 kWh or less $4.20
> Next 42 kWh @ $.09
> Next 46 kWh @ $.08
> Over 100 kWh @ $.069

4 Given the reading at the beginning and at the end of the period, use the above rates in each of the following to find the cost of the electricity used during the period.

	READING	
	At Beginning	At End
1.	4726	4773
2.	6175	6266
3.	1592	1704
4.	3204	3381
5.	9887	0069
6.	8153	8163
7.	2261	2345
8.	0095	0506
9.	7949	8482
10.	5838	6097

5 At the average cost of 8.6¢ per kilowatt hour (1 kilowatt hour = 1,000 watts/h), find the hourly cost of operating each of the following:

1. Toaster, 1,000 watts   **2.** Dryer, 4,000 watts   **3.** 100-watt light bulb

4. Heater, 1,500 watts   **5.** Iron, 1,200 watts   **6.** Radio, 40 watts

6 In many communities water bills are issued every 3 months. The reading on a certain water meter on February 1 was 43582, on May 1 it was 46106, on August 1 it was 51263, and on November 1 it was 54091.

1. How many gallons of water were used during the 3-month period ending:
    a. May 1?   **b.** August 1?   **c.** November 1?

2. Find the cost of the water consumed during each of these periods using the rates: First 2,500 gallons or less, $7.58; Over 2,500 gallons, $1.77 for every 1,000 gallons.

7 During the year Mr. Ross bought 210 gallons of fuel oil at $1.249 per gallon, 185 gallons at $1.219 each, 235 gallons at $1.226 each, 220 gallons at $1.235 each, and 190 gallons at $1.253 each. How many gallons of fuel oil did he buy? What was the total cost of his fuel oil for the year?

7-7 Owning an Automobile

There are many costs involved in owning an automobile. The variable costs, like gasoline, tires, and repairs, are dependent on how many miles you drive. Costs like insurance and registration fees are fixed and usually remain the same from year to year. Depreciation, which is the decrease in value of an automobile on a yearly basis, is another fixed cost.

PRACTICE PROBLEMS

1 **1.** Mr. Delano bought a new car for $8,250. After using it for 4 years, he sold it for $1,725. What was the average yearly depreciation of the car?

2. At the end of the first year the odometer read 9425 miles, at the end of the second year 21132 miles, at the end of the third year 33051 miles, and at the end of the fourth year 47009 miles. How many miles did Mr. Delano drive during the second year? During the third year? During the fourth year?

3. During the first year his expenses in addition to the average yearly depreciation were: gasoline, 620 gallons at $1.59 per gal; oil, 30 quarts at $2.10 per qt; insurance, $385 per yr; license fees, $36 per yr; parking, $65 per yr; miscellaneous expenses, $125.

    a. How much did it cost Mr. Delano to run his car for the first year?

    b. What was the average cost per month? **c.** Per week? **d.** Per mile?

2 How far can a car go on a tankful of gasoline if it averages 15 miles on a gallon and the tank holds 22 gallons?

3 How long will it take a person, driving at an average speed of 50 m.p.h., to travel from Washington to Boston, a distance of 460 miles? If the car can average 14 miles on a gallon of gasoline, how many gallons are required to make the trip?

4 Mrs. Potter insures her car costing $9,000 against fire and theft for 80% of its value at the annual rate of 85¢ per $100. If the charge per year for property damage insurance is $81.75, for liability insurance is $68.50, and for collision insurance is $76, what is her total annual premium?

READING A ROAD MAP

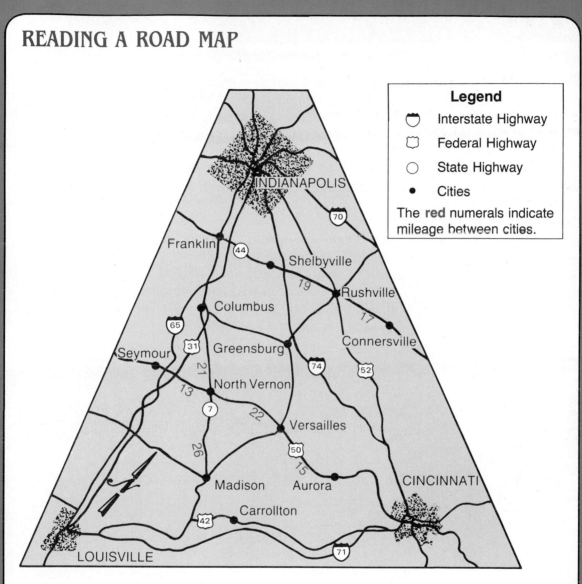

Legend

⬡ Interstate Highway

⬚ Federal Highway

◯ State Highway

● Cities

The **red** numerals indicate mileage between cities.

EXERCISES

Listing highway markers and towns, what route would you take to drive by car from:

1. Indianapolis to Cincinnati?

2. Louisville to Indianapolis?

3. Cincinnati to Louisville?

4. What alternate routes could you take for each of the above?

How far is it by highway from:

5. Shelbyville to Connersville?   **6.** Columbus to Madison?   **7.** Seymour to Aurora?

319

7-8 Installment Buying

Installment buying, or buying on credit, is a purchase plan by which the customer usually pays a certain amount of money, called the down payment, at the time of purchase and then pays, in equal installments at regular intervals, usually months, both the balance and a finance charge on this balance.

This finance charge, also called a carrying charge, is the interest which usually is applied on the entire balance as being owed for the full time of payment. The carrying or finance charge is the difference between the cash price and the total amount paid.

Many stores use deferred payment purchase plans. In these plans a specified monthly payment is made to reduce the balance owed. Sometimes these payments are fixed amounts. Other times these payments decrease as the bracket of the balance owed falls. However, a finance charge, usually $1\frac{1}{4}\%$ or $1\frac{1}{2}\%$ per month, is applied to the previous monthly unpaid balance and is added to this previous balance.

PRACTICE PROBLEMS

1 Mrs. Smith bought an automatic dishwasher, paying $25 down and $46.50 a month for 8 months. How much more did the dishwasher cost on the installment plan if the cash price was $375?

2 Lisa's mother can purchase a refrigerator for the cash price of $489 or $39 down and 10 equal monthly payments of $48.38 each. How much is the carrying charge? How much can be saved by paying cash?

320

3 Find the carrying charge when each of the following articles is purchased on the installment plan:

	Cash Price	Down Payment	Monthly Payment	Number of Monthly Payments
1. Computer	$950	$50	$63.00	15
2. Stereo	$175	$25	$18.00	9
3. Clothes dryer	$259	$19	$15.10	18
4. Air conditioner	$245	$45	$27.50	8
5. Typewriter	$179	$24	$16.43	10

4 Find the monthly payment if the amount financed is $960 and the finance charge is $180, both to be paid in 30 equal installments.

5 A TV set can be purchased for the cash price of $395 or $35 down and the balance and finance charge to be paid in 12 equal monthly payments. If 6% was charged on the full balance, how much is the monthly payment?

6 Find the monthly payment when each of the following articles is purchased on the installment plan and the finance-charge interest rate is on the full balance:

	Cash Price	Down Payment	Finance Charge Interest Rate	Number of Monthly Payments
1. Camera	$118	$10	6%	12
2. VCR	$595	$55	6%	12
3. Bedroom set	$869	$59	8%	18
4. Sewing machine	$275	$35	$8\frac{1}{2}$%	6
5. Microwave oven	$425	$25	$7\frac{1}{2}$%	10
6. Fur coat	$585	$75	6%	12
7. CB radio	$190	$15	8%	6
8. Drum set	$250	$25	6%	8
9. Desk	$220	$20	9%	12

7 A department store uses the following deferred payment schedule:

Monthly Balance	Monthly Payments	Monthly Balance	Monthly Payments
$0 to $10	Full Balance	$300.01 to $350	$25
$10.01 to $200	$10	$350.01 to $400	$30
$200.01 to $250	$15	$400.01 to $500	$40
$250.01 to $300	$20	Over $500	$\frac{1}{10}$ of balance

If finance charges are computed at the monthly rate of $1\frac{1}{4}$% on the previous month's balance, find the monthly payment due on each of the following previous month's balances. First compute the finance charge, add it to the month's balance, and then find in the above schedule the monthly payment due.

1. $79   **2.** $198   **3.** $600   **4.** $265   **5.** $349

The federal government requires each credit purchase or loan contract to indicate the annual percentage rate, which is the *true* finance charge for the year. Consumers can use this information to compare the cost of credit for various types of financing.

PROCEDURE

To compute the annual percentage rate:
   Multiply the rate per period by the number of periods per year.

EXAMPLE: Find the annual percentage rate on a finance charge of $1\frac{1}{2}$% per month.

$$12 \times 1\frac{1}{2}\% = 18\%$$

ANSWER: 18% annual percentage rate

PRACTICE PROBLEMS

1 Find the annual percentage rate if the monthly finance charge rate is:

1. 1%   **2.** $1\frac{3}{4}$%   **3.** $2\frac{1}{2}$%   **4.** $1\frac{2}{3}$%   **5.** $2\frac{1}{4}$%

2 Find the annual percentage rate if the monthly late payment finance charge rate on past due amounts is:

1. 2%   **2.** $\frac{3}{4}$ of 1%   **3.** $\frac{2}{3}$ of 1%   **4.** $\frac{7}{8}$ of 1%   **5.** $\frac{5}{6}$ of 1%

Most stores use Annual Percentage Rate Tables to determine the rates.

PROCEDURE

To find the annual percentage rate using a table when the amount to be financed, the finance charge for this amount, and the number of monthly payments are known:
(1) Multiply the given finance charge by 100 and divide this product by the amount financed.
(2) Locate the number of monthly payments on the table.
(3) Read across the table until you find the number nearest to the amount determined in step 1.
(4) The rate at the top of this column is the annual percentage rate.

EXAMPLE: Find the annual percentage rate if the amount to be financed is $1,500, the finance charge is $151.65, and the number of monthly payments is 15.
$151.65 \times 100 = \$15,165$ $\$15,165 \div \$1,500 = 10.11$

ANSWER: 14.75%

Number of Payments	Annual Percentage Rate										
	14.00%	14.25%	14.50%	14.75%	15.00%	15.25%	15.50%	15.75%	16.00%	17.00%	18.00%
(FINANCE CHARGE PER $100 OF AMOUNT FINANCED)											
1	1.17	1.19	1.21	1.23	1.25	1.27	1.29	1.31	1.33	1.42	1.50
2	1.75	1.78	1.82	1.85	1.88	1.91	1.94	1.97	2.00	2.13	2.26
3	2.34	2.38	2.43	2.47	2.51	2.55	2.59	2.64	2.68	2.85	3.01
4	2.93	2.99	3.04	3.09	'3.14	3.20	3.25	3.30	3.36	3.57	3.78
5	3.53	3.59	3.65	3.72	3.78	3.84	3.91	3.97	4.04	4.29	4.54
6	4.12	4.20	4.27	4.35	4.42	4.49	4.57	4.64	4.72	5.02	5.32
7	4.72	4.81	4.89	4.98	5.06	5.15	5.23	5.32	5.40	5.75	6.09
8	5.32	5.42	5.51	5.61	5.71	5.80	5.90	6.00	6.09	6.48	6.87
9	5.92	6.03	6.14	6.25	6.35	6.46	6.57	6.68	6.78	7.22	7.65
10	6.53	6.65	6.77	6.88	7.00	7.12	7.24	7.36	7.48	7.96	8.43
11	7.14	7.27	7.40	7.53	7.66	7.79	7.92	8.05	8.18	8.70	9.22
12	7.74	7.89	8.03	8.17	8.31	8.45	8.59	8.74	8.88	9.45	10.02
13	8.36	8.51	8.66	8.81	8.97	9.12	9.27	9.43	9.58	10.20	10.81
14	8.97	9.13	9.30	9.46	9.63	9.79	9.96	10.12	10.29	10.95	11.61
15	9.59	9.76	9.94	10.11	10.29	10.47	10.64	10.82	11.00	11.71	12.42
16	10.20	10.39	10.58	10.77	10.95	11.14	11.33	11.52	11.71	12.46	13.22
17	10.82	11.02	11.22	11.42	11.62	11.82	12.02	12.22	12.42	13.23	14.04
18	11.45	11.66	11.87	12.08	12.29	12.50	12.72	12.93	13.14	13.99	14.85
19	12.07	12.30	12.52	12.74	12.97	13.19	13.41	13.64	13.86	14.76	15.67
20	12.70	12.93	13.17	13.41	13.64	13.88	14.11	14.35	14.59	15.54	16.49
21	13.33	13.58	13.82	14.07	14.32	14.57	14.82	15.06	15.31	16.31	17.32
22	13.96	14.22	14.48	14.74	15.00	15.26	15.52	15.78	16.04	17.09	18.15
23	14.59	14.87	15.14	15.41	15.68	15.96	16.23	16.50	16.78	17.88	18.98
24	15.23	15.51	15.80	16.08	16.37	16.65	16.94	17.22	17.51	18.66	19.82
25	15.87	16.17	16.46	16.76	17.06	17.35	17.65	17.95	18.25	19.45	20.66
26	16.51	16.82	17.13	17.44	17.75	18.06	18.37	18.68	18.99	20.24	21.50
27	17.15	17.47	17.80	18.12	18.44	18.76	19.09	19.41	19.74	21.04	22.35
28	17.80	18.13	18.47	18.80	19.14	19.47	19.81	20.15	20.48	21.84	23.20
29	18.45	18.79	19.14	19.49	19.83	20.18	20.53	20.88	21.23	22.64	24.06
30	19.10	19.45	19.81	20.17	20.54	20.90	21.26	21.62	21.99	23.45	24.92
31	19.75	20.12	20.49	20.87	21.24	21.61	21.99	22.37	22.74	24.26	25.78
32	20.40	20.79	21.17	21.56	21.95	22.33	22.72	23.11	23.50	25.07	26.65
33	21.06	21.46	21.85	22.25	22.65	23.06	23.46	23.86	24.26	25.88	27.52
34	21.72	22.13	22.54	22.95	23.37	23.78	24.19	24.61	25.03	26.70	28.39
35	22.38	22.80	23.23	23.65	24.08	24.51	24.94	25.36	25.79	27.52	29.27
36	23.04	23.48	23.92	24.35	24.80	25.24	25.68	26.12	26.57	28.35	30.15
37	23.70	24.16	24.61	25.06	25.51	25.97	26.42	26.88	27.34	29.18	31.03
38	24.37	24.84	25.30	25.77	26.24	26.70	27.17	27.64	28.11	30.01	31.92
39	25.04	25.52	26.00	26.48	26.96	27.44	27.92	28.41	28.89	30.85	32.81
40	25.71	26.20	26.70	27.19	27.69	28.18	28.68	29.18	29.68	31.68	33.71

PROCEDURE

To find the finance charge and the monthly payment when the amount to be financed, the annual percentage rate, and number of months are known:
(1) Locate the annual percentage rate in the top row of the table.
(2) Read down that column until you reach the appropriate number of monthly payments.
(3) Multiply the amount found in step 2 by the amount to be financed, then divide this product by 100 to get the finance charge.
(4) Add the finance charge found in step 3 to the amount to be financed and then divide by the number of monthly payments to find the monthly payment.

EXAMPLE: Find the finance charge and the monthly payment when $1,500 is financed at an annual percentage rate of 14.75% for 15 months.

$10.11 \times \$1,500 = \$15,165$ $\$151.65 + \$1,500 = \$1,651.65$
$\$15,165 \div 100 = \$151.65$, finance charge $\$1,651.65 \div 15 = \$110.11$, monthly payment

ANSWER: $151.65 finance charge; $110.11 monthly payment

PRACTICE PROBLEMS

1 Using the table, find the annual percentage rate when:

	Total Amount Financed	Finance Charge on Total Amount	Number of Monthly Payments
1.	$100	$6.65	10
2.	$100	$14.82	21
3.	$400	$120.60	36
4.	$750	$93.45	16
5.	$575	$33.35	8

2 Using the table, find the finance charge for each of the following amounts financed at the given annual percentage rates. Also find the monthly payment.

	Amount Financed	Annual Percentage Rate	Number of Monthly Payments
1.	$100	15.00%	9
2.	$100	18.00%	5
3.	$900	14.25%	18
4.	$1,600	15.75%	24
5.	$4,500	14.50%	30

7-10 Fire Insurance

Fire insurance is a plan by which persons share risks so that each person is protected against financial loss. The insured person is charged a sum of money called a premium. The insurance company uses the fund created by these premiums to pay the insured persons who suffer losses.

The written contract between the insured person and the insurance company is called the policy. The amount of insurance specified in the policy is called the face value of the policy. The length of time the insurance is in force is called the term of the policy.

PROCEDURE

To find the premium:
(1) For a 1-year policy, multiply the number of $100's of insurance by the yearly rate per $100.
(2) For a pre-paid 3-year policy, multiply the number of $100's of insurance by 2.7 times the yearly rate per $100.
(3) For a 3-year policy paid in yearly installments, multiply the amount found in step 2 above by 35% to find each year's installment.

EXAMPLE: Find the yearly premium on an $80,000 policy at a yearly rate of $.18 per $100 for:

a. a 1-year policy $80,000 ÷ $100 = 800   800 × $.18 = $144, premium
b. a 3-year pre-paid policy $144 × 2.7 = $388.80, premium
c. a 3-year policy with yearly installments $388.80 × 35% = $136.08, premium

ANSWERS: $144, annual; $388.80, 3-year pre-paid; $136.08, 3-year installment

PRACTICE PROBLEMS

1 Find the premium for each of the following 1-year policies:

	1.	2.	3.	4.
Face value of policy	$25,000	$69,000	$14,000	$107,500
Yearly rate per $100	$.14	$.35	$.25	$.30

2 Find for each of the following 3-year policies the total premium and the yearly premium if paid by yearly installments:

	1.	2.	3.	4.
Face value of policy	$46,000	$11,000	$50,000	$95,000
Yearly rate per $100	$.16	$.32	$.28	$.26

3 How much does Mr. Collins save by buying a 3-year policy instead of three 1-year policies if he insures his house for $42,500 at the yearly rate of $.24 per $100?

Life insurance offers financial protection to the dependents of an insured person in the event of the insured person's death. The person who receives the money when the insured person dies is called the beneficiary. There are four main types of life insurance.

Term insurance—The person is insured for a specified period of time. Premiums are paid only during that time. The beneficiary receives face value of policy if the insured person dies within the specified time.

Ordinary life insurance—The person is insured until death. Premiums are paid until death of the insured person. The beneficiary receives face value of policy when the insured person dies.

Limited-payment life insurance—The person is insured until death. Premiums are paid for a specified period of time. The beneficiary receives face value of policy when the insured person dies.

Endowment insurance—The person is insured for a specified period of time. Premiums are paid only during that time. The beneficiary receives face value of policy if the insured person dies within the specified time. The insured person, or policyholder, receives face value of policy if alive at the end of the specified time.

Annual Life Insurance Premiums Per $1,000

Age	5-Year Term		Ordinary Life		20-Payment Life		20-Year Endowment	
	Male	Female	Male	Female	Male	Female	Male	Female
18	3.77	3.74	13.48	12.95	23.97	23.22	45.17	45.17
20	4.01	4.01	14.20	13.62	24.91	24.11	45.30	45.30
22	4.20	4.20	15.00	14.35	25.89	25.05	45.46	45.46
24	4.40	4.40	15.86	15.15	26.94	26.03	45.64	45.64
25	4.50	4.50	16.34	15.59	27.49	26.55	45.75	45.75
26	4.60	4.60	16.83	16.04	28.06	27.09	45.86	45.85
28	4.83	4.83	17.90	17.02	29.26	28.21	46.14	46.10
30	5.09	5.05	19.09	18.11	30.54	29.42	46.48	46.40
35	6.08	5.86	22.71	21.43	34.24	32.88	47.72	47.49
40	8.03	7.44	27.20	25.56	38.49	36.85	49.45	49.00
45	11.10	10.13	32.79	30.73	43.34	41.40	51.81	51.06
50	16.00	14.45	40.07	37.50	49.28	46.94	55.40	54.18

PROCEDURE

To find the premium:
(1) Select from the table the type of insurance, and whether male or female.
(2) Read down the column until you reach the appropriate age.
(3) Multiply this rate by the number of $1,000's of insurance of your policy.

EXAMPLE: Find the annual premium on an ordinary life insurance policy for $90,000 issued to a woman at age 35.

$90,000 ÷ $1,000 = 90$ $90 × $21.43 = $1,928.70$

ANSWER: $1,928.70 annual premium

PRACTICE PROBLEMS

1 Find the annual premium on each of the following policies:

  1. 20-payment life for $10,000 issued to a woman at age 30.
  2. 5-year term for $50,000 issued to a man at age 40.
  3. 20-year endowment for $5,000 issued to a woman at age 25.
  4. Ordinary life for $60,000 issued to a man at age 20.
  5. 5-year term for $40,000 issued to a woman at age 22.
  6. Ordinary life for $15,000 issued to a man at age 26.
  7. 20-payment life for $80,000 issued to a woman at age 18.
  8. 20-year endowment for $25,000 issued to a woman at age 35.
  9. Ordinary life for $100,000 issued to a man at age 24.
  10. 20-year endowment for $30,000 issued to a man at age 28.

2 How much more is the male's life insurance premium per $1,000 than the female's rate on each of the following policies:

  1. 20-year endowment, issued at age 24?   **2.** Ordinary life, issued at age 30?
  3. 5-year term, issued at age 45?   **4.** 20-payment life, issued at age 28?
  5. Ordinary life, issued at age 20?

3 Ms. Myers bought a 20-payment life insurance policy for $20,000 when she was 35 years old. What is the total amount of premium she must pay before the policy is fully paid up?

4 At the age of 25 Mr. Ryan bought an ordinary life policy for $10,000.

  1. What annual premium does he pay?
  2. How much will his beneficiary receive when he dies?

5 Mrs. Miller purchased a 20-year endowment policy for $5,000 when she was 30 years old. She is now 50 years old.

  1. How much will she receive?   **2.** What was the total premium she paid?

6 When Mr. Walker was 40 years old he purchased a 5-year term policy for $25,000. He died after paying 3 annual premiums. How much more did the beneficiary receive than Mr. Walker paid?

7 Rosa's brother, now 25 years old, wishes to buy either a 20-payment life policy or an ordinary life policy.

  1. What will the difference be in total premiums paid per $1,000 insurance for these policies when he becomes 45 years of age?
  2. 55 years of age?
  3. 65 years of age?

A tax on buildings and land is called a real property or real estate tax. The value that tax officers place on property for tax purposes is assessed value of the property.

The tax rate on property may be expressed as: cents per $1.00, mills per $1.00, dollars and cents per $100, dollars and cents per $1,000, and percent.

> 1 mill = one-thousandth of a dollar = $.001
> 10 mills = one cent = $.01

PROCEDURE

To find the amount of taxes on property:
(1) Express the assessed value so that the appropriate tax rate can be applied.
(2) Multiply by the tax rate.

EXAMPLE: Find the taxes on a house assessed for $19,800 when the tax rate is $2.70 per $100.

$19,800 ÷ $100 = 198

$2.70 tax rate per $100
× 198 number of $100's
2160
2430
270
$534.60 taxes

ANSWER: $534.60, taxes

PRACTICE PROBLEMS

1 Write as a decimal part of a dollar:
1. 7 mills　　2. 19 mills　　3. 50 mills　　4. 132 mills
5. 84 mills　　6. 175 mills　　7. 60 mills　　8. 92 mills

2 Write as cents:
1. 20 mills　　2. 9 mills　　3. 35 mills　　4. 14 mills
5. 146 mills　　6. 180 mills　　7. 58 mills　　8. 250 mills

3 Write as mills:
1. $.04　　2. $.08　　3. $.005　　4. $.026
5. $.061　　6. $.105　　7. $.070　　8. $.095

4 Write as cents per $1:
1. $4 per $100　　2. 28 mills per $1　　3. $43 per $1,000　　4. 3.4%

5 Write as mills per $1:
1. $5 per $100　　2. $33 per $1,000　　3. 4.5%　　4. $.19 per $1

6 Write as dollars and cents per $100:
1. $.08 per $1　　2. 5.3%　　3. 36 mills per $1　　4. $19 per $1,000

7 Write as dollars and cents per $1,000:

1. $4.50 per $100   **2.** 2.9%   **3.** 42 mills per $1   **4.** $.25 per $1

8 Write as percent:

1. $4 per $100   **2.** $24.50 per $1,000   **3.** $.07 per $1   **4.** 61 mills per $1

9 A community requires $1,000,000 in tax money for a year. The assessed valuation of the property in the community is $50,000,000.

1. What percent of the assessed valuation is the tax needed?
2. Express this tax rate in dollars per $100.
3. In cents per $1.   **4.** In mills per $1.

10 How much must you pay for taxes on a house assessed for $10,200 if the tax rate is $5.60 per $100?

11 If the tax rate is 43 mills per $1, how much must you pay for taxes on a building assessed for $28,500?

Find the amount of taxes on property having the following assessed valuations and tax rates:

12

	1.	2.	3.	4.
Assessed valuation	$16,000	$9,500	$54,000	$41,700
Tax rate per $100	$2.90	$4.25	$3.80	$5.45

13

	1.	2.	3.	4.
Assessed valuation	$14,800	$8,600	$45,000	$79,500
Tax rate per $1	$.04	$.10	$.058	$.07$\frac{1}{2}$

14

	1.	2.	3.	4.
Assessed valuation	$20,000	$7,500	$43,200	$59,800
Tax rate per $1	26 mills	41 mills	75 mills	69 mills

15

	1.	2.	3.	4.
Assessed valuation	$8,000	$32,000	$15,600	$70,500
Tax rate per $1,000	$25	$53	$91	$33

16

	1.	2.	3.	4.
Assessed valuation	$5,000	$28,300	$66,500	$85,000
Tax rate	3%	5%	4.6%	2.8%

17 Mr. Moore's house is assessed for 75% of its cost. The tax rate is $2.45 per $100. How much property tax does he pay if the house cost $42,400?

18 **1.** The assessed valuation of the Santini's house is $19,700. The tax rate is $3.40 per $100. What is the yearly property tax?

2. A 3% discount is allowed if the tax is paid within the first quarter of the year. How much will the Santini family save and how much tax will they pay if they take advantage of this discount?

When borrowing money to buy a house, a condominium, or space in a cooperative, the borrower gives the lender, generally a bank, a written claim, called a mortgage, to the property until full payment is made. *To find the amount of the mortgage, subtract the down payment from the selling price.*

The banks charge interest on the mortgage for the use of their money. This amount of interest depends on the period of the mortgage, the amount of the mortgage, and the present interest rate for that bank. The interest rate can vary from bank to bank.

Monthly Payments on a $1,000 Loan						
Period of Loan	**Yearly Interest Rate**					
	10%	**11%**	**$11\frac{1}{2}$%**	**12%**	**13%**	**14%**
20 years	9.66	10.33	10.67	11.01	11.72	12.44
25 years	9.09	9.81	10.17	10.53	11.28	12.04
30 years	8.78	9.53	9.91	10.29	11.07	11.85
35 years	8.60	9.37	9.77	10.16	10.96	11.76
40 years	8.50	9.29	9.69	10.09	10.90	11.72

PROCEDURE

To find the monthly payment for a loan:
(1) Find the amount of the mortgage.
(2) Express the amount of the mortgage in $1,000's.
(3) Find monthly payment per $1,000 at the given rate of interest for period of loan.
(4) Multiply this amount by the number of $1,000's.

EXAMPLE: The Tungs bought a house for $81,500 with a down payment of $16,300 and a loan for the balance on a 30-year mortgage at a yearly rate of 12%. How much were their monthly payments for the loan?

Amount of mortgage $81,500 − $16,300 = $65,200

$65,200 ÷ $1,000 = 65.2   Monthly payment per $1,000

Monthly payment   65.2 × $10.29 = $670.91

ANSWER: $670.91, monthly payment

PRACTICE PROBLEMS

1 If no more than 25% of a person's income should be spent for the rent of a house or apartment, find the highest monthly rent someone earning each of the following incomes can afford to pay:

1. $18,500 per year

2. $750 per month

3. $440 per week

2 If a family should not buy a house costing more than $2\frac{1}{2}$ times its annual income, what is the highest price a family can afford to pay if its income is:

1. $27,250 per year?   **2.** $1,250 per month?   **3.** $250 per week?

3 A woman bought a house for $65,495. She paid $16,495 down and gave a mortgage for the balance. What is the amount of her mortgage?

What is the monthly payment on each of the following mortgage loans:

4 A $50,000 loan for a period of 20 years at 14% yearly interest?

5 A $68,500 loan for a period of 35 years at $11\frac{1}{2}$% yearly interest?

6 The Mugridges bought a house for $107,000 with a down payment of $21,400 and a loan for the balance on a 25-year mortgage at a yearly rate of 13%. How much were their monthly payments for the loan?

7 Todd's parents bought a condominium for $47,500, paying $7,500 down. They are charged 13% interest on a loan for the balance. The condominium is assessed for $20,600 and the tax rate is $5.20 per $100. The condominium is insured for 80% of its cost at $.28 per $100. The mortgage is to be paid off in 20 years in monthly installments. The down payment of $7,500 was formerly invested at $8\frac{1}{2}$% interest.

1. How much interest is due on the mortgage for the first year?

2. What is the property tax?   **3.** What does the insurance cost?

4. How much interest do Todd's parents lose on money used as down payment?

5. How much does the ownership of the condominium cost Todd's parents for the first year? What is the monthly cost?

6. How much is the annual installment due on the mortgage?

7. How much is the monthly payment on the loan?

8 Compare the costs of renting and buying a house for the first year when the house can be rented for $385 per month or can be purchased for $39,900 with a down payment of $4,900. It is assessed for $22,500 with a tax rate of $3.85 per $100. Insurance costs $325 for a year. The principal used as a down payment could be invested at 9% interest. The interest on the mortgage is 14%. Repairs average $400 per year.

Career
Applications

Spending too much for insurance could ruin your budget, but not having enough insurance coverage could be much worse. *Insurance agents* sell insurance. An insurance agent can analyze a customer's needs and plan a package with the appropriate insurance on the customer's life, health, car, and home or apartment.

Most insurance agencies will hire high-school graduates who have sales ability. An agent must pass a state examination on insurance laws.

1. A unit of life insurance has a face value of $1000. A customer purchases 50 units. The annual premium per unit is $14.30. What is the total annual premium?

2. A customer's employer pays 85% of the cost of medical insurance. If the total monthly premium is $64.80, how much does the employer pay?

3. An insurance agent explains that a customer can pay quarterly premiums (four times a year) instead of one large annual premium. The quarterly premium is 25.5% of the annual premium. If the quarterly premium is $191.25, what is the annual premium?

4. To find the annual premium for automobile insurance, an agent multiplies the base premium by the customer's driver-rating factor. One customer's driver-rating factor is 1.85. She chooses insurance with a base premium of $115.40. What is her annual premium?

5. A homeowner decides to insure his house for $83,700. The replacement value of the house is $93,000. What percent of the replacement value is covered by insurance?

6. Because an apartment house has smoke detectors, the annual premium for renter's insurance is 8% less than the base premium. The base premium is $110. How much does the tenant save?

CHAPTER REVIEW

Vocabulary

annual percentage rate p. 322	finance charge p. 320	policy p. 325
assessed value p. 328	fire insurance p. 325	premium p. 325
carrying charge p. 320	installment buying p. 320	rate of discount p. 308
deferred payment p. 320	life insurance p. 326	real estate tax p. 328
depreciation p. 318	limited payment life	real property tax p. 328
discount p. 308	insurance p. 326	sale price p. 308
down payment p. 320	list price p. 308	sales tax p. 314
endowment insurance p. 326	mortgage p. 330	term insurance p. 326
face value of the policy	net price p. 308	term of the policy p. 325
p. 325	ordinary life insurance p. 326	unit price p. 305

The numerals in the boxes indicate Lessons where help may be found.

1. What is the cost of 3 lb of tomatoes at $.79 per pound? 7-1

2. Find the unit price (cost per pound) of a 12 oz package of cheese costing $1.92. 7-2

3. Which is the better buy: 3 cans of soup for 87¢ or 7 cans for $2? 7-2

4. A VCR marked $798 was sold at a reduction of 20%. What was the sale price? 7-3

5. Find the regular price of a computer that sold for $1,149 at a 25% reduction sale. 7-3

6. Antonio purchased a shirt for $17.95 and a pair of shoes for $34.75. How much change should he get from $60? 7-4

7. If the sales tax rate is 8%, what would the tax be on a radio that sells for $37.50? 7-5

8. The meter reading at the beginning of the month is 5924 and at the end of the month is 6419. At the average cost of 9.8¢ per kilowatt hour, find the cost of electricity used. 7-6

9. Linda's automobile had a full tank of gasoline when the odometer read 23869. The next time she stopped for gasoline, it took 12 gallons to fill the tank and the odometer read 24115. How many miles per gallon did the automobile average on a gallon of gasoline? 7-7

10. Kim's mother can purchase a microwave oven for the cash price of $449 or $49 down and 12 equal monthly payments of $39.35 each. How much can she save by paying cash? 7-8

11. Find the annual percentage rate if the monthly finance charge rate is $1\frac{1}{2}$%. 7-9

12. What is the cost of $20,000 fire insurance for 1 year at the rate of $.22 per $100? 7-10

13. If the annual life insurance premium per $1,000 is $25.56, how much will a $25,000 policy cost per year? 7-11

14. If the tax rate is $4.90 per $100, how much must you pay for taxes on a house assessed for $37,000? 7-12

15. Kwon's father bought a house for $53,000. He paid $13,000 cash and gave a mortgage bearing $12\frac{1}{2}$% interest for the balance. What is his annual interest? 7-13

COMPETENCY CHECK TEST

The numerals in the boxes indicate the Lessons where help may be found.

Solve each problem and select the letter corresponding to your answer.

1. The unit cost (cost per pound) of an 8-oz jar of peanuts costing $1.19 is: 7-2
   a. $.595   b. $1.98   c. $2.38   d. $2.49

2. When you buy 1 gallon of paint at $14.99 instead of 4 individual quarts of the same paint at $4.99 each, you save: 7-2
   a. $10.99   b. $10.00   c. $8.98   d. $4.97

3. A ski jacket that regularly sells for $69 is reduced 25%. The sale price is: 7-3
   a. $51.75   b. $44.00   c. $17.25   d. $52.50

4. When a computer with a list price of $1,200 is sold for $960, the rate of discount is: 7-3
   a. 80%   b. 10%   c. 20%   d. 40%

5. If a purchase is $17.09, the change from a $20 bill should be: 7-4
   a. $3.09   b. $3.91   c. $2.91   d. $2.09

6. If the rate is 5%, the sales tax on a TV set selling for $479 is: 7-5
   a. $47.90   b. $4.79   c. $23.95   d. $474

7. The total cost of using an electric heater of 1,500 watts for 8 hours at the average cost of 9.5¢ per kilowatt hour is: 7-6
   a. $1.58   b. $1.14   c. .76¢   d. $9.58

8. If the driver's reaction time before applying the brakes is one second, the distance the car will go in one second at the speed of 50 m.p.h. is: 7-7
   a. 50 ft   b. $175\frac{1}{2}$ ft   c. 10 ft   d. $73\frac{1}{3}$ ft

9. Carmen's brother can buy an automobile for the cash price of $8,969 or $369 down and 48 equal payments of $264.75. By paying cash, he saves: 7-8
   a. $2,593   b. $4,108   c. $681.75   d. Answer not given

10. If the finance charge is $1\frac{1}{3}$% per month, the annual percentage rate is: 7-9
    a. 12%   b. 15%   c. 16%   d. 18%

11. The cost of $50,000 fire insurance for 1 year at $.22 per $100 is: 7-10
    a. $220   b. $110   c. $50.22   d. $122

12. At $19.09 per $1,000, the premium for $100,000 life insurance per year is: 7-11
    a. $1,909   b. $19.09   c. $190.90   d. $1,019.09

13. If the tax rate is $5.75 per $100, the property tax on a house costing $72,500 and assessed for $60,000 is: 7-12
    a. $12,500   b. $5,750   c. $3,450   d. $4,168.75

14. Mr. Chin bought a house for $72,500. He paid 20% down and gave a 13% mortgage for the balance. The amount of interest he pays per year is: 7-13
    a. $7,540   b. $2,600   c. $3,300   d. Answer not given

CHAPTER **8** **Managing Money**

The following forms are used in banking:

DEPOSIT SLIP

A deposit slip is used to deposit money in a savings account or checking account.

EXAMPLE: On December 15, 1985, Nancy Mugridge made a deposit to her account, numbered 999-123456, consisting of 2 $10-bills, 3 $5-bills, a check #85-24/260 in the amount of $128.63, and a check #23-847/105 in the amount of $79.82.

CENTRAL BANK	December 15 19 85		DOLLARS	CENTS
Pondfield, CA 98765	DATE	CASH ▶	35	00
ACCOUNT NUMBER 9 9 9 • 1 2 3 4 5 6		CHECKS BY BANK NO.		
		1 85-24/260	128	63
DEPOSIT TO THE ACCOUNT OF:		2 23-847/105	79	82
Name Nancy Mugridge		3		
		4		
		5		
		6		
Teller Validation		TOTAL ▶	243	45

DEPOSIT

WITHDRAWAL SLIP

A withdrawal slip is used to withdraw cash from a bank account.

EXAMPLE: On January 18, 1986 Stephen Scott wrote a withdrawal slip for the sum of $125 to be withdrawn from his account, numbered 129-345786.

CENTRAL BANK
Pondfield, CA 98765

JANUARY 18 19 86   1 2 9 • 3 4 5 7 8 6
DATE   ACCOUNT NUMBER

PASSBOOK WITHDRAWAL (Non-Negotiable)

PAY TO STEPHEN SCOTT _____ OR BEARER, THE SUM OF $ 125 | 00

ONE HUNDRED, TWENTY-FIVE AND ———————— 00/100 DOLLARS

If this is a Joint Account, I represent that both depositors are now living:

Stephen Scott
DEPOSITOR SIGNS HERE — DO NOT PRINT

	#	☐ M.O.	$
	#	☐ Check	$
PASSBOOK MUST ACCOMPANY THIS ORDER	#	☐ Redeposit	$
INTEREST CREDIT $ _____ 19 ____ DATE OF NOTICE		☐ Cash	$

TELLER'S STAMP

PASSBOOK WITHDRAWAL (Non-Negotiable)

CHECK

A check is used to transfer funds from a checking account to persons or business establishments. When a check is deposited, it must be signed (endorsed) on its back by the payee.

EXAMPLE: Lisa Todd wrote a check on June 18, 1986 payable to the Town Department Store in the amount of $87.35 for clothing.

CHECK STUB

The check stub is used if attached to the check to keep a record of the checks written and the deposits and withdrawals made.

EXAMPLE: Assume that Lisa Todd has a bank balance of $826.59 and wrote the check above. She would note the transaction on the corresponding check stub as shown to the right.

CHECK REGISTER

A check register is used in place of check stubs. When a check is written or a deposit or withdrawal is made, the transaction is recorded in the check register.

EXAMPLE: Assume that Lisa Todd has the same bank balance as above ($826.59). She would record the transaction in her check register as shown below.

RECORD OF ACTIVITY		Please be sure to deduct charges that affect your account.						BALANCE FORWARD	
CHECK NUMBER	DATE	ISSUED TO OR DESCRIPTION OF DEPOSIT	AMOUNT OF PAYMENT		OTHER DEDUCT.	AMOUNT OF DEPOSIT		826	59
101	6-18-86	TO *Town Department Store*	87	35				87	35
		FOR *Clothing*						739	24
		TO							
		FOR							
		TO							
		FOR							
		TO							
		FOR							
		TO							

BANK STATEMENT

Banks send, usually at the end of each month, a statement showing the deposits, withdrawals, cleared checks, bank charges, previous monthly balance, and closing monthly balance of the account. Thus, check stubs or check register records may be reconciled with the bank records.

PRACTICE PROBLEMS

Use deposit slips, checks, and either check stubs or a check register to record the following. If these forms are not available, copy them as illustrated.

1 On August 1, you have a bank balance of $1,593.07. Write this as the balance brought forward on the first check stub or balance forward in the check register.

2 On August 2, you deposit 4 $20 bills, 6 $5 bills, 12 $1 bills and a check numbered 83-91/218 drawn on the Valley Bank for $284.75. Fill in a deposit slip. Write the total deposit on a check stub or a check register.

3 On August 5, you send a check to Andrew Jaffe for $508.40 for purchases of appliances. Fill in the first check stub or use check register. Calculate the account balance. Write the check.

4 On August 12, you send a check to the Stephen Furniture Co. in the amount of $241.95 for the furniture purchased. Fill in the second check stub or use check register. Calculate the account balance. Write the check.

5 On August 16, you deposit 16 $10 bills, 9 $5 bills, 23 $1 bills, a check numbered 85-106/295 drawn on the First City Bank for $305.50 and a check numbered 6-39/318 drawn on the Second National Bank for $193.84. Fill in the deposit slip. Write the total deposit on the third check stub or use check register.

6 On August 22, you send a check to the Scott Department Store for $485.69 for merchandise purchased. Fill in the third check stub or use check register. Calculate the account balance. Write the check.

7 On August 29, you send a check to Peter Todd, Inc. for $94.95 for a desk purchased. Fill in the fourth check stub or use check register. Calculate the account balance. Write the check.

Simple Interest

Interest is money paid for the use of money. The money borrowed or invested, and on which interest is paid, is called principal. Interest paid on the principal only is called simple interest. The interest charged is generally expressed as a percent of the principal. This percent is called the rate of interest. The rate of interest is usually understood as the rate per year unless specified otherwise. The sum of the principal and the interest is called the amount.

PROCEDURE

To find the interest:
(1) Multiply the principal by the rate of interest per year.
(2) Then multiply the product found in step 1 by the time expressed in years.

$$\text{Interest } (i) = \text{principal } (p) \times \text{rate } (r) \times \text{time } (t)$$
$$i = prt$$

EXAMPLE: Find the interest on $900 for 3 yr 6 mo at 12%, using the formula.

$p = \$900$

$r = 12\% = \frac{12}{100}$

$t = 3$ yr 6 mo or $3\frac{1}{2}$ yr

$i = ?$

$i = prt$

$i = \$900 \times \frac{12}{100} \times 3\frac{1}{2}$

$i = \$378$

$\frac{\overset{9}{\cancel{900}}}{1} \times \frac{\overset{6}{\cancel{12}}}{\underset{1}{\cancel{100}}} \times \frac{7}{\underset{1}{2}} = 378$

ANSWER: $378 interest

To find the amount:
(1) Find the interest.
(2) Add the interest to the principal.

EXAMPLE: Find the amount due on $625 borrowed for 7 yr at 5%.

$625 principal
× .05 rate
$31.25 interest for 1 year
× 7
$218.75 interest for 7 years

$625.00 principal
+ 218.75 interest
$843.75 amount

ANSWER: $843.75 amount

To find the annual (yearly) rate of interest:
(1) Divide the annual interest by the principal.
(2) Express the quotient as a percent.

EXAMPLE: What is the annual rate of interest if the annual interest on a principal of $250 is $20?

$$\frac{\$20}{\$250} = \$20.00 \div \$250 = .08 = 8\%$$

ANSWER: 8% annual rate of interest

PRACTICE PROBLEMS

Find the interest on:
  1. 1. $200 for 1 yr at 15%   **2.** $2,800 for 1 yr at 10.25%

  2 1. $1,450 for 8 yr at 14%   **2.** $2,000 for 6 yr at $9\frac{3}{4}$%

  3 1. $900 for $3\frac{1}{2}$ yr at 10%   **2.** $1,600 for $2\frac{1}{4}$ yr at $13\frac{1}{2}$%

  4 1. $700 for 3 mo at 11%   **2.** $4,800 for 11 mo at 4%

  5 1. $50 for 1 yr 6 mo at 15%   **2.** $3,000 for 4 yr 1 mo at $9\frac{1}{2}$%

  6 Find the semi-annual (6 mo.) interest on:
    1. $400 at 7%   **2.** $11,000 at $6\frac{1}{4}$%   **3.** $40,000 at 9.383%

  7 Find the quarterly (3 mo.) interest on:
    1. $900 at 20%   **2.** $8,000 at $10\frac{3}{4}$%   **3.** $1,600 at 8.64%

  8 Find the monthly interest on:
    1. $1,200 at 8%   **2.** $900 at $12\frac{1}{2}$%   **3.** $12,000 at 9.791%

  9 Using 1 year = 360 days, 1 month = 30 days, find the interest on:
    1. $800 for 30 da. at 6%   **2.** $675 for 90 da. at 4%   **3.** $2,400 for 60 da. at 15%
    4. $4,000 for 120 da. at 8%   **5.** $120 for 15 da. at 10%   **6.** $1,000 for 45 da. at 20%

  10 Find the exact interest (1 year = 365 days) on:
    1. $425 for 73 da. at 7%   **2.** $2,920 for 15 da. at 12%   **3.** $2,400 for 30 da. at 9%

  11 Find the interest and amount on the following loans:
    1. $300 for 1 yr at 15%   **2.** $8,000 for 6 yr at $10\frac{1}{2}$%
    3. $2,700 for 9 yr 2 mo at 8.85%   **4.** $12,000 for 1 yr 11 mo at 6%

  12 Mr. Sanchez owns a $1,000 bond bearing 12.95% interest. How much interest does he receive every six months?

  13 What is the amount due on $420 borrowed at 15% and repaid at the end of 2 yr 8 mo?

  14 Find the annual rates of interest when the interest for
    1. 1 yr on $150 is $18   **2.** 3 yr on $2,000 is $600   **3.** 4 yr on $1,600 is $288

  15 What sum, invested at 8%, will earn $20,000 per year?

Discount Loan and Add-On-Interest Loan

The discount loan and the add-on-interest loan are the two types of loans most generally used.

When a discount loan is made:

a. The lending company immediately deducts the full interest from the principal (amount of note) and the borrower receives only the difference, called the net proceeds.

b. The borrower, however, repays the full principal in a specified number of equal monthly payments.

When an add-on-interest loan is made:

a. The borrower receives the full principal.

b. The interest is added on to this principal and the borrower repays the full amount (principal plus interest) in a specified number of equal monthly payments.

EXAMPLE: The Elliots borrowed $1,800 at 12% annual interest and are required to repay the loan in 24 equal monthly payments over 2 years. How much money do they receive and what is their monthly payment if they take a discount loan? If they take an add-on-interest loan?

Discount Loan

$1,800 principal
−   432 interest (12% × $1,800 × 2 yr)
$1,368 amount received

    $75 monthly payment
24)$1,800 amount to repay

ANSWER: $1,368, amount received
        $75, monthly payment

Add-On-Interest Loan

$1,800 amount received
  $1,800 principal
+   432 interest
  $2,232 amount to repay

     $93 monthly payment
24)$2,232

ANSWER: $1,800, amount received
       $93, monthly payment

PRACTICE PROBLEMS

What is the annual percentage rate if the rate of interest charged on a loan:

1 Per month is: **1.** 1%?   **2.** $\frac{1}{4}$%?   **3.** $1\frac{1}{2}$%?   **4.** .6%?   **5.** $\frac{3}{4}$%?

2 Per week is: **1.** 1%?   **2.** $\frac{1}{2}$%?   **3.** $1\frac{3}{4}$%?   **4.** .2%?   **5.** $\frac{1}{4}$%?

3 Per day is: **1.** 1%   **2.** .1%?   **3.** $\frac{3}{4}$%?   **4.** $1\frac{1}{2}$%?   **5.** $\frac{1}{2}$%?

4 What is the annual rate of interest you are paying when you:

   1. Borrow $10 and pay back $12.50 at the end of a month?
   2. Borrow $1 and pay back $1.05 at the end of a week?
   3. Borrow $5 and pay back $5.10 the next day?

5 What is the actual amount of money (net proceeds) the borrower receives and how much is the monthly payment in each of the following discount loans?

	Amount of Note	Interest Rate	Period of Loan
1.	$600	6%	12 months
2.	$1,500	10%	24 months
3.	$9,000	15%	18 months
4.	$2,700	$12\frac{1}{2}$%	36 months
5.	$1,800	8%	9 months
6.	$4,000	16%	30 months
7.	$7,200	13%	48 months
8.	$10,000	$14\frac{1}{2}$%	12 months
9.	$14,700	$6\frac{1}{4}$%	42 months
10.	$3,200	$7\frac{3}{4}$%	60 months

6 What is the actual amount of money the borrower receives and how much is the monthly payment in each of the following add-on-interest loans?

	Principal	Interest Rate	Period of Loan
1.	$600	6%	12 months
2.	$7,200	10%	36 months
3.	$3,000	20%	24 months
4.	$900	12%	30 months
5.	$1,800	8%	48 months
6.	$6,000	$9\frac{1}{2}$%	18 months
7.	$2,400	15%	6 months
8.	$7,500	18%	54 months
9.	$5,000	$12\frac{1}{4}$%	60 months
10.	$12,000	$8\frac{3}{4}$%	42 months

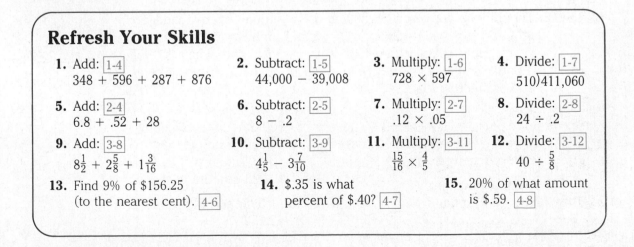

Refresh Your Skills

1. Add: 1-4
348 + 596 + 287 + 876

2. Subtract: 1-5
44,000 − 39,008

3. Multiply: 1-6
728 × 597

4. Divide: 1-7
510)411,060

5. Add: 2-4
6.8 + .52 + 28

6. Subtract: 2-5
8 − .2

7. Multiply: 2-7
.12 × .05

8. Divide: 2-8
24 ÷ .2

9. Add: 3-8
$8\frac{1}{2} + 2\frac{5}{8} + 1\frac{3}{16}$

10. Subtract: 3-9
$4\frac{1}{5} - 3\frac{7}{10}$

11. Multiply: 3-11
$\frac{15}{16} \times \frac{4}{5}$

12. Divide: 3-12
$40 \div \frac{5}{8}$

13. Find 9% of $156.25 (to the nearest cent). 4-6

14. $.35 is what percent of $.40? 4-7

15. 20% of what amount is $.59. 4-8

Financing a Car

Persons who finance the purchase of automobiles usually pay by monthly installments. The finance charge is the difference between the cash price and the total amount paid. The table below indicates some of these monthly payments.

Amount to be Borrowed	Amount of Each Monthly Payment		
	24 mo	30 mo	36 mo
$1,000	$45.83	$37.50	$31.94
1,200	55.00	45.00	38.33
1,500	68.75	56.25	47.91
1,800	82.50	67.50	57.50
2,000	91.66	75.00	63.88
2,500	114.58	93.75	79.86
3,000	137.50	112.50	95.83
3,500	160.41	131.25	111.80

PRACTICE PROBLEMS

Use the above table in each of the following problems:

1 What is the monthly payment to finance a car if you borrow:
- **1.** $1,200 for 30 months?
- **2.** $2,000 for 36 months?
- **3.** $1,500 for 24 months?
- **4.** $3,000 for 30 months?

2 Find the total payment if you borrow:
- **1.** $1,800 for 24 months.
- **2.** $2,500 for 30 months.

3 Find the total finance charge if you borrow:
- **1.** $3,500 for 24 months.
- **2.** $1,500 for 36 months.
- **3.** $2,000 for 30 months.
- **4.** $3,000 for 36 months.

4 How much more is the total finance charge on the 36-month loan of $2,500 than on the 24-month loan?

5 How much less is the total finance charge when you borrow $3,500 for 30 months instead of 36 months?

8-5 Amortizing a Loan

When a loan is made to purchase a house or condominium, each monthly payment to pay off the loan reduces the principal, and interest is paid on the *new reduced balance* of the principal owed each month.

This repayment of a mortgage loan is sometimes referred to as amortizing the loan. This type of repayment is different from those found with discount loans and add-on-interest loans where the borrower pays interest on the full amount of the loan for the entire period of the loan.

The following partial schedule indicates the monthly payments necessary to amortize a loan that was made at 13% annual interest.

Term Amount

Term	$5,000	$10,000	$15,000	$20,000	$25,000	$30,000	$35,000	$40,000	$45,000	$50,000
15 yr	63.27	126.53	189.79	253.05	316.32	379.58	442.84	506.10	569.36	632.63
20 yr	58.58	117.16	175.74	234.32	292.90	351.48	410.06	468.64	527.21	585.79
25 yr	45.12	112.79	169.18	225.57	281.96	338.36	394.75	451.14	507.53	563.92
30 yr	44.25	110.62	165.93	221.24	276.55	331.86	387.17	442.48	497.79	553.10
35 yr	43.81	109.52	164.28	219.04	273.80	328.56	383.32	438.08	492.84	547.60
40 yr	43.59	108.96	163.43	217.91	272.38	326.86	381.33	435.81	490.29	544.76

PRACTICE PROBLEMS

Use the above table of monthly payments in the following problems:

1 If a person borrowed $40,000 at 13% annual interest, what is the monthly payment when the loan is to be amortized in:

1. 25 years? **2.** 40 years? **3.** 15 years? **4.** 30 years? **5.** 20 years? **6.** 35 years?

2 If a loan at 13% annual interest is to be amortized in 25 years, what is the monthly payment when the amount of this loan is:

1. $50,000? **2.** $35,000? **3.** $25,000? **4.** $40,000? **5.** $15,000? **6.** $45,000?

3 Find the total interest that was paid on each of the following loans at 13% annual interest when the loans were amortized in the specified terms:

Amount	Term		Amount	Term		Amount	Term
1. $10,000	15 years		**2.** $30,000	40 years		**3.** $50,000	35 years
4. $45,000	20 years		**5.** $20,000	25 years		**6.** $35,000	30 years

4 **1.** Find the total payment over the full term required on a loan of $45,000 at 13% annual interest when it is amortized in 30 years. **2.** When it is amortized in 40 years. **3.** For which term is the interest less? **4.** How much less?

5 Mrs. Rizzo purchased a house for $42,500. She paid $7,500 in cash and obtained a mortgage loan at 13% annual interest for the balance. What is the monthly payment due if it is to be amortized in 35 years?

Compound Interest

Savings and loan associations and some savings banks advertise that savings left with them earn interest compounded semiannually (every 6 months) or quarterly (every 3 months). Some banks even advertise interest compounded daily on passbook savings or on savings certificates.

Compound interest is interest paid on both the principal and the interest earned previously.

The table on page 346 shows how much $1 will amount to based on the given rate and the given number of years.

PROCEDURE

To find how much a given principal will amount to at a given rate for a certain period of time:

If compounded annually:
(1) Read down the column for the given rate until you reach the given period (years) of time.
(2) Multiply the amount found in step 1 by the given principal.

If compounded semiannually:
(1) Read down the column for one-half the given rate until you reach two times as many periods as the given number of years.
(2) Multiply the amount found in step 1 by the given principal.

If compounded quarterly:
(1) Read down the column for one-fourth the given rate until you reach four times as many periods as the given number of years.
(2) Multiply the amount found in step 1 by the given principal.

If compounded monthly:
(1) Read down the column for one-twelfth the given rate until you reach twelve times as many periods as the given number of years.
(2) Multiply the amount found in step 1 by the given principal.

> *To find the compound interest only*, subtract the principal from the amount.

EXAMPLE: Find the amount and the interest earned on $800 deposited for 3 years at 12% when the interest is:

Compounded Annually;
Rate, 12%; Periods, 3.

$$\begin{array}{r} 1.40492 \\ \times \quad \$800 \\ \hline \$1{,}123.93600 = \$1{,}123.94 \text{ amount} \end{array}$$

$$\begin{array}{r} \$1{,}123.94 \text{ amount} \\ - \quad 800.00 \text{ principal} \\ \hline \$323.94 \text{ interest} \end{array}$$

ANSWER: $1,123.94 amount, $323.94 interest

Compounded Quarterly;
Rate, 3%; Periods, 12.

$$\begin{array}{r} 1.42575 \\ \times \quad \$800 \\ \hline \$1{,}140.60000 = \$1{,}140.60 \text{ amount} \end{array}$$

$$\begin{array}{r} \$1{,}140.60 \text{ amount} \\ - \quad 800.00 \text{ principal} \\ \hline \$340.60 \text{ interest} \end{array}$$

ANSWER: $1,140.60 amount, $340.60 interest

Compound Interest Table						
Showing How Much $1 Will Amount to at Various Rates						
Periods	1%	2%	3%	4%	5%	6%
1	1.01000	1.02000	1.03000	1.04000	1.05000	1.06000
2	1.02010	1.04040	1.06090	1.08160	1.10250	1.12360
3	1.03030	1.06120	1.09272	1.12486	1.15762	1.19101
4	1.04060	1.08243	1.12550	1.16985	1.21550	1.26247
5	1.05101	1.10408	1.15927	1.21665	1.27628	1.33822
6	1.06152	1.12616	1.19405	1.26531	1.34009	1.41851
7	1.07213	1.14868	1.22987	1.31593	1.40709	1.50362
8	1.08285	1.17165	1.26676	1.36856	1.47745	1.59384
9	1.09368	1.19509	1.30477	1.42330	1.55132	1.68947
10	1.10462	1.21899	1.34391	1.48024	1.62889	1.79084
11	1.11566	1.24337	1.38423	1.53945	1.71033	1.89829
12	1.12682	1.26823	1.42575	1.60102	1.79585	2.01219
13	1.13809	1.29360	1.46852	1.66506	1.88564	2.13292
14	1.14947	1.31947	1.51258	1.73167	1.97992	2.26089
15	1.16096	1.34586	1.55796	1.80093	2.07892	2.39654
16	1.17257	1.37277	1.60469	1.87297	2.18286	2.54034
17	1.18429	1.40023	1.65283	1.94789	2.29201	2.69276
18	1.19614	1.42823	1.70242	2.02580	2.40661	2.85432
19	1.20810	1.45680	1.75349	2.10683	2.52694	3.02558
20	1.22018	1.48593	1.80610	2.19111	2.65328	3.20712
Periods	8%	10%	12%	14%	16%	20%
1	1.08000	1.10000	1.12000	1.14000	1.16000	1.20000
2	1.16640	1.21000	1.25440	1.29960	1.34560	1.44000
3	1.25971	1.33100	1.40492	1.48154	1.56089	1.72800
4	1.36048	1.46410	1.57351	1.68896	1.81063	2.07360
5	1.46932	1.61051	1.76234	1.92541	2.10034	2.48832
6	1.58687	1.77156	1.97382	2.19497	2.43638	2.98598
7	1.71382	1.94871	2.21067	2.50226	2.82621	3.58318
8	1.85092	2.14358	2.47596	2.85258	3.27841	4.29981
9	1.99900	2.35794	2.77307	3.25194	3.80295	5.15977
10	2.15892	2.59374	3.10584	3.70721	4.41143	6.19173
11	2.33163	2.85311	3.47854	4.22622	5.11725	7.43008
12	2.51816	3.13842	3.89597	4.81789	5.93601	8.91609
13	2.71961	3.45226	4.36348	5.49240	6.88578	10.6993
14	2.93718	3.79749	4.88710	6.26133	7.98750	12.8391
15	3.17216	4.17724	5.47355	7.13792	9.26550	15.4070
16	3.42593	4.59496	6.13038	8.13723	10.7479	18.4884
17	3.70000	5.05446	6.86602	9.27644	12.4676	22.1860
18	3.99600	5.55990	7.68995	10.5751	14.4624	26.6232
19	4.31568	6.11589	8.61274	12.0556	16.7764	31.9479
20	4.66094	6.72748	9.64627	13.7434	19.4606	38.3375

PRACTICE PROBLEMS

1 Find the amount and the interest earned when the interest is compounded annually on:

1. $100 for 6 yr at 14%   **2.** $900 for 11 yr at 6%   **3.** $1,400 for 3 yr at 16%
4. $10,000 for 15 yr at 5%   **5.** $8,500 for 7 yr at 8%   **6.** $2,750 for 20 yr at 10%

2 Find the amount and the interest earned when the interest is compounded semiannually on:

1. $600 for 1 yr at 12%   **2.** $2,300 for 7 yr at 16%   **3.** $9,000 for 4 yr at 6%
4. $15,000 for 10 yr at 4%   **5.** $7,450 for 8 yr at 8%   **6.** $20,000 for 5 yr at 10%

3 Find the amount and the interest earned when the interest is compounded quarterly on:

1. $350 for 2 yr at 12%   **2.** $2,000 for 4 yr at 4%   **3.** $1,500 for 1 yr at 8%
4. $10,000 for 5 yr at 20%   **5.** $6,300 for 3 yr at 8%   **6.** $50,000 for 4 yr at 16%

A bank advertises an 8% interest rate compounded quarterly on all deposits.

4 **1.** What would a deposit of $500 amount to at the end of 3 months?   **2.** 6 months?
3. 9 months?   **4.** 1 year?   **5.** 15 months?   **6.** 18 months?   **7.** 2 years?   **8.** 30 months?
9. 3 years?   **10.** 42 months?   **11.** 4 years?   **12.** 51 months?   **13.** 5 years?

5 **1.** What is the interest earned on a $1,000 deposit at the end of 3 months?
2. 6 months?   **3.** 9 months?   **4.** 1 year?   **5.** 18 months?   **6.** 21 months?   **7.** 2 years?
8. 30 months?   **9.** 3 years?   **10.** 45 months?   **11.** 4 years?   **12.** 54 months?
13. 5 years?

At the interest rate of 12% compounded monthly (12 times a year):

6 What will each of the following deposits amount to at the end of one year?
1. $700   **2.** $1,250   **3.** $4,800   **4.** $10,000

7 How much interest is earned at the end of 1 year on each of the following deposits?
1. $250   **2.** $980   **3.** $3,600   **4.** $5,000

8 What will each of the following deposits amount to?

1. A deposit of $75 at the end of 8 months
2. A deposit of $400 at the end of 3 months
3. A deposit of $1,800 at the end of 1 month

9 How much interest is earned on:

1. A deposit of $60 at the end of 9 months?
2. A deposit of $840 at the end of 6 months?
3. A deposit of $3,000 at the end of 2 months?

10 Find the interest earned on a $5,000 deposit at the end of 1 year at each of the following rates:

1. 12% compounded quarterly   **2.** 12% compounded annually
3. 12% compounded semiannually   **4.** 12% compounded monthly
5. Which rate brings the greatest amount of interest?
6. Which rate brings the smallest amount of interest?

STOCKS

A person buying a share of **stock** becomes a part-owner of the business and sometimes receives dividends from the profits earned by the business. Stock quotations are in terms of dollars.

$$36\tfrac{1}{2} \text{ means } \$36.50$$

The P.E. (price to earning) ratio column heading in the stock quotations refers to an approximate ratio comparing the closing price of a share of a stock to its annual earning.

EXAMPLE:

If the P.E. ratio is 9, then the closing price of the stock is approximately 9 times its annual earnings.

To find the annual earnings, divide the closing price by the P.E. ratio number.

Fees, or commissions, are paid to stock brokers for the service of buying or selling stocks. These fees are subject to change and at present there is no one set of fixed fees for all brokers. Below is a typical fee schedule.

Cost of Stock	Fee Paid to Broker
$100 but under $2,500	1.3% of cost of stock plus $16
$2,500 but under $20,000	0.9% of cost of stock plus $28
$20,000 but under $30,000	0.6% of cost of stock plus $90
Over $30,000	0.4% of cost of stock plus $152

STOCK QUOTATIONS

P.E.	Sales 100s	High	Low	Close	Net Chg.	Yearly High	Yearly Low	Stock	Div.	Yld. %	P.E.	Sales 100s	High	Low	Close	Net Chg.	Yearly High
7	1234	21 3/4	16 1/8	17 5/8	....	17 1/8	14	INA In	1.92	12.	...	11	16 1/4	15 7/8	16	+ 1/8	47 3/8
13	97	32 1/8	25 1/4	29 1/8	+ 1/8	47 3/8	20 5/8	ITT Cp	1	3.8	7	3548	26 5/8	26 1/8	26 1/2	+ 1/4	156
6	1135	127 1/4	119 1/8	122 1/8	+ 3/8	25 5/8	15 1/4	IU Int	1.20	7.7	21	294	15 3/4	15 1/2	15 1/2	- 1/8	35 1/2
12	87	38 1/4	36 3/4	38 1/8	- 1/8	37 7/8	30 3/4	Idaho P.	3.28	9.0	7	98	36 3/4	36 3/4	36 1/2	...	53 5/8
17	294	15 3/4	15 1/2	15 1/2	...	23 3/4	17 5/8	Ill Powr	2.64	12.	6	1309	22 7/8	22 1/2	22 3/4	- 1/8	97 1/8
7	101	40 1/4	39 7/8	40 5/8	- 1/8	55 3/4	35 1/2	Inger R	2.60	6.3	...	412	41 3/8	40 5/8	41 1/8	+ 1/4	59 3/4
...	1237	15 3/4	15 1/8	15 5/8	+ 1/2	32 3/4	20 5/8	Inld Stl	.50	2.3	...	1061	21 5/8	21 1/4	21 1/2	...	14 3/8
...	412	36 3/4	36 3/8	36 1/2	...	128 1/2	99	IBM	4.40	3.6	12	10668	121 3/8	119 5/8	121	- 1/4	13 1/8
6	98	22 3/8	20 3/4	21 1/2	+ 1/2	14 3/4	5 1/8	Int Harv	...	...	...	7163	8 1/4	7 3/8	8 1/4	+ 1/8	27 3/4
7	3548	27 5/8	27 1/8	26 1/2	+ 1/4	49	32 7/8	Int Min	2.60	6.8	11	114	38 1/8	37 3/4	38 1/8	+ 1/4	31 1/8
.·.	7163	8 1/8	7 3/4	8	+ 1/8	60	46	Int Paper	2.40	4.7	11	958	51 1/4	49 7/8	51 1/4	+ 1	19 1/8
12	10668	121 1/8	120 7/8	121 1/8	+ 1/4	18 1/8	14 1/4	Iowa El	1.90	11.	7	101	17 7/8	17 5/8	17 5/8	...	14 7/8
7	101	13 1/8	12 3/4	13 1/8	...	46	34	IRT	1.08	3.8	...	293	44 5/8	44 1/8	44 1/2	+ 1/8	25 5/8

PRACTICE PROBLEMS

1 Using the lowest prices listed in the stock quotations, find the cost of each of the following purchases, excluding the broker's fee.

1. 100 shares of INA In   **2.** 100 shares of IU Int   **3.** 100 shares of Int Paper
4. 40 shares of Iowa El   **5.** 75 shares of ITT Cp   **6.** 25 shares of IBM
7. 35 shares Ill Powr   **8.** 200 shares of Int Min   **9.** 50 shares of Int Harv

2 Using the highest prices listed in the stock quotations, find the total cost of each of the following purchases, including the broker's fee.

1. 100 shares of Int Paper   **2.** 100 shares of ITT Cp   **3.** 100 shares of Ill Powr
4. 100 shares of Int Harv   **5.** 100 shares of Int Min   **6.** 50 shares of Iowa El
7. 75 shares of Idaho P   **8.** 200 shares of INA In   **9.** 60 shares of Inld Stl

3 Using the closing prices listed in the stock quotations, find the net amount due (selling price of stock less broker's fee) on each sale of stock.

1. 100 shares of Int Min   **2.** 100 shares of Iowa El   **3.** 100 shares of Idaho P
4. 100 shares of Ill Powr   **5.** 100 shares of Int Paper   **6.** 200 shares of ITT Cp
7. 30 shares of IBM   **8.** 50 shares of IU Int   **9.** 70 shares of INA In

4 How many points above the year low and below the year high is each of the following stock's closing price?

1. Idaho P   **2.** ITT Cp   **3.** Int Harv   **4.** Int Min   **5.** IBM   **6.** Iowa El

5 The dividend listed for each stock in the stock quotations is for one year. Determine for each stock the amount of the quarterly dividend.

6 Find the annual earnings (to nearest cent) for each of the stocks listed in the stock quotations on page 348.

7 Include the broker's fee in each of the following and find the amount of profit:

1. 100 shares of XYZ, purchased at 24, sold for 32
2. 100 shares of EIS, purchased at $18\frac{1}{2}$, sold for $27\frac{1}{4}$
3. 50 shares of LJL, purchased at $33\frac{3}{4}$, sold for 40
4. 85 shares of MSL, purchased at $45\frac{7}{8}$, sold for $51\frac{3}{4}$
5. 30 shares of ASP, purchased at $37\frac{5}{8}$, sold for $46\frac{1}{8}$

8 Include the broker's fee in each of the following and find the amount of loss:

1. 100 shares of ZZZ, purchased at 47, sold for 38
2. 100 shares of EBS, purchased at $25\frac{1}{4}$, sold for 20
3. 60 shares of SHJ, purchased at $18\frac{5}{8}$, sold for $12\frac{1}{2}$
4. 25 shares of NMN, purchased at $41\frac{3}{8}$, sold for $29\frac{3}{4}$
5. 78 shares of PQR, purchased at $62\frac{1}{8}$, sold for $47\frac{7}{8}$

BONDS

A bond is a written promise of a private corporation or of a local, state, or the national government to pay a given rate of interest at stated times, and to repay the face value (original value) of the bond at a specified time (date of maturity).

A person buying a bond is lending money to the business or government and receives interest on the face value of the bond.

The value at which a bond sells at any given time is called market value.

Bond quotations are in terms of percent.

$101\frac{1}{4}$ means $101\frac{1}{4}\%$

EXAMPLE:

Xerox $10\frac{5}{8}$s 93 $94\frac{1}{2}$

In the above bond quotation:

$10\frac{5}{8}$s represents a $10\frac{5}{8}\%$ annual rate of interest.

93 is the date of maturity, 1993.

$94\frac{1}{2}$ is the market price, indicating $94\frac{1}{2}\%$ of the face value of the bond.

To find the rate of income or yield of a bond (column marked Cur. Yld.), find what percent the annual interest is of the market price of the bond.

Fees that are paid to brokers for buying and selling bonds may vary.

BOND QUOTATIONS

Low	Close	Net Chg.
99 1/2	101	....
100	102	+ 1/4
98 7/8	100 3/8	+ 1/2
97 3/8	99 7/8	− 1/8
96 5/8	98 1/4	− 1
102 3/8	104	− 1/2
89	95 1/4	+ 1/2
96 3/4	98	+ 1
93	94 1/8	.....
94 3/4	95 3/4	− 1/8
102 1/8	103 1/4	− 1/8
101 5/8	102	+ 1
98 1/2	100 5/8	+ 1/2
100 3/8	101 7/8	+ 1/4
94 1/4	96 1/2	− 1/8
86 3/8	91 1/4	+ 7/8
87	90 3/8	+ 1
102	102 1/4	− 3/8

Bonds	Cur. Yld.	Vol.	High	Low	Close	Net Chg.
ATT 13 1/4 91	13.	388	104 5/8	104 1/2	104 5/8	+ 1/2
Ames 8 1/2 09	cv*	105	102	100 1/2	100 1/2	− 1 1/2
Citicp 12s 90	12.	201	100	99 5/8	100	+ 1/4
Deere 9s 08	cv*	5	110	110	110	− 1
Digit 8s 09	cv*	61	98 1/4	97 1/2	97 1/2	− 1/2
GTE 10 1/8 95	11.	4	88 1/2	88 1/2	88 1/2	+ 1/2
GaPw 13 1/8 12	13.	5	99	99	99	+ 1/2
Ill Bel 12 1/4 17	12.	25	99 1/4	99 1/4	99 1/4	− 1/4
InTT 12 5/8 05	13.	5	100 3/8	100 3/8	100 3/8	− 7/8
MCI 15s 00	15.	12	102	101 7/8	101 7/8	− 7/8
McKess 6s 94	4.9	122	99 1/4	98	98	− 1
Memrx 5 1/4 90	7.0	75	101	97	99	+ 1/8
MerLy 11 5/8 87	12.	101	89	87	88	− 1/4
MGM 9 1/25 00	10.	94	102 3/8	99	101	.....
NCNB 9.40s 95	12.	81	101 1/2	100	100	− 1/4
NEngT 8.2s 04	12.	68	102	100	100	− 1/8
NYTel 11 5/8 19	13.	91	103	99	101	− 5/8
NiaM 10.20 05	12.	84	89	85	88	+ 1 7/8

*The abbreviation "cv" indicates convertible bonds.

Bonds	Cur. Yld.
Telex 11 3/45 96	13.
Tennc 12 1/8 05	13.
TVA 7.40s 97D	11.
Texaco 8 1/2 06	12.
TxCap 13s 91	13.
TideInc 7 3/4 05	12.
Towle 9 1/2s 00	12.
TRW 9 7/8s 00	12.
UCarb 14 1/2 91	14.
UOCal 8 3/8s 85	8.5
USHm 5 1/2s 96	6.7
USLeas 9s 07	8.
UTech 9 3/8s 04	12.
UtahP 8 3/8s 06	12.
WellsF 7 7/8s 97	11.
WestUn 16s 91	18.
WestgE 9s 09	7.5
WillEl 12 3/4 96	17.

PRACTICE PROBLEMS

1 Find the date of maturity and the interest rate for each of the following bonds:

1. PGE 5s 89   **2.** Beth Steel 9s 00   **3.** Gulf Oil $8\frac{1}{2}$s 95

4. Rydr $11\frac{1}{2}$s 90   **5.** Dow 6.7s 98   **6.** Duke P $7\frac{3}{4}$s 02

2 Find the cost of each of the following purchases, excluding commissions, using the highest prices in the bond quotations.

1. One GTE $10\frac{1}{8}$s 95 bond, face value $1,000

2. Ten Ames $8\frac{1}{2}$s 09 bonds, face value $1,000 each

3. Forty Ill Bel $12\frac{1}{4}$s 17 bonds, face value $1,000 each

4. Five Deere 9s 08 bonds, face value $1,000 each

5. Twenty Digit 8s 09 bonds, face value $1,000 each

6. Fifty MCI 15s 00 bonds, face value $1,000 each

3 Determine how much interest the purchaser will receive semi-annually on the purchases above.

4 For each bond listed above in the bond quotations, find the current yield (percent the annual interest is of the closing market price) to the nearest tenth of a percent, and check with the current yield column (Cur. Yld.). The face value of each bond is $1,000.

MUTUAL FUNDS

A mutual fund is an investment company. It obtains money for investment by selling shares of stock of the company. Mutual fund quotations are expressed as a Net Asset Value (NAV) and an offer price. When we *buy* shares, we pay the offer price of each share. When we *sell* shares, we receive the Net Asset Value (NAV) of each share.

N.L. means no load or no commission charged for buying shares. When N.L. is indicated, the offer price is the same as the NAV.

MUTUAL FUNDS

	N.A.V.	Offer Price		N.A.V.	Offer Price
Bullock Fd.	15.73	17.19	Dreyfus Fd.	12.26	13.40
Delaware	18.21	19.90	Phila Fd.	8.52	9.31
Fidelity Fd.	14.97	N.L.	Pilgrim Fd.	13.18	14.21
Ivest	15.02	N.L.	Value Line	10.53	N.L.
Daily Fd.	15.02	N.L.	Kidder Fd.	13.40	N.L.
Muni Fd.	13.84	15.13	Vanguard	17.32	N.L.
Rowe Price	9.53	10.32	Lehman Fd.	15.13	16.79
Dean Witter	11.07	N.L.	Fidelity Fd.	12.78	13.52
DBL Fd.	13.37	14.70	Municipal	14.97	N.L.

PRACTICE PROBLEMS

Use the mutual fund quotations above to do the following:

1 Find the cost of buying:

   1. 100 shares of Pilgrim Fd   **2.** 100 shares of Ivest

   3. 50 shares of Phila Fd   **4.** 75 shares of Dreyfus Fd

2 How much money should you receive when you sell each of the following:

   1. 100 shares of Bullock Fd   **2.** 40 shares of Value Line

   3. 100 shares of Fidelity Fd   **4.** 85 shares of Delaware

GOVERNMENT BONDS; TREASURY BILLS, NOTES, AND BONDS

The United States government borrows money by selling savings bonds and Treasury bills (T bills) (3 mo, 6 mo, or 1 yr), notes (2 yr to 10 yr), or bonds (over 10 yr).

The EE savings bonds mature in 10 years and sell at 50% of face amount, starting with a $50 bond mininum. The HH bonds mature in 10 years and sell in multiples of $500. The rates of interest for both of these bonds may vary.

The Treasury bills, notes, and bonds can be sold at any time. The rates of interest may vary. Treasury bills are discounted. Investors pay the **auction price** (face amount less interest) but receive the full amount at maturity.

PRACTICE PROBLEMS

1 Mrs. Berk owns three $10,000 Treasury notes that pay an annual rate of $10\frac{3}{4}$%. How much interest does she receive semi-annually?

2 Andy bought a $100 EE bond. How much did he pay for it?

3 The interest rate on a 6-month Treasury bill was announced at 9.5%. If you purchase a $5,000 T bill at auction for $4,985 and the interest is also deducted, what is your final cost?

MONEY RATES—
CERTIFICATES OF DEPOSIT

Prime Rate: The prime rate is the basic interest rate commercial banks charge their most valued customers. In recent years this rate has ranged from 8% to $21\frac{1}{2}$%.

Passbook Interest Rates: Passbook interest on savings is paid by savings institutions at the fixed annual rate of 5.5%. Commercial banks pay 5.25%.

Money Market Rates: Banks offer money-market accounts with a minimum deposit of $2,500 paying rates of interest which change weekly.

Certificates of Deposit: Banks and savings and loan associations issue savings certificates called certificates of deposit (CD) paying varying rates of interest depending on the term of the certificate.
The interest rates on money-market accounts and on certificates of deposit change weekly. These rates can change as little as a **basis point,** which is one-hundredth of one percent (.01%). It takes 100 basis points to make 1 percent.

EXERCISES

1. A money market account pays 8.962% annually on an average monthly balance of $15,000. How much interest is earned for the month?

In the table below, the annual yield shows the actual interest rate being earned due to the daily compounding of the savings in a certificate of deposit.

Interest Rate		Annual Yield
9.90%	6-month certificate	10.556%
10.15%	1-year certificate	10.838%
10.37%	2-year certificate	11.085%

2. Using the annual yield rates above, find the total interest earned by a 1-year certificate of deposit for $10,000.

3. Find the total interest earned by a 2-year certificate of deposit for $25,000.

The following are typical certificate of deposit rates for a minimum investment of $100,000. These rates change daily.

30– 59 days	8.50%
60– 89 days	8.75%
90–119 days	9.00%
120–179 days	9.40%
180–360 days	9.90%

Use these rates to find the interest earned on a $100,000 CD purchased for:

4. 30 days **5.** 90 days

6. 75 days **7.** 160 days

8. If a 9.45% interest rate is increased by 25 basis points, what is the new rate?

Customers go to banks to make deposits and withdrawals, cash checks, buy savings bonds and travelers' checks, and apply for loans. *Bank tellers* handle the money and paperwork for all these transactions. In most banks they use computers to do calculations and keep records.

Banks hiring tellers look for high-school graduates who are careful, accurate, and courteous. Tellers learn their jobs in training programs run by banks, and by watching experienced tellers.

1. A teller receives a deposit of $23.50 in cash plus checks for $189.27 and $46.75. What is the total deposit?

2. A customer has $463.12 in her account. She withdraws $275. How much money is left in her account?

3. A teller explains that the cost of a savings bond is 50% of its face value. If the cost is $37.50, what is the face value?

4. A bank will grant a loan for 80% of the cost of a new car. The car costs $11,782.65. What is the amount of the loan?

5. A money order is a safe way to send money through the mail. A customer wants a money order for $119.95 to buy a camera. The bank charges a service fee of $1.55. What is the total cost of the money order?

6. A customer wants to buy $500 worth of travelers' checks. The bank charges a 1% service fee. What is the total cost of the travelers' checks?

CHAPTER REVIEW

Vocabulary

add-on-interest loan p. 341
amortizing the loan p. 344
amount p. 339
auction price p. 352
basis point p. 353
bond p. 350
certificate of deposit (CD) p. 353
check p. 336
check register p. 336
check stub p. 336
compound interest p. 345

deposit slip p. 336
discount loan p. 341
finance charge p. 343
interest p. 339
market value p. 350
money market rate p. 353
mutual fund p. 351
net asset value (NAV) p. 351
net proceeds p. 341
no load (NL) p. 351
offer price p. 351

passbook interest rate p. 353
price to earnings ratio
  (P.E.) p. 348
prime rate p. 353
principal p. 339
rate of income p. 350
rate of interest p. 339
rate of yield p. 350
simple interest p. 339
stock p. 348
withdrawal slip p. 336

The numerals in boxes indicate Lessons where help may be found.

1. On August 5, 1986, Frank Peres withdrew the cash sum of $108.40 from his account #925-416-8. Write the withdrawal slip. 8-1

2. On October 16, 1986, Antonia Ricci had a bank balance of $1,209.80 in her checking account #626-914-3. That day she wrote a check payable to the Computer Store in the amount of $416.49 for the purchase of a monitor. Write this check and record the transaction on the check stub. 8-1

3. Mr. Carroll receives $150 interest each year on an investment of $4,000. He also receives $180 semiannual interest on an investment of $9,000. On which investment does he receive a higher rate of interest? 8-2

4. Roberto's mother owns a $10,000 T-note bearing 10.75% interest. How much interest does she receive every 6 months? 8-2

5. What is the actual amount of money the borrower receives and how much is the monthly payment if the amount of the note of a discount loan is $1,000, the interest rate is 15%, and the period of the loan is 12 months? 8-3

6. What is the actual amount of money the borrower receives and how much is the monthly payment if the principal of an add-on-interest loan is $5,000, the interest rate is 12%, and the period of the loan is 24 months? 8-3

7. Using the table on p. 343, find the total finance charge if you borrow $2,500 for 36 months to finance your car. 8-4

8. Using the table on p. 344, find the total interest that is paid on a loan of $30,000 at 13% annual interest and amortized in 20 years. 8-5

9. Using the table on p. 346, find the amount and the interest earned on $3,000 deposited for 2 years at 8% compounded semiannually. 8-6

10. Excluding broker's fees, how much profit do you make if you buy 100 shares of stock at $17\frac{7}{8}$ and sell them at $21\frac{3}{4}$? 8-7

11. How much interest do you receive semiannually if you buy eight bonds listed as "Gen Elect 14s 90," with a face value of $1,000 each? 8-7

12. What is the cost of the above bonds, excluding broker's fee, when they are quoted at $106\frac{1}{2}$? 8-7

COMPETENCY CHECK TEST

The numerals in boxes indicate Lessons where help may be found.

Solve each problem and select the letter corresponding to your answer.

1. Luis Gomez paid the Eastern Telephone Co. by check in the amount of $27.45 on September 26, 1986.

```
┌──────────────────────────────────────────────────────┐
│         WESTERN STATE BANK              650            │
│                                                        │
│                               _____19____   │
│   PAY TO THE                                           │
│   ORDER OF_____ $_____   │
│   ①                                        /DOLLARS    │
│   _____  │
│                                                        │
│   MEMO_____        _____   │
│   ⑆781543365⑆ 54 207654 3⑈                             │
└──────────────────────────────────────────────────────┘
```

On line 1 should be written: 8-1

  a. Eastern Telephone Co.   **b.** Twenty-seven and 45/100
  c. September 26, 1986   **d.** Luis Gomez

2. The interest on $7,500 for 2 years at $12\frac{1}{2}$% annual rate is: 8-2

  a. $937.50   **b.** $3,750   **c.** 25%   **d.** $1,875

3. The amount due on $4,000 borrowed at $11\frac{3}{4}$% and repaid at the end of 3 years is: 8-2

  a. $1,410   **b.** $4,575   **c.** $5,410   **d.** Answer not given

4. If $6,000 is borrowed at 12% for 12 months, the monthly payment on an add-on-interest loan is: 8-3

  a. $560   **b.** $720   **c.** $500   **d.** $600

5. If $10,000 is borrowed at 15% for 12 months on a discount loan, the amount of money the borrower receives is: 8-3

  a. $15,000   **b.** $1,500   **c.** $8,500   **d.** $11,500

6. If you borrow $3,000 for 30 months to buy a car and pay $125 per month, the total finance charge is: 8-4

  a. $3,750   **b.** $750   **c.** $375   **d.** Answer not given

7. If the monthly payment to amortize a loan of $40,000 in 30 years at 13% annual interest is $442.48, the total interest paid is: 8-5

  a. $119,292.80   **b.** $40,442.48   **c.** $96,406.28   **d.** $12,472.48

8. A sum of $1 compounded annually at 10% interest amounts to $1.61051 in 5 years and compounded semiannually amounts to $1.62889. When the sum of $20,000 is invested at each rate, the difference in the interest earned is: 8-6

  a. $323.94   **b.** $967.38   **c.** $1,200   **d.** $367.60

9. If you purchase 100 shares of *xyz* at $20\frac{3}{4}$ and sell at $24\frac{1}{8}$, your profit, excluding commissions, is: 8-7

  a. $555.80   **b.** $410.25   **c.** $337.50   **d.** $219.47

10. Mrs. Murphy owns a $5,000 Treasury note that pays an annual interest rate of $10\frac{3}{4}$%. She receives semiannually: 8-7

  a. $500   **b.** $537.50   **c.** $1,075   **d.** $268.75

Achievement Test

Solve each problem and select the letter corresponding to your answer.

5-1   Problem: Joe's mother bought a wall unit that is 9 feet long. If the sales tax is $21.80, find the total cost of the wall unit.

   1. The question asked in the above problem is:
     a. What is the length of the wall unit?   **b.** What is the dealer's cost?
     c. What is the cost excluding sales tax?   **d.** What is the cost including sales tax?

   2. The missing fact needed to solve the above problem is:
     a. The dealer's cost   **b.** The sales tax
     c. The purchase price excluding tax   **d.** The height of the wall unit

   3. The fact not needed to solve the above problem is:
     a. The purchase price   **b.** The length of the wall unit
     c. The sales tax   **d.** Answer not given

5-2   Problem: A bus travels 240 miles in 5 hours. What is its average rate of speed?

   4. The operation needed to solve the above problem is:
     a. Addition   **b.** Subtraction   **c.** Multiplication   **d.** Division

5-3   **5.** The nearest estimate to the quotient of $7,088 \div 69$ is:
     a. 1,000   **b.** 500   **c.** 250   **d.** 100

5-4   **6.** The answer 925 is reasonable if the estimate is:
     a. 10   **b.** 100   **c.** 1,000   **d.** 10,000

5-5
6-1   **7.** The weekly earnings of a person who works 40 hours at $5.65 per hour are:
     a. $216   **b.** $224.60   **c.** $242.30   **d.** $226

5-6
6-2   **8.** Esther receives $185 per week plus 2% commission on sales. One week her sales totaled $8,509. Her total earnings that week were:
     a. $202.02   **b.** $287.48   **c.** $355.18   **d.** $306.52

6-3   **9.** At what price should a dealer sell a watch costing $96 to make a profit of 40% on the selling price?
     a. $134.40   **b.** $384   **c.** $136   **d.** $160

6-4   **10.** When the withholding tax is $25.40, Social Security tax is 7.05% of earnings,
6-5   and the city wage tax is $1\frac{1}{2}$% of earnings, the take-home pay on weekly wages of
6-6 $275 is:
6-7   **a.** $208.35   **b.** $86.04   **c.** $188.96   **d.** $226.09

6-8   **11.** If your Federal taxes for the year are $2,059 and your withholding tax payments are $1,775, the amount of tax you still owe is:
     a. $384   **b.** $3,834   **c.** $374   **d.** $284

7-1
7-2
12. Which is the best buy of the same kind of dinner rolls?

7-2
    a. 6 for 59¢   **b.** 8 for 75¢   **c.** 12 for $1   **d.** 20 for $1.69

7-2   **13.** What is the unit price (cost per kilogram) of a jar of applesauce which weighs 1.25 kg and costs $1?

    a. $1.25   **b.** $.80   **c.** $.25   **d.** $2.25

7-3   **14.** A camera which regularly sells for $62.50 is reduced 18%. Its sale price is:

    a. $44.50   **b.** $51.25   **c.** $62.32   **d.** Answer not given

7-4   **15.** If a purchase is $9.27, the change from a $20 bill should be:

    a. $11.83   **b.** $10.83   **c.** $10.73   **d.** $11.73

7-5   **16.** If the rate is 6%, the sales tax on a radio selling for $49.50 is:

    a. $6.00   **b.** $2.97   **c.** $4.95   **d.** $29.70

5-7
7-6
17. At the average cost of 6.5¢ per kilowatt hour, the weekly cost of operating 4 hours each day a TV that uses 200 watts of power is:

    a. 36.4¢   **b.** $3.64   **c.** 52¢   **d.** $5.20

7-7   **18.** A car leaves New York at 6 P.M. and arrives in Baltimore, 296 km away, at 10 P.M. The average speed of this car is:

    a. 75 km/h   **b.** 98 km/h   **c.** 74 km/h   **d.** 85 km/h

7-8   **19.** Andrew's uncle can purchase a clothes washer for the cash price $389.95 or $25 down and 12 equal monthly payments of $34.06. By paying cash he can save:

    a. $34.06   **b.** $37.00   **c.** $55.89   **d.** $43.77

7-9   **20.** When the monthly finance charge rate is $1\frac{1}{4}$%, the annual rate is:

    a. 10%   **b.** 12%   **c.** 15%   **d.** 18%

7-10 **21.** The cost of $40,000 fire insurance for 1 year at $.28 per $100 is:

    a. $28   **b.** $112   **c.** $280   **d.** $400

7-11 **22.** At the annual rate of $22.71 per $1,000 for life insurance, the total annual premium for $75,000 insurance is:

    a. $1,022   **b.** $1,703.25   **c.** $74,000   **d.** $75,022.71

7-12 **23.** If the tax rate is $7.50 per $100, the property tax on a house costing $104,000 and assessed for $70,000 is:

    a. $1,040   **b.** $7,800   **c.** $5,250   **d.** $2,650

7-13 **24.** Joe's mother bought a house for $75,000. She paid 20% down and gave a mortgage for the remainder. The amount of the mortgage is:

    a. $60,000   **b.** $33,000   **c.** $67,000   **d.** $15,000

8-1 **25.** Elaine Bauman paid the Plaza Furniture Co. by check in the amount of $59.75 on July 18, 1986.

```
┌─────────────────────────────────────────────────────────────┐
│              FIRST STATE BANK                  650           │
│                                    _____19____         │
│   PAY TO THE                                                 │
│   ORDER OF_____ $_____         │
│   ①_____/DOLLARS       │
│                                                              │
│                                                              │
│   MEMO_____      _____                  │
│   ⑆956461718⑆ 36 ⑈ 113298 7⑈                                 │
└─────────────────────────────────────────────────────────────┘
```

On line ① should be written:

a. Elaine Bauman   **c.** July 18, 1986

b. Plaza Furniture Co.   **d.** Fifty nine and $\frac{75}{100}$

8-2 **26.** The interest on $2,600 for 3 years at $7\frac{1}{2}$% annual rate is:

a. $585   **b.** $260   **c.** $195   **d.** $695

8-3 **27.** What is the monthly payment on an add-on-interest loan, if $6,000 is borrowed at 10% for 12 months?

a. $600   **b.** $550   **c.** $500   **d.** $720

8-4 **28.** If you borrow $3,000 for 24 months to buy a car and pay $137.50 per month, the finance charge is:

a. $137.50   **b.** $350   **c.** $300   **d.** $287.50

8-5 **29.** If the monthly payment to amortize a loan of $30,000 at 13% annual interest rate for 20 years is $351.48, the total interest paid on the loan is:

a. $54,355.20   **b.** $7,029.60   **c.** $37,029.60   **d.** $3,900.00

8-6 **30.** By compounding interest semiannually at the annual rate of 12%, each $1 is worth $2.01219 at the end of 6 years.

The total interest earned on $1,000 compounded semiannually at 12% annual rate at the end of 6 years is:

a. $2,012.79   **b.** $3,013.19   **c.** $1,012.19   **d.** Answer not given

8-7 **31.** If you purchase 100 shares of TEL at 24 and sell at $30\frac{1}{2}$, your profit, excluding commissions, is:

a. $750   **b.** $700   **c.** $650   **d.** $550

32. Ms. Costello owns a $10,000 Treasury Note that pays an annual interest rate of $9\frac{1}{4}$%. She receives semiannually:

a. $1,000   **b.** $925   **c.** $462.50   **d.** $437.50

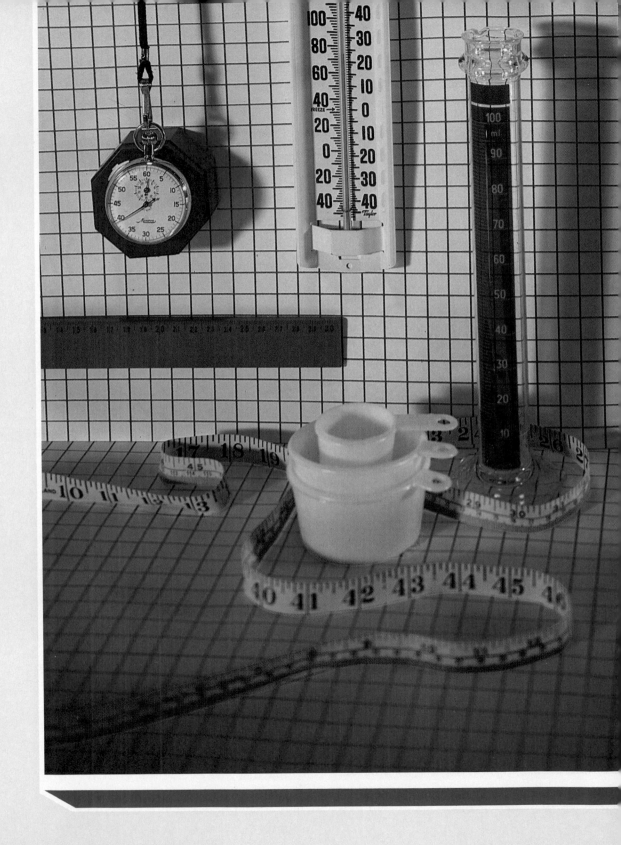

UNIT 3

Measurement, Graphs, Statistics, and Probability

Inventory Test

9-1 Change: **1.** 42 cm to mm   **2.** 690 m to km

9-2 Change: **3.** 5,500 mg to g   **4.** 1.8 kg to g

9-3 Change: **5.** 7 L to cL   **6.** 655 mL to L

9-4 Change: **7.** 800 m^2 to cm^2 **8.** 49 km^2 to hectares

9-5 Change: **9.** 8.3 m^3 to cm^3 **10.** 9,250 cm^3 to dm^3

       11. 6.75 L of water occupies _____ cm^3 and weighs _____ kg.

10-1 **12.** How many feet are in 264 inches?

10-2 **13.** What part of a pound is 12 ounces?

10-3 **14.** Find the number of pints in 9 gallons.

10-4 **15.** How many bushels are in 96 pecks?

10-5 **16.** Change 108 square inches to square feet.

10-6 **17.** How many cubic feet are in 7 cubic yards?

10-7 **18.** What volume in cubic inches do 16 gallons of water occupy?

10-8 **19.** What part of a year is 10 months?

10-9 **20.** Using the train schedule on p. 400, find at what times train 179 leaves New London and arrives in Newark.

10-10 **21.** Change 72 km/h to m/min.   **10-11** **22.** Which is colder: 15°C or 20°F?

11-1 **23.** Construct a bar graph showing the elevations of the following mountain peaks in the United States:

Mt. Rainier, 4,395 meters   Mt. Washington, 1,918 meters

Mt. McKinley, 6,198 meters   Mt. Hood, 3,427 meters

        Guadalupe Peak, 2,669 meters

11-2 **24.** Construct a line graph showing the average number of days per month that Seattle and New Orleans have precipitation.

	Jan.	Feb.	Mar.	Apr.	May	June	July	Aug.	Sept.	Oct.	Nov.	Dec.
Seat.	20	16	18	14	10	10	5	6	9	14	18	21
N.O.	10	10	9	7	7	10	15	13	10	6	7	10

11-3 **25.** Construct a circle graph showing the average annual outlays of the Federal government in recent years:

Education, 5%; Health, 8%; Interest, 7%; Defense, 33%; Income Security, 28%; Other, 19%

11-4 **26.** Make a frequency distribution table for the following list of scores:

72, 66, 75, 67, 73, 74, 69, 72, 68, 71, 70, 65,

66, 71, 68, 74, 65, 71, 66, 74, 70, 67, 75, 68,

74, 69, 68, 70, 72, 69, 74

11-5 For the following scores—39, 42, 39, 48, 41, 36, 49—find:

     27. Mean   **28.** Median   **29.** Mode   **30.** Range

11-6 **31.** What is the probability of drawing at random on the first draw a yellow ball from a box containing 12 white balls and 18 yellow balls?

     32. What are the odds?

CHAPTER 9　Metric System

INTRODUCTION TO THE METRIC SYSTEM

The United States monetary system is a decimal system in which the dollar is the basic unit. In the metric system, also a decimal system, the meter (m) is the basic unit of length, the gram (g) is the basic unit of weight or mass, and the liter (L) is the basic unit of capacity (dry and liquid measures).

Other metric units of length, weight, or capacity are found by combining the following prefixes with the basic units of measure.

Prefix	Symbol	Value
mega-	M	million (1,000,000)
kilo-	k	thousand (1,000)
hecto-	h	hundred (100)
deka-	da	ten (10)
deci-	d	one-tenth (.1 or 1/10)
centi-	c	one-hundredth (.01 or 1/100)
milli-	m	one-thousandth (.001 or 1/1,000)
micro-	μ	one-millionth (.000001 or 1/1,000,000)
nano-	n	one-billionth (.000000001 or 1/1,000,000,000)
pico-	p	one-trillionth (.000000000001 or 1/1,000,000,000,000)

In the chart below, each unit of United States money, length, weight, or capacity is ten (10) times as great as the unit to the right. Therefore, 10 of any unit is equivalent to one (1) unit of the next larger size.

It should be noted that just as 3 dollars 8 dimes 4 cents may be written as a single numeral, 3.84 *dollars* or $3.84; so 3 meters 8 decimeters 4 centimeters may also be written as a single numeral, 3.84 *meters*.

Abbreviations or symbols are written without periods after the last letter. The same symbol is used for both one or more quantities. Thus, "cm" is the symbol for centimeter *or* centimeters.

	1,000	100	10	1	.1	.01	.001
Decimal Place Value	thousands	hundreds	tens	ones	tenths	hundredths	thousandths
United States Money	$1,000 bill	$100 bill	$10 bill	dollar	dime	cent	mill
Metric Length	kilometer km	hectometer hm	dekameter dam	meter m	decimeter dm	centimeter cm	millimeter mm
Metric Weight	kilogram kg	hectogram hg	dekagram dag	gram g	decigram dg	centigram cg	milligram mg
Metric Capacity	kiloliter kL	hectoliter hL	dekaliter daL	liter L	deciliter dL	centiliter cL	milliliter mL

The number of units of the smaller denomination that is equivalent to one unit of the larger denomination is called the *conversion factor*.

Each conversion factor is some power of 10, so that the short method of computation may be used.

9-1 Measures of Length—Metric

Vocabulary

The **millimeter**, **centimeter**, **meter**, and **kilometer** are the units commonly used to measure length in the metric system.

Examine the following section of a centimeter ruler.

One Decimeter

Each subdivision represents 1 millimeter (mm).
Ten millimeters equal 1 centimeter (cm).

Centimeter rulers and meter sticks are commonly used to measure length. The kilometer is used to measure longer distances, such as those between cities.

10 millimeters (mm) = 1 centimeter (cm)	10 meters (m) = 1 dekameter (dam)
10 centimeters (cm) = 1 decimeter (dm)	10 dekameters (dam) = 1 hectometer (hm)
10 decimeters (dm) = 1 meter (m)	10 hectometers (hm) = 1 kilometer (km)
1 meter (m) = 100 centimeters (cm) = 1,000 millimeters(mm)	
1 kilometer (km) = 1,000 meters (m)	

PROCEDURE

To change a given metric unit of length to another:
(1) Find the conversion factor.
(2) Multiply by this conversion factor when changing to a smaller unit.
(3) Divide by this conversion factor when changing to a larger unit.

EXAMPLES:

Change 23 centimeters to millimeters.
  1 cm = 10 mm
23 cm = 23 × 10 = 230 mm

ANSWER: 230 millimeters

Change 9,000 meters to kilometers.
  1 km = 1,000 m
9,000 m = 9,000 ÷ 1,000 = 9 km

ANSWER: 9 kilometers

Express 9 km 3 hm 5 m 2 dm in meters.

9 km = 9 × 1,000 = 9,000   m
3 hm = 3 × 100   =   300   m
5 m   5   m
2 dm = 2 ÷ 10   =   ____.2 m
                              9,305.2 m

ANSWER: 9,305.2 meters

Shortcut: Use place value.

9 km 3 hm 0 dam 5 m   2 dm
 ↓   ↓   ↓   ↓   ↓
 9   3   0   5   .   2

PRACTICE EXERCISES

1 What metric unit of length does each of these symbols represent?

1. km   **2.** cm   **3.** m   **4.** mm   **5.** dm   **6.** dam   **7.** hm

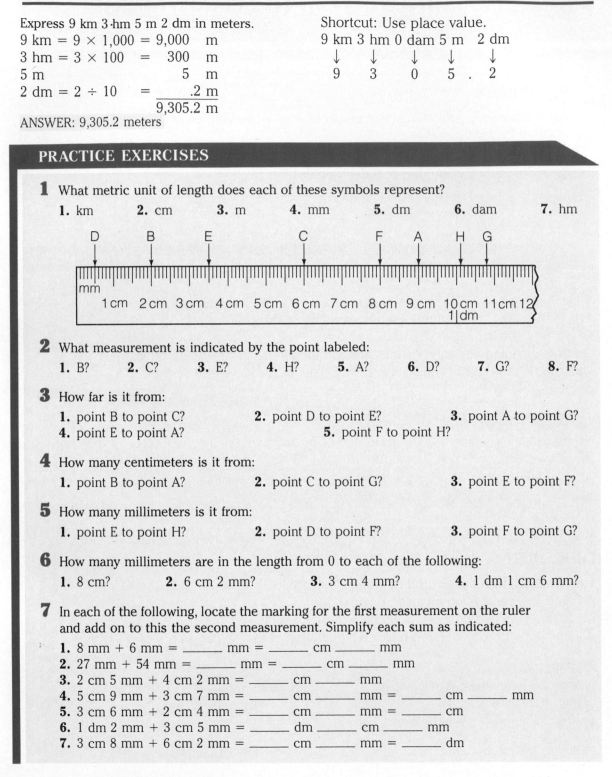

2 What measurement is indicated by the point labeled:

1. B?   **2.** C?   **3.** E?   **4.** H?   **5.** A?   **6.** D?   **7.** G?   **8.** F?

3 How far is it from:

1. point B to point C?   **2.** point D to point E?   **3.** point A to point G?
4. point E to point A?   **5.** point F to point H?

4 How many centimeters is it from:

1. point B to point A?   **2.** point C to point G?   **3.** point E to point F?

5 How many millimeters is it from:

1. point E to point H?   **2.** point D to point F?   **3.** point F to point G?

6 How many millimeters are in the length from 0 to each of the following:

1. 8 cm?   **2.** 6 cm 2 mm?   **3.** 3 cm 4 mm?   **4.** 1 dm 1 cm 6 mm?

7 In each of the following, locate the marking for the first measurement on the ruler and add on to this the second measurement. Simplify each sum as indicated:

1. 8 mm + 6 mm = _____ mm = _____ cm _____ mm
2. 27 mm + 54 mm = _____ mm = _____ cm _____ mm
3. 2 cm 5 mm + 4 cm 2 mm = _____ cm _____ mm
4. 5 cm 9 mm + 3 cm 7 mm = _____ cm _____ mm = _____ cm _____ mm
5. 3 cm 6 mm + 2 cm 4 mm = _____ cm _____ mm = _____ cm
6. 1 dm 2 mm + 3 cm 5 mm = _____ dm _____ cm _____ mm
7. 3 cm 8 mm + 6 cm 2 mm = _____ cm _____ mm = _____ dm

8 Find the 108-mm mark on the metric ruler. Subtract from it a measurement of 4.3 cm. What measurement does the mark you reach indicate?

Complete each of the following:

9 1. _____ mm = 1 cm   2. _____ cm = 1 dm   3. _____ dm = 1 m
    4. _____ m = 1 dam   5. _____ dam = 1 hm   6. _____ hm = 1 km
    7. _____ cm = 1 mm   8. _____ dm = 1 cm   9. _____ m = 1 dm
    10. _____ dam = 1 m   11. _____ hm = 1 dam   12. _____ km = 1 hm

10 1. _____ mm = 1 m   2. _____ cm = 1 m   3. _____ dm = 1 m
    4. _____ mm = 1 cm   5. _____ mm = 1 dm   6. _____ m = 1 km
    7. _____ km = 1 m   8. _____ m = 1 cm   9. _____ m = 1 mm
    10. _____ m = 1 dm
    11. 1 m = _____ dm = _____ cm = _____ mm
    12. 1 km = _____ hm = _____ dam = _____ m
    13. 1 km = _____ m = _____ dm = _____ cm = _____ mm
    14. 1 mm = _____ m = _____ km

11 Express each of the following in meters:
    1. 7 km 5 hm 8 dam 3 m = _____ m   2. 8 m 2 dm 6 cm 5 mm = _____ m
    3. 6 m 4 cm 9 mm = _____ m   4. 3 km 5 m 1 cm 6 mm = _____ m
    5. 9 km 4 hm 2 dam 7 m 3 dm 7 cm 5 mm = _____ m

12 Change each of the following to <u>millimeters</u>:
    1. 34 cm   2. 28 m   3. 4 km   4. 8.5 cm   5. 7.6 m

13 Change each of the following to <u>centimeters</u>:
    1. 4.7 dm   2. 6.4 m   3. 851 mm   4. 8 km   5. 9.3 m

14 Change each of the following to <u>decimeters</u>:
    1. 439 m   2. 760 cm   3. 6,875 mm   4. 16.7 m   5. 2.6 km

15 Change each of the following to <u>meters</u>:
    1. 78 dam   2. 4,206 cm   3. 3.9 hm   4. 60.4 dm   5. 569 mm
    6. .57 km   7. 21.4 dam   8. 5 cm   9. 12,600 mm   10. 9.625 km

16 Change each of the following to <u>dekameters</u>:
    1. 46 m   2. 53 hm   3. 7.2 km   4. 38.9 m   5. 3,625 cm

17 Change each of the following to <u>hectometers</u>:
    1. 60 km   2. 16.4 m   3. 400 dam   4. 225.38 m   5. 2.6 km

18 Change each of the following to <u>kilometers</u>:

    1. 81 hm   **2.** 5,328 m   **3.** 794,000 cm   **4.** 362.3 dam   **5.** 9,270,000 mm

19 Find the missing equivalent measurements:

	km	hm	dam	m	dm	cm	mm
1.	?	?	?	700	?	?	?
2.	3	?	?	?	?	?	?
3.	?	?	?	?	?	5,000	?
4.	?	?	?	?	?	?	6,800,000

Complete each of the following:

20   **1.** 7 cm 4 mm = _____ mm   **2.** 5 m 9 cm = _____ cm   **3.** 8 km 620 m = _____ m
     4. 3 m 7 mm = _____ mm   **5.** 1 cm 8 mm = _____ mm   **6.** 6 m 39 cm = _____ cm

21   **1.** 4 m 8 dm 2 cm = _____ cm   **2.** 2 m 6 cm 5 mm = _____ mm
     3. 1 km 9 m 3 cm = _____ cm   **4.** 7 m 5 cm 4 mm = _____ mm

22   **1.** 3 cm 8 mm = _____ cm   **2.** 5 km 2 m = _____ km   **3.** 5 m 62 mm = _____ m
     4. 2 cm 9 mm = _____ cm   **5.** 4 km 218 m = _____ km   **6.** 6 m 7 cm = _____ m

23   **1.** 6 m 9 cm 4 mm = _____ m   **2.** 4 km 17 m 52 cm = _____ km
     3. 8 km 6 m 9 cm = _____ km   **4.** 5 m 2 cm 6 mm = _____ m

24   **1.** 6 m 1 cm 8 mm = _____ cm   **2.** 7 km 4 m 3 cm = _____ m
     3. 3 cm 7 mm = _____ m   **4.** 92 m 6 cm = _____ km

25 If each curtain panel requires 125 cm of fabric and you need 8 panels, how many meters of fabric should you buy?

26 Sound travels at a speed of 1,450 meters per second in water. How many kilometers does it travel in water in 15 seconds?

27 How much longer is a metal strip that measures 1.2 meters than one that measures 894 millimeters?

28 Which measurement is greater:

    1. 9,000 mm or 7 m?   **2.** 4 km or 3,000 m?   **3.** 4,000 cm or 12 m?
    4. 800 mm or 6 m?   **5.** 5.1 cm or 72 mm?   **6.** 4.5 m or 450 cm?

29 Which measurement is smaller:

    1. 7 m or 9 km?   **2.** 40 cm or 3 m?   **3.** 5 km or 260,000 mm?
    4. 73 mm or 6.9 cm?   **5.** 2.7 km or 2,994 m?   **6.** 8.49 m or 849 mm?

30 Arrange the following measurements in order of size (longest first):
    6,800 m   526,000 cm   6.7 km   692,400 mm

31 Arrange the following measurements in order of size (shortest first):
    8.45 km   84.5 mm   845 m   8,450,000 cm

9-2 Measures of Mass or Weight—Metric

Vocabulary

The **kilogram** and **gram** are units commonly used to measure the **mass** in the metric system.

In everyday use, the word "weight" almost always means "mass." Therefore, for our purposes we will use kilogram and gram as units of weight.

10 milligrams (mg) = 1 centigram (cg)	10 grams (g) = 1 dekagram (dag)
10 centigrams (cg) = 1 decigram (dg)	10 dekagrams (dag) = 1 hectogram (hg)
10 decigrams (dg) = 1 gram (g)	10 hectograms (hg) = 1 kilogram (kg)
1,000 kilograms (kg) = 1 metric ton (t)	

PROCEDURE

To change a given metric unit of weight to another:
(1) Find the conversion factor.
(2) Multiply by this conversion factor when changing to a smaller unit.
(3) Divide by this conversion factor when changing to a larger unit.

EXAMPLES:

Change 6 kilograms to grams.
1 kg = 1,000 g
6 kg = 6 × 1,000 = 6,000 g
ANSWER: 6,000 grams

Change 58 milligrams to centigrams.
1 cg = 10 mg
58 mg = 58 ÷ 10 = 5.8 cg
ANSWER: 5.8 centigrams

Change 1 cg 4 mg to centigrams.
1 cg =   1 cg
4 mg = 4 ÷ 10 =   .4 cg
                 ———
                 1.4 cg
ANSWER: 1.4 centigrams

PRACTICE EXERCISES

1 What metric unit of weight does each of these symbols represent?

1. mg   **2.** kg   **3.** dg   **4.** g   **5.** cg   **6.** hg   **7.** dag

Complete each of the following:

2 **1.** ___ mg = 1 cg   **2.** ___ cg = 1 dg   **3.** ___ dg = 1 g   **4.** ___ g = 1 dag

5. ___ dag = 1 hg   **6.** ___ hg = 1 kg   **7.** ___ cg = 1 mg   **8.** ___ dg = 1 cg

9. ___ g = 1 dg   **10.** ___ dag = 1 g   **11.** ___ hg = 1 dag   **12.** ___ kg = 1 hg

3
1. ___ mg = 1 g   2. ___ cg = 1 g   3. ___ g = 1 cg
4. ___ g = 1 mg   5. ___ g = 1 kg   6. ___ kg = 1 g
7. 1 kg = ___ g = ___ mg   8. 1 mg = ___ g = ___ kg   9. ___ hg = 1 kg
10. ___ dag = 1 kg   11. ___ dg = 1 kg   12. ___ cg = 1 kg
13. ___ mg = 1 kg   14. ___ kg = 1 hg   15. ___ kg = 1 dag
16. ___ kg = 1 dg   17. ___ kg = 1 cg   18. ___ kg = 1 mg

4 Change each of the following to <u>milligrams</u>:
1. 9 cg   2. 16 g   3. 4.7 cg   4. 2.36 kg   5. 0.841 g

5 Change each of the following to <u>centigrams</u>:
1. 39 mg   2. 5.74 dg   3. 62 g   4. 1.4 kg   5. 28.9 g

6 Change each of the following to <u>decigrams</u>:
1. 25 g   2. 78 cg   3. 3.1 kg   4. 4,500 mg   5. 9.4 g

7 Change each of the following to <u>grams</u>:
1. 60 dag   2. 530 dg   3. 9 kg   4. 25.8 cg   5. 750 mg

8 Change each of the following to <u>dekagrams</u>:
1. 3,000 g   2. 68 kg   3. 76.7 cg   4. 5.63 hg   5. 9,100 mg

9 Change each of the following to <u>hectograms</u>:
1. 533 dag   2. 7,000 g   3. 40 kg   4. 451.9 g   5. 6.5 kg

10 Change each of the following to <u>kilograms</u>:
1. 7,864 g   2. 85 hg   3. 961,000 cg   4. 423.6 dag   5. 7,400,000 mg

11 Change each of the following to <u>metric tons</u>:
1. 3,000 kg   2. 570 kg   3. 6,300,000 g   4. 19,000 hg   5. 82,600 kg

12 Find the missing equivalent weights:

	kg	g	cg	mg
1.	?	5,000	?	?
2.	7.3	?	?	?
3.	?	?	2,900	?
4.	?	?	?	800,000
5.	?	456	?	?

Complete each of the following:

13
1. 5 cg 6 mg = _____ mg   2. 8 kg 130 g = _____ g   3. 7 g 25 cg = _____ cg
4. 2 g 84 mg = _____ mg   5. 3 cg 9 mg = _____ mg   6. 6 kg 370 g = _____ g

14
1. 4 g 8 dg 5 cg _____ cg   2. 3 g 6 cg 9 mg = _____ mg
3. 7 kg 2 g 8 cg = _____ cg   4. 9 kg 1 g 5 mg = _____ mg

15
1. 5 cg 2 mg = _____ cg   2. 15 kg 260 g = _____ kg   3. 9 g 4 cg = _____ g
4. 7 g 1 mg = _____ g   5. 8 kg 7 g = _____ kg   6. 19 g 58 mg = _____ g

16
1. 8 g 5 cg 2 mg = _____ g   2. 5 kg 9 g 4 cg = _____ kg
3. 6 kg 7 g 1 mg = _____ kg   4. 3 g 6 cg 6 mg = _____ g

Vocabulary

The **liter** and **milliliter** are units commonly used to measure capacity in the metric system.

The units that are used to measure capacity are also used to measure volume.

The difference between volume and capacity is that volume is the amount of space within a three-dimensional figure and capacity is the amount of substance that the figure will hold.

10 milliliters (mL) = 1 centiliter (cL)	10 liters (L) = 1 dekaliter (daL)
10 centiliters (cL) = 1 deciliter (dl.)	10 dekaliters (daL) = 1 hectoliter (hL)
10 deciliters (dL) = 1 liter (L)	10 hectoliters (hL) = 1 kiloliter (kL)

PROCEDURE

To change a given metric unit of capacity to another:
(1) Find the conversion factor.
(2) Multiply by this conversion factor when changing to a smaller unit.
(3) Divide by this conversion factor when changing to a larger unit.

EXAMPLES:

Change 4 liters to centiliters.

1 L = 100 cL
4 L = 4 × 100 = 400 cL

ANSWER: 400 centiliters

Change 7,250 milliliters to liters.

1 L = 1,000 mL
7,250 mL = 7,250 ÷ 1,000 = 7.250 L

ANSWER: 7.25 liters

Change 1 liter 8 centiliters to liters.

1 L   1 L
8 cL = 8 ÷ 100   .08 L
                 ‾‾‾‾‾‾
                 1.08 L

ANSWER: 1.08 liters

Shortcut: Use place value.

1 L   0 dL 8 cL
 ↓   ↓   ↓
 1 . 0   8

PRACTICE EXERCISES

1 What metric unit of capacity does each symbol represent?

    1. kL   **2.** L   **3.** mL   **4.** cL   **5.** hL   **6.** dL   **7.** daL

Complete each of the following:

2 **1.** _____ mL = 1 cL   **2.** _____ cL = 1 dL   **3.** _____ dL = 1 L   **4.** _____ L = 1 daL

    5. _____ daL = 1 hL   **6.** _____ hL = 1 kL   **7.** _____ cL = 1 mL   **8.** _____ dL = 1 cL

    9. _____ L = 1 dL   **10.** _____ daL = 1 L   **11.** _____ hL = 1 daL   **12.** _____ kL = 1 hL

3 **1.** _____ mL = 1 L   **2.** _____ cL = 1 L   **3.** _____ L = 1 kL   **4.** _____ kL = 1 L

5. _____ L = 1 cL   **6.** _____ L = 1 mL

7. 1 kL = _____ L = _____ mL   **8.** 1 mL = _____ L = _____ kL

4 Change each of the following to <u>milliliters</u>:

1. 7 cL   **2.** 4.6 L   **3.** 53 dL   **4.** 375 L   **5.** 80.4 cL

5 Change each of the following to <u>centiliters</u>:

1. 85 L   **2.** 9 dL   **3.** 18 mL   **4.** 6.7 L   **5.** 44.2 mL

6 Change each of the following to <u>deciliters</u>:

1. 31 L   **2.** 68 cL   **3.** 2.25 L   **4.** 520 mL   **5.** 767 cL

7 Change each of the following to <u>liters</u>:

1. 9 dL   **2.** 60 hL   **3.** 21.6 kL   **4.** 400 cL   **5.** 8,405 mL

8 Change each of the following to <u>dekaliters</u>:

1. 29 hL   **2.** 108 L   **3.** 4.7 hL   **4.** 16.4 L   **5.** 332 dL

9 Change each of the following to <u>hectoliters</u>:

1. 7.9 kL   **2.** 5.2 daL   **3.** 6,008 L   **4.** 21.57 kL   **5.** 825.1 L

10 Change each of the following to <u>kiloliters</u>:

1. 6,582 L   **2.** 43 hL   **3.** 875 L   **4.** 79.5 hL   **5.** 200 daL

11 Find the missing equivalent capacities:

	L	dL	cL	mL
1.	7	?	?	?
2.	?	645	?	?
3.	?	?	?	900
4.	3.8	?	?	?
5.	?	?	545.2	?

Complete each of the following:

12 **1.** 7 cL 4 mL = _____ mL   **2.** 1 L 6 cL = _____ cL   **3.** 5 L 8 mL = _____ mL

4. 8 kL 600 L = _____ L   **5.** 4 L 91 mL = _____ mL   **6.** 13 L 25 cL = _____ cL

13 **1.** 6 L 1 cL 9 mL = _____ mL   **2.** 1 L 57 cL 8 mL = _____ mL

14 **1.** 3 cL 7 mL = _____ cL   **2.** 5 L 5 cL = _____ L   **3.** 4 kL 46 L = _____ kL

4. 8 L 2 mL = _____ L   **5.** 9 L 54 cL = _____ L   **6.** 6 cL 9 mL = _____ cL

15 **1.** 8 L 3 cL 7 mL = _____ L   **2.** 3 L 6 cL 8 mL = _____ L

16 **1.** 4 L 1 cL 4 mL = _____ cL   **2.** 5 L 9 cL 6 mL = _____ cL

17 Frank mixed 15 cL of a weed-killer solution with 485 cL of water. How many liters of the mixture did he have?

18 Which capacity is less:

1. 40 cL or 65 mL?   **2.** 6.8 L or 2,000 cL?   **3.** 54.8 mL or 1 L?

4. 2.6 cL or 3 mL?   **5.** 65 dL or 65 daL?   **6.** 8.1 L or 9,700 mL?

19 Arrange the following capacities in order of size (largest first):

63.8 cL   637 mL   .639 L   6.39 dL

20 Arrange the following capacities in order of size (smallest first):

19 L   6,235 mL   89.4 dL   622 cL

Computer Activities

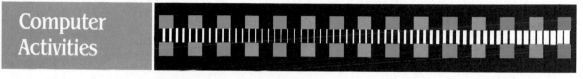

The computer program below converts any number of *kilometers* to the *equivalent* number of: hectometers, dekameters, or meters.

```
10 PRINT "HOW MANY KILOMETERS";
20 INPUT K
30 PRINT K;" KILOMETERS IS EQUAL TO:"
40 PRINT K*10;" HECTOMETERS. . .OR"
50 PRINT K*100;" DEKAMETERS. . .OR"
60 PRINT K*1000;" METERS"
```

a) ENTER the program and then RUN it to make sure it works.

b) Modify the program so that it converts any number of kilograms to the equivalent number of: hectograms, dekagrams, or grams.

c) Modify the program so that it converts any number of kiloliters to the equivalent number of: hectoliters, dekaliters, or liters.

d) Write a program which converts meters to: decimeters, centimeters, or millimeters.

e) Modify the program so that it converts grams to: decigrams, centigrams, or milligrams.

f) Modify the program so that it converts liters to: deciliters, centiliters, or milliliters.

Refresh Your Skills

1. Add: 1-4
93,514
6,648
87,529
896
9,388

2. Subtract: 1-5
109,625
 87,592

3. Multiply: 1-6
1,658
 973

4. Divide: 1-7
$5,280)\overline{237,600}$

5. Add: 2-4
.008 + .8 + .08

6. Subtract: 2-5
.97 − .3

7. Multiply: 2-7
.09 × .09

8. Divide: 2-8
$4 ÷ $.05

9. Add: 3-8
$1\frac{7}{12} + 2\frac{1}{6} + 4\frac{2}{3}$

10. Subtract: 3-9
$1\frac{3}{4} - \frac{5}{6}$

11. Multiply: 3-11
$\frac{2}{3} × 1\frac{4}{5}$

12. Divide: 3-12
$5\frac{1}{4} ÷ \frac{3}{10}$

13. Find $13\frac{1}{4}$% of $5,000 4-6

14. $1.25 is what percent of $10? 4-7

15. $4.90 is 10% of what amount? 4-8

SIGNIFICANT DIGITS

Digits are significant in an approximate number when they indicate the precision (closeness to the true measurement) which is determined by the value of the place of the last significant digit on the right.

SIGNIFICANT DIGITS
I All nonzero digits.
II All zeros located between significant digits.
III All zeros at the end of a decimal or mixed decimal.
IV All digits of the first factor when a number is expressed in scientific notation.
V Underscored or specified zeros of a whole number ending in zeros.

DIGITS THAT ARE NOT SIGNIFICANT
I Zeros at the end of a whole number (unless specified to be significant).
II Zeros following the decimal point in a number between 0 and 1.

EXAMPLES:
684 has three significant digits: 6, 8, and 4.

5,006 has four significant digits: 5, 0, 0, and 6.

86.190 has five significant digits: 8, 6, 1, 9, and 0.

7.32×10^9 has three significant digits: 7, 3, and 2.

51,000 has two significant digits: 5 and 1.

51,0̲0̲0 has four significant digits: 5, 1, 0, and 0.

.0070 has two significant digits: 7 and 0

2.00960 has six significant digits: 2, 0, 0, 9, 6, and 0.

.003 has one significant digit: 3.

In 64.75, the significant digit 5 indicates precision to the nearest hundredth.

In 83,0̲0̲0, the underlined zero indicates precision to the nearest ten.

EXERCISES

In each of the following, determine the number of significant digits and name them.

1. 28	**2.** 600	**3.** 4,271	**4.** 37,000	**5.** 9
6. .08	**7.** .0095	**8.** .013	**9.** .0130	**10.** 9,050,000
11. 23.05	**12.** .0001000	**13.** 87,0̲0̲0	**14.** 350,0̲00	**15.** 6.3×10^5
16. 8.49×10^7	**17.** 2,500	**18.** .0007	**19.** 6,810	**20.** 51.0
21. 3.1416	**22.** 7,060	**23.** 95,280	**24.** 16,003	**25.** 40,0̲00
26. .000000015	**27.** 9.00000027	**28.** 8,259,078	**29.** 4.0060	**30.** 203,0̲0̲0

9-4 Measures of Area—Metric

Vocabulary

Square centimeters, **square meters**, and **square kilometers** are the units commonly used to measure area in the metric system. **Centare** may be used instead of square meter and **hectare** may be used instead of square hectometer.

To measure area, we use square units. In square measure, 100 of any metric unit is equivalent to 1 of the next higher unit. The exponent ^{2} is used to replace the word square.

1 cm

1 cm

Area: 1 cm^2

$$100 \text{ square millimeters (mm}^2) = 1 \text{ square centimeter (cm}^2)$$
$$100 \text{ square centimeters (cm}^2) = 1 \text{ square decimeter (dm}^2)$$
$$100 \text{ square decimeters (dm}^2) = 1 \text{ square meter (m}^2)$$
$$100 \text{ square meters (m}^2) = 1 \text{ square dekameter (dam}^2)$$
$$100 \text{ square dekameters (dam}^2) = 1 \text{ square hectometer (hm}^2)$$
$$100 \text{ square hectometers (hm}^2) = 1 \text{ square kilometer (km}^2)$$
$$100 \text{ centares} = 1 \text{ are (a)}$$
$$100 \text{ ares} = 1 \text{ hectare (ha)}$$
$$100 \text{ hectares} = 1 \text{ square kilometer}$$

PROCEDURE

To change a given metric unit of area to another:
(1) Find the conversion factor.
(2) Multiply by this conversion factor when changing to a smaller unit.
(3) Divide by this conversion factor when changing to a larger unit.

EXAMPLES:

Change 9 square kilometers to square meters.
$1 \text{ km}^2 = 1,000,000 \text{ m}^2$
$9 \text{ km}^2 = 9 \times 1,000,000 = 9,000,000 \text{ m}^2$

ANSWER: 9,000,000 square meters

Change 670 square millimeters to square centimeters.

$1 \text{ cm}^2 = 100 \text{ mm}^2$
$670 \text{ mm}^2 = 670 \div 100 = 6.70 \text{ cm}^2$

ANSWER: 6.7 square centimeters

Change 900 hectares to square kilometers.
$1 \text{ km}^2 = 100 \text{ ha}$
$900 \text{ ha} = 900 \div 100 = 9 \text{ km}^2$

ANSWER: 9 square kilometers

PRACTICE EXERCISES

1 What metric unit of square measure does each of the symbols represent?

1. cm² **2.** km² **3.** m² **4.** mm² **5.** ha **6.** dm² **7.** dam²

2 **1.** How many square centimeters are in 1 square meter?
  2. How many square millimeters are in 1 square meter?
  3. How many square meters are in 1 square kilometer?
  4. How many square meters are in 1 hectare?
  5. How many hectares are in 1 square kilometer?

3 Change each of the following to <u>square millimeters</u>:

  1. 6 cm² **2.** 2.15 m² **3.** 1.7 dm² **4.** 41.9 cm² **5.** 0.85 m²

4 Change each of the following to <u>square centimeters</u>:

  1. 37 dm² **2.** 8.4 m² **3.** 700 mm² **4.** 0.66 m² **5.** 5,300 mm²

5 Change each of the following to <u>square decimeters</u>:

  1. 819 cm² **2.** 30 m² **3.** 600,000 mm² **4.** 7.5 m² **5.** 240.7 cm²

6 Change each of the following to <u>square meters (or centares)</u>:

  1. 45 km² **2.** 19 dm² **3.** 20,000 cm² **4.** 809,000 mm² **5.** 0.259 km²

7 Change each of the following to <u>square dekameters (or ares)</u>:

  1. 8,000 m² **2.** 31 hectares **3.** 2.77 m² **4.** 6.8 hectares **5.** 12,300 m²

8 Change each of the following to <u>hectares (or square hectometers)</u>:

  1. 75,000 m² **2.** 78.2 dam² **3.** 14 km² **4.** 387.5 m² **5.** 4.3 km²

9 Change each of the following to <u>square kilometers</u>:

  1. 500 hectares **2.** 72.6 hectares
  3. 3,825,000 m² **4.** 609,220 m²

10 A farm contains 40 hectares of land. How many square meters does it measure? What part of a square kilometer is it?

9-5 Measures of Volume—Metric

Vocabulary

Cubic centimeters and cubic meters are the units commonly used to measure volume in the metric system. Volume is related to capacity, so milliliters and liters can also be used to measure volume.

To measure volume, we use cubic units. In cubic measure, 1,000 of any metric unit is equivalent to 1 of the next higher unit. The exponent ^{3} is used to replace the word cubic.

The volume of 1 cubic decimeter has the same capacity as 1 liter.

The volume of 1 cubic centimeter has the same capacity as 1 milliliter.

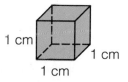

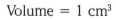

Volume $= 1$ cm^3

1,000 cubic millimeters (mm³) = 1 cubic centimeter (cm³)
1,000 cubic centimeters (cm³) = 1 cubic decimeter (dm³)
1,000 cubic decimeters (dm³) = 1 cubic meter (m³)
1 liter = 1 cubic decimeter = 1,000 cubic centimeters
1 milliliter = 1 cubic centimeter

Metric units of weight and volume (or capacity) are related as follows:

1,000 cubic centimeters (or 1 liter) of water weighs 1 kilogram at 4° Celsius.
1 cubic centimeter (or 1 milliliter) of water weighs 1 gram at 4° Celsius.

PROCEDURE

To change a given metric unit of volume to another:
(1) Find the conversion factor.
(2) Multiply by this conversion factor when changing to a smaller unit.
(3) Divide by this conversion factor when changing to a larger unit.

EXAMPLES:

Change 7 cubic decimeters to cubic centimeters.

1 dm³ = 1,000 cm³
7 dm³ = 7 × 1,000 = 7,000 cm³

ANSWER: 7,000 cubic centimeters

Change 490,000 cubic decimeters to cubic meters.

1 m³ = 1,000 dm³
490,000 dm³ = 490,000 ÷ 1,000 = 490 m³

ANSWER: 490 cubic meters

2 liters of liquid occupies how many cubic centimeters?

1 L = 1,000 cm³
2 L = 2 × 1,000 = 2,000 cm³
ANSWER: 2,000 cubic centimeters

At 4° Celsius, 8 liters of water weighs about how many kilograms?

1 L weighs 1 kg
8 L weighs 8 × 1 = 8 kg
ANSWER: 8 kilograms

PRACTICE EXERCISES

1 What metric unit of cubic measure does each of the following symbols represent?
   1. m³   **2.** mm³   **3.** km³   **4.** cm³   **5.** dm³   **6.** hm³   **7.** dam³

2 **1.** How many cubic millimeters are in 1 cubic decimeter?
   2. How many cubic millimeters are in 1 cubic meter?
   3. How many cubic centimeters are in 1 cubic meter?

3 Change each of the following to <u>cubic millimeters</u>:
   1. 8 cm³   **2.** 15 m³   **3.** 6.1 dm³   **4.** 25.8 cm³   **5.** 53.6 m³

4 Change each of the following to <u>cubic centimeters</u>:
   1. 27 m³   **2.** 4.95 dm³   **3.** 6,226 mm³   **4.** 8.5 m³   **5.** 740 mm³

5 Change each of the following to <u>cubic decimeters</u>:
   1. 63 m³   **2.** 2,930 cm³   **3.** 17.9 m³   **4.** 73,250 mm³   **5.** 86.4 cm³

6 Change each of the following to <u>cubic meters</u>:
   1. 17,000 dm³   **2.** 36,800 cm³   **3.** 9,400,000 mm³   **4.** 575 dm³   **5.** 600,000 cm³

7 **1.** 21 milliliters of liquid will fill how many cubic centimeters?
   2. 9 liters of liquid will fill how many cubic centimeters of space?
   3. A space of 430 cubic centimeters will hold how many milliliters of liquid?
   4. A space of 18.6 cubic decimeters will hold how many liters of liquid?
   5. A space of 7,900 cubic centimeters will hold how many liters of liquid?

Assume for Exercises 6-16 the water is at a temperature of 4° Celsius.

   6. 5 liters of water weigh approximately how many kilograms?
   7. 67 milliliters of water weigh how many grams?
   8. 82 centiliters of water weigh how many grams?
   9. 9.1 liters of water weigh how many kilograms?
   10. 0.615 liter of water weighs how many grams?
   11. How many liters of water are in a container if the water weighs 4 kilograms?
   12. How many milliliters of water weigh 75 grams?
   13. How many liters of water are in a container if the water weighs 545 grams?
   14. How many cubic decimeters do 26 kilograms of water occupy?
   15. Water weighing 19 grams occupies a space of how many cubic centimeters?
   16. Water weighing 8.8 kilograms fills a space of how many cubic centimeters?

8 How much space is occupied by each of the following capacities?
In <u>cubic centimeters (cm³):</u>
   1. 7 mL   **2.** 43.6 mL   **3.** 61 cL   **4.** 4 dL   **5.** 0.75 L

In <u>cubic decimeters (dm³):</u>
   6. 6 L   **7.** 84 dL   **8.** 257 cL   **9.** 14.2 L   **10.** 8.25 L

9 Find the capacity that will fill each of the following volumes:
In <u>milliliters (mL):</u>
   1. 13 cm³   **2.** 2.5 cm³   **3.** 800 mm³   **4.** 7 dm³   **5.** 0.09 m³

In <u>centiliters (cL):</u>
   6. 8 cm³   **7.** 3.18 m³   **8.** 2.6 dm³   **9.** 5.07 cm³   **10.** 9,100 mm³

In <u>liters (L):</u>
   11. 7 dm³   **12.** 28.3 dm³   **13.** 5.9 m³   **14.** 6,000 cm³   **15.** 43,000 mm³

10 Find the weight of each of the following volumes or capacities of water at 4°C:
In <u>kilograms (kg):</u>
   1. 5 dm³   **2.** 79 L   **3.** 196 cm³   **4.** 4.81 L   **5.** 6,300 cm³

In <u>grams (g):</u>
   6. 11 cm³   **7.** 0.225 L   **8.** 8.7 mL   **9.** 6.08 cL   **10.** 5,600 mm³

11 Find the volume or capacity occupied by each of the following weights of water at 4°C:
In cubic <u>centimeters (cm³):</u>
   1. 6 kg   **2.** 30 g   **3.** 0.6 kg   **4.** 950 mg   **5.** 8.74 g

In <u>liters (L):</u>
   6. 15 kg   **7.** 4,100 g   **8.** 520 g   **9.** 0.004 kg   **10.** 7,090 mg

In <u>milliliters (mL):</u>
   11. 7 g   **12.** 138.51 g   **13.** 1.8 kg   **14.** 6,270 mg   **15.** 9,500 cg

In <u>centiliters (cL):</u>
   16. 26 g   **17.** 5.7 kg   **18.** 615 g   **19.** 17,400 cg   **20.** 9,000 mg

12 A tank holds 800 liters of oil. How many cubic decimeters of space are in the tank?

13 It takes 190,000 liters of water to fill a swimming pool. What is the weight of the water when the pool is nine-tenths full?

14 How many liters of water will fill an aquarium if it occupies a space of 6,600 cubic centimeters? What is the weight of the water when the aquarium is full?

15 A water tank has a volume of 7.54 cubic meters. How many liters of water can the tank hold? What is the weight of the water when the tank is full?

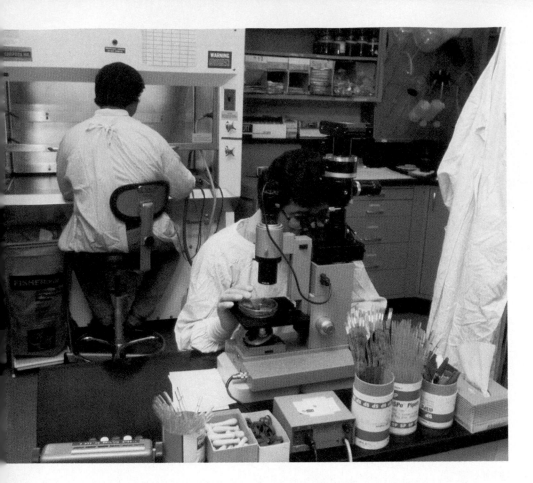

Laboratory tests are important in the diagnosis and treatment of many diseases. A large laboratory may employ a variety of workers. Medical laboratory assistants help other workers and do some work on their own. A medical laboratory assistant may weigh and measure samples, use a microscope to look for abnormal cells, do blood tests, and prepare solutions for tests.

After high-school graduation, medical laboratory assistants take a one-year training program at a college or hospital.

1. A medical laboratory assistant needs to measure the weight of a sample. The weight of the empty container is 126.83 g. The weight of the container and the sample is 174.36 g. What is the weight of the sample?

2. A medical laboratory assistant looks at a parasite egg under a microscope. The microscope field is divided into units .0023 mm long. The egg is 8 units long. How long is the egg in millimeters?

3. A medical laboratory assistant plans to test 23 5-mL blood samples. What is the total volume of the samples?

4. To test blood for sickle cell anemia, a medical laboratory assistant uses a solution of 2 g of sodium metabisulfate in 100 mL of water. Sodium metabisulfate is available as .2 g tablets. How much water should be used for one tablet?

5. A solution is to be 3.8% sodium citrate by weight. How much sodium citrate is needed for 500 g of solution?

CHAPTER REVIEW

Vocabulary

centare p. 375
centimeter p. 365
cubic centimeter p. 377
cubic meter p. 377
gram p. 369
hectare p. 375

kilogram p. 369
kilometer p. 365
liter p. 371
mass p. 369
meter p. 365
milliliter p. 371

millimeter p. 365
square centimeter p. 375
square kilometer p. 375
square meter p. 375

The numerals in boxes indicate Lessons where help may be found.

Change: [9-1] **1.** 85 mm to cm **2.** 4.6 km to m

3. Which measurement is shorter: 2.1 m or 205 cm? [9-1]

4. Arrange according to size (longest measurement first): [9-1]
   a. 560 m **b.** 490,000 mm **c.** 7.3 km **d.** 72,900 cm

5. Andrew has 2 pieces of copper tubing, one measuring 1 m 5 cm and the other 4 times as long. What is the total length of the 2 pieces of tubing? [9-1]

Change: [9-2] **6.** 375 g to kg **7.** 15.6 cg to mg

8. Which is heavier: 3.25 kg or 3,500 g? [9-2]

9. Arrange according to weight (lightest first): [9-2]
   a. 590 cg **b.** .042 kg **c.** 67.3 g **d.** 30,000 mg

10. How many 50-mg tablets will be equivalent to a dose of 1 gram? [9-2]

Change: [9-3] **11.** 10.4 L to cL **12.** 392 mL to dL

13. Which capacity is larger: 819 cL or 8.2 L? [9-3]

14. Arrange according to size (smallest capacity first): [9-3]
   a. 2,500 mL **b.** 3.4 L **c.** 36.8 dL **d.** 329 cL

15. A tank holds 2.3 kiloliters of water. How long will it take to fill if water flows in at the rate of 100 liters per minute? [9-3]

Change: [9-4] **16.** 9.8 m^2 to cm^2 **17.** 290 mm^2 to cm^2

18. A farm contains 65 hectares of land. How many square meters does it measure? [9-4]

Change: [9-5] **19.** 11.2 cm^3 to mm^3 **20.** 5,300 dm^3 to m^3

21. 23 milliters of liquid will occupy a space of how many cubic centimeters? [9-5]

22. 8.7 liters of water weigh how many kilograms? [9-5]

23. A space of 7,500 cubic centimeters will hold how many liters of liquid? [9-5]

24. 49 centiliters of water weigh how many grams? [9-5]

25. Sound travels in water at a speed of 1,450 meters per second. How many kilometers does sound travel in water in 5 seconds? [9-1]

COMPETENCY CHECK TEST

The numerals in boxes indicate Lessons where help may be found.

Solve each problem and select the letter corresponding to your answer.

1. What measurement does point *A* indicate on the ruler? 9-1

    a. 3.2 cm   **b.** 35 mm   **c.** 37 mm   **d.** 3.8 cm

2. 53 cm changed to mm is: 9-1
    a. 5.3 mm   **b.** 53 mm   **c.** 530 mm   **d.** 5,300 mm

3. The shortest measurement in the following is: 9-1
    a. 6.47 m   **b.** 38.9 cm   **c.** 0.006 km   **d.** 2,755 mm

4. 3.8 kg changed to g is: 9-2
    a. 38 g   **b.** 380 g   **c.** 3,800 g   **d.** .38 g

5. The heaviest weight in the following is: 9-2
    a. 0.002 kg   **b.** 97.8 mg   **c.** 2,500 cg   **d.** 395 g

6. 420 cL changed to L is: 9-3
    a. 4,200 L   **b.** 42 L   **c.** 4.2 L   **d.** .042 L

7. The smallest capacity in the following is: 9-3
    a. 500 cL   **b.** 4.5 L   **c.** 37.6 dL   **d.** 1,890 mL

8. 7 m^2 changed to cm^2 is: 9-4
    a. .7 cm^2 **b.** 700 cm^2 **c.** 7,000 cm^2 **d.** 70,000 cm^2

9. 6,000 cm^3 changed to dm^3 is: 9-5
    a. 60,000 dm^3 **b.** 600 dm^3 **c.** 60 dm^3 **d.** 6 dm^3

10. 12 milliliters of water occupy a space of: 9-5
    a. 120 mm^3 **b.** 12 cm^3 **c.** 1.2 dm^3 **d.** .012 m^3

11. 5 liters of water weigh: 9-5
    a. 500 g   **b.** 50 g   **c.** .05 kg   **d.** 5 kg

12. A tank holds 40 liters of gasoline. The number of cubic decimeters of space in the tank is: 9-5
    a. .04 dm^3 **b.** 4 dm^3 **c.** 40 dm^3 **d.** 400 dm^3

13. If the water weighs 13 kilograms, the number of liters of water in the container is: 9-5
    a. 13 L   **b.** 130 L   **c.** 13,000 L   **d.** .0013 L

14. Lisa mixed 87 cL of warm water with 63 cL of cold water. The total number of liters of water she mixed is: 9-3
    a. 15 L   **b.** 150 L   **c.** .15 L   **d.** 1.5 L

CHAPTER 10 Customary System, Time, and Temperature

10-1 Measures of Length—Customary

Vocabulary

The **inch**, **foot**, **yard**, and **mile** are basic units commonly used to measure the length of an object. Numbers expressed in terms of units of measure are called **denominate numbers**.

In the customary system, there are many different conversion factors. Once the particular conversion factor is known, the same procedures as we used in the metric system are used to change from one unit to another.

1 foot (ft) = 12 inches (in.)	1 statute mile (mi) = .8684 nautical mile
1 yard (yd) = 3 feet (ft)	= 320 rods (rd)
= 36 inches (in.)	= 1,760 yards (yd)
1 rod (rd) = $16\frac{1}{2}$ feet (ft)	= 5,280 feet (ft)
= $5\frac{1}{2}$ yards (yd)	1 nautical mile = 6,080 feet (ft)
	= 1.1515 statute miles (mi)

PROCEDURE

To change a given customary unit of length to another:
(1) Find the conversion factor.
(2) Multiply by this conversion factor when changing to a smaller unit.
(3) Divide by this conversion factor when changing to a larger unit.

EXAMPLES:

Change 6 feet to inches.

1 ft = 12 in.
6 ft = 6 × 12 = 72 in.

ANSWER: 72 inches

Change 2,464 yards to miles.

1 mi = 1,760 yd
2,464 yd = 2,464 ÷ 1,760 = 1.4 mi

ANSWER: 1.4 miles

To add or subtract denominate numbers:
(1) Arrange the numbers in columns.
(2) Add or subtract each column.
(3) Simplify the answer.

EXAMPLES:

Add: 6 ft 7 in. + 1 ft 9 in.

$$\begin{array}{r} 6 \text{ ft } 7 \text{ in.} \\ + 1 \text{ ft } 9 \text{ in.} \\ \hline 7 \text{ ft } 16 \text{ in.} = 8 \text{ ft } 4 \text{ in.} \end{array}$$

16 in. = 1 ft 4 in.

ANSWER: 8 ft 4 in.

Subtract: 3 mi 200 yd − 1 mi 650 yd

1 mi = 1,760 yd ⟶ 3 mi 200 yd = 2 mi 1,960 yd
$$\begin{array}{r} 1 \text{ mi } 650 \text{ yd} = 1 \text{ mi } 650 \text{ yd} \\ \hline 1 \text{ mi } 1,310 \text{ yd} \end{array}$$

ANSWER: 1 mi 1,310 yd

To multiply or divide denominate numbers:
(1) Multiply or divide each unit by the multiplier or divisor.
(2) In multiplication, simplify the product.
In division, if there is a remainder with any unit, change the remainder to the next smaller unit and combine with the number of these units.

EXAMPLES:

Multiply: 5 yd 11 in. × 4

   5 yd 11 in.
×   4   44 in. = 1 yd 8 in.
20 yd 44 in. = 21 yd 8 in.

ANSWER: 21 yd 8 in.

Divide: 6 yd 1 ft 6 in. by 3

      2 yd 0 ft 6 in.
3)6 yd 1 ft 6 in.
  6 yd 0
      1 ft 6 in. = 18 in.
                18 in.

ANSWER: 2 yd 6 in.

PRACTICE EXERCISES

Find the number of <u>inches</u> in:

1	**1.** 8 ft	**2.** 6.25 ft	**3.** $3\frac{1}{2}$ ft	**4.** 50 ft
2	**1.** 9 yd	**2.** 1.75 yd	**3.** $4\frac{2}{3}$ yd	**4.** 16 yd
3	**1.** 5 ft 7 in.	**2.** 9 ft 3 in.	**3.** 1 yd 11 in.	**4.** 5 yd 21 in.

Find the number of <u>feet</u> in:

4	**1.** 6 yd	**2.** 1.4 yd	**3.** $\frac{3}{8}$ yd	**4.** $15\frac{1}{2}$ yd
5	**1.** 8 mi	**2.** 6.5 mi	**3.** $23\frac{1}{4}$ mi	**4.** 39 mi
6	**1.** 4 rd	**2.** 3.25 rd	**3.** $5\frac{3}{4}$ rd	**4.** 26 rd
7	**1.** 5 yd 2 ft	**2.** 47 yd 1 ft	**3.** 11 mi 1,000 ft	**4.** 3 rd 10 ft
8	**1.** 24 in.	**2.** 192 in.	**3.** 57 in.	**4.** 588 in.

Find the number of <u>yards</u> in:

9	**1.** 7 mi	**2.** 13.1 mi	**3.** $4\frac{1}{2}$ mi	**4.** 38 mi
10	**1.** 9 rd	**2.** 4.6 rd	**3.** $6\frac{1}{4}$ rd	**4.** 25 rd
11	**1.** 4 mi 200 yd	**2.** 15 mi 800 yd	**3.** 3 rd 5 yd	**4.** 13 rd 3 yd
12	**1.** 180 in.	**2.** 756 in.	**3.** 50.4 in.	**4.** 576 in.
13	**1.** 21 ft	**2.** 57 ft	**3.** 89 ft	**4.** 7.5 ft

Find the number of <u>rods</u> in:

14	**1.** 10 mi	**2.** 4.2 mi	**3.** $9\frac{3}{4}$ mi	**4.** 39 mi
15	**1.** 99 ft	**2.** $8\frac{1}{4}$ ft	**3.** 198 yd	**4.** $41\frac{1}{4}$ yd

Find the number of <u>miles</u> in:

16	**1.** 15,840 ft	**2.** 36,960 ft	**3.** 3,300 ft	**4.** 132,000 ft
17	**1.** 8,800 yd	**2.** 70,400 yd	**3.** 4,840 yd	**4.** 6,000 yd

18 What part of a <u>foot</u> is:
    1. 4 in.   **2.** 9 in.   **3.** 6 in.   **4.** 1 in.   **5.** 3 in.   **6.** 11 in.   **7.** 10 in.

19 What part of a <u>yard</u> is:
    1. 21 in.   **2.** 30 in.   **3.** 24 in.   **4.** 7 in.   **5.** 1 ft   **6.** 2 ft   **7.** $1\frac{1}{2}$ ft

20 What part of a <u>mile</u> is:
    1. 440 yd   **2.** 2,640 ft   **3.** 1,100 yd   **4.** 280 rd   **5.** 1,320 yd   **6.** 1,980 ft?

21 Add and simplify: **22** Subtract:

    1. 2 ft 9 in.   **2.** 3 yd 27 in.   **1.** 9 yd 7 in.   **2.** 10 ft
       <u>11 ft 2 in.</u>   <u>5 yd 19 in.</u>   <u>4 yd 5 in.</u>   <u>7 ft 8 in.</u>

23 Multiply and simplify:
    1. 5 ft 2 in.   **2.** 4 mi 300 yd   **3.** 2 yd 9 in.   **4.** 4 yd 1 ft 11 in.
         <u>4</u>   <u>9</u>   <u>8</u>   <u>6</u>

24 Divide:
    1. $4\overline{)12 \text{ ft } 8 \text{ in.}}$ **2.** $7\overline{)17 \text{ yd } 4 \text{ in.}}$ **3.** $5\overline{)6 \text{ mi } 100 \text{ ft}}$ **4.** $3\overline{)7 \text{ yd } 2 \text{ ft } 6 \text{ in.}}$

Change:

25 To <u>feet</u>:   **26** To <u>yards</u>:   **27** To <u>statute miles</u>:
    1. 6 naut. mi   **1.** 11 naut. mi   **1.** 7 naut. mi
    2. 2.8 naut. mi   **2.** 8.75 naut. mi   **2.** 9.3 naut. mi
    3. $4\frac{1}{4}$ naut. mi   **3.** $\frac{7}{8}$ naut. mi   **3.** $36\frac{1}{2}$ naut. mi
    4. 40 naut. mi   **4.** 145 naut. mi   **4.** 145 naut. mi

Change to <u>nautical miles</u>:

28 **1.** 30,400 ft   **2.** 18,240 ft   **3.** 10,000 ft   **4.** 152,000 ft
29 **1.** 42,560 yd   **2.** 4,000 yd   **3.** 11,500 yd   **4.** 54,720 yd
30 **1.** 5 stat. mi   **2.** 61.25 stat. mi   **3.** $28\frac{3}{4}$ stat. mi   **4.** 130 stat. mi

Applications

Solve each problem and select the letter corresponding to your answer.

1. If the floor boards are 2 inches wide, how many will be required to cover a floor 17 feet wide?
    a. 34 boards   **b.** 102 boards   **c.** 84 boards   **d.** Answer not given

2. How many yards of ribbon are needed to make 30 school officer badges if it takes 9 inches of ribbon to make one badge?
    a. $27\frac{1}{2}$ yards   **b.** $7\frac{1}{2}$ yards   **c.** 12 yards   **d.** $6\frac{2}{3}$ yards

3. How many miles high is Mount Everest if its elevation is 29,028 feet? Find the answer correct to the nearest tenth.
    a. 5.5 miles   **b.** 5.8 miles   **c.** 54.9 miles   **d.** Answer not given

Vocabulary

The **pound**, **ounce**, and **ton** are basic units commonly used to measure the weight of an object.

In the customary system, the pound, ounce, and ton are units of weight, not mass. There are two weights commonly called tons, the short ton and the long ton.

1 pound (lb) = 16 ounces (oz)
1 short ton (T) = 2,000 pounds (lb)
1 long ton (l. ton) = 2,240 pounds (lb)

PROCEDURE

To change a given customary unit of weight to another:
(1) Find the conversion factor.
(2) Multiply by this conversion factor when changing to a smaller unit.
(3) Divide by this conversion factor when changing to a larger unit.

EXAMPLES:

Change $3\frac{1}{2}$ pounds to ounces.

1 lb = 16 oz

$3\frac{1}{2}$ lb = $3\frac{1}{2} \times 16 = \frac{7}{2} \times 16 = 56$ oz

ANSWER: 56 ounces

Change 2,800 pounds to short tons.

1 T = 2,000 lb

2,800 lb = 2,800 ÷ 2,000 = 1.4 T

ANSWER: 1.4 short tons

PRACTICE EXERCISES

Find the number of <u>ounces</u> in:

1 **1.** 6 lb **2.** 2.2 lb **3.** $5\frac{3}{4}$ lb **4.** 37 lb

2 **1.** 1 lb 2 oz **2.** 4 lb 8 oz **3.** 9 lb 7 oz **4.** 18 lb 13 oz

Find the number of <u>pounds</u> in:

3 **1.** 7 T **2.** 8.75 T **3.** $4\frac{1}{2}$ T **4.** 148 T

4 **1.** 59 l. ton **2.** 9.5 l. ton **3.** $17\frac{3}{4}$ l. ton **4.** 31 l. ton

5 **1.** 6 T 400 lb **2.** 9 l. ton 500 lb **3.** 20 T 1,200 lb

6 **1.** 64 oz **2.** 176 oz **3.** 26 oz **4.** 368 oz

7 Find the number of <u>short tons</u> in:
1. 6,000 lb
2. 24,000 lb
3. 7,000 lb

8 Find the number of <u>long tons</u> in:
1. 6,720 lb
2. 17,920 lb
3. 4,000 lb

9 What part of a <u>pound</u> is:
1. 12 oz
2. 8 oz
3. 14 oz
4. 5 oz

10 What part of a <u>short</u> ton is:
1. 1,200 lb
2. 500 lb
3. 750 lb
4. 1,800 lb

11 What part of a <u>long</u> ton is:
1. 560 lb
2. 1,120 lb
3. 840 lb
4. 1,960 lb

12 Add and simplify:

1. 3 lb 6 oz
   1 lb 8 oz

2. 2 T 1,000 lb
   5 T 1,200 lb

3. 4 lb 10 oz
   7 lb 9 oz
   6 lb 13 oz

4. 2 l. ton 1,200 lb
   1 l. ton   850 lb
   5 l. ton 1,000 lb

13 Subtract:

1. 6 lb 13 oz
   3 lb 4 oz

2. 9 lb 5 oz
   8 lb 14 oz

3. 4 T   700 lb
   2 T 1,900 lb

4. 5 lb
   3 lb 6 oz

14 Multiply and simplify:

1. 3 lb 2 oz
   × 6

2. 7 T 400 lb
   × 5

3. 1 lb 8 oz
   × 9

4. 6 lb 4 oz
   × 8

15 Divide:

1. 5)10 lb 15 oz

2. 4)5 l. ton 160 lb

3. 6)14 lb 4 oz

4. 8)10 T 200 lb

Applications

Solve each problem and select the letter corresponding to your answer.

1. What is the cost of a chicken weighing 5 lb 12 oz at $.96 per pound?
   a. $5.24   **b.** $5.48   **c.** $5.52   **d.** $6.52

2. At $2.88 per pound, how many ounces of cake should you get for $1.80?
   a. 9 oz   **b.** 10 oz   **c.** 11 oz   **d.** 12 oz

3. How many 3-oz boxes of cookies can be made from a 9-lb box?
   a. 36   **b.** 40   **c.** 44   **d.** 48

4. How much is saved by buying a 1-lb box of cereal at $1.67 instead of two 8-oz boxes at $.89 each?
   a. $.08   **b.** $.09   **c.** $.10   **d.** $.11

10-3 Liquid Measure—Customary

Vocabulary

The **ounce**, **pint**, **quart**, and **gallon** are basic units commonly used to measure liquid capacity.

There are two systems for measuring capacity in the customary system, one for dry measure and one for liquid measure.

1 pint (pt) = 16 fluid ounces (oz)	1 gallon (gal.) = 4 quarts (qt)
1 quart (qt) = 2 pints (pt)	= 8 pints (pt)
= 32 fluid ounces (oz)	= 128 ounces (oz)

PROCEDURE

To change a given customary unit of liquid measure to another:
(1) Find the conversion factor.
(2) Multiply by this conversion factor when changing to a smaller unit.
(3) Divide by this conversion factor when changing to a larger unit.

EXAMPLES:

Change 3 gallons to pints.
1 gal. = 8 pt
3 gal. = 3 × 8 = 24 pt

ANSWER: 24 pints

Change 224 fluid ounces to quarts.
1 qt = 32 oz
224 oz = 224 ÷ 32 = 7 qt

ANSWER: 7 quarts

PRACTICE EXERCISES

Find the number of <u>fluid ounces</u> in:

1 **1.** 9 pt | **2.** 4.75 pt | **3.** $5\frac{1}{2}$ pt | **4.** 17 pt
2 **1.** 5 qt | **2.** 1.1 qt | **3.** $\frac{4}{5}$ qt | **4.** 24 qt
3 **1.** 7 gal. | **2.** .25 gal. | **3.** $6\frac{3}{4}$ gal. | **4.** 39 gal.
4 **1.** 2 pt 8 oz | **2.** 5 qt 12 oz | **3.** 1 gal 48 oz

Find the number of <u>pints</u> in:

5 **1.** 3 qt | **2.** 5.5 qt | **3.** $4\frac{1}{4}$ qt | **4.** 18 qt
6 **1.** 6 gal. | **2.** 16.8 gal. | **3.** $7\frac{1}{2}$ gal. | **4.** 40 gal.
7 **1.** 5 qt 1 pt | **2.** 7 qt 3 pt | **3.** 9 qt 2 pt
8 **1.** 48 oz | **2.** 72 oz | **3.** 256 oz | **4.** 352 oz

Find the number of <u>quarts</u> in:

9   **1.** 8 gal.   **2.** 10.25 gal.   **3.** $17\frac{1}{2}$ gal.   **4.** 100 gal.

10   **1.** 6 gal. 3 qt   **2.** 7 gal. 1 qt   **3.** 1 gal. 2 qt

11   **1.** 14 pt   **2.** 6.5 pt   **3.** 9 pt   **4.** 27 pt

12   **1.** 96 oz   **2.** 320 oz   **3.** 192 oz   **4.** 700 oz

Find the number of <u>gallons</u> in:

13   **1.** 20 qt   **2.** 76 qt   **3.** 6 qt   **4.** 35 qt

14   **1.** 24 pt   **2.** 56 pt   **3.** 18 pt   **4.** 31 pt

15   **1.** 256 oz   **2.** 768 oz   **3.** 192 oz   **4.** 96 oz

16   What part of a <u>pint</u> is:

   1. 8 oz   **2.** 12 oz   **3.** 7 oz   **4.** 4 oz   **5.** 6 oz   **6.** 10 oz

17   What part of a <u>quart</u> is:

   1. 1 pt   **2.** $1\frac{1}{2}$ pt   **3.** 16 oz   **4.** $\frac{1}{2}$ pt   **5.** 24 oz   **6.** 4 oz

18   What part of a <u>gallon</u> is:

   1. 2 qt   **2.** 3 qt   **3.** 32 oz   **4.** 1 pt   **5.** 96 oz   **6.** $3\frac{1}{2}$ qt

19   Add and simplify: **20** Subtract:

   1. 5 gal. 2 qt   **2.** 2 qt 14 oz   **1.** 5 qt 1 pt   **2.** 1 pt 8 oz
      3 gal. 1 qt   1 qt 7 oz   2 qt 1 pt   14 oz

21   Multiply and simplify:

   1. 2 qt 1 pt   **2.** 3 qt 8 oz   **3.** 1 pt 7 oz   **4.** 5 gal. 3 qt
        5   8   4   6

22   Divide:

   1. 2)8 gal. 2 qt   **2.** 3)9 pt 12 oz   **3.** 6)8 qt 2 oz   **4.** 9)20 gal. 1 qt

Applications

How many ounces of water must be added to:

1. A 6-oz can of frozen lemon concentrate to make 1 quart of lemonade?

2. A 27-oz can of frozen orange concentrate to make 1 gallon of orange juice?

Solve each problem and select the letter corresponding to your answer.

3. How much is saved by buying one gallon of paint for $13.99 instead of four 1-qt cans at $4.50 each?
   a. $5.01   **b.** $4.51
   c. $4.01   **d.** Answer not given

4. Find the difference in price per liquid ounce when a 1-pt 14 oz bottle of beverage sells for $.48 and a 10-oz bottle sells for $.25.
   a. 1¢   **b.** 0.9¢
   c. 0.8¢   **d.** 1.2¢

10-4 Dry Measure—Customary

Vocabulary

The **pint**, **quart**, and **bushel** are basic units commonly used to measure dry capacity.

These units of capacity are used to measure corn, wheat, fruit, and other nonliquid quantities. In dry measure, 2 pints is also 1 quart.

1 quart (qt) = 2 pints (pt)	1 bushel (bu) = 4 pecks (pk)
1 peck (pk) = 8 quarts (qt)	= 32 quarts (qt)
= 16 pints (pt)	= 64 pints (pt)

PROCEDURE

To change a given customary unit of dry measure to another:
(1) Find the conversion factor.
(2) Multiply by this conversion factor when changing to a smaller unit.
(3) Divide by this conversion factor when changing to a larger unit.

EXAMPLES:

Change $4\frac{1}{2}$ pecks to quarts.

1 pk = 8 qt

$4\frac{1}{2}$ pk = $4\frac{1}{2} \times 8 = \frac{9}{2} \times 8 = 36$ qt

ANSWER: 36 quarts

Change 76 pints to pecks.

1 pk = 16 pt

76 pt = $76 \div 16 = 4\frac{3}{4}$ pk

ANSWER: $4\frac{3}{4}$ pecks

PRACTICE EXERCISES

Find the number of:

1 <u>pints</u> in: **1.** 4 qt   **2.** 30 qt   **3.** $3\frac{3}{4}$ qt   **4.** 9 qt 1 pt

2 <u>quarts</u> in: **1.** 15 pk   **2.** $2\frac{1}{4}$ pt   **3.** 12 pt   **4.** 5 pt

3 <u>pecks</u> in: **1.** 9 bu   **2.** $\frac{3}{4}$ bu   **3.** 2 bu 1 pk   **4.** 22 qt

4 <u>bushels</u> in: **1.** 36 pk   **2.** 92 pk   **3.** 23 pk   **4.** 108 qt

5 What part of a <u>bushel</u> is: **1.** 2 pk   **2.** 1 pk   **3.** 3 pk   **4.** 4 qt

6 What part of a <u>peck</u> is: **1.** 6 qt   **2.** 4 qt   **3.** 5 qt   **4.** 6 pt

7 Add and simplify:

7 bu 1 pk
4 bu 3 pk

8 Subtract:

12 pk 7 qt
 8 pk 2 qt

9 Multiply:

6 bu 3 pk
    6

10 Divide:

1. 4)12 pk 4 qt

2. 5)13 bu 3 pk

391

10-5 Measures of Area—Customary

Vocabulary

The **square inch**, **square foot**, **square yard**, **square mile**, and **acre** are basic units commonly used to measure area.

Just as in the metric system, square units are used in the customary system to measure area.

1 square foot (ft²) = 144 square inches (in.²) 1 square yard (yd²) = 9 square feet (ft²) = 1,296 square inches (in.²) 1 square rod (rd²) = 30.25 square yards (yd²) = 272.25 square feet (ft²)	1 square mile (mi²) = 640 acres 1 acre = 160 square rods (rd²) = 4,840 square yards (yd²) = 43,560 square feet (ft²)

PROCEDURE

To change a given customary unit of area to another:
(1) Find the conversion factor.
(2) Multiply by this conversion factor when changing to a smaller unit.
(3) Divide by this conversion factor when changing to a larger unit.

EXAMPLES:

Change 10 square feet to square inches.

$1 \text{ ft}^2 = 144 \text{ in.}^2$
$10 \text{ ft}^2 = 10 \times 144 = 1{,}440 \text{ in.}^2$

ANSWER: 1,440 square inches

Change 9,680 square yards to square rods.

$1 \text{ rd}^2 = 30.25 \text{ yd}^2$
$9{,}680 \text{ yd}^2 = 9{,}680 \div 30.25 = 320 \text{ rd}^2$

ANSWER: 320 square rods

PRACTICE EXERCISES

Change:

1 To square inches: **1.** 16 ft²   **2.** 5 yd²   **3.** 14.75 ft²   **4.** $\frac{2}{3}$ yd²

2 To square feet: **1.** 1,584 in.²   **2.** 12 yd²   **3.** 40 rd²   **4.** 5 acres

3 To square yards: **1.** 45 ft²   **2.** 9,072 in.²   **3.** 7 acres   **4.** 3.25 mi²

4 To square rods:
   1. 23 acres   **2.** 847 yd²   **3.** 1,452 yd²   **4.** 16.875 acres

5 To acres:
   1. 800 rd²   **2.** 3.6 mi²   **3.** 24,200 yd²   **4.** 13,068 ft²

6 To square miles:
   1. 10,880 acres   **2.** 18,585,600 yd²   **3.** 14,720 acres

7 What part of a square foot is:
  1. 72 in.²   **2.** 126 in.²   **3.** 81 in.²

8 What part of a square yard is:
  1. 8 ft²   **2.** 6 ft²   **3.** 648 in²

9 What part of a square mile is:
  1. 240 acres   **2.** 61,952 yd²   **3.** 278,784 ft²

10 What part of an acre is:
  1. 120 rd²   **2.** 1,210 yd²   **3.** 27,225 ft²

 Applications

1. A kitchen has 108 ft² of floor space. How many square yards of linoleum are needed to cover the entire floor?

2. The air pressure at sea level is 14.7 lb per in.² Find the force on 2 ft² of surface at sea level.

Computer Activities

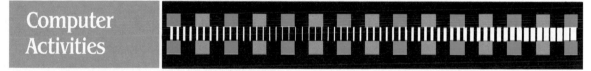

The computer program below converts any number of gallons into the equivalent number of: quarts, pints, or fluid ounces.

```
10 PRINT "HOW MANY GALLONS";
20 INPUT G
30 PRINT G;" GALLONS IS EQUAL TO:"
40 PRINT G*4;" QUARTS. . .OR"  ←————— 4 quarts in a gallon
50 PRINT G*4*2;" PINTS. . .OR"  ←————— 2 pints in a quart
60 PRINT G*4*2*16;" OUNCES"  ←————— 16 ounces in a pint
```

a) ENTER the program and then RUN it to make sure it works.

b) Modify the program so that it converts any number of bushels to: pecks, quarts, or pints.

c) Modify the program so that it converts any number of ounces to: pints, quarts, or gallons.

d) Modify the program so that it converts any number of pints to: quarts, pecks, or bushels.

Vocabulary

The **cubic inch**, **cubic foot**, and **cubic yard** are basic units commonly used to measure volume.

Just as in the metric system, cubic units are used in the customary system to measure area.

> 1 cubic foot (ft³) = 1,728 cubic inches (in.³)
> 1 cubic yard (yd³) = 27 cubic feet (ft³) = 46,656 cubic inches (in.³)

PROCEDURE

To change a given customary unit of volume to another:
(1) Find the conversion factor.
(2) Multiply by this conversion factor when changing to a smaller unit.
(3) Divide by this conversion factor when changing to a larger unit.

EXAMPLES:

Change 4.2 cubic yards to cubic feet.

$1 \text{ yd}^3 = 27 \text{ ft}^3$
$4.2 \text{ yd}^3 = 4.2 \times 27 = 113.4 \text{ ft}^3$

ANSWER: 113.4 cubic feet

Change 55,987.2 cubic inches to cubic yards.

$1 \text{ yd}^3 = 46,656 \text{ in.}^3$
$55,987.2 \text{ in.}^3 = 55,987.2 \div 46,656 = 1.2 \text{ yd}^3$

ANSWER: 1.2 cubic yards

PRACTICE EXERCISES

Change:

1 To cubic inches:　　1. 38 ft³　　2. $\frac{3}{4}$ yd³　　3. 7.25 ft³

2 To cubic inches:　　1. 17 yd³　　2. $2\frac{1}{2}$ ft³　　3. 9.375 yd³

3 To cubic feet:　　1. 40 yd³　　2. 6.5 yd³　　3. 5,184 in.³

4 To cubic feet:　　1. 29,376 in.³　　2. 432 in.³　　3. $7\frac{2}{3}$ yd³

5 To cubic yards: **1.** 135 ft³ **2.** 93,312 in.³ **3.** 63 ft³

6 To cubic yards: **1.** 233,280 in.³ **2.** 688.5 ft³ **3.** 793,152 in.³

7 What part of a cubic foot is:
1. 432 in.³ **2.** 756 in.³ **3.** 1,440 in.³

8 What part of a cubic yard is:
1. 9 ft³ **2.** $6\frac{3}{4}$ ft³ **3.** 21 ft³

9 What part of a cubic foot is:
1. 576 in.³ **2.** 108 in.³ **3.** 1,152 in.³

10 What part of a cubic yard is:
1. 15 ft³ **2.** 6 ft³ **3.** $13\frac{1}{2}$ ft³

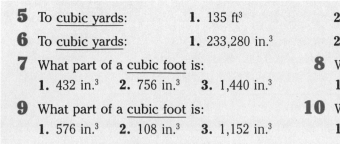

Applications

Solve each problem and select the letter corresponding to your answer.

1. Lead weighs .41 lb per in.³ Find the weight of one cubic foot of lead.
a. 59.04 lb **b.** 4.92 lb
c. 4,100 lb **d.** 708.48 lb

2. A certain truck can hold 2 yd³ of dirt. If a cubic foot of dirt weighs 100 lb, what is the weight of a truckload of dirt?
a. 50 lb **b.** 2,700 lb
c. 5,400 lb **d.** Answer not given

Refresh Your Skills

1. Add: 1-4
693 + 1,229 + 588

2. Subtract: 1-5
4,000 − 986

3. Multiply: 1-6
96 × 385

4. Divide: 1-7
72)15,048

5. Add: 2-4
6.2 + .47

6. Subtract: 2-5
.076 − .0105

7. Multiply: 2-7
1.8 × .009

8. Divide: 2-8
.5)60

9. Add: 3-8
$5\frac{7}{8} + 1\frac{2}{3}$

10. Subtract: 3-9
$8\frac{5}{12} - 4\frac{3}{4}$

11. Multiply: 3-11
$6\frac{1}{4} \times \frac{4}{5}$

12. Divide: 3-12
$24 : 2\frac{1}{4}$

13. Find 5% of 927.
4-6

14. What percent of 40 is 32?
4-7

15. 54 is 18% of what number?
4-8

10-7 Volume, Capacity, and Weight Relationships—Customary

1 gallon (gal.) = 231 cubic inches (in.³)	1 cubic foot of fresh water weighs $62\frac{1}{2}$ pounds.
1 cubic foot (ft³) = $7\frac{1}{2}$ gallons (gal.)	
1 bushel (bu) = $1\frac{1}{4}$ cubic feet (ft³)	1 cubic foot of sea water weighs 64 pounds.
= 2,150.42 cubic inches (in.³)	

PROCEDURE

To change a given customary unit of volume, capacity, or weight to another equivalent unit:
(1) Find the conversion factor.
(2) Multiply or divide by this conversion factor as required.

EXAMPLES:

Find the volume in cubic inches of 9 gallons.
1 gal. = 231 in.³

9 gal. = 9 × 231 = 2,079 in.³
ANSWER: 2,079 in.³

Find the volume in cubic feet of 1,000 pounds of fresh water.

$$1 \text{ ft}^3 = 62\frac{1}{2} \text{ lb}$$

$$1,000 \text{ lb} = 1,000 \div 62\frac{1}{2} = 16 \text{ ft}^3$$
ANSWER: 16 cubic feet

PRACTICE EXERCISES

Find the equivalent:

1 Capacity in gallons: **1.** 1,848 in.³ **2.** 21 ft³ **3.** 4.9 ft³

2 Capacity in bushels: **1.** 80 ft³ **2.** $12\frac{1}{2}$ ft³ **3.** 35 yd³

3 Volume in cubic feet: **1.** 45 gal. **2.** 900 gal. **3.** 206 bu

4 Volume in cubic inches: **1.** 7 gal. **2.** 15.8 gal. **3.** $4\frac{1}{2}$ gal.

5 Volume in cubic feet: **1.** 4,800 lb of sea water **2.** 6,000 lb of fresh water

6 Weight in pounds: **1.** 75 ft³ of fresh water **2.** 120 ft³ of sea water

Applications

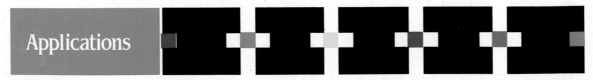

1. A submarine displaces 1,400 tons (1 long ton = 2,240 lb) of sea water when submerged. How many cubic feet of sea water does it displace?

2. The normal flow of water over Niagara Falls is 500,000 tons (1 short ton = 2,000 lb) of fresh water a minute. Find the rate of flow in gallons per minute.

10-8 Measuring Time

Vocabulary

The **second**, **minute**, **hour**, **day**, **month**, and **year** are basic units commonly used to measure time.

The units of time are the same in both the metric and customary systems.

1 hour (h) = 60 minutes (min)	1 week (wk) = 7 days (da)	1 year = 12 months (mo)
= 3,600 (s)	1 day (da) = 24 hours (h)	= 52 weeks (wk)
1 minute (min) = 60 seconds (s)	= 1,440 minutes (min)	− 365 days (da)

PROCEDURE

To change a unit of time to another:
(1) Find the conversion factor.
(2) Multiply by this conversion factor when changing to a smaller unit.
(3) Divide by this conversion factor when changing to a larger unit.
Note: When changing several years to days, use 366 days for every fourth year to account for leap year.

EXAMPLES:

Change 6 years to days.

1 yr = 365 da
6 yr = 5 × 365 + 1 × 366 = 2,191 da
ANSWER: 2,191 days

Change 4,500 seconds to hours.

1 h = 3,600 s
4,500 s = 4,500 ÷ 3,600 = $1\frac{1}{4}$ h
ANSWER: $1\frac{1}{4}$ hours

PRACTICE EXERCISES

1 Find the number of <u>seconds</u> in: **1.** 24 min **2.** 6.25 h **3.** $1\frac{1}{2}$ h

2 Find the number of <u>minutes</u> in: **1.** $\frac{3}{4}$ h **2.** 13 h 52 min **3.** 1,800 s

3 Find the number of <u>hours</u> in: **1.** 30 da **2.** 5 da 13 h **3.** 3,180 min

4 Find the number of <u>days</u> in: **1.** $6\frac{3}{7}$ wk **2.** 3 yr 200 da **3.** 192 h

5 Find the number of <u>months</u> in: **1.** $2\frac{1}{2}$ yr **2.** 1 yr 6 mo **3.** 12 yr 5 mo

6 Find the number of <u>weeks</u> in: **1.** 3 yr **2.** $5\frac{3}{4}$ yr **3.** 2 yr 8 wk

7 Find the number of <u>years</u> in: **1.** 300 mo **2.** 260 wk **3.** 1,095 da

8 What part of a <u>minute</u> is: **1.** 45 s **2.** 18 s **3.** 5 s **4.** 36 s

9 What part of an <u>hour</u> is: **1.** 40 min **2.** 15 min **3.** 30 min **4.** 54 s

10 What part of a <u>day</u> is: **1.** 4 h   **2.** 16 h   **3.** 20 h   **4.** 15 h

11 What part of a <u>week</u> is: **1.** 2 da   **2.** 6 da   **3.** $3\frac{1}{2}$ da   **4.** 5 da

12 What part of a <u>year</u> is: **1.** 9 mo   **2.** 13 wk   **3.** 146 da   **4.** 4 mo

13 Add and simplify: **14** Subtract: **15** Multiply: **16** Divide:

    17 h 22 min   10 yr 4 mo   5 yr 3 mo   6)10 da 2 h

     5 h 38 min   7 mo   3

17 Which is greater: **18** Which is less:

    1. A half-hour or 25 minutes?   **1.** A third of a minute or 16 seconds?

    2. 8 months or three-quarters of a year?   **2.** 10 hours or a half-day?

19 Write as A.M. or P.M. time:

    1. 9 o'clock at night.   **2.** 11 o'clock in the morning.

    3. Half past two in the afternoon.   **4.** Quarter after six in the morning.

20 On a clock, how long does it take the minute hand to move from:

    1. 4 to 5?   **2.** 2 to 7?   **3.** 1 to 9?   **4.** 8 to 3?

21 Find the length of time: **1.** From 2 A.M. to 6 P.M. the same day.

                          2. From 10:45 A.M. to 7:15 P.M. the same day.

                          3. From 12:26 P.M. one day to 3:10 A.M. the following day.

22 Look at your calendar. On what date this month does:

    1. The second Monday fall?   **2.** The fourth Thursday fall?   **3.** The third Saturday fall?

23 How many <u>days</u> are in the month of: **1.** October   **2.** April

24 Find the exact number of <u>days</u> from:

    1. March 15 to May 15.   **2.** April 3 to October 16.   **3.** July 16 to December 9

    4. June 12 of one year to January 8 of the following year.

25 How old will Lisa be on her next birthday if she was born on June 3, 1972?

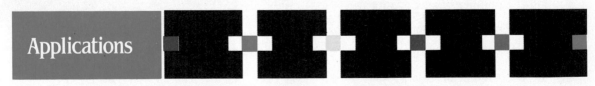

Applications

Solve each problem and select the letter corresponding to your answer.

1. Lisa parked her car at the airport. When she entered the parking lot, her ticket read 30 NOV 7:35 A.M. How long was her car in the parking lot if she picked it up on December 2 at 6:10 P.M.?

    a. 3 da 4 h 35 min   **b.** 2 da 10 h 35 min

    c. 2 da 1 h 25 min   **d.** Answer not given

2. A bus leaves the station every 17 minutes for the center city. If you missed the 10:56 A.M. bus, what time is the next scheduled bus?

    a. 11:12 A.M.   **b.** 11:10 A.M.

    c. 11:11 A.M.   **d.** 11:13 A.M.

TIME ZONES

The four standard time belts in the Continental United States are Eastern (EST), Central (CST), Mountain (MST), and Pacific (PST). Central time is one hour earlier than Eastern, Mountain is one hour earlier than Central, and Pacific is one hour earlier than Mountain.

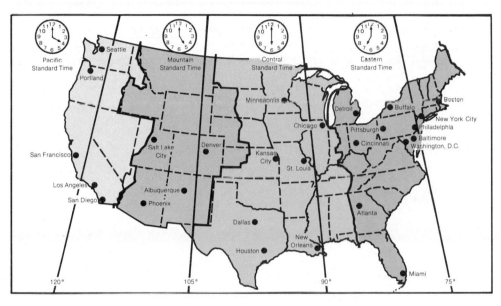

Alaska has four time zones, Pacific, Yukon, Alaska-Hawaii, and Bering. Hawaii uses the Alaska-Hawaii time zone which is two hours earlier than Pacific. Parts of Canada are in the Atlantic Standard time zone, which is one hour later than Eastern.

EXERCISES

1. In what time zone is your city located?

If it is 11 A.M. in the Mountain Time Zone, what time is it in:

2. the Pacific Zone?   **3.** the Eastern Zone?   **4.** the Central Zone?

When it is 8 P.M. in Dallas, what time is it in:

5. Phoenix?   **6.** Washington, D.C.?   **7.** St. Louis?   **8.** San Francisco?

9. Atlanta?   **10.** Kansas City?   **11.** Denver?   **12.** Seattle?

13. Detroit?   **14.** New York?   **15.** Houston?   **16.** Your city?

If a nationwide program is telecast at 10 P.M. from Los Angeles, what time is it seen in:

17. Pittsburgh?   **18.** Albuquerque?   **19.** Minneapolis?

20. Miami?   **21.** Salt Lake City?   **22.** Baltimore?

10-9 Reading a Timetable

Vocabulary

A **timetable** is a schedule that gives arrivals and departures for trains, buses, or airplanes.

The train timetable lists in columns the times of arrival or departure of the train at each stop.

km	mi.	Train Number		123	177	179	193	195
		Operation		Daily	Daily	Daily	Daily	Daily
0	0	Boston, MA	Dp		1 45 P	4 00 P	5 30 P	7 00 P
19	12	Route 128, MA			2 03 P	4 18 P	5 48 P	7 18 P
71	44	Providence, RI			2 39 P	4 54 P	6 25 P	7 54 P
114	71	Kingston, RI				5 19 P		8 20 P
171	106	New London, CT			3 44 P	6 02 P	7 25 P	9 01 P
253	157	New Haven, CT	Ar		4 48 P	7 07 P	8 17 P	10 00 P
			Dp		4 58 P	7 17 P	8 27 P	10 10 P
280	174	Bridgeport, CT			5 18 P	7 37 P		10 30 P
315	196	Stamford, CT			5 41 P	8 00 P	9 10 P	10 55 P
373	232	New York, NY	Ar		6 40 P	8 50 P	10 00 P	11 47 P
			Dp	5 30 P	7 00 P	9 00 P	10 10 P	12 01 A
389	242	Newark, NJ		5 45 P	7 15 P	9 17 P	10 25 P	12 17 A

The letter A indicates A.M.
The letter P indicates P.M.

PRACTICE EXERCISES

1 At what time does train 195 leave Boston? At what time does it arrive in New Haven? In Newark?

2 At what time does train 193 leave Providence? At what time does it arrive in New Haven? In New York?

3 How far (in kilometers) is it by rail from Boston to Bridgeport? How long does it take train 177 to go from Boston to Bridgeport? What is the average rate of speed during this trip?

4 Compare the time it takes train 193 to travel from Providence to New York to that of train 179. What is the difference in time?

The airline schedule gives separate listings for each city. Notice that all arrival and departure times are given in local time.

Leave	Arrive	Flight	Stops	Meals
From: **DALLAS-FORT WORTH** (CST)				
TO: **CHICAGO** (CST)				
6 30a	9 15a	242	One-Stop	M
8 20a	10 15a	54	Non-Stop	M
10 20a	2 10p	248	Two-Stops	S
2 40p	5 30p	140	One-Stop	S
3 20p	5 20p	44	Non-Stop	S
5 20p	7 20p	36	Non-Stop	M
7 15p	10 45p	120/156	Kansas City	M
TO: **DENVER** (MST)				
9 25a	10 05a	62	Non-Stop	M
11 25a	10 05a	68	Non-Stop	M
4 25p	5 30p	66	One-Stop	S
6 25p	7 05p	78	Non-Stop	M

Leave	Arrive	Flight	Stops	Meals
From: **DALLAS-FORT WORTH** (CST)				
TO: **MIAMI** (EST)				
8 35a	12 00n	79	Non-Stop	M
12 05p	3 25p	63	Non-Stop	M
5 00p	9 10p	169	One-Stop	M
8 10p	12 15a	405	One-Stop	S
TO: **NEW ORLEANS** (CST)				
8 35a	9 45a	157	Non-Stop	M
11 30a	12 40p	235	Non-Stop	M
4 40p	5 50p	133	Non-Stop	S
TO: **SEATTLE-TACOMA** (PST)				
11 30a	1 10p	95	Non-Stop	M
6 40p	8 20p	182	Non-Stop	M
7 45p	10 20p	184	One-Stop	M

The letter a indicates A.M.
The letter p indicates P.M.
M—Meal S—Snack

400

PRACTICE EXERCISES

1 At what time does Flight 54 leave Dallas-Fort Worth for Chicago? At what time does it arrive in Chicago? How long does the flight take? Are there any stops on the way?

2 What is the number of the flight leaving Dallas-Fort Worth for New Orleans at 11:30 A.M.? When does it arrive in New Orleans? Is a meal or a snack served?

3 How long does Flight 63 take from Dallas-Fort Worth to Miami?

4 How much longer does it take Flight 66 to fly from Dallas-Fort Worth to Denver than Flight 78?

5 How long does it take Flight 95 to fly from Dallas-Fort Worth to Seattle-Tacoma? How much faster is it than Flight 184?

The bus timetable lists the times of arrival and departure for buses at each town. Read down on the left and up on the right. Notice that not all the towns are in the same time zone.

Times are given in local time.
Red type indicates A.M. time.
Blue type indicates P.M. time.

Read Down							Read Up		
824	822	820		Bus Number			823	825	827
3 30	9 30	3 15	Lv	Atlanta, GA	EST	Ar	7 00	10 55	6 35
4 25	10 25	↓	Lv	Anniston	CST	Ar	3 30	8 05	↑
6 00	11 59	5 05	Ar	Birmingham, AL		Lv	1 45	6 40	2 45
6 30	12 45	6 00	Lv	Birmingham		Ar	1 05	6 00	2 20
10 00	4 35	9 20		Tupelo, MS			9 35	2 15	10 50
12 30	7 20	12 15	Ar	Memphis, TN		Lv	7 00	11 30	8 20
1 10	8 30	1 30	Lv	Memphis		Ar	6 00	10 50	7 00
4 55	12 15	5 15	Ar	Little Rock, AR		Lv	2 15	7 05	3 15
5 15	12 30	6 00	Lv	Little Rock		Ar	1 55	6 40	2 40
9 00	3 40	9 10	Ar	Fort Smith, AR		Lv	10 45	3 20	10 55
9 10	3 55	10 10	Lv	Fort Smith		Ar	10 30	2 55	10 30
12 10	6 30	12 55		McAlester, OK			8 00	↑	↑
3 30	9 30	3 45	Ar	Oklahoma City		Lv	4 15	11 10	5 45
5 00	11 40	4 45	Lv	Oklahoma City		Ar	2 40	10 45	4 40
11 25	6 10	10 35	Ar	Amarillo, TX		Lv	8 15	4 30	11 25
11 55	6 50	11 00	Lv	Amarillo	CST	Ar	7 30	4 15	11 10
4 50	11 45	4 10	Ar	Albuquerque, NM	MST	Lv	12 35	9 00	3 45

PRACTICE EXERCISES

1 At what time does Bus 822 leave Atlanta? At what time does it arrive in Birmingham? In Memphis? In Little Rock?

2 At what time does Bus 827 leave Albuquerque, NM? At what time does it arrive in Forth Smith? In Tupelo, MS? In Atlanta, GA?

3 How long does it take Bus 825 to go from Amarillo to Memphis?

4 At what time does Bus 824 leave Oklahoma City? At what time does it arrive in Albuquerque, NM? How long does it take to make this trip?

5 Compare the time it takes Bus 820 to travel from Atlanta to Amarillo with the time it takes Bus 824 to make the same trip.

10-10 Rates of Speed

Rates of speed are usually expressed per unit of time. The conversion tables given for distance and time are used to convert from one rate to another.

$$1 \text{ knot} = 1 \text{ nautical mile per hour}$$

PROCEDURE

To change from one rate of speed to another:
(1) Convert the distance from that of the first rate to that of the second.
(2) Convert the time from that of the first rate to that of the second.
(3) Simplify the new rate.

EXAMPLES

Change 54 km/h to meters per second.

$$54 \text{ km/h} = \frac{54 \text{ km}}{1 \text{ h}} = \frac{54 \times 1{,}000 \text{ m}}{1 \times 3{,}600 \text{ s}} = 15 \text{ m/s}$$

ANSWER: 15 m/s

Change 60 mi/h to feet per second.

$$60 \text{ mi/h} = \frac{60 \text{ mi}}{1 \text{ h}} = \frac{60 \times 5{,}280 \text{ ft}}{1 \times 3{,}600 \text{ s}} = 88 \text{ ft/s}$$

ANSWER: 88 ft/s

PRACTICE EXERCISES

Change, finding answers to the nearest hundredth whenever necessary:

1 To meters per second: **1.** 400 cm/s   **2.** 150 km/h   **3.** 480 mm/min
2 To meters per minute: **1.** 75 mm/s   **2.** 360 cm/min   **3.** 60 km/h
3 To centimeters per second: **1.** 90 km/h   **2.** 85 m/min   **3.** 540 mm/s
4 To kilometers per hour: **1.** 30 m/s   **2.** 720 cm/s   **3.** 115 knots
5 To knots: **1.** 70 km/h   **2.** 600 m/s   **3.** 250 km/h
6 To feet per second: **1.** 195 stat. m.p.h. **2.** 400 stat. m.p.h. **3.** 29 knots
7 To statute miles per hour: **1.** 352 ft per s   **2.** 90 knots   **3.** 1,800 ft per min
8 To nautical miles per hour: **1.** 16.4 knots   **2.** 50 stat. m.p.h. **3.** 100 ft per s
9 To knots: **1.** 69.2 naut. m.p.h. **2.** 86 stat. m.p.h. **3.** 132 ft per s
10 To feet per minute: **1.** 88 ft per s   **2.** 60 stat. m.p.h. **3.** 145 knots

10-11 Temperature

Vocabulary

The **degree** is the unit used to measure temperature. The three scales used to measure temperature are the Celsius, Fahrenheit, and Kelvin scales.

On the Celsius scale, there are 100 divisions, each called a degree, between the freezing point of water (0°C) and the boiling point of water (100°C). On the Fahrenheit scale, there are 180 divisions, each also called a degree, between the freezing point (32°F) and boiling point (212°F) of water. One degree Celsius is equal to 1 kelvin, the name used to mean degree Kelvin, on the Kelvin scale. A temperature in kelvins is about 273 more than its reading in degrees Celsius.

PROCEDURE

To change a Celsius temperature to a kelvin reading, use the formula: $K = C + 273$
To change a kelvin temperature to a Celsius reading, use the formula: $C = K - 273$
To change a Celsius temperature to a Fahrenheit reading, use the formula:
$$F = \frac{9}{5}C + 32 \text{ or } F = 1.8C + 32$$
To change a Fahrenheit temperature to a Celsius reading, use the formula: $C = \frac{5}{9}(F - 32)$

EXAMPLES

Change 40°C to degrees Fahrenheit.

$F = \frac{9}{5}C + 32$

$F = \frac{9}{5}(40) + 32 = 104$

$F = 104°$

ANSWER: 104°F

Change 77°F to degrees Celsius.

$C = \frac{5}{9}(F - 32)$

$C = \frac{5}{9}(77 - 32) = 25$

$C = 25°$

ANSWER: 25°C

PRACTICE EXERCISES

1 Change each Celsius reading to a corresponding kelvin reading:
   1. 50°C   **2.** 96°C   **3.** 8°C   **4.** 0°C   **5.** 160°C

2 Change each kelvin reading to a corresponding Celsius reading:
   1. 280 K   **2.** 346 K   **3.** 405 K   **4.** 318 K   **5.** 371 K

3 Change each Fahrenheit reading to a corresponding Celsius reading:
   1. 86°F   **2.** 32°F   **3.** 113°F   **4.** 212°F   **5.** 60°F

4 Change each Celsius reading to a corresponding Fahrenheit reading:
   1. 60°C   **2.** 0°C   **3.** 100°C   **4.** 85°C   **5.** 18°C

Career
Applications

You have probably admired the beautiful rings and necklaces in a jewelry store. A *jeweler* knows much more about jewelry than its price. Jewelers may work in factories and repair shops as well as stores. They may make jewelry by casting metal in molds or shaping it with hand tools. Some jewelers are gemologists, experts on diamonds and other precious stones. A jeweler needs good eyesight and skillful hands. Most jewelers learn their skills as apprentices or in technical schools.

1. Pure gold is very soft. Gold used in jewelry is an alloy, a mixture of gold and other metals. 24-carat gold is 100% pure gold. What percent gold does 18-carat gold contain?

2. The melting point of pure gold is 1064.43°C. The melting point of pure silver is 961.93°C. How much higher is the melting point of gold?

3. Gold can be made into sheets $\frac{1}{250,000}$ of an inch thick. How thick would a pile of 400,000 of these sheets be?

4. After repairing a silver cuff link with silver solder, a jeweler cleans it in pickling solution. To make pickling solution, the jeweler dissolves a tablespoon of pickling powder in a pint of water. How much pickling powder would be needed for a gallon of water?

5. A jeweler begins making a ring at 10:30 A.M. and finishes at 1:15 P.M. How long did the job take?

CHAPTER REVIEW

Vocabulary

acre p. 392
bushel p. 391
cubic foot p. 394
cubic inch p. 394
cubic yard p. 394
day p. 397
degree p. 403
denominate numbers p. 384
fluid ounce p. 389
foot p. 384

gallon p. 389
hour p. 397
inch p. 384
mile p. 384
minute p. 397
month p. 397
ounce p. 387
pint pp. 389, 391
pound p. 387
quart pp. 389, 391

rate of speed p. 402
second p. 397
square foot p. 392
square inch p. 392
square mile p. 392
square yard p. 392
timetable p. 400
ton p. 387
yard p. 384
year p. 397

EXERCISES

The numerals in boxes indicate Lessons where help may be found.

1. What part of a yard is 30 inches? 10-1

2. How many ounces are in $3\frac{1}{4}$ pounds? 10-2

3. How many quarts are in 5 gallons? 10-3

4. What part of a bushel is 3 pecks? 10-4

5. Change 720 in.² to ft². 10-5

6. Change 81 ft³ to yd³. 10-6

7. From 4 h 30 min subtract 2 h 19 min 30 sec. 10-8

8. Change 100 m/h to m/s. 10-10

9. Is a temperature of 41°F warmer than a temperature of 5°C? 10-11

10. Use the bus schedule on page 401 to find how long it takes Bus 823 to go from Albuquerque to Little Rock. 10-9

11. Find the capacity in gallons equal to a volume of 26 ft³. 10-7

COMPETENCY CHECK TEST

The numerals in boxes indicate Lessons where help may be found.

Solve each problem and select the letter corresponding to your answer.

1. In $4\frac{1}{4}$ feet there are: 10-1
   a. 17 in.   **b.** $1\frac{1}{2}$ yd   **c.** 51 in.   **d.** $1\frac{5}{8}$ yd

2. 10 ounces is what part of a pound? 10-2
   a. $\frac{3}{4}$ lb   **b.** $\frac{5}{8}$ lb   **c.** $\frac{2}{5}$ lb   **d.** $\frac{1}{2}$ lb

3. In a half-gallon there are: 10-3
   a. 128 oz   **b.** 64 oz   **c.** 32 oz   **d.** 16 oz

4. In 5 bushels there are: 10-4
   a. 40 pk   **b.** 30 pk   **c.** 20 pk   **d.** 10 pk

5. What part of an hour is 40 minutes? 10-8
   a. $\frac{1}{2}$ hour   **b.** $\frac{7}{10}$ hour   **c.** $\frac{4}{5}$ hour   **d.** $\frac{2}{3}$ hour

6. In 54 square feet there are: 10-5
   a. 18 yd^2 **b.** 162 yd^2 **c.** 6 yd^2 **d.** 9 yd^2

7. In $1\frac{1}{2}$ cubic feet there are: 10-6
   a. 2,160 in.^{3} **b.** 2,400 in.^{3} **c.** 2,592 in^3 **d.** 3,296 in.^{3}

8. The volume of 10 cubic feet can hold a capacity of: 10-7
   a. 25 gallons   **b.** 40 quarts   **c.** 75 gallons   **d.** 60 quarts

9. Using the airplane schedule on page 400, the time it takes Flight 182 to fly from Dallas-Fort Worth to Seattle-Tacoma is: 10-9
   a. 1 h 40 min   **b.** 2 h 40 min   **c.** 3 h 40 min   **d.** 1 h 80 min

10. The speed of 30 miles per hour changed to feet per second is: 10-10
    a. 88 ft/s   **b.** 44 ft/s   **c.** 22 ft/s   **d.** 11 ft/s

11. The temperature reading of 15°C corresponds to a reading of: 10-11
    a. 59°F   **b.** 68°F   **c.** 50°F   **d.** 77°F

12. An airplane flying at 26,400 feet is at a height of: 10-1
    a. 6 mi   **b.** $4\frac{1}{2}$ mi   **c.** $3\frac{3}{4}$ mi   **d.** 5 mi

13. The normal body temperature for humans is 98.6°F. The corresponding Celsius temperature is: 10-11
    a. 37°C   **b.** 42°C   **c.** 60.5°C   **d.** 72.4°C

14. The number of pints of water that must be added to a pint can of frozen orange-juice concentrate to make a half-gallon of orange juice is: 10-3
    a. 7 pints   **b.** 5 pints   **c.** 3 pints   **d.** 1 pint

CHAPTER 11 Graphs, Statistics, and Probability

11-1 Bar Graphs

Vocabulary

A **bar graph** uses vertical or horizontal bars to compare sizes of quantities, generally statistical information.

This bar graph shows how many people rode a bus during the first six months of one year in one town.

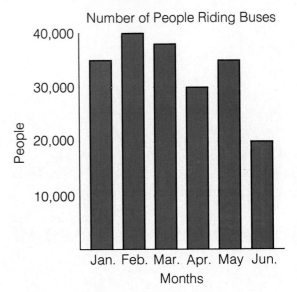

Number of People Riding Buses

PROCEDURE

To construct a bar graph:

(1) Draw a horizontal guide line on the bottom of the squared paper and a vertical guide line on the left.

(2) Select a convenient scale for the numbers that are being compared, first rounding large numbers. For a vertical bar graph, write the number scale along the vertical guide line; for a horizontal bar graph, use the horizontal guide line. Label the scale.

(3) Print names of the items opposite alternate squares along the other guide line. Label the items.

(4) Mark off the height corresponding to the given number for each item. Draw lines to complete the bars. All bars should have the same width.

(5) Select and print an appropriate title.

EXAMPLES:

In the graph above, in which month did the most people ride the bus?

ANSWER: February

In the graph above, about how many people rode the bus in January?

ANSWER: 35,000

Construct a bar graph showing the following information:

Of the first ten presidents, 6 were born in Virginia, 2 in Massachusetts, 1 in South Carolina, and 1 in New York.

ANSWER:

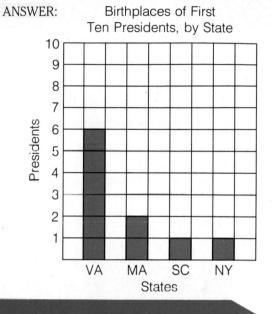

Birthplaces of First Ten Presidents, by State

PRACTICE EXERCISES

1 Use the graph on the right to answer the following questions regarding the lengths of some foreign rivers.

   1. How many miles does the side of a small square represent in the vertical scale of the bar graph? $\frac{1}{4}$ of the side of the square?

   2. Which river is the longest? The shortest?

   3. Find the approximate length of each river.

2 Construct bar graphs showing the following data:

   1. Marilyn's final grades for the year:
     English, 92   Mathematics, 87
     Social Studies, 85   Science, 90
     Art, 86   Music, 82
     Physical and Health Education, 78

   2. The lengths of the channel spans of some famous suspension bridges in the United States:
     Golden Gate, 4,200 ft   Delaware River, 1,750 ft
     Bear Mountain, 1,632 ft   Brooklyn, 1,595 ft
     George Washington, 3,500 ft

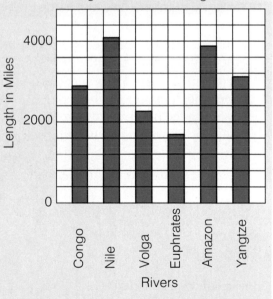

Lengths of Some Foreign Rivers

11-2 Line Graphs

This line graph shows daily highs in Tucson during the first five days in July one year.

PROCEDURE

To construct a line graph:

(1) Draw a horizontal guide line on the bottom of the squared paper and a vertical guide line on the left.

(2) Select a convenient scale for the related numbers, first rounding large numbers. Write the number scale along one of the guide lines. Label the scale.

(3) Print names of the items below the other guide line. Label the items.

(4) On each of these lines mark a dot to show the location of the value corresponding to the given number.

(4) Connect the successive dots with straight lines.

(5) Select and print an appropriate title.

EXAMPLES:

What was the temperature on July 3?

ANSWER: 101°

On which days was it 104°?

ANSWER: July 1 and July 4

Construct a line graph showing the data:

At Barbara's Appliances in January, José sold 18 appliances, Robin sold 22, Marti sold 19, and Curt sold 25.

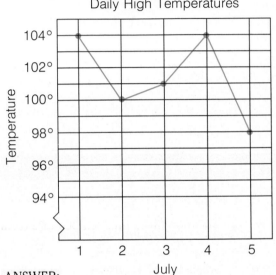

Daily High Temperatures

ANSWER:

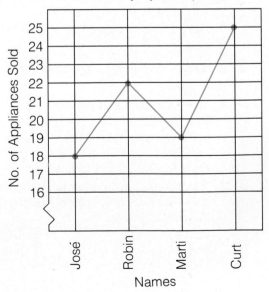

Appliances Sold at Barbara's Appliances in January by Salesperson

410

PRACTICE EXERCISES

1 Use the line graph to the right to answer the following questions regarding lateness at the Wilson School.

1. How many pupils late does the side of a small square indicate in the vertical scale of the line graph? $\frac{1}{2}$ of the side of a square?

2. During what month was the punctuality poorest? best?

3. How many pupils were late each month?

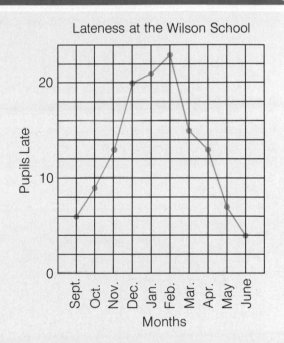

Lateness at the Wilson School

2 Construct line graphs showing the data below:

1. In twelve arithmetic progress tests, each containing 20 examples, George solved 8, 10, 13, 15, 12, 11, 14, 17, 16, 19, 17, and 20 examples correctly.

2. The monthly normal temperature (Fahrenheit) in some United States cities:

	Jan.	Feb.	Mar.	Apr.	May	June	July	Aug.	Sept.	Oct.	Nov.	Dec.
Atlanta	43	45	52	61	70	76	78	77	72	63	52	45
Chicago	24	26	35	47	58	67	72	72	65	54	40	29
Cincinnati	30	33	41	52	63	71	75	74	67	56	42	33
Los Angeles	55	56	58	59	62	66	70	71	69	65	61	57
New Orleans	54	57	63	69	75	81	82	82	79	71	62	56
New York	31	31	38	49	61	69	74	73	67	56	44	35
Philadelphia	33	34	41	52	63	71	76	75	68	58	46	36
St. Paul	13	16	29	46	58	67	72	69	61	49	32	19
Seattle	41	43	46	51	57	61	66	65	60	54	47	43

Vocabulary

A **circle graph** is used to show the relationship of the parts to a whole and to each other.

This circle graph at the right shows the uses of electricity in the home.

PROCEDURE

To construct a circle graph:

(1) Make a table showing: **a.** the given facts, **b.** the fractional part or percent each quantity is of the whole, **c.** the number of degrees representing each fractional part or percent, obtained by multiplying 360° by the fraction or percent.

(2) Draw a convenient circle. With a protractor construct successive central angles, using the number of degrees representing each part.

(3) Label each part.

(4) Select and print a title for the graph.

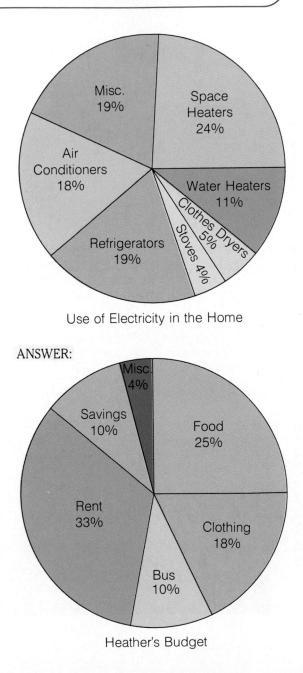

Use of Electricity in the Home

Heather's Budget

EXAMPLES:

Which use of electricity above was most common?

ANSWER: for space heaters

What percent of electricity above was used for refrigerators and stoves combined?
19% + 4% = 23%

ANSWER: 23%

Construct a circle graph to show Heather's budget.

Food, 25%; Clothing, 18%; Bus, 10%; Rent, 33%; Savings, 10%; Misc., 4%

PRACTICE EXERCISES

1 Use the graph on the right to answer the following questions regarding the family budget.

  1. What fractional part of its entire income does the Harris family plan to spend for food? for shelter?

  2. Compare the percent to be spent for shelter to the percent of savings.

  3. If the family income is $18,600 per year, what are the amounts to be spent for each item per year? per month? per week?

2 Construct circle graphs showing the following data:

  1. A certain large city planned to spend its income as follows: schools, 30%; interest on debt, 25%; safety, 15%; health and welfare, 12%; public works, 10%; other services, 8%.

  2. Charles spends his time as follows: 9 hours for sleep, 6 hours for school, 2 hours for study, 3 hours for recreation, and 4 hours for miscellaneous activities.

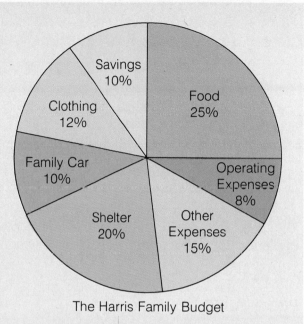

The Harris Family Budget

  3. A mathematics test showed the following distribution of marks: **A**, 6 pupils; **B**, 8 pupils; **C**, 12 pupils; **D**, 10 pupils.

Refresh Your Skills

1. Add: 1-4
3,586
2,975
4,598
6,284

2. Subtract: 1-5
305,008
214,069

3. Multiply: 1-6
690
504

4. Divide: 1-7
365)75,920

5. Add: 2-4
$.54 + $9.89 + $24.06

6. Subtract: 2-5
$30 − $1.98

7. Multiply: 2-7
.006 × .02

8. Divide: 2-8
.1).586

9. Add: 3-8
$8\frac{1}{4} + 5\frac{1}{3} + 2\frac{5}{12}$

10. Subtract: 3-9
$\frac{9}{10} - \frac{3}{5}$

11. Multiply: 3-11
$1\frac{3}{4} \times 1\frac{3}{5}$

12. Divide: 3-12
$9\frac{3}{4} \div 13$

13. Find $11\frac{1}{2}$% of $9,500
4-6

14. What percent of 50 is 75?
4-7

15. $1.62 is 60% of what amount?
4-8

Vocabulary

Statistics is the study of collecting, organizing, analyzing, and interpreting data. The following terms are used:

Frequency: The number of times a score or datum occurs.

Frequency Distribution: Arrangement of data in table form.

Histogram: Frequency distribution shown in a bar graph.

Frequency Polygon: Frequency distribution shown in a line graph.

The data studied in statistics may be scores, measurements, or other numerical facts. Data are recorded by tallying.

PROCEDURE

To make a histogram:

(1) Make a frequency distribution by listing the classes of data. Then tally the data and list the frequency of each class.

(2) Make a bar graph showing the scores and frequencies.

To make a frequency polygon:

(1) Make a frequency distribution as described above.

(2) Make a line graph showing the score and the frequency of each score.

EXAMPLE:

Make a frequency distribution table, a histogram, and a frequency polygon for the following data.

A class of 34 pupils made the following scores on a test:

80, 95, 70, 80, 85, 65, 90, 75, 60, 90, 100, 65, 85, 75, 80, 95, 85, 65, 80, 70, 90, 70, 80, 85, 80, 60, 75, 80, 85, 95, 80, 70, 90, 85

ANSWER:

Frequency Distribution Table		
Score	Tally	Frequency
100	I	1
95	III	3
90	IIII	4
85	IHT I	6
80	IHT III	8
75	III	3
70	IIII	4
65	III	3
60	II	2
		Total 34

ANSWER:

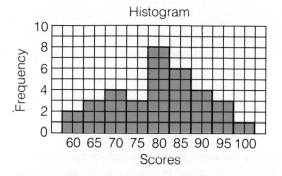

ANSWER:

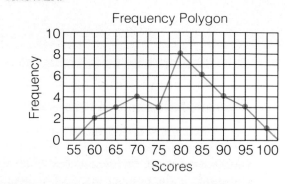

PRACTICE EXERCISES

Using the lists of scores given below, make for each:

1 A frequency distribution table

2 A histogram

3 A frequency polygon

1. 85 60 75 40 85 70 90 100 95 85 75 80 70 70 85 90 60 65 75 80
90 75 90 80 95

2. 9 2 7 6 4 5 8 8 6 9 7 10 5 7 2 8 3 7 4 6 5 9 1 8 7 6 9 7 5 8

3. 45 43 44 48 45 52 47 48 46 50 45 49 49 47 51 49 52 50 43 45
48 44 47 49 45 51 50 43 45 51 46 49 50 45 44

4. 17 14 12 15 17 16 12 15 19 14 15 15 12 17 14 19 13 12 15 14
13 17 12 11 18 13 19 14 16 14 18 11 10 14 18 14 16 19 13 18 17

5. 6 8 4 2 3 9 6 7 5 8 8 3 10 5 2 6 7 6 8 9 4 9 1 7 4 6 9 2
5 5 5 7 4 9 2 8 8 3 5 10 6 7 4 8 9 6 8 10 8 7

Refresh Your Skills

1. Add: 1-4
6,109 + 56,274 + 398

2. Subtract: 1-5
81,040 − 2,951

3. Multiply: 1-6
9,008 × 406

4. Divide: 1-7
$6{,}080\overline{)54{,}720}$

5. Add: 2-4
.724 + 5.66 + 20.8

6. Subtract: 2-5
45 − 3.9

7. Multiply: 2-7
.03 × .0008

8. Divide: 2-8
2.8 ÷ .007

9. Add: 3-8
$3\frac{2}{3} + 4\frac{1}{2} + 2\frac{3}{5}$

10. Subtract: 3-9
$8\frac{1}{6} - 5\frac{1}{3}$

11. Multiply: 3-11
$12 \times 4\frac{5}{8}$

12. Divide: 3-12
$7\frac{1}{2} \div 2\frac{1}{2}$

13. Find $8\frac{3}{4}$% of $20,000
4-6

14. 48 is what percent of 60?
4-7

15. 9 is 36% of what number?
4-8

11-5 Averages—Measures of Central Tendency

> The **mean**, **median**, and **mode** are three commonly used measures of central tendency.
> The **arithmetic mean** (or simply **mean**) is the average.
> The **median** is the middle number when a group of numbers is arranged in order of size.
> The **mode** is the number that occurs most frequently in a group of numbers. There may be more than one mode.
> The **range** is the difference between the highest and lowest numbers in the group.

Mean, median, mode, and range are descriptions of a set of data that tell us about the data—their average, which one is in the middle, which one occurs most frequently, and how spread out they are.

PROCEDURE

To find the mean:
(1) Find the sum of the numbers. (To find the sum using a frequency distribution table, multiply each score by its frequency. Add these products.)
(2) Divide the sum by the number of items.

EXAMPLE:
Find the arithmetic mean of the scores 78, 83, 91, 82, and 86.

$$\frac{78 + 83 + 91 + 82 + 86}{5} = \frac{420}{5} = 84$$

ANSWER: The mean is 84.

PROCEDURE

To find the median:
(1) Arrange the numbers by order of size.
(2) Count from either end to find the middle number. If the number in the group of numbers is even, divide the sum of the two middle numbers by 2.

EXAMPLE:
Find the median of the scores 78, 85, 83, 81, 92, 86, and 90.

78   81   83   85   86   90   92

ANSWER: The median is 85.

PROCEDURE

To find the mode: Find the number that occurs most frequently.

EXAMPLE:

Find the mode of the numbers 80, 75, 93, 81, 98, 93, and 57.   93 appears 2 times.

ANSWER: The mode is 93.

PROCEDURE

To find the range:
(1) Identify the highest and lowest numbers in the set.
(2) Subtract to find the range.

EXAMPLE:

Find the range of the following scores: 78, 83, 91, 82, 86.
$91 - 78 = 13$

ANSWER: The range is 13.

EXAMPLE:

Arrange the following group of scores in a frequency distribution table; then
find the mean, median, mode, and range:
5, 7, 9, 8, 5, 8, 6, 5, 9, 8, 7, 8, 7, 5, 9, 5, 8, 5, 9

ANSWERS:

Frequency Distribution Table		
Score	Tally	Frequency
9	IIII	4
8	ЖІ	5
7	III	3
6	I	1
5	ЖІ I	6
		Total 19

Arithmetic Mean:
$9 \times 4 = 36$
$8 \times 5 = 40$
$7 \times 3 = 21$ $\frac{133}{19} = 7$
$6 \times 1 = 6$
$5 \times 6 = 30$
Sum = 133
Arithmetic mean = 7

Median or Middle Score:
10th score from top or
bottom = 7

Mode or most frequent
score = 5

Range: $9 - 5 = 4$

PRACTICE EXERCISES

1 Find the mean for each of the following lists of scores:
   1. 83, 89, 80, 81, 84, 86, 85
   2. 9, 5, 8, 4, 10, 3, 7, 8, 6, 7, 9, 9, 7, 3, 8, 9
   3. 47, 43, 50, 48, 42, 47, 49, 43, 45, 47, 48, 46, 47, 44, 40, 44, 47, 49, 42, 44, 50, 47, 46, 49
   4. 17, 24, 12, 28, 19, 16, 25, 28, 92, 17, 21, 18, 15, 24, 19, 22, 20, 26, 29, 16, 23, 25, 18, 14, 90, 23, 28, 86, 15, 19, 20, 18, 24

2 Find the median for each of the following lists of scores:

1. 8, 6, 7, 5, 9, 10, 3, 0, 5
2. 4.2, 5.6, 3.8, 4.7, 6.1, 4.5
3. 65, 85, 70, 95, 75, 80, 70, 85, 80, 85, 95, 60, 75, 80, 75, 95, 85, 60, 75, 90, 85
4. 9, 6, 8, 3, 4, 9, 2, 8, 3, 7, 7, 6, 8, 10, 4, 5, 9, 6, 7, 8, 8, 5, 6, 8, 10, 8, 5, 9, 6, 7, 6, 6, 9, 10, 10, 9, 5, 2, 6, 8, 7, 9, 4, 10

3 Find the mode for each of the following lists of scores:

1. 11, 15, 10, 19, 15, 12, 11, 17, 14, 15, 13, 12
2. 57, 49, 64, 53, 58, 55, 53, 60, 58, 52, 53, 64, 58, 49, 60
3. 5, 4, 9, 7, 6, 8, 6, 4, 5, 9, 7, 7, 5, 6, 9, 8, 3, 2, 5, 7, 4, 9, 3, 6, 2, 7, 8
4. 65, 25, 95, 80, 30, 85, 65, 30, 45, 60, 95, 70, 80, 95, 45, 90, 70, 45, 25, 85, 95, 75, 70, 45, 85, 70, 65, 80, 45, 60, 65, 80, 70, 45, 55

4 Find the range for each of the following lists of scores:

1. 31, 26, 29, 43, 34, 39, 47, 28, 45
2. 85, 90, 75, 65, 35, 80, 55, 85, 60, 75, 40, 80
3. 12, 7, 9, 15, 8, 4, 11, 6, 17, 9, 5, 10, 2, 16, 3
4. 23, 41, 37, 16, 27, 32, 19, 49, 36, 52, 28, 46, 59, 44, 27, 30

5 Tally and arrange each of the following lists of scores in a frequency distribution table; then find the mean, median, mode, and range:

1. 6, 4, 8, 3, 7, 8, 6, 9, 8, 5, 7, 9, 7, 4, 9, 6, 7, 9, 8, 5, 8, 9, 8, 6, 9
2. 48, 47, 50, 43, 42, 47, 49, 43, 45, 47, 48, 46, 47, 44, 47, 44, 40, 49, 48
3. 18, 15, 12, 19, 14, 16, 14, 18, 13, 17, 16, 19, 16, 12, 14, 17, 15, 16, 20, 18, 11, 14, 19, 15, 17, 20, 19, 17, 18, 13, 14
4. 58, 62, 64, 59, 57, 61, 60, 59, 63, 58, 64, 60, 63, 59, 59, 56, 64, 61, 58, 62, 64, 60, 57, 59, 63, 55, 61, 56, 60, 59, 61, 58, 62, 59, 57, 62
5. 9, 6, 2, 5, 8, 3, 7, 6, 5, 9, 4, 4, 3, 1, 8, 10, 9, 6, 9, 4, 2, 5, 7, 3, 6, 8, 5, 9, 7, 2, 8, 4, 7, 5, 4, 6, 9, 8, 8, 3, 9, 10, 5, 7, 3, 9, 5

Computer Activities

The computer program below computes the *arithmetic mean* of any 5 numbers.

```
1 LET T = Ø
1Ø PRINT "PLEASE TYPE A NUMBER AFTER
EACH ?-MARK"
2Ø FOR I = 1 TO 5
3Ø INPUT X
4Ø LET T = T + X
5Ø NEXT I
6Ø PRINT "THE ARITHMETIC MEAN IS: "; T/5
```

a) Enter the program and then RUN it to make sure it works.

b) Modify the program to compute the arithmetic mean of any *10* numbers.

11-6 Probability and Odds

Vocabulary

Probability and **odds** are numerical measures used to indicate the likelihood of an event happening. They are expressed as ratios in the form of a fraction, a decimal, or a percent. An **outcome** is the way something happens in an experiment or activity; as tossing a coin.

When tossing a coin, there are two outcomes: heads and tails. If the desired outcome is heads, there is one favorable outcome—heads, and one unfavorable outcome—tails.

The probability of a particular outcome is a numerical measure of the chance, or likelihood, that the outcome or event will occur.

$$\text{Probability} = \frac{\text{Number of favorable outcomes}}{\text{Total number of outcomes}}$$

If there are no favorable outcomes, the probability is 0. If all are favorable outcomes, the probability is 1.

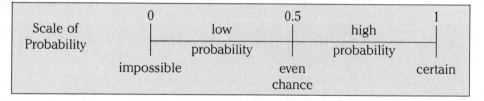

PROCEDURE

To find the probability of an outcome:
Write the ratio of favorable outcomes to the total number of outcomes.

EXAMPLES:

Find the probability of getting a head when tossing a coin.
There are 2 outcomes: heads and tails. $\frac{\text{Number of favorable outcomes}}{\text{Total number of outcomes}} = \frac{1}{2}$
There is 1 favorable outcome: heads
ANSWER: $\frac{1}{2}$

There are 3 blue marbles in a bag containing 12 marbles. What is the probability of drawing a blue marble if one is drawn at random?
$\frac{\text{Number of favorable outcomes}}{\text{Number of outcomes}} = \frac{3}{12} = \frac{1}{4}$
Write: P(blue marble) $= \frac{1}{4}$ Read: The probability of a blue marble is $\frac{1}{4}$.
ANSWER: P(blue marble) $= \frac{1}{4}$.

> The odds for an event happening are: $\dfrac{\text{Number of favorable outcomes}}{\text{Number of unfavorable outcomes}}$
>
> The odds against an event happening are: $\dfrac{\text{Number of unfavorable outcomes}}{\text{Number of favorable outcomes}}$

PROCEDURE

To find the odds for an event happening:
Write the ratio of the number of favorable outcomes to the number of unfavorable outcomes.

To find the odds against an event happening:
Write the ratio of the number of unfavorable outcomes to the number of favorable outcomes.

Example:

There are 3 blue marbles and 9 marbles of other colors in a bag. Find the odds for drawing a blue marble from the bag.

$$\frac{\text{Number of favorable outcomes}}{\text{Number of unfavorable outcomes}} = \frac{3}{9} = \frac{1}{3} \text{ or } 1:3$$

ANSWER: The odds of selecting a blue marble are 1 to 3.

PRACTICE EXERCISES

1 What is the probability of drawing at random on the first draw:

   1. A green ball from a box containing 9 balls of which 4 are green?

   2. A marked card from a hat containing 15 cards of which 9 are marked?

   3. A red checker from a box containing 20 checkers of which 8 are red?

   4. A blue bead from a bag containing 48 beads of which 30 are blue?

   5. A yellow jelly bean from a bag containing 54 jelly beans of which 36 are yellow?

2 What is the probability of drawing at random on the first draw:

   1. A blue marble from a bag containing 7 blue marbles and 14 yellow marbles?

   2. A black checker from a box containing 12 red checkers and 12 black checkers?

   3. An orange gumdrop from a bag containing 18 green gumdrops and 24 orange gumdrops?

   4. A brown bead from a bag containing 42 brown beads and 28 purple beads?

   5. A white ball from a box containing 56 yellow balls and 40 white balls?

3 **1.** A die is a cube each of whose six faces is marked with a number of dots, from one through six. The plural of die is dice.

   When a die is rolled, what are the odds that the face with six dots will be up?

   2. A hat contains 16 cards of which 6 are marked and the rest are blank. What are the odds of selecting at random on the first draw a blank card?

3. A spinner has a circle with twelve sectors each of the same size numbered 1 through 12. What are the odds that the pointer on the first spin does not stop on a sector named by a number divisible by 4?

4. What are the odds against selecting at random on the first try a yellow ball from a box containing 15 yellow balls and 30 white balls?

5. What are the odds of selecting at random on the first try a nickel from your pocket when it contains 9 nickels and 12 quarters?

4 What is the probability that you will win the prize if you hold 50 raffle tickets of the 1,500 raffle tickets sold at a charity affair? What are the odds that you will not win the prize?

5 A standard deck of 52 playing cards consists of an ace, king, queen, jack, 10, 9, 8, 7, 6, 5, 4, 3, and 2 in four different suits: clubs, diamonds, hearts, and spades. Diamond and heart suits are in red; club and spade suits are in black. The king, queen, and jack cards are called picture cards.

From a well-shuffled deck drawn at random,

1. What is the probability that the first draw will be:

    (1) a red card?   **(2)** a picture card?

    (3) a king?   **(4)** the queen of spades?

2. What is the probability that the first draw will not be:

    (1) an ace?   **(2)** a black card?

    (3) a picture card?   **(4)** a red 7?

3. What are the odds in favor of the first draw being:

    (1) a black 10?   **(2)** a red card?

    (3) the jack of diamonds?   **(4)** a picture card?

4. What are the odds against the first draw being:

    (1) a black picture card?   **(2)** the ace of clubs?

    (3) a queen?   **(4)** a spade card?

Career Applications

Keeping a forest healthy, beautiful, and productive requires a lot of work. *Forestry technicians* inspect trees for disease, prevent fires and flood damage, and plant new trees. They work in national and state parks to help maintain forest areas for hiking and camping, and teach visitors about park rules.

Most forestry technicians study forestry for one or two years after high school. Others learn their skills by working on-the-job training.

Histogram

No. of Trees — Types of Trees

Distribution of Campers

1. A forestry technician counted the trees in a certain area and made the graph at the right. How many more birch trees than spruce trees were counted?

2. The circle graph shows the distribution of campers at a park. If there were 5,394 tent campers this season, how many campers were there in all?

3. A forestry technician counted the diseased trees in fifteen plots. The numbers were: 53, 64, 52, 44, 39, 41, 60, 58, 49, 42, 44, 55, 59, 36, and 63. What are the mean, median, and mode of the data?

CHAPTER REVIEW

Vocabulary

bar graph p. 408
circle graph p. 410
frequency p. 414
frequency distribution p. 414
frequency polygon p. 414

histogram p. 414
line graph p. 410
mean p. 416
median p. 416
mode p. 416

odds p. 419
outcome p. 419
probability p. 419
range p. 416
statistics p. 414

EXERCISES

The numerals in boxes indicate lessons where help may be found.

1. Construct a bar graph showing the lengths of the world's major tunnels: 11-1

   Mont Blanc, 7.2 miles; St. Gotthard, 10.01 miles; Great St. Bernard, 3.6 miles; Baltimore Harbor, 1.4 miles; Arlberg, 8.7 miles; and Brooklyn-Battery, 1.73 miles.

2. Construct a line graph showing the gross savings and investments in the United States for five years: 11-2

   1st yr, $375 billion;   2nd yr, $423 billion;   3rd yr, $406 billion;   4th yr, $484 billion;   and   5th yr, $406 billion

3. Construct a circle graph showing how a family spends its monthly income of $3,000. It spends $900 for food, $600 for housing, $390 for clothing, $270 for transportation, $540 for miscellaneous expenses, and allows $300 for savings. 11-3

For the following list of scores—6, 4, 5, 2, 1, 6, 8, 5, 2, 7, 3, 9, 5, 6, 7, 4, 2, 3, 8, 5, 1, 4, 7, 4, 6, 8, 5, 2, 8, 3, 5, 6—make: 11-4

4. A frequency distribution table
5. A histogram
6. A frequency polygon

Tally and arrange the following list of scores in a frequency distribution table—19, 16, 18, 17, 14, 16, 17, 16, 19, 15, 20, 17, 16, 19, 18, 16, 14, 16, 19, 17, 19, 16, 18, 16, 17. Determine: 11-5

7. The mean
8. The median
9. The mode
10. The range

Suppose Mr. O'Leary has 8 one-dollar bills, 7 five-dollar bills, and 5 ten-dollar bills mixed in his wallet. 11-6

11. What is the probability that he will select at random on the first try a one-dollar bill?
12. What are the odds against his selecting a ten-dollar bill on the first try?

COMPETENCY CHECK TEST

The numerals in boxes indicate Lessons where help may be found.

Solve each problem and select the letter corresponding to your answer.

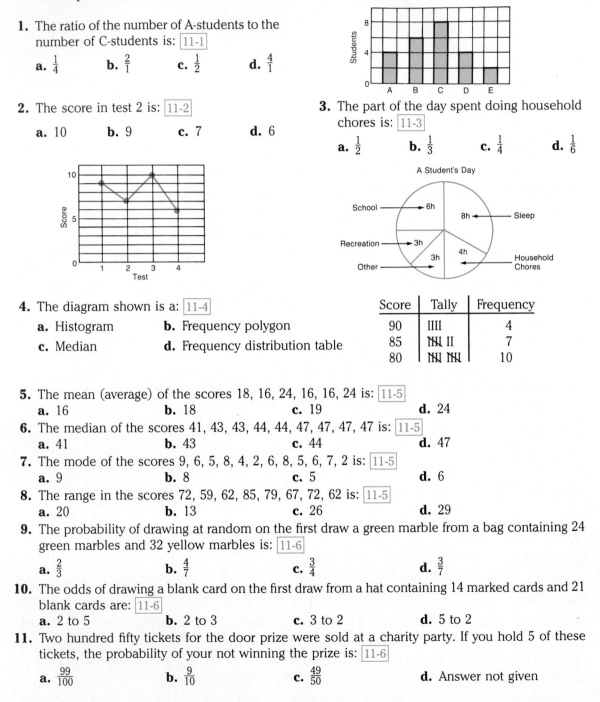

1. The ratio of the number of A-students to the number of C-students is: 11-1
   a. $\frac{1}{4}$ b. $\frac{2}{1}$ c. $\frac{1}{2}$ d. $\frac{4}{1}$

2. The score in test 2 is: 11-2
   a. 10   b. 9   c. 7   d. 6

3. The part of the day spent doing household chores is: 11-3
   a. $\frac{1}{2}$ b. $\frac{1}{3}$ c. $\frac{1}{4}$ d. $\frac{1}{6}$

4. The diagram shown is a: 11-4
   a. Histogram
   c. Median
   b. Frequency polygon
   d. Frequency distribution table

Score	Tally	Frequency
90	IIII	4
85	ℕℕ II	7
80	ℕℕ ℕℕ	10

5. The mean (average) of the scores 18, 16, 24, 16, 16, 24 is: 11-5
   a. 16   b. 18   c. 19   d. 24

6. The median of the scores 41, 43, 43, 44, 44, 47, 47, 47, 47 is: 11-5
   a. 41   b. 43   c. 44   d. 47

7. The mode of the scores 9, 6, 5, 8, 4, 2, 6, 8, 5, 6, 7, 2 is: 11-5
   a. 9   b. 8   c. 5   d. 6

8. The range in the scores 72, 59, 62, 85, 79, 67, 72, 62 is: 11-5
   a. 20   b. 13   c. 26   d. 29

9. The probability of drawing at random on the first draw a green marble from a bag containing 24 green marbles and 32 yellow marbles is: 11-6
   a. $\frac{2}{3}$ b. $\frac{4}{7}$ c. $\frac{3}{4}$ d. $\frac{3}{7}$

10. The odds of drawing a blank card on the first draw from a hat containing 14 marked cards and 21 blank cards are: 11-6
    a. 2 to 5   b. 2 to 3   c. 3 to 2   d. 5 to 2

11. Two hundred fifty tickets for the door prize were sold at a charity party. If you hold 5 of these tickets, the probability of your not winning the prize is: 11-6
    a. $\frac{99}{100}$ b. $\frac{9}{10}$ c. $\frac{49}{50}$ d. Answer not given

Achievement Test

9-1   Change: **1.** 63.8 m to cm   **2.** 419 mm to cm

9-2   Change: **3.** 12.4 kg to g   **4.** 35 mg to g

9-3   Change: **5.** 759 mL to L   **6.** 12 L to cL

9-4   Change: **7.** 26.45 cm^2 **8.** 31.7 km^2 to hectares

9-5   Change: **9.** 1,495 cm^3 to dm^3 **10.** 83.9 m^3 to dm^3

        11. 33.4 liters of water occupy _____ dm^3 and weighs _____ kg.

10-1   **12.** How many yards are in $6\frac{2}{3}$ miles?

10-2   **13.** Change $2\frac{5}{8}$ pounds to ounces.

10-3   **14.** How many liquid ounces are in 3 quarts?

10-4   **15.** Change $10\frac{1}{2}$ bushels to pecks.

10-5   **16.** What part of a square foot is 48 square inches?

10-6   **17.** Change 297 cubic feet to cubic yards.

10-7   **18.** What is the weight in pounds of 100 cubic feet of fresh water?

10-8   **19.** What part of a minute is 45 seconds?

10-9   **20.** Using the bus schedule on p. 401, find at what times Bus 820 leaves Memphis and arrives in Oklahoma City.

10-10   **21.** Change 54 km/h to m/s.

10-11   **22.** Is a temperature of 50°F warmer than 30°C?

11-1   **23.** Construct a bar graph showing the heights of the world's tallest buildings: World Trade Center, 1,350 ft; CNR Tower, 1,815 ft; Centerpoint Tower, 1,065 ft; Eiffel Tower, 984 ft; and Sears Tower, 1,454 ft.

11-2   **24.** Construct a line graph showing the U.S. national income for five recent years: 1st yr, $1,760 billion; 2nd yr, $1,967 billion; 3rd yr, $2,117 billion; 4th yr, $2,373 billion; 5th yr, $2,450 billion.

11-3   **25.** Construct a circle graph showing a family's budget: Housing, 25%; Food, 30%; Transportation, 20%; Recreation, 10%; Savings, 5%; Other, 10%.

        For the following list of scores—5, 9, 6, 4, 5, 8, 7, 3, 9, 6, 2, 4, 5, 9, 7, 4, 8, 3, 5, 7, 4, 6, 8, 9, 7, 5, 8, 4, 6, 9, 4, 3, 8, 6, 7:

11-4   **26.** Make a frequency distribution table.

11-5   Find: **27.** The mean   **28.** The median   **29.** The mode   **30.** The range

11-6   A spinner has ten equal sectors numbered from 1 through 10.

        31. What is the probability that the pointer on the spinner will stop on an even-numbered sector?

        32. What are the odds of the pointer *not* stopping on an even-numbered sector?

UNIT **4**

Geometry

Inventory Test

12-1 Read, or write in words, each of the following:

1. $\overline{BC}$ **2.** $\overleftrightarrow{NR}$ **3.** $\overrightarrow{AT}$ **4.** $\overleftrightarrow{CF}$ **5.** $\overline{SY}$ **6.** $\overrightarrow{PQ}$

Name each of the following and express them symbolically:

7. M ●————● S   **8.** ←——— D ●——● E ———→   **9.** ←——— C ●——● J ———→

Draw a representation of:

10. a curved line   **11.** a straight line   **12.** a broken line

Draw a representation of a line in:

13. a slanting position   **14.** a vertical position   **15.** a horizontal position

Draw a representation of a pair of:

16. intersecting lines   **17.** parallel lines   **18.** perpendicular lines

12-2 **19.** Name the angle below in 3 ways.

12-3 **20.** Draw an obtuse angle.

21. Draw a right angle.

22. Draw an acute angle.

12-4 **23.** Using a ruler, measure the following line segment: ●————————●

12-5 **24.** If the distance from R to A is 50 miles, what is the distance from A to B?

R ● A ●————————● B

12-6 **25.** What kind of triangle is:

12-7 **26.** What is the figure below called?

12-8 **27.** With a protractor, draw an angle of 125°.

12-9 **28.** If two angles of a triangle measure 49° and 75°, what is the measure of the third angle?

12-10 **29.** Find the measure of the fourth angle of a quadrilateral when the other three angles measure 95°, 88°, and 43°.

12-11 **30.** What is the complement of an angle measuring 69°?

31. What is the supplement of an angle measuring 108°?

32. If $m\angle 1 = 39°$ and $m\angle 4 = 105°$, find the measure of $\angle 2$. Of $\angle 3$.

33. If $m\angle 3 = 67°$, what is the measure of $\angle 1$? $\angle 2$? $\angle 4$?

428

12-12 34. Parallel lines *AB* and *CD* are cut by transversal *EF*. If $m\angle 6 = 95°$, what is the measure of $\angle 1$? $\angle 2$? $\angle 3$? $\angle 4$? $\angle 5$? $\angle 7$? $\angle 8$?

12-13 35. Which pair of triangles below are congruent?

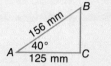 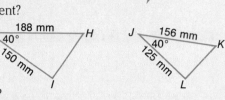

12-14 36. Which pair of triangles above are similar?

13-1 37. Make a copy of the following line segment: ●————————————●

13-2 38. Draw an angle of 30° with a protractor, then copy the angle using a compass.

13-3 39. Construct a triangle with the sides measuring $1\frac{5}{8}$ in., $2\frac{1}{8}$ in., and $1\frac{3}{4}$ in.

40. Construct a triangle with sides measuring 5 cm and 4.5 cm and an included angle of 75°.

41. With a scale of $\frac{1}{4}$ in. = 1 ft, construct a triangle with angles measuring 30° and 60° and an included side measuring 9 ft.

13-4 42. Draw any line. Use a compass to construct a perpendicular at a point on the line.

13-5 43. Draw any line. Use a compass to construct a perpendicular from a point outside the line.

13-6 44. Draw any line segment. Use a compass to bisect this segment.

13-7 45. Draw any angle. Use a compass to bisect this angle.

13-8 46. Draw any line. Locate a point outside this line. Through this point construct a line parallel to the first line.

14-1 47. Find the square of 27.

14-2 48. Find $\sqrt{6,400}$

14-3 49. Find $\sqrt{50}$ by the estimate, divide, and average method.

14-4 50. Find the base of a right triangle if the hypotenuse is 146 meters and the altitude is 96 meters.

51. Find the hypotenuse of a right triangle if the altitude is 112 feet and the base is 384 feet.

52. Find the altitude of a right triangle if the base is 45 centimeters and the hypotenuse is 117 centimeters.

14-5 53. Find the height of a flagpole that casts a shadow of 450 feet at a time when a girl, $5\frac{1}{2}$ feet tall, casts a shadow of 33 feet.

14-6 In right triangle *ABC*:

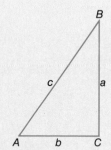

54. Find side *a* when angle *A* = 89° and side *b* = 250 centimeters.

55. Find side *a* when angle *A* = 71° and side *c* = 800 yards.

56. Find side *b* when angle *A* = 47° and side *c* = 500 meters.

15-1 **57.** Find the perimeter of a rectangle 138 meters long and 57 meters wide.

15-2 **58.** Find the perimeter of a square whose side measures 106 millimeters.

15-3 **59.** Find the perimeter of a triangle with sides measuring $8\frac{9}{16}$ inches, $7\frac{5}{8}$ inches, and $6\frac{3}{4}$ inches.

15-4 Find the circumference of:

60. A circle whose radius is 49 feet.

61. A circle whose diameter is 64 meters.

15-5 **62.** Find the area of a rectangle 293 meters long and 185 meters wide.

15-6 **63.** Find the area of a square whose side measures 78 kilometers.

15-7 **64.** Find the area of a parallelogram with an altitude of 98 feet and a base of 107 feet.

15-8 **65.** Find the area of a triangle whose altitude is $10\frac{1}{2}$ inches and base is 9 inches.

15-9 **66.** Find the area of a trapezoid with bases of 87 meters and 41 meters and a height of 62 meters.

15-10 Find the area of:

67. A circle whose radius is 3.8 kilometers.

68. A circle whose diameter is 56 inches.

15-11 **69.** Find the total area of the outside surface of a rectangular solid $7\frac{1}{2}$ inches long, 6 inches wide, and $2\frac{3}{4}$ inches high.

15-12 **70.** Find the total area of the outside surface of a cube whose side measures 9.3 meters.

15-13 Find the total area of the outside surface of:

71. A right circular cylinder 20 centimeters in diameter and 34 centimeters high.

72. A sphere whose diameter is 84 millimeters.

15-14 **73.** Find the volume of a rectangular solid 109 centimeters long, 97 centimeters wide, and 63 centimeters high.

15-15 **74.** Find the volume of a cube whose edge measures 2 feet 8 inches.

15-16 **75.** Find the volume of a right circular cylinder with a radius of 8 centimeters and a height of 10 centimeters.

15-17 Find the volume of:

76. A sphere whose diameter is 4.5 meters.

77. A right circular cone 10 inches in diameter and 9 inches high.

78. A square pyramid 16 yards on each edge of the base and 13 yards high.

CHAPTER **12** **Lines, Angles, and Triangles**

Vocabulary

A **point** is an exact location in space.

A **line** is a collection of points that extends indefinitely in opposite directions.

A point separates a line into two **half-lines**.

A **line segment**, or **segment**, is a definite part of a line consisting of two endpoints and all the points between.

An **interval** is a definite part of a line excluding its endpoints.

A **ray** is a half-line which includes one endpoint.

Parallel lines (∥) are two lines in the same plane (See Lesson 12-6) that never meet.

Intersecting lines are two lines in the same plane (See Lesson 12-6) that meet.

Perpendicular lines (⊥) are two intersecting lines that form a right angle.

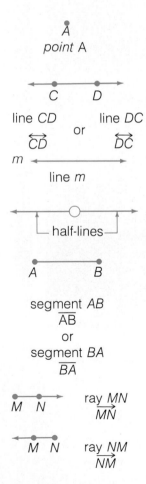

A point has no size, nor can it be seen. Any dot we use to show a point simply represents the point. We name points with capital letters.

The arrowheads show that a line is endless in both directions. A line has an infinite number of points but no endpoints. The "lines" we draw are only representations of geometric lines. We cannot see a geometric line. We name a line by using two labeled points on it or a single lowercase letter.

A half-line extends indefinitely in one direction and does not include the point from which it extends. It separates the line into two half-lines.

We name a line segment by its endpoints. A segment has length, but no width or thickness.

To name a ray, we write the letter naming the endpoint first and then the letter that names any other point on the ray.

432

In geometry, the word *line* means "straight line." There are also curved and broken lines.

line   curve   broken line

Lines may be in three positions.

vertical line   horizontal line   slanting line

Intersecting lines, also called concurrent lines, meet in one point.

$\overleftrightarrow{AC}$ intersects $\overleftrightarrow{DE}$ at point *B*.

Parallel lines will never meet.

m is parallel to *n*.
$m \parallel n$

Two intersecting lines or rays or segments that form a right angle are said to be perpendicular to each other.

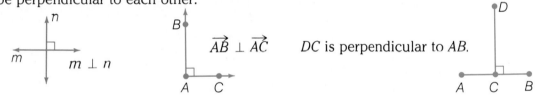

$m \perp n$ $\overrightarrow{AB} \perp \overrightarrow{AC}$ *DC* is perpendicular to *AB*.

EXAMPLES:

Name and then express symbolically each of the following:

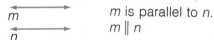

ANSWER: Line *AB*; $\overleftrightarrow{AB}$ ANSWER: Ray *AB*; $\overrightarrow{AB}$ ANSWER: $\overleftrightarrow{AB}$ is parallel to $\overleftrightarrow{CD}$; $\overleftrightarrow{AB} \parallel \overleftrightarrow{CD}$

PRACTICE EXERCISES

Name and then express symbolically each of the following:

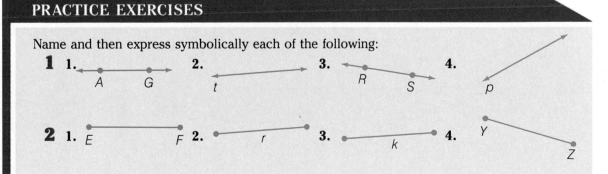

1 1. *A G*   2. *t*   3. *R S*   4. *p*

2 1. *E F*   2. *r*   3. *k*   4. *Y Z*

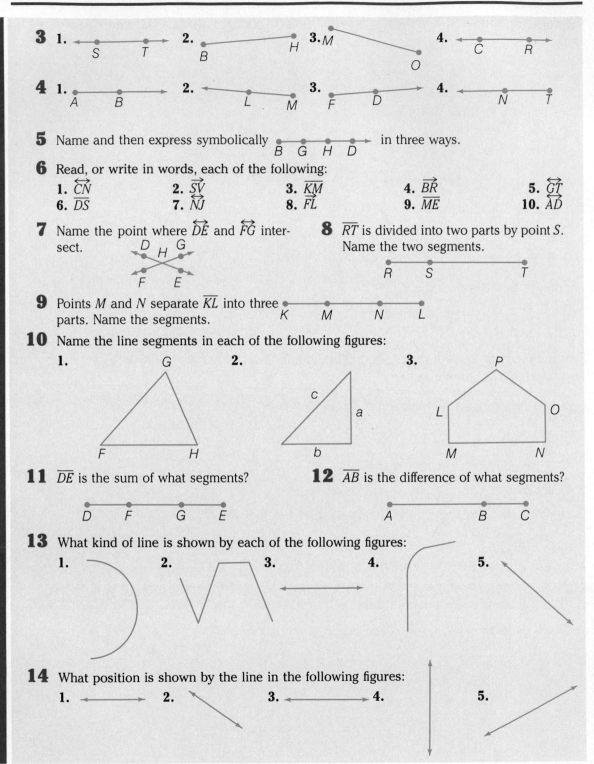

3 1. [S T] 2. [B H] 3. [M O] 4. [C R]

4 1. [A B] 2. [L M] 3. [F D] 4. [N T]

5 Name and then express symbolically [B G H D] in three ways.

6 Read, or write in words, each of the following:

1. $\overleftrightarrow{CN}$ 2. $\overrightarrow{SV}$ 3. $\overline{KM}$ 4. $\overrightarrow{BR}$ 5. $\overleftrightarrow{GT}$
6. $\overline{DS}$ 7. $\overleftrightarrow{NJ}$ 8. $\overleftrightarrow{FL}$ 9. $\overline{ME}$ 10. $\overrightarrow{AD}$

7 Name the point where $\overleftrightarrow{DE}$ and $\overleftrightarrow{FG}$ intersect.

8 $\overline{RT}$ is divided into two parts by point *S*. Name the two segments.

[R S T]

9 Points *M* and *N* separate $\overline{KL}$ into three parts. Name the segments.

[K M N L]

10 Name the line segments in each of the following figures:

1.   2.   3.

11 $\overline{DE}$ is the sum of what segments?

[D F G E]

12 $\overline{AB}$ is the difference of what segments?

[A B C]

13 What kind of line is shown by each of the following figures:

1.   2.   3.   4.   5.

14 What position is shown by the line in the following figures:

1.   2.   3.   4.   5.

15 **1.** Which of the following are intersecting lines or rays?
2. Which of the following are parallel lines?
3. Which of the following are perpendicular lines or rays?

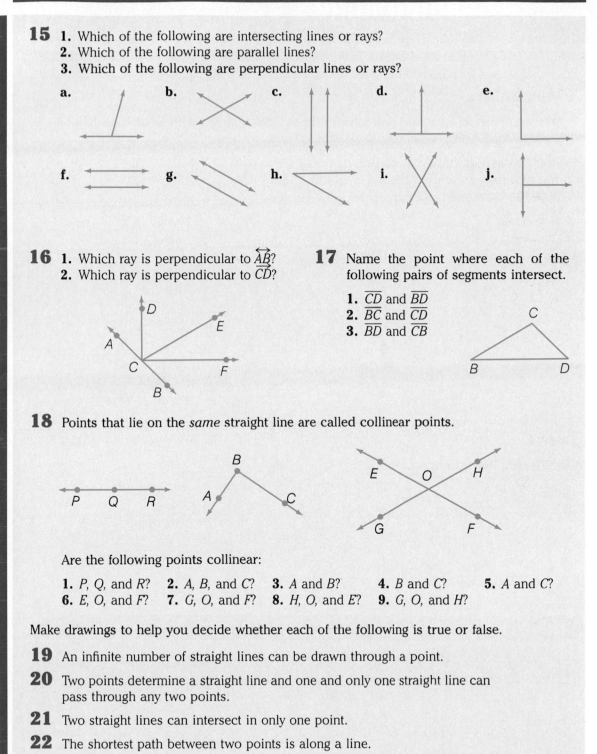

a. **b.** **c.** **d.** **e.**

f. **g.** **h.** **i.** **j.**

16 **1.** Which ray is perpendicular to $\overleftrightarrow{AB}$?
2. Which ray is perpendicular to $\overleftrightarrow{CD}$?

17 Name the point where each of the following pairs of segments intersect.

1. $\overline{CD}$ and $\overline{BD}$
2. $\overline{BC}$ and $\overline{CD}$
3. $\overline{BD}$ and $\overline{CB}$

18 Points that lie on the *same* straight line are called collinear points.

Are the following points collinear:

1. P, Q, and R? **2.** A, B, and C? **3.** A and B? **4.** B and C? **5.** A and C?
6. E, O, and F? **7.** G, O, and F? **8.** H, O, and E? **9.** G, O, and H?

Make drawings to help you decide whether each of the following is true or false.

19 An infinite number of straight lines can be drawn through a point.

20 Two points determine a straight line and one and only one straight line can pass through any two points.

21 Two straight lines can intersect in only one point.

22 The shortest path between two points is along a line.

12-2 Naming Angles

Vocabulary

An **angle** (symbol $\angle$) is the figure formed by two different rays with a common endpoint.

The common endpoint is called the vertex of the angle and the two rays are called the sides of the angle.

side

vertex — side

Angles can be named in several different ways.

angle *A*	angle 1	angle *ABC* or angle *CBA*
$\angle A$	$\angle 1$	$\angle ABC$ or $\angle CBA$
		The middle letter is always the vertex.

EXAMPLE:

Name the angle in three ways.

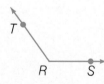

ANSWER: $\angle TRS$, $\angle SRT$, $\angle R$

PRACTICE EXERCISES

1 Name the sides and vertex of the following angle:

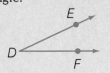

2 Name the following angle in two ways:

F

2

3 Name each of the following angles:

1.

4

2.

B   C

O

3.

N

4.

3

5.

E   F

D

6.

R   T   S

7.

A   K

L

4 Name ∠1, ∠2, and ∠3 in the figure below, using three letters.

E

G

2

1   3

F

H

5 Name each angle of the triangle below in four ways.

M

2

1   3

K   L

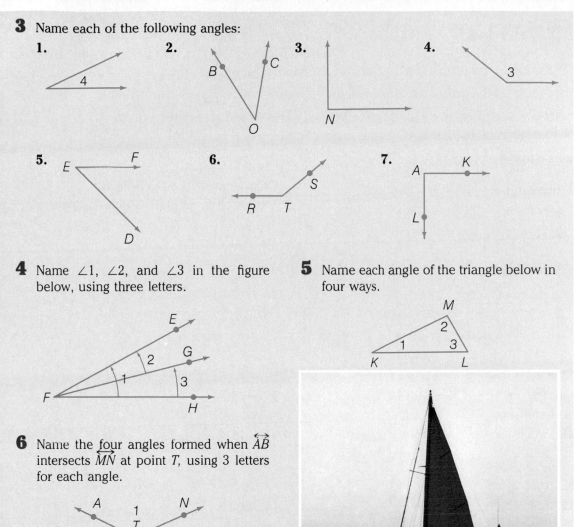

6 Name the four angles formed when $\overleftrightarrow{AB}$ intersects $\overleftrightarrow{MN}$ at point T, using 3 letters for each angle.

A   N

1

T

2   4

3

M   B

7 In the following triangle, find the angle opposite to:

1. Side *PR*   **2.** Side *RQ*   **3.** Side *PQ*

R

P   Q

MEASURE OF ANGLES

A complete rotation of a ray about its fixed endpoint is 360°. A degree (symbol °) is the unit of measure of angles and arcs and is $\frac{1}{360}$ part of a complete rotation. A degree is divided into 60 equal parts called minutes (symbol ′). A minute is divided into 60 equal parts called seconds (symbol ″).

EXAMPLES:

Change 56° to minutes.

$56° = 56 \times 60 = 3,360'$

ANSWER: 3,360′

Change 54,000″ to degrees.

$54,000'' = 54,000 \div 3,600 = 15°$

                        ↑
                    └──60 × 60

ANSWER: 15°

EXERCISES

Change to minutes:

1. 45°   **2.** 32°6′   **3.** 2,940″   **4.** 13°

Change to seconds:

5. 38′   **6.** 5°   **7.** 12′45″   **8.** 2°18′50″

Change to degrees:

9. 180′   **10.** 4,320′   **11.** 10,800″   **12.** 36,000″

Add and simplify:

13. 14° 26′
      52° 19′

14. 63° 27′ 40″
      39° 18′ 20″

15.   44° 21′ 19″
      27° 53′ 24″
    102° 39′ 43″

16. 73° 35′ 45″
      32° 18′ 53″
      65° 5′ 22″

Subtract:

17. 120° 48′
       67° 19′

18. 90°
      38° 45′

19. 78° 6′ 37″
      56° 15′ 20″

20. 180°
      95° 24′ 31″

Multiply and simplify:

21. 24° 13′
         4

22. 12° 32′
         6

23. 28° 15′ 24″
         5

24. 13° 49′ 30″
         15

Divide:

25. 6)42° 18′   **26.** 15)8° 45′   **27.** 4)29° 33′ 48″   **28.** 15)128° 41′ 30″

29. An angle measuring 112° is to be divided into 4 equal angles. What is the measure of each angle?

30. Find the sum of measures of two angles when one angle measures 27° and the other measures 145°

31. Angle C is 3 times the size of angle D. Angle D measures 19° 35′. What is the measure of angle C?

32. One angle measures 50°. A second angle measures 37° 54′. How much greater is the measure of the first angle?

Angles are measured in degrees. There is 360° in a complete rotation. Angles can be classified by the number of degrees or amount of rotation.

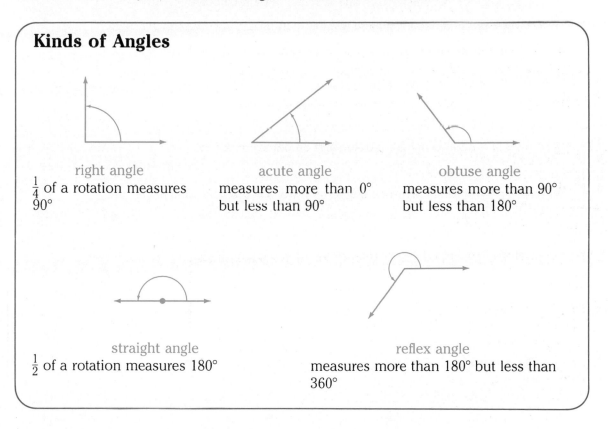

Kinds of Angles

right angle
$\frac{1}{4}$ of a rotation measures 90°

acute angle
measures more than 0° but less than 90°

obtuse angle
measures more than 90° but less than 180°

straight angle
$\frac{1}{2}$ of a rotation measures 180°

reflex angle
measures more than 180° but less than 360°

EXAMPLES:

Classify each angle as acute, right, obtuse, straight, or reflex.

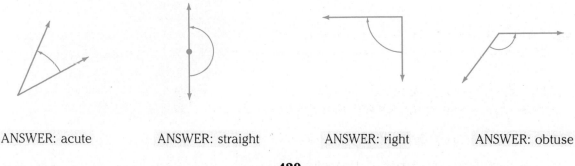

ANSWER: acute   ANSWER: straight   ANSWER: right   ANSWER: obtuse

PRACTICE EXERCISES

1 What unit is used to measure angles? What symbol is written for it?

2 One complete rotation measures _____°.

3 A right angle measures _____° and is _____ of a rotation.

4 The measure of an acute angle is greater than _____° and less than _____°.

5 The measure of an obtuse angle is greater than _____° and less than _____°.

6 A straight angle is _____ of a rotation and measures _____°.

7 Which of the following are measures of an acute angle?
   1. 67°   **2.** 98°   **3.** 89°   **4.** 155°   **5.** 90°

8 Which of the following are measures of an obtuse angle?
   1. 39°   **2.** 126°   **3.** 94°   **4.** 88°   **5.** 191°

9 Indicate by writing corresponding letters which of the following angles are:
   1. right angles   **2.** acute angles   **3.** obtuse angles   **4.** straight angles

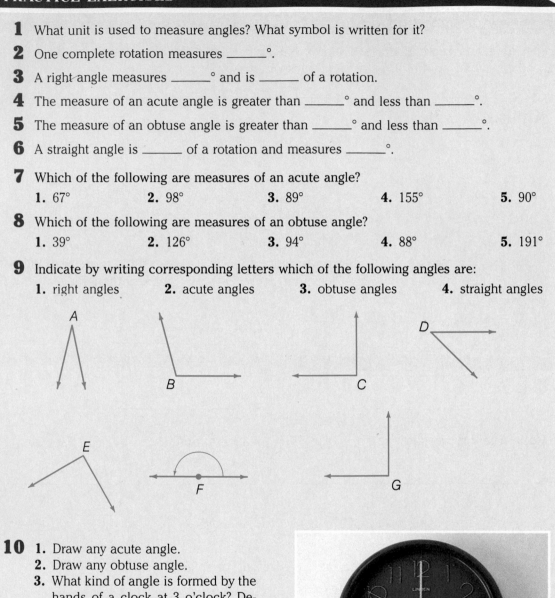

10 **1.** Draw any acute angle.
    2. Draw any obtuse angle.
    3. What kind of angle is formed by the hands of a clock at 3 o'clock? Describe only the smaller of the two angles formed.

12-4 Drawing and Measuring Line Segments

Vocabulary

A **straightedge** and a **ruler** are used to draw and measure segments.
A **compass** is an instrument also used in measuring.

A straightedge is used to draw line segments.
A ruler is a straightedge with calibrated measurements. It is used to measure line segments and can be used to draw them also.

EXAMPLES:

Draw a line segment $1\frac{3}{4}$ in. long.

ANSWER: _____

Measure the length of the segment.

_____

ANSWER: $1\frac{7}{8}$ in.

Use a compass to draw a segment $1\frac{1}{2}$ in. long.

Draw a working line with a straightedge.

Open the compass to measure $1\frac{1}{2}$ in.

With the compass set at $1\frac{1}{2}$ in., mark the length on the working line.

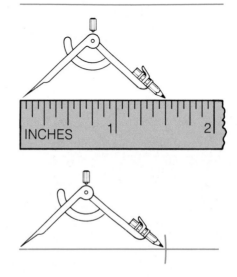

ANSWER: _____

441

PRACTICE EXERCISES

1 Using a ruler, draw line segments having the following lengths:

1. $4\frac{1}{2}$ in.   **2.** $2\frac{3}{4}$ in.   **3.** $3\frac{5}{8}$ in.   **4.** $4\frac{13}{16}$ in.

2 Measure the length of each of the following line segments. Use a ruler for segments 1, 2, and 3, and a compass and ruler for segments 4 and 5.

1. ―――――――――――   **2.** ――――――

3. ―――――――――――   **4.** ―――――――――

5. ――――――――――――

3 Using a metric ruler, draw line segments having the following lengths:

1. 6 cm   **2.** 49 mm   **3.** 13 cm   **4.** 108 mm

4 Draw a line segment 15 centimeters long. Using a compass, mark in succession on this segment lengths of 2.8 centimeters, 4.9 centimeters, and 3.6 centimeters. How long is the remaining segment?

5 The symbol $m\overline{BR}$ is read "the measure of line segment BR" and represents the length of the segment.

1. Find $m\overline{BR}$; $m\overline{AD}$; $m\overline{ST}$; $m\overline{NH}$ **2.** Does $m\overline{AD} = m\overline{NH}$?   **3.** Does $m\overline{ST} = m\overline{BR}$?

6 Draw any line segment EF. Label a point not on $\overline{EF}$ as A. Draw a line segment AG equal in length to $\overline{EF}$.

7 Draw a line segment twice as long as $\overline{BR}$ in exercise 5.

Refresh Your Skills

1. Add: 1-4
53,284
68,596
37,230
81,957
49,866

2. Subtract: 1-5
750,000
169,508

3. Multiply: 1-6
5,280
165

4. Divide: 1-7
1,728)25,920

5. Add: 2-4
6 + .666 + .66

6. Subtract: 2-5
8.5 − .42

7. Multiply: 2-7
.25 × 8.04

8. Divide: 2-8
1.5).75

9. Add: 3-8
$1\frac{2}{3} + \frac{5}{6} + 1\frac{7}{8}$

10. Subtract: 3-9
$16 - 3\frac{7}{10}$

11. Multiply: 3-11
$3\frac{3}{5} \times 5\frac{1}{3}$

12. Divide: 3-12
$20 \div 1\frac{1}{4}$

13. Find 18.4% of $9,500
4-6

14. What percent of $15 is $.30?
4-7

15. $10 is 5% of what amount?
4-8

PRECISION AND ACCURACY

Measurement is never exact—it is only approximate. Precision is the closeness of a measurement to the true measurement. *The smaller the unit of measure, the more precise the measurement.*

The same length measured on each of the following scales shows a measurement of:

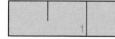 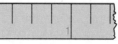  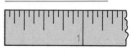

$1\frac{1}{2}''$ unit of measure: $\frac{1}{2}''$ $1\frac{1}{4}''$ unit of measure: $\frac{1}{4}''$ $1\frac{3}{8}''$ unit of measure: $\frac{1}{8}''$ $1\frac{5}{16}''$ unit of measure: $\frac{1}{16}''$

       least precise   most precise

The greatest possible error for a measurement is $\frac{1}{2}$ the unit of measure being used.
The relative error is the ratio of the greatest possible error to the measurement.
The smaller the relative error, the more accurate the measurement.

EXAMPLE:

For the measurement $2\frac{3}{8}$ in., find the precision and accuracy.

ANSWER: The measurement is precise to the nearest $\frac{1}{8}$ in.

     The greatest possible error is $\frac{1}{2} \times \frac{1}{8} = \frac{1}{16}$ in.

     The relative error is the ratio:

$$\frac{1}{16} \text{ to } 2\frac{3}{8} = \frac{1}{16} \div 2\frac{3}{8} = \frac{1}{16} \times \frac{8}{19} = \frac{1}{38}$$

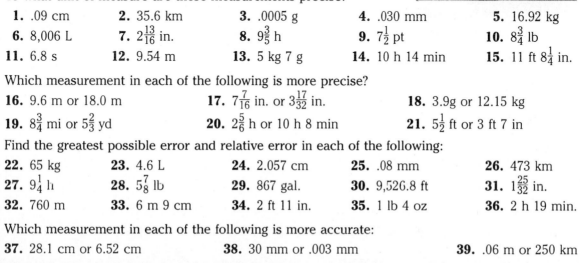

EXERCISES

To what unit of measure are these measurements precise?

1. .09 cm	**2.** 35.6 km	**3.** .0005 g	**4.** .030 mm	**5.** 16.92 kg
6. 8,006 L	**7.** $2\frac{13}{16}$ in.	**8.** $9\frac{3}{5}$ h	**9.** $7\frac{1}{2}$ pt	**10.** $8\frac{3}{4}$ lb
11. 6.8 s	**12.** 9.54 m	**13.** 5 kg 7 g	**14.** 10 h 14 min	**15.** 11 ft $8\frac{1}{4}$ in.

Which measurement in each of the following is more precise?

16. 9.6 m or 18.0 m   **17.** $7\frac{7}{16}$ in. or $3\frac{17}{32}$ in.   **18.** 3.9g or 12.15 kg

19. $8\frac{3}{4}$ mi or $5\frac{2}{3}$ yd   **20.** $2\frac{5}{6}$ h or 10 h 8 min   **21.** $5\frac{1}{2}$ ft or 3 ft 7 in

Find the greatest possible error and relative error in each of the following:

22. 65 kg	**23.** 4.6 L	**24.** 2.057 cm	**25.** .08 mm	**26.** 473 km
27. $9\frac{1}{4}$ h	**28.** $5\frac{7}{8}$ lb	**29.** 867 gal.	**30.** 9,526.8 ft	**31.** $1\frac{25}{32}$ in.
32. 760 m	**33.** 6 m 9 cm	**34.** 2 ft 11 in.	**35.** 1 lb 4 oz	**36.** 2 h 19 min.

Which measurement in each of the following is more accurate:

37. 28.1 cm or 6.52 cm   **38.** 30 mm or .003 mm   **39.** .06 m or 250 km

12-5 Scale

A scale like 1 in. = 8 ft means 1 scale inch represents 8 actual feet. The scale 1:96 means 1 unit of scale represents 96 units in the actual dimensions. The scale 1 in. = 8 ft can also be written as $\frac{1}{8}$ in. = 1 ft or as 1:96, since 8 ft = 96 in.

PROCEDURE

To find the actual distance when the scale and scale distance are known:
  Multiply the scale distance by the scale value of a unit.

EXAMPLE:

If the scale is 1 cm = 30 km, what actual distance is represented by 4.7 cm?

  Scale: 1 cm = 30 km   Scale distance: 4.7 cm = ? km
$$4.7 \times 30 = 141.0$$

ANSWER: 141 km

To find the scale distance when the scale and the actual distance are known:
  Divide the actual distance by the scale value of a unit.

EXAMPLE:

If the scale is 1 in. = 16 ft, how many inches represent 78 ft?

  Scale: 1 in. = 16 ft   Scale distance: ? in. = 78 ft
$$78 \div 16 = 4\frac{7}{8}$$

ANSWER: $4\frac{7}{8}$ in.

To find the scale when the actual and scale distances are known:
  Divide the actual distance by the scale distance.

EXAMPLE:

Find the scale when the actual distance is 60 km and the scale distance is 15 mm.

  $60 \div 15 = 4$ Each scale millimeter represents 4 actual kilometers.

ANSWER: 1 mm = 4 km, or 1:4,000,000

PRACTICE EXERCISES

Find the actual distances using the given scale in each of the following:

1 Scale: 1 mm = 15 mm   **1.** 7 mm   **2.** 13 mm   **3.** 65 mm   **4.** 41 mm

2 Scale: 1 cm = 30 km   **1.** 6 cm   **2.** 23 cm   **3.** 8.5 cm   **4.** 17.1 cm

3 Scale: 1 in. = 48 mi   **1.** 5 in.   **2.** $6\frac{1}{2}$ in.   **3.** $2\frac{3}{4}$ in.   **4.** $3\frac{5}{8}$ in.

4 Scale: $\frac{1}{4}$ in. = 1 ft   **1.** 3 in.   **2.** $4\frac{1}{2}$ in.   **3.** $7\frac{3}{4}$ in.   **4.** $5\frac{9}{16}$ in.

5 Scale: 1:24   **1.** 2 in.   **2.** $6\frac{1}{2}$ in.   **3.** $9\frac{1}{4}$ in.   **4.** $3\frac{7}{8}$ in.

6 Scale: 1:600   **1.** 4 m   **2.** 27 cm   **3.** 9.3 cm   **4.** 154 mm

7 Scale: 1:2,500,000   **1.** 18 mm   **2.** 7 mm   **3.** 1 cm   **4.** 15.6 cm

Find the scale distances, using the given scale in each of the following:

8 Scale: 1 in. = 64 mi   **1.** 192 mi   **2.** 160 mi   **3.** 304 mi   **4.** 372 mi

9 Scale: $\frac{1}{8}$ in. = 1 ft   **1.** 48 ft   **2.** 126 ft   **3.** 29 ft   **4.** 167 ft

10 Scale: $\frac{1}{4}$ in. = 1 ft   **1.** 20 ft   **2.** 3 ft   **3.** 117 ft   **4.** $30\frac{1}{2}$ ft

11 Scale: 1 cm = 25 m   **1.** 75 m   **2.** 325 m   **3.** 110 m   **4.** 42.5 m

12 Scale: 1:2,000,000   **1.** 8 km   **2.** 116 km   **3.** 98 km   **4.** 47 km

Find the scale when:

13 The scale length of 7 mm represents an actual distance of 420 km.

14 The scale length of $6\frac{5}{8}$ in. represents an actual distance of 106 ft.

15 The actual distance of 870 km is represented by the scale length of 5.8 cm.

16 The actual distance of 750 mi. is represented by the scale length of $4\frac{11}{16}$ in.

17 The scale length of 10.7 cm represents an actual distance of 192.6 m.

18 The scale length of $3\frac{7}{8}$ in. represents an actual distance of 186 ft.

19 Using the scale 1 cm = 100 km, draw line segments representing:

   1. 300 km   **2.** 60 km   **3.** 140 km   **4.** 290 km   **5.** 450 km

20 Using the scale $\frac{1}{8}$ inch = 1 foot, draw line segments representing:

   1. 24 ft   **2.** 10 ft   **3.** 19 ft   **4.** 35 ft   **5.** $61\frac{1}{2}$ ft

21 What are the actual dimensions of a warehouse floor if plans drawn to the scale of 1:200 shows dimensions of 12 cm by 9 cm?

22 Draw a floor plan of a schoolroom 33 feet long and 18 feet wide, using the scale 1 inch = 12 feet.

23 Using the scale of *miles*

0 12 24 36 48

find the distance represented by each of the following line segments:

1. _____   2. _____
3. _____   4. _____
5. _____

24 Using the scale of *kilometers*

0   10   20   30   40   50

find the distance represented by each of the following line segments:

1. _____   2. _____
3. _____   4. _____
5. _____

25 If the distance from *A* to *B* is 56 miles, what is the distance from *G* to *H*?

_____   _____
A   B   G   H

26 If the distance from *D* to *E* is 70 kilometers, what is the distance from *M* to *N*?

_____   _____
D   E   M   N

Computer Activities

The computer program below computes the *actual distance* when the *scale* and *scale distance* are known.

```
1Ø PRINT "SCALE: 1 CM = ? KM"

2Ø PRINT "HOW MANY KILOMETERS (KM)";

3Ø INPUT K

4Ø PRINT "WHAT IS THE SCALE DISTANCE
(IN CM)";

5Ø INPUT D

6Ø PRINT "THE ACTUAL DISTANCE IS:"

7Ø PRINT K*D;" KILOMETERS"
```

a) Enter the program and then RUN it to make sure it works

b) Modify the program so that
   (1) the scale is—
       SCALE: 1 IN. = ? FT
   (2) the scale distance is expressed in *inches*.
   (3) the actual distance is expressed in *feet*.

Vocabulary

A geometric **plane** is an endless flat surface consisting of a collection of points.
Coplanar points are points that are in the same plane.
Coplanar lines are lines that are in the same plane.
A **polygon** is a simple closed curve made up of line segments called *sides*.
A **vertex** is a point where a pair of intersecting sides of a polygon meet.
A **diagonal** is a line segment connecting two nonadjacent vertices of a polygon.
A **regular polygon** is a polygon with all sides equal in length and all angles equal in measure.
A **circle** is a collection of points in a plane which are equidistant (the same distance) from a fixed point in the plane called the *center*.

A plane extends indefinitely far in any direction. We represent and name a plane three ways:

plane *EFG*   By three capital letters which name three points not on the same line in a plane.

plane *RS*   By two letters written outside opposite corners.

plane *Q*   By one letter written inside a corner.

A, *B*, *C*, and *D* are coplanar points.
$\overleftrightarrow{AB}$ and $\overleftrightarrow{CD}$ are coplanar lines.

Types of Polygons			
Polygon	**Number of Sides**	**Polygon**	**Number of Sides**
Triangle	3	Septagon (Heptagon)	7
Quadrilateral	4	Octagon	8
Pentagon	5	Decagon	10
Hexagon	6	Dodecagon	12

Triangles can be classified by the length of the sides or by the measure of the angles.

Equilateral Triangle
Three equal sides.

Isosceles Triangle
Two equal sides.

Scalene Triangle
No equal sides.

447

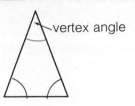

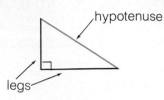

Equiangular Triangle
Three equal angles.
The base is generally the side on which a triangle rests.

Isosceles Triangle
Two equal base angles.
The vertex angle is the angle formed by the two equal sides.

Right Triangle
One right angle
The hypotenuse is the side opposite the right angle. The other two sides are called legs.

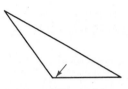

Obtuse Triangle
One obtuse angle.

Acute Triangle
All acute angles.

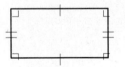

When working with triangles, there are three (3) line segments that are commonly used: An altitude is a perpendicular segment from any vertex to the opposite side or extension of that side.
A median is a segment connecting any vertex to the midpoint of the opposite side. An angle bisector is a segment that is drawn from any vertex to the opposite side and that bisects the angle at that vertex.

Quadrilaterals can be described by their sides and their relation to each other, and by their angles.

Rectangle
2 pairs of opposite sides are equal.
4 right angles.

Square
4 equal sides.
4 right angles.

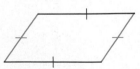

Parallelogram
2 pairs of opposite sides equal and parallel.

Trapezoid
1 pair of opposite sides parallel.

The square is a special rectangle. The rectangle and square are special parallelograms.

A circle has various line segments and curves associated with it.

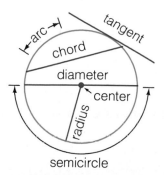

A radius is a segment with one endpoint on the circle and the other at the center of the circle.

A chord is a segment that has both its endpoints on the circle.

A diameter is a chord that passes through the center of the circle.

An arc is part of a circle.

A semicircle is an arc that is one-half a circle.

A tangent to a circle is a line that has one and only one point in common with the circle.

PRACTICE EXERCISES

1 Name each of the following planes:

1. •A •C •B

2. •P •R •Q •S

3. M L

4. T

2 What is the name of each of the following figures?

1.   **2.**   **3.**   **4.**

5.   **6.**

7.   **8.**

9.   **10.**

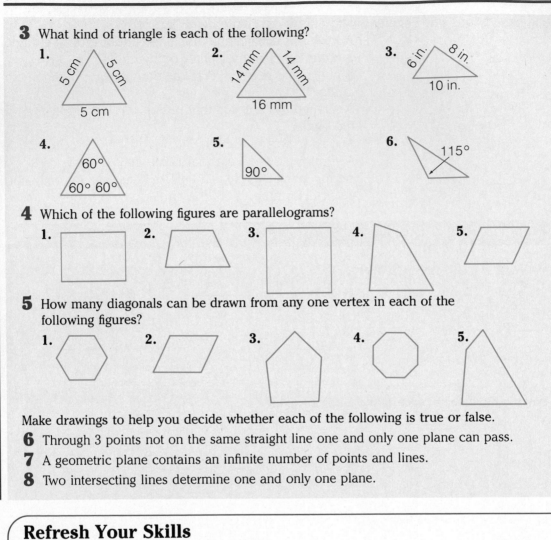

3 What kind of triangle is each of the following?

1. 5 cm, 5 cm, 5 cm

2. 14 mm, 14 mm, 16 mm

3. 6 in., 8 in., 10 in.

4. 60°, 60° 60°

5. 90°

6. 115°

4 Which of the following figures are parallelograms?

1.   2.   3.   4.   5.

5 How many diagonals can be drawn from any one vertex in each of the following figures?

1.   2.   3.   4.   5.

Make drawings to help you decide whether each of the following is true or false.

6 Through 3 points not on the same straight line one and only one plane can pass.

7 A geometric plane contains an infinite number of points and lines.

8 Two intersecting lines determine one and only one plane.

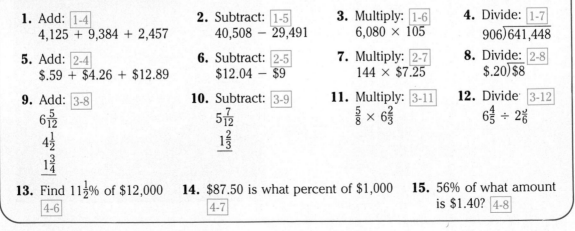

Refresh Your Skills

1. Add: 1-4
4,125 + 9,384 + 2,457

2. Subtract: 1-5
40,508 − 29,491

3. Multiply: 1-6
6,080 × 105

4. Divide: 1-7
906)641,448

5. Add: 2-4
$.59 + $4.26 + $12.89

6. Subtract: 2-5
$12.04 − $9

7. Multiply: 2-7
144 × $7.25

8. Divide: 2-8
$.20)$8

9. Add: 3-8
$6\frac{5}{12}$
$4\frac{1}{2}$
$1\frac{3}{4}$

10. Subtract: 3-9
$5\frac{7}{12}$
$1\frac{2}{3}$

11. Multiply: 3-11
$\frac{5}{8} \times 6\frac{2}{3}$

12. Divide 3-12
$6\frac{4}{5} \div 2\frac{5}{6}$

13. Find $11\frac{1}{2}$% of $12,000
4-6

14. $87.50 is what percent of $1,000
4-7

15. 56% of what amount is $1.40? 4-8

12-7 Geometric Figures—Solid

Vocabulary

Space is the infinite set of all points.

A **polyhedron** is a closed geometric figure consisting of four or more polygons and their interiors, all in different planes.

The **faces** of a polyhedron are the polygons and their interiors.

The **edges** of a polyhedron are line segments where the faces intersect.

The **vertices** of a polyhedron are points where the edges intersect.

The length, width, and height of space are endless.

Common polyhedra are the rectangular solid (right rectangular prism), the cube, and the pyramid.

Rectangular Solid
Six rectangular faces.

Cube
Six square faces.
All edges are equal in length.

Pyramid
Base is a polygon.
Triangular faces meet in a common vertex.

The relationship among the number of faces, edges, and vertices of a polyhedron is expressed by Euler's formula, $F + V - E = 2$. This formula tells us that the number of faces (F) plus the number of vertices (V) minus the number of edges (E) is equal to two for *any* polyhedron.

Other solid geometric figures are the cylinder, sphere, and cone.

A cylinder has two equal and parallel circular bases and a lateral curved surface.

A sphere has a curved surface in space on which every point is equidistant from a fixed point.

A cone has a circular base and a curved surface that comes to a point at the vertex.

451

PRACTICE EXERCISES

1 Classify each of the following figures.

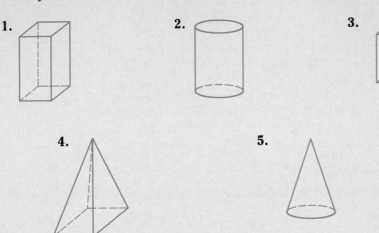

1.   2.   3.

4.   5.   6.

2 1. How many faces (*F*) does a rectangular solid have?
2. How many vertices (*V*)?
3. How many edges (*E*)?
4. Does $F + V - E = 2$?

3 1. How many faces (*F*) does a pyramid with a square base have?
2. How many vertices (*V*)?
3. How many edges (*E*)?
4. Does $F + V - E = 2$?

Make drawings to help you decide whether each of the following is true or false.

4 When two different planes intersect, their intersection is a straight line.

5 The intersection of a plane and a line not in the plane is one and only one point.

6 A line that is perpendicular to each of two intersecting lines at their point of intersection is perpendicular to the plane in which these lines lie.

Vocabulary

A **protractor** is an instrument used to measure and draw angles.

The measure of an angle ($m\angle$) does not depend on the lengths of the sides.

PROCEDURE

To measure an angle:
(1) Place the protractor on the angle so that its center is at the vertex and its zero line lies along a ray.
(2) Read the measure of the angle on the scale which has zero on the side of the angle.

EXAMPLE: Measure $\angle RST$.

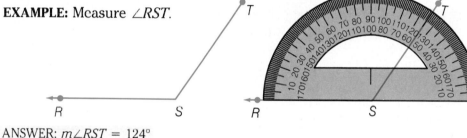

ANSWER: $m\angle RST = 124°$

PROCEDURE

To draw an angle:
(1) Draw a ray to represent one side of the angle.
(2) Place the protractor so that its straight edge falls on this ray and its center mark is on the vertex of the angle.
(3) Use the scale whose zero line lies along the ray and mark the correct number of degrees.
(4) Remove the protractor and draw a ray from the vertex through the dot.

EXAMPLE:

Draw an angle of 75°.

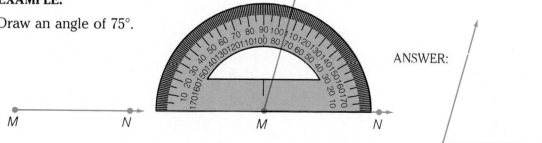

ANSWER:

PRACTICE EXERCISES

1 Estimate the size of each of the following angles. Then measure each angle with a protractor.

1.   **2.**   **3.**

4.   **5.**   **6.**

2 **1.** Which of the following angles is smaller?
   2. The size of an angle does not depend upon
      the length of the _____ .

r   *s*

3 How many degrees are in the angle formed by the hands of a clock:

   1. at 5 o'clock?   **2.** at 3 o'clock?
   3. at 2 o'clock?   **4.** at 6 o'clock?

4 Through how many degrees does the minute hand of a clock turn in:

   1. 30 minutes?   **2.** 10 minutes?   **3.** 15 minutes?
   4. 25 minutes?   **5.** 40 minutes?   **6.** 1 hour?

For each of the following, first draw a ray with the endpoint either on the left or on the right as required, then with a protractor draw the angle of the given measure.

5 Left endpoint as the vertex:
   1. 30° **2.** 75° **3.** 83° **4.** 120° **5.** 270° **6.** 145° **7.** 100° **8.** 58° **9.** 225° **10.** 305°

6 Right endpoint as the vertex:
   1. 80° **2.** 45° **3.** 9° **4.** 165° **5.** 330° **6.** 25° **7.** 124° **8.** 67° **9.** 96° **10.** 200°

7 With a protractor, measure each of the following angles, then draw an angle equal to it.

   1.   **2.** *B*   **3.** *C*

   A

With a protractor draw:
8 a right angle
9 a straight angle

Draw any triangle. Measure its three angles.
Find the sum of the measures of these three angles.
Draw a second triangle. Find the sum of the measures of its three angles.
Find the sum of the measures of the three angles of a third triangle.
   You should find that the sum of the measures of the angles in each of the three triangles is 180°.

The sum of the measures of the angles of any triangle is 180°.

PROCEDURE

To find the measure of an angle of a triangle given the measures of the other two angles:
(1) Add the two given measures.
(2) Subtract the sum from 180°.

EXAMPLE:

Find the measure of the third angle of a triangle if the other two angles have measures of 33° and 101°.

$$33° + 101° = 134° \quad 180° - 134° = 46°$$

ANSWER: 46°

PRACTICE EXERCISES

1 In each of the following, find the measure of the third angle of a triangle when the other two angles have measures of:
1. 29° and 87°   **2.** 63° and 72°   **3.** 106° and 51°
4. 90° and 45°   **5.** 18° and 121°   **6.** 133° and 6°

2 What is the measure of each angle of an equiangular triangle?

3 If each of the equal angles of an isosceles triangle measures 59°, find the measure of the third angle.

4 If the vertex angle of an isosceles triangle measures 70°, find the measure of each of the other two angles.

5 Find the measure of the third angle in each of the following:

1.

48°
90° ?

2.

88°
42° ?

3.

?
128°
15°

12-10 Sum of the Measures of the Angles of Other Polygons

Draw a parallelogram and a trapezoid. Measure the four angles of each. Find the sum of the measures of the angles of the parallelogram. Of the trapezoid.

Draw any other quadrilateral. Find the sum of the measures of the angles of the quadrilateral.

You should find that the sum of the measures of the angles for each of the three figures is 360°.

> The sum of the measures of the angles of any quadrilateral is 360°.

Draw a pentagon and a hexagon. What is the sum of the measures of the angles of each of these polygons? Check whether your sum of measures of the angles of each of these figures matches the angle measure determined by substituting the number of sides in the polygon for n in the expression $180(n - 2)$ and performing the required operations.

PROCEDURE

To find the sum of the measures of the angles of any polygon of n sides:
(1) Subtract 2 from the number of sides (n).
(2) Multiply this difference by 180°.

$$180°(n - 2)$$

EXAMPLE:

Find the sum of the measures of the angles of an octagon.

$180°(n - 2) = 180°(8 - 2) = 180°(6) = 1,080°$

ANSWER: 1,080°

PRACTICE EXERCISES

1 What is the sum of the measures of the angles of a rectangle? Of a square?

2 In each of the following, find the measure of the fourth angle of a quadrilateral when the other three angles measure:

    1. 45°, 85°, and 97°   **2.** 119°, 90°, and 90°   **3.** 76°, 102°, and 161°

3 The opposite angles of a parallelogram are equal. If one angle measures 65°, find the measures of the other three angles.

4 Three angles of a trapezoid measure 90°, 90°, and 108°. What does the fourth angle measure?

5 Find the number of degrees in each of the following polygons:

    1. heptagon   **2.** decagon   **3.** dodecagon

Vocabulary

Complementary angles are two angles whose sum of measures is 90°.
Supplementary angles are two angles whose sum of measures is 180°.

∠A and ∠B are complementary.   ∠S and ∠T are supplementary.

PROCEDURE

To find the complement of an angle, subtract its measure from 90°.
To find the supplement of an angle, subtract its measure from 180°.

EXAMPLES:

Find the complement of an
angle whose measure is 24°.

$$90° - 24° = 66°$$

ANSWER: 66°

Find the supplement of
an angle whose measure is 72°.

$$180° - 72° = 108°$$

ANSWER: 108°

Draw two intersecting lines, forming four angles. Measure a pair of angles
that are directly opposite to each other. Are the measures equal? Measure the
other pair of opposite angles. Are the measures equal?

You should find that the angles opposite each other are equal in measure.

> When two straight lines intersect, the opposite (vertical) angles are
> equal in measure.

EXAMPLE:

The measure of ∠1 is 115° and the measure of ∠2 is 65°. Find the
measures of angles 3 and 4.

$m\angle 3 = m\angle 2 = 65°$ because angles 2 and 3 are **vertical** angles.
$m\angle 4 = m\angle 1 = 115°$ because angles 1 and 4 are **vertical** angles.

ANSWER: $m\angle 3 = 65°, m\angle 4 = 115°$

By extending any side of a triangle, an exterior angle is formed.

Angle 4 is an exterior angle of triangle *ABC* at the right.

Measure angles 1, 2, and 4.

> The measure of an exterior angle of a triangle is equal to the sum of the measures of the two opposite interior angles.

EXAMPLE:

Find the measure of $\angle 4$ in the drawing above if $m\angle 1 = 35°$ and $m\angle 2 = 83°$.

$m\angle 4 = m\angle 1 + m\angle 2 = 35° + 83° = 118°$

ANSWER: 118°

PRACTICE EXERCISES

1 Which of the following pairs of angles are complementary?
1. $m\angle E = 50°$, $m\angle G = 40°$ 2. $m\angle R = 37°$, $m\angle T = 53°$
3. $m\angle 6 = 22°$, $m\angle 3 = 68°$ 4. $m\angle a = 39°$, $m\angle b = 41°$
5. $m\angle D = 80°$, $m\angle H = 100°$ 6. $m\angle 4 = 76°$, $m\angle 5 = 14°$

2 Find the measure of the angle that is the complement of each of the following angles.
1. $m\angle 2 = 21°$ 2. $m\angle L = 87°$ 3. $m\angle B = 6°$
4. $m\angle 8 = 33°$ 5. $m\angle x = 43° \, 18'$ 6. $m\angle P = 68° \, 51' \, 15''$

3 Angle *ABC* is a right angle.
1. If $m\angle 2 = 26°$, find the measure of $\angle 1$.
2. If $m\angle 1 = 53°$, find the measure of $\angle 2$.
3. If $m\angle 2 = 14°$, find the measure of $\angle 1$.
4. If $m\angle 1 = 67°$, find the measure of $\angle 1$.
5. If $m\angle 2 = 42°30'$, find the measure of $\angle 1$.

4 In the figure $\overrightarrow{FB} \perp \overrightarrow{FD}$ and $\overrightarrow{FC} \perp \overleftrightarrow{AE}$.
1. If $m\angle 4 = 55°$, find the measure of $\angle 1$. Of $\angle 2$. Of $\angle 3$.
2. If $m\angle 1 = 61°$, find the measure of $\angle 2$. Of $\angle 3$. Of $\angle 4$.
3. If $m\angle 3 = 74°$, find the measure of $\angle 1$. Of $\angle 2$. Of $\angle 4$.
4. If $m\angle 2 = 19°$, find the measure of $\angle 3$. Of $\angle 1$. Of $\angle 4$.
5. If $m\angle 4 = 25° \, 45'$, find the measure of $\angle 3$. Of $\angle 2$. Of $\angle 1$.

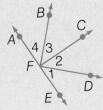

5 Why is the complement of an acute angle also an acute angle?

6 Which of the following pairs of angles are supplementary?

    1. $m\angle A = 134°, m\angle N = 46°$ **2.** $m\angle 3 = 85°, m\angle 4 = 105°$
    3. $m\angle B = 79°, m\angle D = 91°$ **4.** $m\angle c = 127°, m\angle e = 53°$
    5. $m\angle T = 32°, m\angle F = 148°$ **6.** $m\angle G = 63°, m\angle K = 27°$

7 Find the measure of the angle that is the supplement of each of the following angles.

    1. $m\angle H - 44°$ **2.** $m\angle a = 117°$
    3. $m\angle 5 = 89°$ **4.** $m\angle R = 8°$
    5. $m\angle Q = 96° \ 20'$ **6.** $m\angle E = 170° \ 42' \ 59''$

8 Measure $\angle CDF$ and $\angle EDF$. Is the sum of their measures 180°? Do you see that *when one straight line meets another, the adjacent angles, which have the same vertex and a common side, are supplementary?*

    1. What is the measure of $\angle CDF$ when $m\angle EDF = 32°$?
    2. Find the measure of $\angle EDF$ when $m\angle CDF = 116°$.
    3. Find the measure of $\angle CDF$ when $m\angle EDF = 90°$.
    4. What is the measure of $\angle EDF$ when $m\angle CDF = 145°$?
    5. Find the measure of CDF when $m\angle EDF = 76° \ 30'$.

9 Is the supplement of an obtuse angle also an obtuse angle? Explain your answer.

10 In the drawing at the right, what angle is opposite to $\angle 1$? Does $\angle 1 = \angle 3$? What angle is opposite to $\angle 4$? Does $\angle 4 = \angle 2$?

11 If $m\angle 3 = 68°$, what is the measure of
    1. $\angle 1$?   **2.** $\angle 2$?   **3.** $\angle 4$?

12 If $m\angle 4 = 110°$, what is the measure of
    1. $\angle 2$?   **2.** $\angle 1$?   **3.** $\angle 3$?

13 If $m\angle 1 = 17°$, what is the measure of
    1. $\angle 4$?   **2.** $\angle 2$?   **3.** $\angle 3$?

14 If $m\angle 2 = 145°$, what is the measure of
    1. $\angle 3$?   **2.** $\angle 4$?   **3.** $\angle 1$?

15 If $m\angle 1 = 42°$ and $m\angle 2 = 69°$, find the measure of $\angle 4$. Of $\angle 3$.

16 If $m\angle 3 = 53°$ and $m\angle 1 = 75°$, find the measure of $\angle 2$. Of $\angle 4$.

17 If $m\angle 2 = 38°$ and $m\angle 3 = 81°$, find the measure of $\angle 1$. Of $\angle 4$.

18 If $m\angle 4 = 110°$ and $m\angle 1 = 26°$, find the measure of $\angle 2$. Of $\angle 3$.

19 If $m\angle 2 - 45°$ and $m\angle 4 = 131°$, find the measure of $\angle 3$. Of $\angle 1$.

20 Can an exterior angle of a triangle have the same measure as one of the angles of the triangle? As one of the opposite interior angles of the triangle? Explain your answers.

Vocabulary

Alternate interior angles and **corresponding angles** are pairs of angles formed when parallel lines are cut by a line called a **transversal**.

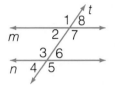

m is parallel to *n*, and *m* and *n* are cut by transversal *t* (a line that intersects two or more lines, each at a different point).
Angles 1 and 3 are corresponding angles.
Angles 2 and 6 are alternate interior angles.

EXAMPLES:

Name other pairs of corresponding angles in the figure above.
ANSWER: ∠2 and ∠4; ∠6 and ∠8; ∠5 and ∠7

Name another pair of alternate interior angles in the figure above.
ANSWER: ∠3 and ∠7

When parallel lines are cut by a transversal, the measures of corresponding angles are equal, and the measures of alternate interior angles are equal.

When lines are cut by a transversal and the measures of a pair of corresponding or alternate interior angles are equal, the lines are parallel.

EXAMPLE:

Are *m* and *n* parallel?

m   35°
35°
n

ANSWER: Yes, the measures of a pair of alternate interior angles are equal.

PRACTICE EXERCISES

Use the figure below for all problems. Complete each statement as required in problems 1 through 6.

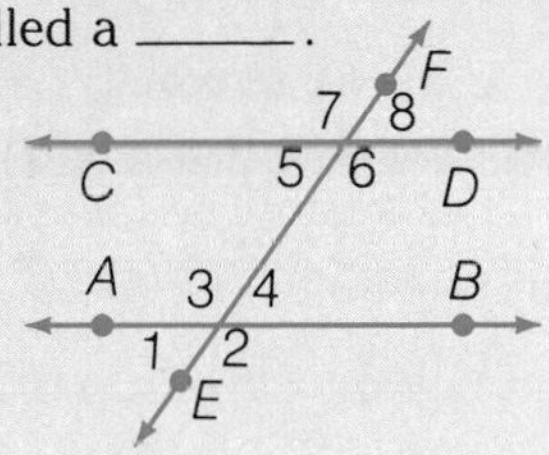

1 1. If lines *AB* and *CD* are cut by a third line *EF*, the line *EF* is called a _____ .
  2. ∠6 and ∠3 are _____ angles.
  3. ∠5 and ∠4 are _____ angles.
  4. ∠5 and ∠1 are _____ angles.
  5. ∠8 and ∠4 are _____ angles.

2 If two parallel lines are cut by a transversal, the alternate-interior angles are _____ . Therefore, ∠3 = _____ , ∠5 = _____ .

3 If two parallel lines are cut by a transversal, the corresponding angles are _____ .
Therefore, ∠2 = _____ , ∠7 = _____ , ∠1 = _____ , ∠8 = _____ .

4 What is the sum of the measures of ∠5 and ∠7?
They are a pair of _____ angles.

5 What is the sum of the measures of ∠1 and ∠2?
They are a pair of _____ angles.

6 ∠5 and ∠8 are a pair of _____ angles.

7 If line *AB* is parallel to line *CD* in the figure above,
  1. Show that ∠2 = ∠7.
  2. Show that ∠1 and ∠6 are supplementary angles.
  3. Show that ∠8 = ∠1.
  4. Show that ∠3 and ∠5 are supplementary angles.
  5. Show that ∠8 and ∠2 are supplementary angles.

8 If line *AB* is parallel to line *CD* in the figure above, what is the measure of:
  1. ∠6 if *m*∠2 = 130°?   2. ∠7 if *m*∠6 = 106°?
  3. ∠5 if *m*∠4 = 75°?   4. ∠2 if *m*∠7 = 93°?
  5. ∠3 if *m*∠8 = 47°?   6. ∠8 if *m*∠1 = 64°?
  7. ∠2 if *m*∠5 = 83°?   8. ∠4 if *m*∠6 = 115°?
  9. all other angles when *m*∠8 = 48°?   10. all other angles when *m*∠3 = 125°?

9 If two lines are cut by a transversal making a pair of corresponding angles or a pair of alternate interior angles equal, the lines are _____ .

10 Are lines *AB* and *CD* parallel in the figure above when:
  1. *m*∠1 = 70° and *m*∠5 = 70°?   2. *m*∠7 = 120° and *m*∠3 = 120°?
  3. *m*∠7 = 135° and *m*∠4 = 45°?   4. *m*∠2 = 108° and *m*∠8 = 82°?
  5. *m*∠8 = 52° and *m*∠1 = 52°?   6. *m*∠4 = 81° and *m*∠6 = 99°?
  7. *m*∠1 = 89° and *m*∠6 = 89°?   8. *m*∠5 = 69° and *m*∠2 = 111°?
  9. *m*∠3 = 94° and *m*∠7 = 86°?

12-13 Congruent Triangles

Vocabulary

Congruent triangles are triangles whose corresponding sides are equal in length and whose corresponding angles are equal in size.

The symbol ≅ means "is congruent to."

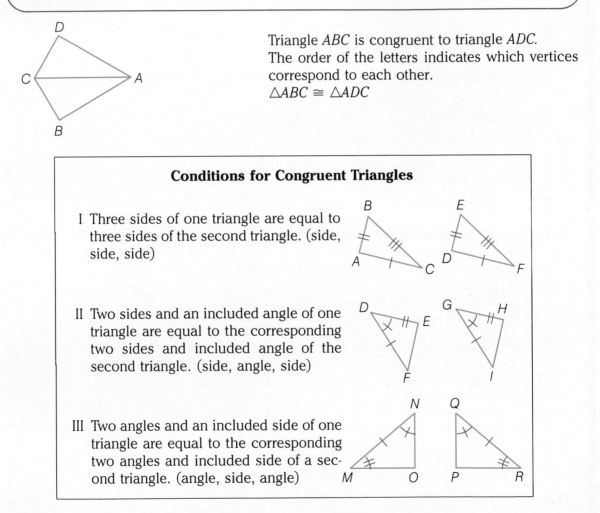

Triangle *ABC* is congruent to triangle *ADC*.
The order of the letters indicates which vertices correspond to each other.
△*ABC* ≅ △*ADC*

Conditions for Congruent Triangles

I Three sides of one triangle are equal to three sides of the second triangle. (side, side, side)

II Two sides and an included angle of one triangle are equal to the corresponding two sides and included angle of the second triangle. (side, angle, side)

III Two angles and an included side of one triangle are equal to the corresponding two angles and included side of a second triangle. (angle, side, angle)

PROCEDURE

To determine if two triangles are congruent:
(1) Compare the sides and angles of one triangle to those of a second.
(2) If one of the conditions above exists, the triangles are congruent.

EXAMPLES:

State why each pair of triangles is congruent.

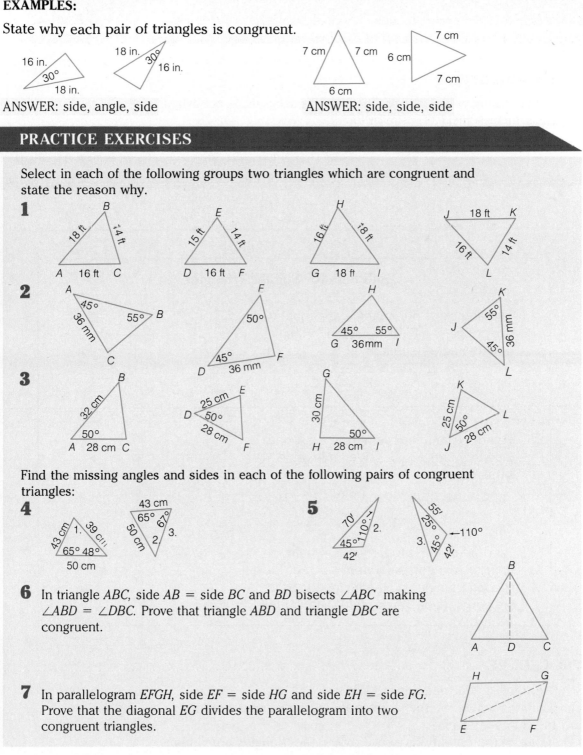

16 in.
18 in.
30°
30°
16 in.
30°
18 in.

ANSWER: side, angle, side

7 cm
7 cm
6 cm
6 cm
7 cm
7 cm

ANSWER: side, side, side

PRACTICE EXERCISES

Select in each of the following groups two triangles which are congruent and state the reason why.

1

2

3

Find the missing angles and sides in each of the following pairs of congruent triangles:

4

5

6 In triangle *ABC*, side *AB* = side *BC* and *BD* bisects ∠*ABC* making ∠*ABD* = ∠*DBC*. Prove that triangle *ABD* and triangle *DBC* are congruent.

7 In parallelogram *EFGH*, side *EF* = side *HG* and side *EH* = side *FG*. Prove that the diagonal *EG* divides the parallelogram into two congruent triangles.

Vocabulary

Similar triangles are triangles that have the same shape but differ in size. The symbol ~ means "is similar to."

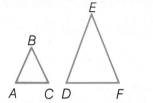

Triangle *ABC* is similar to △*DEF*. The order of the letters indicates the corresponding sides and angles.

△*ABC* ~ △*DEF*

Conditions for Similar Triangles

I Two angles of one triangle are equal to two angles of the second triangle. (angle, angle—similarity)

II The three ratios of pairs of corresponding sides are equal. (side, side, side—similarity)

III Two sides are proportional (equal ratios) to two corresponding sides of the second triangle and the included angles are equal. (side, angle, side—similarity)

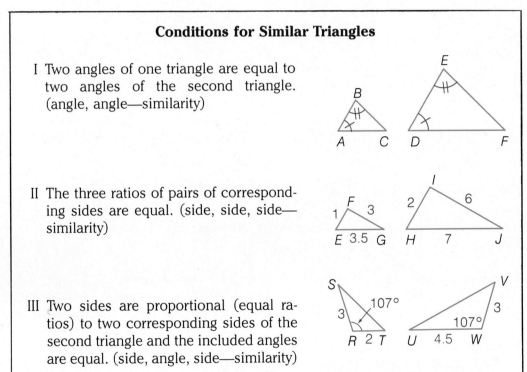

PROCEDURE

To determine if two triangles are similar:
(1) Find the ratios of the corresponding sides and compare any angle measures known.
(2) If one of the conditions above exists, the triangles are similar.

EXAMPLES:

State why the triangles of each pair are similar.

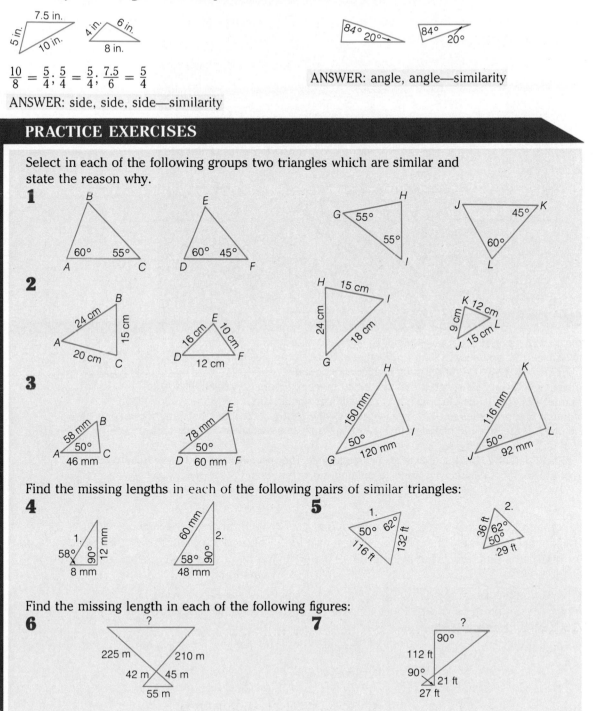

$\dfrac{10}{8} = \dfrac{5}{4}; \dfrac{5}{4} = \dfrac{5}{4}; \dfrac{7.5}{6} = \dfrac{5}{4}$

ANSWER: side, side, side—similarity

ANSWER: angle, angle—similarity

PRACTICE EXERCISES

Select in each of the following groups two triangles which are similar and state the reason why.

1

2

3

Find the missing lengths in each of the following pairs of similar triangles:

4   **5**

Find the missing length in each of the following figures:

6   **7**

Photographers pride themselves in taking excellent photographs that appear in advertisements and books. Their knowledge of types of cameras and lenses and the amounts of light to use makes such high quality possible.

Many photographers specialize in one kind of picture. Portrait photographers take pictures of people, often for yearbooks or weddings. Industrial photographers photograph products and processes to be shown in catalogs and reports. Scientific photographers take pictures of research and new developments. They may use special equipment such as X-ray machines or microscopes.

Almost half of all photographers are self-employed. They usually have no formal training in photography. A person who wants to be a photographer should get as much experience as possible.

1. A photographer takes a picture of a building. One centimeter on the picture represents 5 m on the building. What is the scale of the picture?

2. A photographer takes a picture of a flower. Three inches on the picture represents one inch on the flower. What is the scale of the picture?

A photographer uses a variety of lenses. A wide angle lens "sees" a wide field of view. A telephoto lens can pick out a small detail of a scene. Measure the field of view (angle) of each of these lenses.

3. wide angle

4. normal

5. telephoto

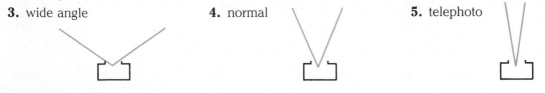

CHAPTER REVIEW

Vocabulary

acute angle p. 439	degree p. 438	minute p. 438	relative error p. 443
acute triangle p. 448	diagonal p. 447	obtuse angle p. 439	right angle p. 439
alternate interior angles p. 460	diameter p. 449	obtuse triangle p. 448	right triangle p. 448
altitude p. 448	edge p. 451	parallel lines p. 432	scale p. 444
angle p. 436	equiangular triangle p. 448	parallelogram p. 448	second p. 438
angle bisector p. 448	Euler's formula p. 451	perpendicular lines p. 432	segment p. 432
arc p. 449	exterior angle p. 438	point p. 432	semicircle p. 449
base p. 448	face p. 451	polygon p. 447	similar triangles p. 464
chord p. 449	geometric plane p. 447	polyhedron p. 451	space p. 451
circle p. 447	greatest possible error p. 443	precision p. 443	sphere p. 451
compass p. 441	half-line p. 432	protractor p. 453	straight angle p. 439
cone p. 451	hypotenuse p. 448	pyramid p. 451	straightedge p. 441
complementary angles p. 457	intersecting lines p. 432	radius p. 449	supplementary angles p. 457
congruent triangles p. 462	interval p. 432	ray p. 432	tangent p. 449
coplanar lines p. 447	isosceles triangle p. 448	rectangle p. 448	transversal p. 460
coplanar points p. 447	leg p. 448	rectangular solid p. 451	trapezoid p. 448
corresponding angles p. 460	line p. 432	reflex angle p. 439	vertex p. 447
cube p. 451	line segment p. 432	regular polygon p. 447	vertex angle p. 436
cylinder p. 451	median p. 448		

The numerals in boxes indicate Lessons where help may be found.

1. Read, or write in words, each of the following: $\overleftrightarrow{RT}$; $\overrightarrow{OM}$; $\overrightarrow{DG}$ ⬚12-1

Draw a representation of: ⬚12-1

2. perpendicular lines **3.** parallel lines **4.** a horizontal line

5. Name the following angle in three ways: ⬚12-2

6. Which is the measure of an actue angle? **a.** 98° **b.** 90° **c.** 89° ⬚12-3

7. If the scale is 1 cm = 100 km, what actual distance is represented by 5.7 cm? ⬚12-5

8. What is the figure ▱ called? ⬚12-7 **9.** Draw an angle measuring 110°. ⬚12-8

10. Find the measure of the third angle of a triangle when the other two measure 67° and 59°. ⬚12-9

11. Find the complement and the supplement of an angle measuring 41°. ⬚12-11

12. *CD* and *EF* intersect forming angles 1, 2, 3, and 4. If $m\angle 2 = 105°$, what is the measure of ∠1? ∠3? ∠4? ⬚12-11

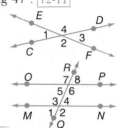

13. Parallel lines *MN* and *OP* are cut by transversal *OR*. If $m\angle 5 = 79°$, what is the measure of ∠4? ∠6? ∠3? ∠2? ∠8? ∠1? ∠7? ⬚12-12

14. State three conditions why two triangles can be congruent. ⬚12-13

15. State three conditions why two triangles can be similar. ⬚12-14

COMPETENCY CHECK TEST

The numerals in the boxes indicate Lessons where help may be found.

Solve each problem and select the letter corresponding to your answer.

1. Which line is in a horizontal position? 12-1
     a.   **b.**   **c.**   **d.**

2. $\overline{TV}$ is named: 12-1
     a. line *TV*   **b.** ray *TV*   **c.** line segment *TV*   **d.** plane *TV*

3. Which pair of lines are parallel? 12-1
     a.   **b.**   **c.**   **d.**

4. Which figure is a square? 12-6
     a.   **b.**   **c.**   **d.**

5. If the scale is 1:2,500,000, what is the actual distance represented by 31 mm? 12-5
     a. 77.5 km   **b.** 6.75 km   **c.** 795 km   **d.** Answer not given

6. Which triangle is isosceles? 12-6
     a. 6 m, 6 m, 6 m   **b.** 8 m, 6 m, 10 m   **c.** 6 m, 5 m, 6m   **d.** 7 m, 6 m, 4 m

7. An angle measuring 48° is a(an): 12-3
     a. right angle   **b.** acute angle   **c.** obtuse angle   **d.** straight angle

8. Which figure illustrates a pyramid? 12-7
     a.   **b.**   **c.**   **d.**

9. The measure of the third angle of a triangle, when the other two angles measure 74° and 39°, is: 12-9
     a. 113°   **b.** 16°   **c.** 90°   **d.** 67°

10. The complement of an angle measuring 53° is: 12-11
     a. 27°   **b.** 37°   **c.** 97°   **d.** 127°

11. Angle *NOP* in the following figure measures: 12-11
     a. 60°   **b.** 100°   **c.** 120°   **d.** 90°

12. Parallel lines *EF* and *GH* are cut by transversal *MN*. If $m\angle 8 = 95°$, the measure of $\angle 1$ is: 12-12
     a. 85°   **b.** 105°
     c. 95°   **d.** 75°

13. The two congruent triangles are: 12-13
     a. I and II   **b.** II and IV
     c. III and IV   **d.** I and III

CHAPTER 13 Constructions

To copy a line segment means to draw a line segment equal in length to the given line segment.

PROCEDURE

To copy a given line segment:

Using a ruler:
(1) Measure the given line segment.
(2) Draw another line segment the same length.

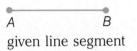

A   *B*

given line segment

Using a compass and a straightedge:
(1) Draw a line segment longer than the given line segment.
(2) Open the compass to the length of the given line segment.
(3) Transfer the compass to the new line segment, placing the metal point on one endpoint.
(4) Draw an arc cutting the new line segment. The intersection is the second endpoint.

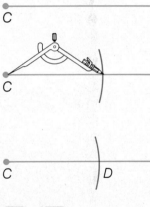

C

C

C   *D*

$$\overline{AB} = \overline{CD}$$

PRACTICE EXERCISES

1 Make a copy of each of the following line segments, using a ruler:

1. ——————— **2.** ——————————— **3.** ————————

2 Make a copy of each of the following line segments, using a compass and a straightedge. Check with ruler.

1. ——————— **2.** ——————————— **3.** ————————

3 Draw any line segment and label the endpoints *D* and *E*. Label a point not on $\overline{DE}$ as *F*. Using a compass and a straightedge, draw a line segment *FG* equal in length to $\overline{DE}$.

470

13-2 Copying An Angle

To copy an angle means to construct an angle equal in size to the given angle.

PROCEDURE

To copy a given angle:

Using a protractor:
(1) Measure the given angle.
(2) Draw another angle of the same size.

Using a compass:
(1) Draw a ray with endpoint P.
(2) Open the compass and place the metal point on vertex A of the given angle. Draw an arc intersecting the sides of the given angle at points B and C.
(3) With the same radius, place the metal point on endpoint P and draw an arc cutting the ray at Q.
(4) Set the radius of the compass equal to BC and draw an arc with center Q intersecting at R the arc drawn in step 3.
(5) Draw $\overrightarrow{PR}$.

given angle

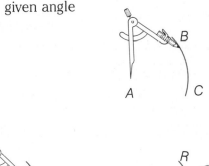

$m\angle BAC = m\angle RPQ$

For each of the following, check your work with a protractor.

1 Draw angles having the following measures, using a protractor. For each angle construct with a compass an angle of equal size.

1. 40°   **2.** 150°   **3.** 65°   **4.** 21°   **5.** 163°   **6.** 90°

2 Draw an acute angle. Construct with a compass an angle of equal size.

3 Draw any obtuse angle. Construct with a compass an angle of equal size.

4 Draw a right angle, using a protractor. Construct with a compass an angle of equal size.

5 Draw any angle. Construct with a compass an angle having the same measure.

A triangle contains three sides and three angles. A triangle may be constructed when any of the following combinations of three parts are known:

I. Three sides
II. Two sides and an included angle.
III. Two angles and an included side.

PROCEDURE I

To construct a triangle when three sides are given:

a ────────

b ────────── given sides

c ──────────

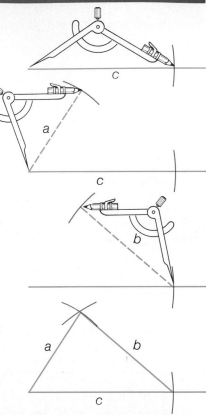

(1) Copy given side c.
(2) Open the compass to a radius equal to given side a and draw an arc using one of the endpoints of side c as the center.
(3) Set the radius of the compass equal to given side b and draw an arc crossing the arc from step 2 using the other endpoint of side c as the center.
(4) Draw line segments from the endpoints of side c to the intersection of the arcs.

PRACTICE EXERCISES

1 Construct triangles having sides that measure:
 1. 39 mm, 43 mm, 27mm **2.** $2\frac{3}{4}$ in., $2\frac{1}{4}$ in., 3 in. **3.** 5.8 cm, 4.6 cm, 6.3 cm

2 Construct an equilateral triangle whose sides are each 3.1 cm long. Measure the three angles. Are their measures equal?

3 Construct an isosceles triangle whose base is $2\frac{3}{16}$ in. long and each of whose two equal sides is $1\frac{5}{8}$ in. long.

4 Using the scale 1 inch = 8 feet, construct triangles with sides:
 1. 16 ft, 22 ft, 9 ft **2.** 30 ft, 24 ft, 15 ft
 3. 12 ft, 12 ft, 12 ft **4.** 18 ft, 10 ft, 10 ft

5 Using the scale 1:2,000,000, construct triangles with sides:
 1. 40 km, 70 km, 58 km **2.** 30 km, 42 km, 38 km
 3. 60 km, 54 km, 72 km **4.** 26 km, 36 km, 48 km

PROCEDURE II

To construct a triangle when two sides and an included angle are given:

given sides and included angle

a ─────────────

b ─────────────

(1) Copy given angle *EDF.*
(2) Open the compass to a radius equal to given side *a* and draw an arc crossing $\overrightarrow{DE}$ using *D* as the center. Then, set the compass to a radius equal to given side *b* and draw an arc crossing $\overrightarrow{DF}$ using *D* as the center.
(3) Draw a line segment connecting the intersections of the arcs found above.

PRACTICE EXERCISES

1 Construct triangles having the following sides and included angles:

1. 61 mm, 48 mm, 55°   **2.** 5.7 cm, 4.6 cm, 70°   **3.** $2\frac{7}{8}$ in., $2\frac{7}{8}$ in., 90°

2 Construct a right triangle having:

1. A base of 6.5 cm and an altitude of 5.9 cm
2. A base of $2\frac{15}{16}$ in. and an altitude of $3\frac{5}{8}$ in.

3 Construct an isosceles triangle in which the equal sides each measure $3\frac{1}{8}$ in. and the vertex angle formed by these sides measures 63°. Measure the angles opposite the equal sides. Are their measures equal?

4 Using the scale 1:500, construct a triangle in which two sides and the included angle measure:

1. 15 m, 10 m, and 50°   **2.** 20 m, 35 m, and 75°   **3.** 18 m, 27 m, and 80°

PROCEDURE III

To construct a triangle when two angles and an included side are given:

given angles and included sides

A ●─────────● B

(1) Copy side *AB.*
(2) At *A*, copy given ∠1.
(3) At *B*, copy given ∠2.
(4) The intersection of the sides of these angles is the third vertex (*C*) of the triangle.

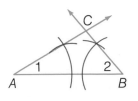

PRACTICE EXERCISES

1 Construct triangles having the following angles and included side that measure:
  1. 90°, 45°, 5 cm   **2.** 105°, 35°, $3\frac{1}{4}$ in.   **3.** 70°, 30°, 58 mm

2 Construct a triangle having two equal angles each measuring 50° and the included side measuring 46 mm. Measure the sides opposite the equal angles. Are their measures equal?

3 Construct a triangle having two equal angles each measuring 60° and the included side measuring $2\frac{11}{16}$ in. Measure the third angle. Measure the other two sides. Is the triangle equiangular? Equilateral?

4 Construct a triangle having two angles and an included side equal to these angles and line segment.

5 Using the scale $\frac{1}{4}$ in. = 1 ft, construct a triangle in which two angles and an included side measure:
  1. 60°, 50°, 16 ft   **2.** 45°, 45°, 24 ft   **3.** 95°, 55°, 22 ft   **4.** 70°, 65°, 19 ft

Refresh Your Skills

1. Add: 1-4
485,296
246,583
627,416
583,529

2. Subtract: 1-5
406,592 − 319,086

3. Multiply: 1-6
240 × 509

4. Divide: 1-7
3,600)86,400

5. Add: 2-4
.859 + .26 + .8

6. Subtract: 2-5
13.9 − .055

7. Multiply: 2-7
1.2 × .06

8. Divide: 2-8
.4 ÷ .08

9. Add: 3-8
$3\frac{11}{12} + 1\frac{5}{6} + 2\frac{3}{4}$

10. Subtract: 3-9
$1\frac{1}{2} - \frac{2}{3}$

11. Multiply: 3-11
$3\frac{1}{3} \times 1\frac{4}{5}$

12. Divide: 3-12
$9\frac{1}{6} \div 2\frac{1}{5}$

13. Find $9\frac{3}{4}$% of $10,000
4-6

14. 96 is what percent of 144?
4-7

15. 21 is 10.5% of what number?
4-8

13-4 Constructing a Perpendicular Through a Point on the Line

Vocabulary

Perpendicular lines (or rays or segments) are two lines (or rays or segments) that meet to form right angles.

The symbol ⊥ means "is perpendicular to."

PROCEDURE

To construct a line perpendicular to a given line at or through a given point on the given line:

(1) Use point C as the center and open the compass to any radius to draw an arc crossing $\overleftrightarrow{AB}$ at points D and E.

(2) With D and E as centers, open the compass to a radius greater than DC and draw arcs intersecting at F.

(3) Draw $\overleftrightarrow{CF}$.

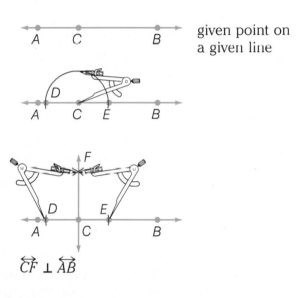

given point on a given line

$\overleftrightarrow{CF} \perp \overleftrightarrow{AB}$

PRACTICE EXERCISES

1 Draw any line. Select a point on this line. Construct a line perpendicular to the line you have drawn at your selected point.

2 Construct a rectangle 43 mm long and 35 mm wide.

3 Construct a square with each side $2\frac{1}{2}$ inches long.

4 Make a plan of a room 33 ft by 21 ft using the scale 1 in. = 8 ft.

5 Draw a circle. Draw any diameter of this circle. At the center construct another diameter perpendicular to the first diameter, dividing the circle into four arcs. Are these four arcs equal in length? Use a compass to check.

Constructing a Perpendicular From a Point Not on the Line

PROCEDURE

To construct a perpendicular to a given line from or through a given point not on the given line:

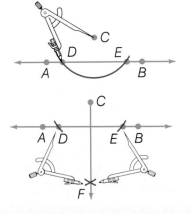

(1) Use point C as the center and draw an arc crossing $\overleftrightarrow{AB}$ at points D and E.

(2) With D and E as centers, open the compass to a radius greater than one-half the distance from D to E and draw arcs intersecting at F.

(3) Draw $\overleftrightarrow{CF}$.

PRACTICE EXERCISES

1 Draw any line. Select a point not on this line. Construct a perpendicular from your selected point to the line you have drawn.

2 Construct an equilateral triangle with each side 6.2 cm long. From each vertex construct a perpendicular to the opposite side. Is each angle bisected? Is each side bisected?

3 Draw any acute triangle. From each vertex construct a perpendicular to the opposite side. What do these perpendicular line segments represent in a triangle? Are they concurrent?

4 Draw any right triangle. Construct the altitude to each side. Are they concurrent? If so, what is the common point?

5 Draw a circle. Draw any chord in this circle except the diameter.

1. Construct a perpendicular from the center of the circle to this chord. Check whether the chord is bisected.

2. Extend the perpendicular line segment so that it intersects the circle. Check with a compass whether the arc corresponding to the chord is also bisected.
Observe that a radius of a circle that is perpendicular to a chord bisects the chord and its corresponding arc.

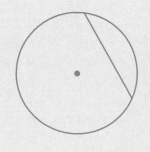

13-6 Bisecting a Line Segment

Vocabulary

The **midpoint** of a line segment is the point that separates the segment into two equal parts. The **perpendicular bisector** of a line segment is a line which both bisects a line segment and is perpendicular to it.

PROCEDURE

given line segment

To bisect a given line segment:
(1) Open the compass so that the radius is more than half the length of $\overline{AB}$.
(2) With A and B as centers, draw arcs which cross above and below the given line segment at points C and D.
(3) Draw $\overleftrightarrow{CD}$ bisecting $\overline{AB}$ at point E.

$\overleftrightarrow{CD} \perp \overline{AB}$
and bisects $\overline{AB}$ at E.

PRACTICE EXERCISES

1 Draw line segments which measure: **1.** 52 mm **2.** $4\frac{3}{8}$ in. **3.** 4.6 cm
Bisect each line segment, using a compass. Check with a ruler.

2 Copy, then using a compass, divide the following segment into four equal parts. Check with a ruler. _____

3 Draw any line segment.
1. Bisect it, using a compass. Then use a protractor to check whether each of the angles formed is a right angle.
2. What do we call a line that bisects and is perpendicular to a segment?

4 Draw any triangle.
1. Bisect each side by constructing the perpendicular bisector of that side.
2. Are these perpendicular bisectors concurrent?
3. Do they meet in a point equidistant from the vertices of the triangle? Check by measuring.
4. Using this common point as the center and the distance from this point to any vertex of the triangle as the radius, draw a circle through the three vertices of the triangle.
   When each side of the triangle is a chord of the circle or each vertex is a point on the circle, we say that *the circle is circumscribed about the triangle or that the triangle is inscribed in a circle.*

5 Draw any triangle.
1. Find the midpoint of each side by constructing the perpendicular bisector of that side.
2. Draw the median from each vertex to the midpoint of the opposite side.
3. Along each median, is the distance from the common point to the vertex twice the distance from the common point to the opposite side?

13-7 Bisecting an Angle

PROCEDURE

To bisect a given angle:

(1) Use a compass with B as the center and any radius to draw an arc cutting $\overrightarrow{BA}$ at D and $\overrightarrow{BC}$ at E.

(2) With D and E as centers and a radius of more than the distance from D to E, draw arcs crossing at F.

(3) Draw $\overrightarrow{BF}$.

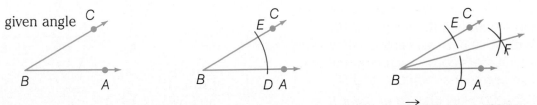

$\overrightarrow{BF}$ bisects $\angle CBA$

PRACTICE EXERCISES

1 Draw angles having the following measures. Bisect each angle using a compass. Check with a protractor.

1. 70° **2.** 130° **3.** 54° **4.** 145° **5.** 43° **6.** 137°

2 Draw any angle. Bisect it by using a compass only. Check by measuring your angle and each bisected angle.

3 Copy, then bisect each of the following angles:

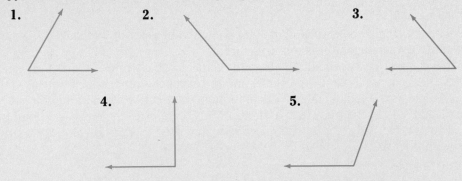

4 Draw any angle. Using a compass only, divide the angle into four equal angles. Check with a protractor.

Vocabulary

Parallel lines are lines in the same plane which do not meet.
The symbol ∥ means "is parallel to."

PROCEDURE

To construct a line parallel to a given line through a given point not on the given line:

(1) Draw a line through point C crossing $\overleftrightarrow{AB}$ at F.

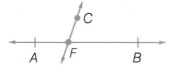

(2) At point C, construct ∠2 equal to ∠1.

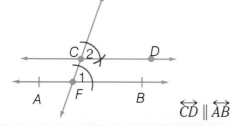

$\overleftrightarrow{CD} \parallel \overleftrightarrow{AB}$

PRACTICE EXERCISES

1 Draw any line. Select a point that is not on this line. Through this point construct a line parallel to the line you have drawn.

2 Construct a parallelogram with a base 75 mm long, a side 58 mm long, and an included angle of 60°.

3 Draw any acute triangle. Bisect one of the sides. Draw a line segment from this midpoint, parallel to one of the other sides, until it intersects the third side.

   1. Does this line segment bisect the third side? Check by measuring.

   2. Also check whether this line segment is one-half as long as the side to which it is parallel.

   3. Draw other triangles and check whether the above findings are true no matter which side is used first or which side is used as the parallel side.

CONSTRUCTING POLYGONS

Although there are different ways to construct some of the regular polygons, in general we use a method that is based on the geometric fact that equal central angles (angles whose vertex is at the center) of a circle intercept equal arcs and equal chords. When we construct regular polygons using this geometric fact, the completed construction will be inscribed within a circle. An inscribed polygon is a polygon whose vertices are points on the circle.

PROCEDURE

To construct a regular polygon:

(1) Draw a circle.

(2) Determine the measure of each central angle by dividing 360° by the number of sides of the polygon.

(3) Draw as many central angles as sides. Use the center of the circle as a common vertex for the angles with their sides (radii) intercepting the circle.

(4) Draw line segments connecting these points of interception to form the polygon.

A regular hexagon and an equilateral triangle may also be drawn by an alternate method:

Draw a circle. With the compass open to the same radius, divide the circle into six equal arcs as shown. Then connect these points of division to form a regular hexagon.

To form an equilateral triangle, simply connect alternate points of division.

EXAMPLE: Hexagon (6 sides)

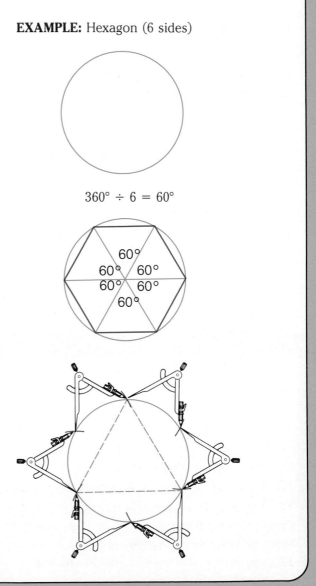

$360° \div 6 = 60°$

A square may be constructed by:

Drawing two diameters of a circle perpendicular to each other. This will divide the circle into four equal arcs.

Then to form a square, connect these points.

Rectangles and parallelograms of specific measurements may be drawn provided enough of these measurements are given.

EXAMPLES:

Draw a rectangle $1\frac{1}{2}$ in. long and $\frac{3}{4}$ in. wide.

Draw $\overline{AB}$ $1\frac{1}{2}$ inches long. Extend AB in both directions. Construct perpendiculars at points A and B. Draw $\overline{BC}$ and $\overline{AD}$ $\frac{3}{4}$ inches each. Then draw $\overline{DC}$ to form rectangle $ABCD$.

Draw a parallelogram with a base 30 mm long, a side 22 mm long, and an included angle of 30°.

Draw $\overline{AB}$ 30 mm long. Extend AB in both directions. Use a protractor to draw a 30° angle at points A and B. Draw $\overline{BC}$ and $\overline{AD}$ 22 mm each. Draw $\overline{DC}$ to form parallelogram $ABCD$.

EXERCISES

Construct each of the following regular polygons:

 1. Pentagon—5 sides
 2. Octagon—8 sides
 3. Decagon—10 sides
 4. Dodecagon—12 sides
 5. Equilateral triangle
 6. Square

Construct a rectangle:

 7. 64 mm long and 41 mm wide
 8. $3\frac{1}{2}$ in. long and $2\frac{3}{8}$ in. wide
 9. 7.6 cm long and 5.3 cm wide
 10. $4\frac{1}{8}$ in. long and $3\frac{9}{16}$ in. wide

Construct a square whose side measures:

11. 5 cm   12. $3\frac{1}{4}$ inches   13. 68 mm   14. $\frac{7}{8}$ inch

Construct a parallelogram with:

15. A base $4\frac{3}{8}$ inches long, a side $3\frac{1}{4}$ inches long, and an included angle of 45°.

16. A base 7 cm long, a side 6.2 cm long, and an included angle of 120°.

Draw a regular hexagon, each side measuring the following. Then draw a circle whose radius has the same measure as each required side.

17. 4 cm   18. 54 mm   19. $2\frac{7}{8}$ inches

Career Applications

When workers are building a satellite, a house, or even a video recorder, they need detailed drawings that show the size and position of every part. The people who prepare these drawings are called *drafters*. The engineers or architects who design a product give drafters sketches and measurements. From these, drafters make accurate, detailed drawings. Drafters use rulers, compasses, protractors, and other drafting devices. Some drafters use computer-aided design systems. They also research or calculate numerical values when necessary.

People who are interested in drafting should have good eyesight and be able to picture objects in three dimensions. The work requires accuracy, attention to detail, and the ability to work with other people. Drafters can learn their skills in high school, by taking college courses, or on the job.

1. Drafters often have to copy part of a drawing. Copy angle *ABC*.
2. A structure is to be in the shape of an equilateral triangle. Construct an equilateral triangle with given base $\overline{HS}$.
3. A tower of a bridge is to be perpendicular to the roadway. Construct a perpendicular to $\overline{BR}$ through point *T*.
4. A knob is to be at the exact center of a radio cabinet. Bisect line segment *RC*.
5. Angle *ROF* represents the roof of a house. Bisect this angle.
6. The edges of an airport runway are to be parallel lines. Construct a line parallel to $\overline{AP}$ through point *R*.

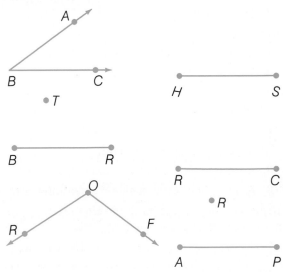

CHAPTER REVIEW

Vocabulary

Midpoint p. 477
Parallel lines (||) p. 479

Perpendicular bisector p. 477
Perpendicular lines (⊥) p. 475

The numerals in the boxes indicate Lessons where help may be found.

1. Use a ruler to measure the following line. ——————————— 13-1

2. Use a compass and a ruler to measure the following line: ——————— 13-1

3. Use a compass and a straightedge to copy this line: ——————— 13-1

4. Use a protractor to measure this angle: 13-2

5. Draw with a protractor an angle of 75°. Then construct with a compass an angle equal to it. Check your copy of the angle with a protractor. 13-2

6. Draw a line segment which measures 6.2 cm. Bisect it using a compass. Check with a ruler. 13-6

7. Draw an angle measuring 68°. Bisect it using a compass. Check with a protractor. 13-7

8. Construct a triangle with two angles measuring 95° and 40° and an included side $2\frac{3}{4}$ inches long. 13-3

9. Construct a triangle with two sides measuring 6.4 cm and 5.8 cm and an included angle measuring 70°. 13-3

10. Using the scale 1:1,000, construct a triangle with sides measuring 45 m, 53 m, and 39 m. 13-3

Draw any line. Using a compass, construct a perpendicular to it:

11. At a point on the line. 13-4   **12.** From a point not on the line. 13-5

13. Draw any line. Select a point outside this line. Through this point construct a line parallel to the line you have drawn. 13-8

14. Draw any equilateral triangle. Select an angle and bisect it. Construct from the vertex of this selected angle the altitude (line segment that is perpendicular to the opposite side). Does this perpendicular line bisect this opposite side? From this same vertex also draw the median to the opposite side. In an equilateral triangle, are all three of these lines (angle bisector, altitude, and median) one and the same line? If this is so, will it be true if each of the other angles is selected? Check 13-6, 13-7

COMPETENCY CHECK TEST

The numerals in boxes indicate Lessons where help may be found.

Solve each problem and select the letter corresponding to your answer.

1. To measure the length of a line, use a: 13-1
   a. protractor   **b.** straightedge   **c.** ruler   **d.** Answer not given

2. To draw an angle measuring 43°, use a: 13-2
   a. ruler   **b.** compass   **c.** protractor   **d.** Answer not given

3. To draw a circle having a diameter of 2 inches, use a: 13-2
   a. protractor   **b.** compass   **c.** ruler   **d.** straightedge

4. To bisect a line segment, use the construction: 13-6
   a.   **b.**   **c.**   **d.**

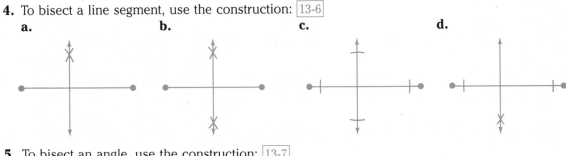

5. To bisect an angle, use the construction: 13-7
   a.   **b.**   **c.**   **d.**

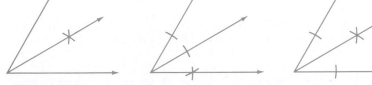

6. To construct a perpendicular to a line from a point not on the line, use the construction: 13-5
   a.   **b.**   **c.**   **d.**

CHAPTER 14 Indirect Measurement

14-1 Squares

Vocabulary

Squaring a number means multiplying the number by itself.

PROCEDURE

To find the square of a number: Multiply the number by itself.

EXAMPLES:

Find the square of 15.

15×15, or $15^2 = 225$

ANSWER: 225

Find the square of .03.

$.03 \times .03$, or $(.03)^2 = .0009$

ANSWER: .0009

Find the square of $\frac{4}{5}$.

$\frac{4}{5} \times \frac{4}{5}$, or $\left(\frac{4}{5}\right)^2 = \frac{16}{25}$

ANSWER: $\frac{16}{25}$

PRACTICE EXERCISES

Find the square of each of the following numbers:

1 **1.** 4 **2.** 1 **3.** 9 **4.** 7 **5.** 10 **6.** 3

2 **1.** 16 **2.** 24 **3.** 39 **4.** 50 **5.** 78 **6.** 100

3 **1.** .2 **2.** .7 **3.** .01 **4.** .09 **5.** .75 **6.** .18

4 **1.** $\frac{1}{2}$ **2.** $\frac{5}{6}$ **3.** $\frac{2}{5}$ **4.** $\frac{1}{3}$ **5.** $\frac{7}{8}$ **6.** $\frac{3}{16}$

5 **1.** $1\frac{1}{2}$ **2.** $1\frac{2}{3}$ **3.** $2\frac{1}{8}$ **4.** $2\frac{3}{4}$ **5.** $3\frac{1}{3}$ **6.** $1\frac{1}{4}$

6 Find the value of each of the following:

1. 6^2 **2.** $(.4)^2$ **3.** $(.009)^2$ **4.** $(8.1)^2$ **5.** $(2.05)^2$ **6.** $\left(\frac{5}{8}\right)^2$ **7.** $\left(3\frac{1}{2}\right)^2$ **8.** 120^2

Applications

To find the area of a square when the length of its side is known, square the given side. Find the area of a square if a side measures:

1. 7 in. **2.** 13 mm **3.** $2\frac{1}{2}$ ft **4.** 25 m **5.** 19 cm

The formula for finding the distance a body falls involves squaring the number of seconds it falls. Square the following numbers representing intervals of time:

6. 5 s **7.** 12 s **8.** 30 s **9.** 27 s **10.** 45 s

The formula for finding the area of a circle involves squaring the radius. Square the following numbers representing radii of circles:

11. 6 mm **12.** 14 cm **13.** $7\frac{1}{2}$ in. **14.** $\frac{3}{4}$ ft **15.** 26 m

Vocabulary

The **square root** (**symbol** $\sqrt{}$) of a number is that number which when multiplied by itself produces the given number.
A perfect square is a number that has an exact square root.

PROCEDURE

To find the square root of a perfect square:
(1) Write all the pairs of factors of the number and look for a pair of equal factors.
(2) The answer is one of the two equal factors.

EXAMPLES:

Find $\sqrt{100}$.

$100 = 1 \times 100; 2 \times 50;$
$ 4 \times 25; 5 \times 20;$
$ 10 \times 10$

ANSWER: 10

Find $\sqrt{144}$.

$144 = 1 \times 144; 2 \times 72;$
$ 3 \times 48; 4 \times 36;$
$ 6 \times 24; 8 \times 18;$
$ 9 \times 16; 12 \times 12$

ANSWER: 12

Find $\sqrt{\dfrac{25}{49}}$.

$25 = 1 \times 25; 5 \times 5$
$49 = 1 \times 49; 7 \times 7$
$\dfrac{25}{49} = \dfrac{5}{7} \times \dfrac{5}{7}$

ANSWER: $\dfrac{5}{7}$

PRACTICE EXERCISES

Find:

1
1. $\sqrt{4}$
2. $\sqrt{49}$
3. $\sqrt{25}$
4. $\sqrt{9}$
5. $\sqrt{81}$
6. $\sqrt{16}$
7. $\sqrt{36}$
8. $\sqrt{1}$
9. $\sqrt{64}$
10. $\sqrt{100}$

2
1. $\sqrt{400}$
2. $\sqrt{2,500}$
3. $\sqrt{100}$
4. $\sqrt{6,400}$
5. $\sqrt{900}$
6. $\sqrt{1,600}$
7. $\sqrt{8,100}$
8. $\sqrt{3,600}$
9. $\sqrt{4,900}$
10. $\sqrt{10,000}$

3
1. $\sqrt{\dfrac{9}{16}}$
2. $\sqrt{\dfrac{4}{9}}$
3. $\sqrt{\dfrac{9}{25}}$
4. $\sqrt{\dfrac{25}{36}}$
5. $\sqrt{\dfrac{16}{49}}$
6. $\sqrt{\dfrac{64}{225}}$
7. $\sqrt{\dfrac{49}{121}}$
8. $\sqrt{\dfrac{81}{100}}$
9. $\sqrt{\dfrac{144}{625}}$
10. $\sqrt{\dfrac{169}{2,500}}$

14-3 Finding Square Roots by Table and Estimation Method

The table below can be used to find square roots of whole numbers between 1 and 99 and perfect squares up to 9,801.

Table of Squares and Square Roots

Number	Square	Square Root	Number	Square	Square Root	Number	Square	Square Root
1	1	1.000	34	1,156	5.831	67	4,489	8.185
2	4	1.414	35	1,225	5.916	68	4,624	8.246
3	9	1.732	36	1,296	6.000	69	4,761	8.307
4	16	2.000	37	1,369	6.083	70	4,900	8.367
5	25	2.236	38	1,444	6.164	71	5,041	8.426
6	36	2.449	39	1,521	6.245	72	5,184	8.485
7	49	2.646	40	1,600	6.325	73	5,329	8.544
8	64	2.828	41	1,681	6.403	74	5,476	8.602
9	81	3.000	42	1,764	6.481	75	5,625	8.660
10	100	3.162	43	1,849	6.557	76	5,776	8.718
11	121	3.317	44	1,936	6.633	77	5,929	8.775
12	144	3.464	45	2,025	6.708	78	6,084	8.832
13	169	3.606	46	2,116	6.782	79	6,241	8.888
14	196	3.742	47	2,209	6.856	80	6,400	8.944
15	225	3.873	48	2,304	6.928	81	6,561	9.000
16	256	4.000	49	2,401	7.000	82	6,724	9.055
17	289	4.123	50	2,500	7.071	83	6,889	9.110
18	324	4.243	51	2,601	7.141	84	7,056	9.165
19	361	4.359	52	2,704	7.211	85	7,225	9.220
20	400	4.472	53	2,809	7.280	86	7,396	9.274
21	441	4.583	54	2,916	7.348	87	7,569	9.327
22	484	4.690	55	3,025	7.416	88	7,744	9.381
23	529	4.796	56	3,136	7.483	89	7,921	9.434
24	576	4.899	57	3,249	7.550	90	8,100	9.487
25	625	5.000	58	3,364	7.616	91	8,281	9.539
26	676	5.099	59	3,481	7.681	92	8,464	9.592
27	729	5.196	60	3,600	7.746	93	8,649	9.644
28	784	5.292	61	3,721	7.810	94	8,836	9.695
29	841	5.385	62	3,844	7.874	95	9,025	9.747
30	900	5.477	63	3,969	7.937	96	9,216	9.798
31	961	5.568	64	4,096	8.000	97	9,409	9.849
32	1,024	5.657	65	4,225	8.062	98	9,604	9.899
33	1,089	5.745	66	4,356	8.124	99	9,801	9.950

PROCEDURE

To find the square root of a number from 1–99 using the table:
(1) Locate the number in the Number column.
(2) Look across to the right to read the corresponding square root in the Square Root column.

To find a square root of a perfect square using the table:
(1) Locate the number in the Square column.
(2) Look across to the left to locate the corresponding square root in the Number column.

EXAMPLES:

Find $\sqrt{12}$. $\sqrt{12} = 3.464$ Find $\sqrt{2{,}704}$. $\sqrt{2{,}704} = 52$

ANSWER: 3.464   ANSWER: 52

The square root of a number may also be found by the estimate/divide/average (E.D.A.) method.

PROCEDURE

(1) Estimate the square root of the given number.
(2) Divide the given number by the estimated square root.
(3) Find the average of the resulting quotient and estimated square root.
(4) Divide the given number by the average from Step 3.
(5) Find the average of the divisor used and quotient found in Step 4.
(6) Continue this process to obtain a greater degree of approximation, as the divisor and quotient will eventually become quite close.

EXAMPLE: Find $\sqrt{12}$ to the nearest thousandth.

Since 12 is between 9 and 16, its square root will be between 3 and 4.

$$3.4\overline{)12.00} \quad \overset{3.5}{} \quad \frac{3.4 + 3.5}{2} = 3.45$$

$$3.45\overline{)12.0000} \quad \overset{3.48}{} \quad \frac{3.45 + 3.48}{2} = 3.465$$

$$3.465\overline{)12.000000} \quad \overset{3.463}{} \quad \frac{3.463 + 3.465}{2} = 3.464$$

ANSWER: 3.464

(1) Use 3.4 as the estimate.
(2) Divide 12 by 3.4 to get the rounded quotient 3.5.
(3) The average of 3.4 and 3.5 is 3.45.
(4) Divide 12 by 3.45 to get the rounded quotient 3.48.
(5) The average of 3.45 and 3.48 is 3.465.
(6) Divide 12 by 3.465 to get the quotient 3.463. The average of 3.465 and 3.463 is 3.464.

PRACTICE EXERCISES

1 Find the square root of each of the following numbers from the table:

1. 19	**2.** 57	**3.** 95	**4.** 73	**5.** 48
6. 29	**7.** 66	**8.** 84	**9.** 33	**10.** 91
11. 324	**12.** 1,225	**13.** 9,216	**14.** 784	**15.** 2,209

2 Find the square root of each number to the nearest thousandth by the E.D.A. method:

1. 6	**2.** 14	**3.** 21	**4.** 59	**5.** 38
6. 75	**7.** 105	**8.** 86	**9.** 41	**10.** 92

SQUARE ROOT ALGORITHM

An algorithm is a special method that can be used to add, subtract, or complete some mathematical operation. There is an algorithm for finding the square root.

Follow the steps below to find the square root of a number.

I Separate the numeral into groups of two figures each, starting at the decimal point. Annex zeros if needed.

II Place the largest possible square under the first group at the left.

III Write the square root of the number in Step II above the first group.

IV Subtract the square from the first group. Annex the next group to the remainder.

V Form the trial divisor by doubling the root already found and annexing a zero that is not written but is used mentally.

VI Divide the dividend (Step IV) by the divisor (Step V). Annex the quotient to the root already found and annex *it* to the trial divisor to form the complete divisor.

VII Multiply the complete divisor by the new figure of the root.

VIII Subtract this product from the dividend.

IX Continue until all groups are used or the answer is of the desired accuracy.

X Place a decimal point in the root directly above the one in the given number.

XI Check by squaring the square root to obtain the given number.

EXAMPLES:

Find the square root of 676.

$$
\begin{array}{r}
2\ \ 6 \\
\sqrt{6\ 76} \\
\hline
4 \\
46\overline{)2\ 76} \\
2\ 76 \\
\hline
\ldots
\end{array}
$$

Check:

$$
\begin{array}{r}
26 \\
26 \\
\hline
156 \\
52 \\
\hline
676
\end{array}
$$

ANSWER: 26

Find the square root of 7 correct to the nearest hundredth:

$$
\begin{array}{r}
2.\ 6\ \ 4\ \ 5 = 2.65 \\
\sqrt{7.00\ 00\ 00} \\
\hline
4 \\
46\overline{)3\ 00} \\
2\ 76 \\
\hline
524\overline{)\ \ 24\ 00} \\
20\ 96 \\
\hline
5285\overline{)\ 3\ 04\ 00} \\
2\ 64\ 25 \\
\hline
39\ 75
\end{array}
$$

ANSWER: 2.65

EXERCISES

Find the square root of each of the following:

1. .04	**2.** .0036	**3.** 625	**4.** 961	**5.** 1,369
6. 7,396	**7.** 23,716	**8.** 85,849	**9.** 398,161	**10.** 175,561
11. 43,264	**12.** 254,016	**13.** 8,826,841	**14.** 3,418,801	**15.** 19,351,201
16. 58,491,904	**17.** 81,018,001	**18.** 49,112,064	**19.** 27,040,000	**20.** 58,982,400

Find, correct to the nearest hundredth, the square root of each of the following:

21. 128   **22.** 1,000   **23.** .59

PROBLEM-SOLVING TECHNIQUES

The strategy below can help you set up and solve geometric problems.

1 READ the problem carefully to find:

a. The facts, dimensions, or values that are given.

b. The fact, dimension, or value that is to be determined.

2 PLAN how to solve the problem.

Decide which formula or formulas can be used that relates the variable representing the value to be found with the variables representing the given values.

3 SOLVE by arranging the solution in three columns.

a. In the left column:
  (1) Draw the geometric figure described in the problem.
  (2) Mark the given dimensions on the figure.
  (3) Below the figure write the variables with corresponding given values and the variable representing the unkown value.

b. In the center column:
  (1) Write the formula that relates the variables.
  (2) Substitute the known given values for the corresponding variables in the formula.
  (3) Perform the necessary operations.
  (4) When necessary, solve the resulting equation.

c. In the right column:
  Perform the necessary arithmetic operations.

4 CHECK the answer directly with the facts, dimensions, or values that are given.

For a sample solution see completely worked-out solution on page 492.

Not all distances can be measured directly. The distances from ships to shore, the distances across lakes and rivers, and others have to be measured indirectly.

One mathematical principle that can be used to find distances is the rule of Pythagoras. The rule of Pythagoras can be used to find the base, altitude, or hypotenuse of a right triangle.

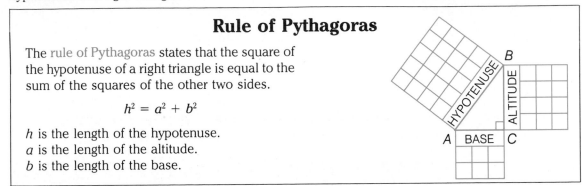

Rule of Pythagoras

The rule of Pythagoras states that the square of the hypotenuse of a right triangle is equal to the sum of the squares of the other two sides.

$$h^2 = a^2 + b^2$$

h is the length of the hypotenuse.
a is the length of the altitude.
b is the length of the base.

If any two sides of a right triangle are known, the third side can be found by using one of the following alternate forms of the rule of Pythagoras.

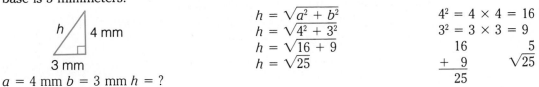

$$h = \sqrt{a^2 + b^2} \qquad a = \sqrt{h^2 - b^2} \qquad b = \sqrt{h^2 - a^2}$$

EXAMPLES:

Find the hypotenuse of a right triangle if the altitude is 4 millimeters and the base is 3 millimeters.

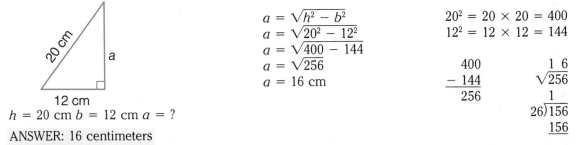

$h = \sqrt{a^2 + b^2}$

$h = \sqrt{4^2 + 3^2}$

$h = \sqrt{16 + 9}$

$h = \sqrt{25}$

$a = 4$ mm $b = 3$ mm $h = ?$

$4^2 = 4 \times 4 = 16$
$3^2 = 3 \times 3 = 9$

$$\begin{array}{r} 16 \\ + \ 9 \\ \hline 25 \end{array} \qquad \begin{array}{r} 5 \\ \sqrt{25} \end{array}$$

ANSWER: 5 millimeters

Find the altitude of a right triangle if the hypotenuse is 20 centimeters and the base is 12 centimeters.

$a = \sqrt{h^2 - b^2}$

$a = \sqrt{20^2 - 12^2}$

$a = \sqrt{400 - 144}$

$a = \sqrt{256}$

$a = 16$ cm

$h = 20$ cm $b = 12$ cm $a = ?$

$20^2 = 20 \times 20 = 400$
$12^2 = 12 \times 12 = 144$

$$\begin{array}{r} 400 \\ - \ 144 \\ \hline 256 \end{array} \qquad \begin{array}{r} 1\ 6 \\ \sqrt{256} \\ \underline{1} \\ 26)\overline{156} \\ \underline{156} \end{array}$$

ANSWER: 16 centimeters

PRACTICE EXERCISES

1 Find the hypotenuse of each right triangle with the following dimensions:

	Altitude	Base
1.	12 m	9 m
2.	8 cm	15 cm
3.	60 km	25 km
4.	33 ft	56 ft
5.	180 m	112 m

2 Find the altitude of each right triangle with the following dimensions:

	Hypotenuse	Base
1.	13 mm	5 mm
2.	35 cm	28 cm
3.	89 ft	80 ft
4.	53 km	45 km
5.	219 ft	144 ft

3 Find the base of each right triangle with the following dimensions:

	Hypotenuse	Altitude
1.	25 cm	24 cm
2.	87 m	63 m
3.	73 mm	48 mm
4.	91 yd	84 yd
5.	153 ft	72 ft

4 By the rule of Pythagoras and by actual measurement:

  1. Find the length of the diagonal (line segment joining opposite corners) of a rectangle 8 cm long and 6 cm wide.

  2. Find the length of the diagonal of a square whose side measures 7 cm.

Applications

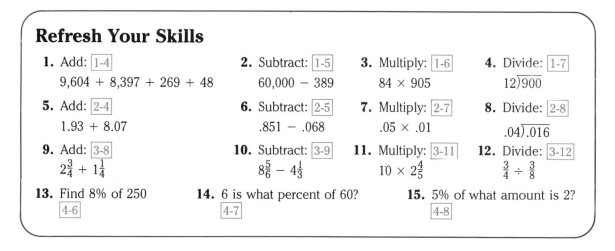

1. What is the shortest distance from first base to third base if the distance between bases is 90 ft?

2. How high up on a wall does a 25-foot ladder reach if the foot of the ladder is 7 feet from the wall?

3. A child lets out 60 m of string in flying a kite. The distance from a point directly under the kite to where the child stands is 36 m. If the child holds the string 1.5 m from the ground, how high is the kite? Disregard any sag.

4. An airplane, flying 252 kilometers from town A due west to town B, drifts off its course in a straight line and is 39 kilometers due south of town B. What distance did the airplane actually fly?

Refresh Your Skills

1. Add: 1-4
9,604 + 8,397 + 269 + 48

2. Subtract: 1-5
60,000 − 389

3. Multiply: 1-6
84 × 905

4. Divide: 1-7
12)‾900

5. Add: 2-4
1.93 + 8.07

6. Subtract: 2-5
.851 − .068

7. Multiply: 2-7
.05 × .01

8. Divide: 2-8
.04)‾.016

9. Add: 3-8
$2\frac{3}{4} + 1\frac{1}{4}$

10. Subtract: 3-9
$8\frac{5}{6} - 4\frac{1}{3}$

11. Multiply: 3-11
$10 \times 2\frac{4}{5}$

12. Divide: 3-12
$\frac{3}{4} \div \frac{3}{8}$

13. Find 8% of 250
4-6

14. 6 is what percent of 60?
4-7

15. 5% of what amount is 2?
4-8

Similar triangles can also be used to find distances indirectly.

The ratios of corresponding sides of similar triangles are equal.

$$\frac{a}{d} = \frac{b}{e} = \frac{c}{f}$$

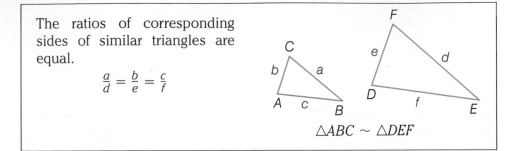

$\triangle ABC \sim \triangle DEF$

PROCEDURE

To find a distance using similar triangles:
(1) Use the given data to draw similar triangles.
(2) Write a proportion using the three known sides and the required distance as the fourth side.
(3) Solve the proportion.

EXAMPLE:

A tree casts a shadow (AB) 27 ft long, while a 4-ft pole (EF) nearby casts a shadow (DE) 3 ft long. What is the tree's height (BC)?

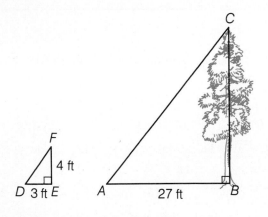

Triangles ABC and DEF are similar. Therefore, we can write a proportion:

$$\frac{\text{side } AB}{\text{side } DE} = \frac{\text{side } BC}{\text{side } EF}$$

Let n = the length of side BC

$$\frac{27}{3} = \frac{n}{4}$$

$3 \times n = 27 \times 4$

$3n = 108$

$n = 36$

$$\begin{array}{r} 27 \\ \times\ 4 \\ \hline 108 \end{array}$$

$$\begin{array}{r} 36 \\ 3\overline{)108} \end{array}$$

ANSWER: The tree is 36 feet high.

494

PRACTICE EXERCISES

1 Triangles *ABE* and *CDE* are similar.

  1. Which side of triangle *ABE* corresponds to side *CE?*

  2. Which side to *DE?*

  3. Which side to *AB?*

  4. Find the length (*AB*) of the lake.

2 Triangles *ABC* and *CDE* are similar.

1. Which side of triangle *ABC* corresponds to side *CE?*

2. Which side to *DE?*

3. What is the distance (*AB*) across the stream?

Enrichment

We can determine missing distances by **Scale Drawings**. First, find the scale (see Lesson 12-5). Then measure the scale lengths and apply the scale to find the actual lengths.

EXAMPLE:

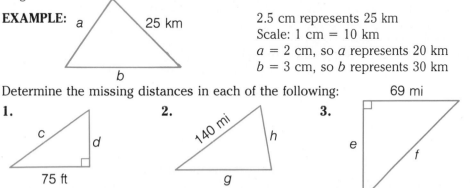

2.5 cm represents 25 km
Scale: 1 cm = 10 km
a = 2 cm, so *a* represents 20 km
b = 3 cm, so *b* represents 30 km

Determine the missing distances in each of the following:

1.

2.

3. 69 mi

4. An airplane, flying due west (270° from north) from starting point *A* at an air speed of 400 km/h, was blown off its course by a 50 km/h wind from the south (180° from north). Using the scale 1 mm = 10 km, make a drawing showing the position of the airplane at the end of one hour. How far was the airplane from its starting point? What course (or track) was it actually flying?

14-6 Numerical Trigonometry

Vocabulary

The ratio of the side opposite an acute angle in a right triangle to the adjacent side is called the **tangent** (abbreviated **tan**) of the angle.

The ratio of the side opposite an acute angle in a right triangle to the hypotenuse is called the **sine** (abbreviated **sin**) of the angle.

The ratio of the adjacent side of an acute angle in a right triangle to the hypotenuse is called the **cosine** (abbreviated **cos**) of the angle.

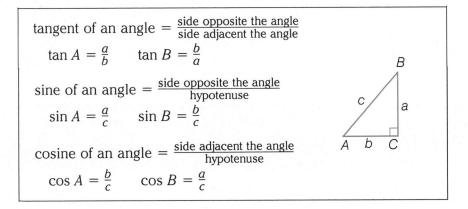

$$\text{tangent of an angle} = \frac{\text{side opposite the angle}}{\text{side adjacent the angle}}$$

$$\tan A = \frac{a}{b} \qquad \tan B = \frac{b}{a}$$

$$\text{sine of an angle} = \frac{\text{side opposite the angle}}{\text{hypotenuse}}$$

$$\sin A = \frac{a}{c} \qquad \sin B = \frac{b}{c}$$

$$\text{cosine of an angle} = \frac{\text{side adjacent the angle}}{\text{hypotenuse}}$$

$$\cos A = \frac{b}{c} \qquad \cos B = \frac{a}{c}$$

PROCEDURE

To solve problems using trigonometric ratios:
(1) Draw a right triangle and label the given dimensions.
(2) Select the proper formula and substitute the given values, using the table of trigonometric values (page 613) when necessary.
(3) Solve the resulting equation.

EXAMPLES:

Find the value of cos 54°.

Use the table. Locate 54° in the angle column and read across to the cosine column.

$$\cos 54° = .5878.$$

ANSWER: .5878

Find the value of tan 38°.

Use the table. Locate 38° in the angle column. Read across to the tangent column.

$$\tan 38° = .7813$$

ANSWER: .7813

In right triangle *ABC*, find side *a* when angle $A = 53°$ and side $b = 100$ ft.

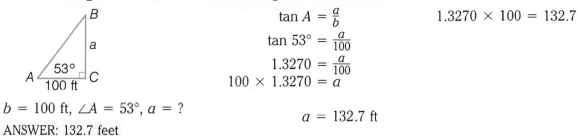

$$\tan A = \frac{a}{b}$$

$$\tan 53° = \frac{a}{100}$$

$$1.3270 = \frac{a}{100}$$

$$100 \times 1.3270 = a$$

$1.3270 \times 100 = 132.7$

$b = 100$ ft, $\angle A = 53°$, $a = ?$

$a = 132.7$ ft

ANSWER: 132.7 feet

In right triangle *ABC*, find side *b* (to nearest tenth) when angle $B = 26°$ and side $c = 45$ meters.

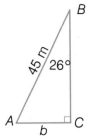

$$\sin B = \frac{b}{c}$$

$$\sin 26° = \frac{b}{45}$$

$$.4384 = \frac{b}{45}$$

$$45 \times .4348 = b$$

$$b = 19.728 = 19.7 \text{ (nearest tenth)}$$

$$\begin{array}{r} .4384 \\ \times \quad 45 \\ \hline 21920 \\ 17536 \\ \hline 19.7280 \end{array}$$

$c = 45$ m, $\angle B = 26°$, $b = ?$

ANSWER: 19.7 meters

An angle measured vertically between the horizontal line and the observer's line of sight to an object is called the angle of depression when the object is below the observer and the angle of elevation when the object is above the observer.

Sometimes the complement of an angle is simpler for computation then the given angle.

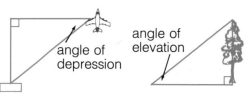

EXAMPLE: The angle of depression from a plane to an object is 13°. The plane is 18 km above the ground. How far is the plane from the object?

$a = 18$ km
$\angle ABC = 77°$ (complement of 13°)
$c = ?$

$$\cos B = \frac{a}{c}$$

$$\cos 77° = \frac{18}{c}$$

$$.2250 = \frac{18}{c}$$

$$.2250c = 18$$

$$c = 80 \text{ km}$$

$$\begin{array}{r} 80. \\ 225\overline{)18000.} \end{array}$$

ANSWER: The plane is 80 km from the object.

PRACTICE EXERCISES

Find the value of:

1 **1.** tan 43°   **2.** tan 81°   **3.** tan 63°   **4.** tan 37°

2 **1.** sin 79°   **2.** sin 7°   **3.** sin 41°   **4.** sin 24°

3 **1.** cos 18°   **2.** cos 56°   **3.** cos 6°   **4.** cos 71°

Find angle A when:

4 **1.** tan A = .3839   **2.** tan A = 3.0777   **3.** tan A = .2493   **4.** tan A = 1.6003

5 **1.** sin A = .9063   **2.** sin A = .3420   **3.** sin A = .7431   **4.** sin A = .9925

6 **1.** cos A = .8988   **2.** cos A = .6293   **3.** cos A = .6018   **4.** cos A = .6691

Find angle B when:

7 **1.** tan B = 19.0811   **2.** tan B = .9004   **3.** tan B = 1.2349   **4.** tan B = .6009

8 **1.** sin B = .6561   **2.** sin B = .9986   **3.** sin B = .5000   **4.** sin B = .9455

9 **1.** cos B = .1045   **2.** cos B = .9998   **3.** cos B = .4540   **4.** cos B = .9986

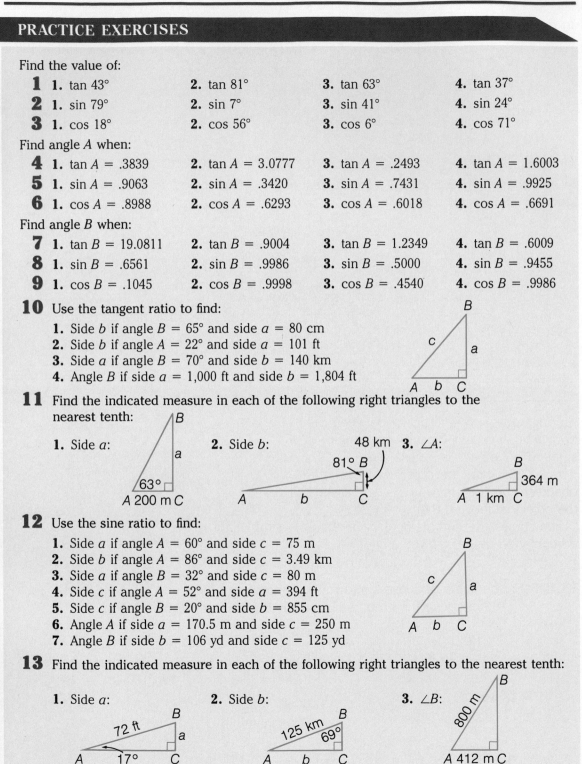

10 Use the tangent ratio to find:

  1. Side b if angle B = 65° and side a = 80 cm
  2. Side b if angle A = 22° and side a = 101 ft
  3. Side a if angle B = 70° and side b = 140 km
  4. Angle B if side a = 1,000 ft and side b = 1,804 ft

11 Find the indicated measure in each of the following right triangles to the nearest tenth:

  1. Side a: **2.** Side b: **3.** ∠A:

12 Use the sine ratio to find:

  1. Side a if angle A = 60° and side c = 75 m
  2. Side b if angle A = 86° and side c = 3.49 km
  3. Side a if angle B = 32° and side c = 80 m
  4. Side c if angle A = 52° and side a = 394 ft
  5. Side c if angle B = 20° and side b = 855 cm
  6. Angle A if side a = 170.5 m and side c = 250 m
  7. Angle B if side b = 106 yd and side c = 125 yd

13 Find the indicated measure in each of the following right triangles to the nearest tenth:

  1. Side a: **2.** Side b: **3.** ∠B:

14 Use the cosine ratio to find:

1. Side b if angle $A = 46°$ and side $c = 100$ m
2. Side a if angle $B = 62°$ and side $c = 250$ yd
3. Side c if angle $A = 70°$ and side $b = 17.1$ km
4. Side c if angle $B = 59°$ and side $a = 515$ cm
5. Angle A if side $b = 23$ mm and side $c = 46$ mm
6. Angle B if side $a = 309$ mi and side $c = 1,000$ mi

15 Find the indicated measure in each of the following right triangles to the nearest tenth:

1. Side b:
2. Side c:
3. $\angle A$:

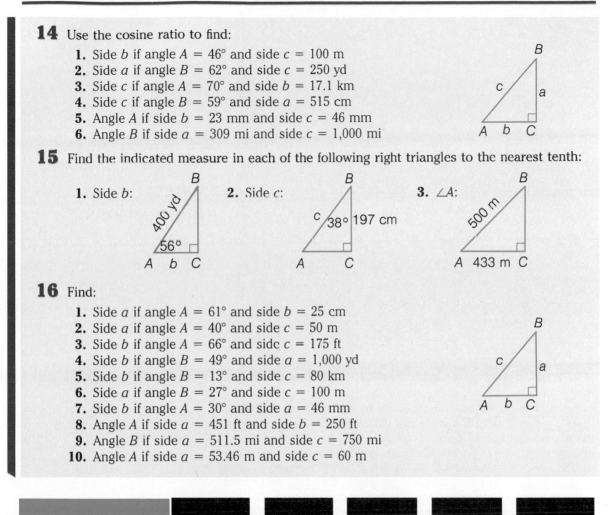

16 Find:

1. Side a if angle $A = 61°$ and side $b = 25$ cm
2. Side a if angle $A = 40°$ and side $c = 50$ m
3. Side b if angle $A = 66°$ and side $c = 175$ ft
4. Side b if angle $B = 49°$ and side $a = 1,000$ yd
5. Side b if angle $B = 13°$ and side $c = 80$ km
6. Side a if angle $B = 27°$ and side $c = 100$ m
7. Side b if angle $A = 30°$ and side $a = 46$ mm
8. Angle A if side $a = 451$ ft and side $b = 250$ ft
9. Angle B if side $a = 511.5$ mi and side $c = 750$ mi
10. Angle A if side $a = 53.46$ m and side $c = 60$ m

Applications

1. From a cliff 300 meters above the sea, the angle of depression of a boat is 87°. How far is the boat from the foot of the cliff?

2. How high is a kite when 160 feet of string is let out and the string makes an angle of 50° with the ground?

3. The altitude of a right triangle is 220 millimeters and the angle opposite the base is 77°. How long is the hypotenuse?

4. What is the elevation of a road if the road rises 364 feet in a horizontal distance of 1,000 feet?

Career Applications

A contractor needs to know the exact size of a building site. A lawyer wants a precise description of the boundaries of a lot. An oil company needs a detailed map of the locations of its oil wells in an area. *Surveyors* are hired to do jobs like these. Surveyors usually work with a team of surveyor technicians. Together they establish the exact locations of points and measure distances and elevations. Their tools range from rods and levels to electronic devices that measure distances.

People interested in surveying should enjoy outdoor work and be able to make accurate measurements. Most surveyors have college training. Surveyor technicians learn their skills on the job. Surveyors must have state licenses because they are legally responsible for the accuracy of their work.

1. A surveyor needs to find distance *JK*, the distance across a river. Triangle *JKL* and triangle *NML* are similar. What is the distance *JK*?

2. A surveyor needs to measure the distance *AB*. Use the rule of Pythagoras to help the surveyor find *AB* to the nearest tenth of a meter.

3. *PQRS* is a scale drawing of a building lot. The scale is 1 in. = 400 ft. What is the approximate length in feet of the actual side *QR*?

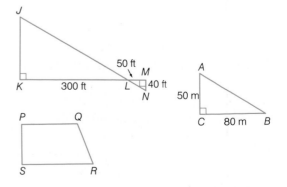

CHAPTER REVIEW

Vocabulary

cosine p. 496	sine p. 496	squaring a number p. 486
perfect square p. 487	square root ($\sqrt{\ }$) p. 487	tangent p. 496
Rule of Pythagoreas p. 492		

The numerals in the boxes indicate Lessons where help may be found.

Find the value of: **1.** 46^2 **2.** $\left(\frac{5}{8}\right)^2$ **3.** $.03^2$ $\boxed{14\text{-}1}$

4. Find $\sqrt{121}$ $\boxed{14\,2}$

5. Find $\sqrt{30}$ to the nearest thousandth by the estimate, divide, and average method. $\boxed{14\text{-}3}$

6. Find the hypotenuse of a right triangle if the altitude is 76 meters and the base is 57 meters. $\boxed{14\text{-}4}$

7. What is the base of a right triangle if the hypotenuse is 377 feet and the altitude is 145 feet? $\boxed{14\text{-}4}$

8. Find the altitude of a right triangle if the base is 66 centimeters and the hypotenuse is 130 centimeters. $\boxed{14\text{-}4}$

9. Show by the rule of Pythagoras that a 30-inch umbrella can be fitted into a suitcase having inside dimensions of 26 inches by 15 inches. $\boxed{14\text{-}4}$

10. Find the height of a building that casts a shadow of 40 feet at the time a 7-foot fence casts a shadow of 5 feet. $\boxed{14\text{-}5}$

Find the indicated parts of the following right triangles: $\boxed{14\text{-}6}$

11. Find side a: **12.** Find side b: **13.** Find side b:

14. The angle of elevation from a ship to the top of a lighthouse, 150 feet high, is 19°. How far from the foot of the lighthouse is the ship? $\boxed{14\text{-}6}$

15. Find the distance across the river (MN) if ST is 75 ft, RT is 60 ft, and MR is 360 ft. $\boxed{14\text{-}5}$

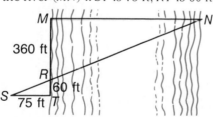

COMPETENCY CHECK TEST

The numerals in the boxes indicate Lessons where help may be found.

Solve each problem and select the letter corresponding to your answer.

1. The square of 16 is: $\boxed{14\text{-}1}$
   a. 256   **b.** 4   **c.** 64   **d.** Answer not given

2. The square root of 9 is: $\boxed{14\text{-}2}$
   a. 81   **b.** 3   **c.** 36   **d.** Answer not given.

3. In a right triangle when the hypotenuse is 34 cm and the base is 16 cm, the altitude is: $\boxed{14\text{-}4}$
   a. 50 cm   **b.** 18 cm   **c.** 544 cm   **d.** 30 cm

4. In a right triangle when the altitude is 40 yd and the hypotenuse is 104 yd, the base is: $\boxed{14\text{-}4}$
   a. 64 yd   **b.** 96 yd   **c.** 144 yd   **d.** 4,160 yd

5. The length of a path running diagonally across a rectangular lot 126 meters long and 102 meters wide is: $\boxed{14\text{-}4}$
   a. 34 m   **b.** 170 m   **c.** 238 m   **d.** Answer not given

6. A flagpole casts a shadow 14 ft long at the time a man, 6 ft tall, casts a shadow 3 ft long. The height of the flagpole is: $\boxed{14\text{-}5}$
   a. 28 ft   **b.** 84 ft   **c.** 42 ft   **d.** Answer not given

7. The distance across the lake is: $\boxed{14\text{-}5}$
   a. 200 ft   **b.** 260 ft

   c. 220 ft   **d.** 320 ft

8. If 200 feet of string is let out and the string makes an angle of 31° with the ground, the height of the kite is:
   a. 231 ft   **b.** 103 ft   **c.** 169 ft   **d.** 156 ft

9. The length of side a is: $\boxed{14\text{-}6}$
   a. 220 m   **b.** 202 m

   c. 170 m   **d.** 305 m

10. The length of side b is: $\boxed{14\text{-}6}$
   a. 225 ft

   b. 725 ft

   c. 500 ft

   d. 950 ft

CHAPTER 15 Perimeter, Area, and Volume

15-1 Perimeter of a Rectangle

Vocabulary

The distance around a polygon is called its **perimeter**.

The perimeter of a polygon is the sum of the lengths of the sides. In a rectangle, pairs of opposite sides are equal in length.

The perimeter of a rectangle is equal to twice its length (l) plus twice its width (w).

Perimeter $= l + w + l + w$

$\qquad = 2l + 2w$

Formula: $p = 2l + 2w$ or $p = 2(l + w)$

EXAMPLE:

Find the perimeter of a rectangle 26 meters long and 17 meters wide.

17 m

26 m

$l = 26$ m, $w = 17$ m; $p = ?$

$p = 2l + 2w$
$p = 2 \times 26 + 2 \times 17$
$p = 52 + 34$
$p = 86$ m

$$\begin{array}{r} 26 \\ \times\ 2 \\ \hline 52 \end{array} \qquad \begin{array}{r} 17 \\ \times\ 2 \\ \hline 34 \end{array} \qquad \begin{array}{r} 52 \\ +\ 34 \\ \hline 86 \end{array}$$

ANSWER: 86 meters

PRACTICE PROBLEMS

1 What is the perimeter of a rectangle if its length is 47 mm and its width is 21 mm?

2 Find the perimeters of the rectangles whose dimensions are:

   1. $l = 16$ cm   **2.** $l = 63$ ft   **3.** $l = 8.375$ km   **4.** $l = 7\frac{3}{8}$ in.   **5.** $l = 2$ ft 7 in.

      $w = 9$ cm $w = 49$ ft $w = 1.875$ km $w = 4\frac{13}{16}$ in. $w = 1$ ft 10 in.

3 How many meters of fencing are required to enclose a rectangular garden 59 m long and 39 m wide? If each 2-meter section costs $11.99, how much will the fencing cost?

4 How many meters of fringe are needed for a border on a bedspread 182 cm by 268 cm?

5 Fred wishes to make a frame for his class picture. The picture measures 25 in. by $12\frac{1}{2}$ in. If he allows $2\frac{1}{4}$ inches extra for each corner, how many feet of molding will he need?

6 How many feet of baseboard are needed for a room 20 ft 6 in. long and 16 ft wide if 5 ft must be deducted for doorways?

The sides of a square are equal in length.

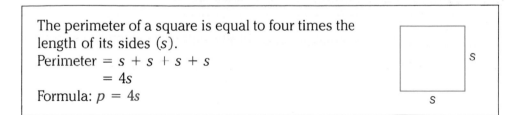

The perimeter of a square is equal to four times the
length of its sides (s).
Perimeter $= s + s + s + s$
$\qquad = 4s$
Formula: $p = 4s$

s

s

EXAMPLE:

Find the perimeter of a square whose side is 19 miles long.

19 mi

$s = 19$ mi
$p = ?$

$p = 4s$
$p = 4 \times 19$
$p = 76$ mi

$$\begin{array}{r} 19 \\ \times\ 4 \\ \hline 76 \end{array}$$

ANSWER: 76 miles

PRACTICE PROBLEMS

1 What is the perimeter of a square whose side is 8 kilometers long?

2 Find the perimeters of the squares whose sides measure:

1. 25 cm

2. 11 km

3. 880 yd

4. 76 mm

5. $20\frac{5}{8}$ in.

6. 5,280 ft

7. .25 m

8. 17.5 cm

9. $8\frac{3}{4}$ ft

10. 2 ft 9 in.

3 If the distance between bases is 90 feet, how many yards does a batter run when he hits a home run?

4 Find the cost of the wire needed to make a fence of 5 strands around a square lot 60 meters by 60 meters if a 400-meter spool of wire costs $34.69.

15-3 Perimeter of a Triangle

The sides of a triangle may or may not be equal in length.
In an equilateral triangle, however, all three sides are the same length.

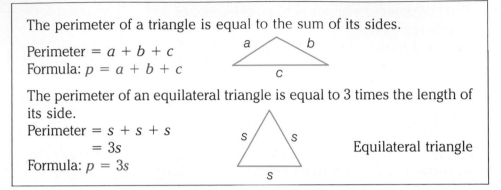

The perimeter of a triangle is equal to the sum of its sides.

Perimeter = $a + b + c$
Formula: $p = a + b + c$

The perimeter of an equilateral triangle is equal to 3 times the length of its side.

Perimeter = $s + s + s$
       = $3s$
Formula: $p = 3s$

Equilateral triangle

EXAMPLE:

Find the perimeter of a triangle with sides measuring 13 cm, 15 cm, and 19 cm.

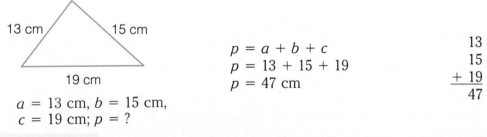

13 cm   15 cm

19 cm

$a = 13$ cm, $b = 15$ cm,
$c = 19$ cm; $p = ?$

$p = a + b + c$
$p = 13 + 15 + 19$
$p = 47$ cm

$$\begin{array}{r} 13 \\ 15 \\ + \ 19 \\ \hline 47 \end{array}$$

ANSWER: 47 cm

PRACTICE PROBLEMS

1 Find the perimeters of triangles with sides measuring:

  1. 18 m, 9 m 15 m   **2.** 21 cm, 17 cm, 29 cm

  3. 6.23 km, 4.7 km, 3.59 km   **4.** $4\frac{1}{4}$ mi, $7\frac{5}{8}$ mi, $6\frac{1}{2}$ mi

2 Find the perimeters of equilateral triangles with sides measuring:

  1. 63 cm   **2.** 14 km   **3.** 4.75 m   **4.** $16\frac{5}{8}$ in.   **5.** 2 ft 4 in.

3 Find the perimeter of an isosceles triangle if each of the equal sides is 15 centimeters and the third side is 8 centimeters.

4 How many meters of hedge are needed to enclose a triangular lot with sides measuring 196 meters, 209 meters, and 187 meters?

15-4 Circumference of a Circle

> ## Vocabulary
> The distance around a circle is called the **circumference**.

The ratio of the circumference (C) to the diameter of a circle is a constant, π, read pi. π is approximately 3.1415927. For computing, use $\pi = \frac{22}{7}$ or $\pi = 3.14$.

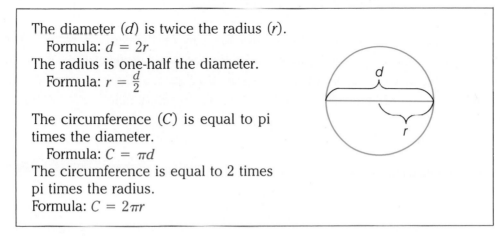

The diameter (d) is twice the radius (r).
  Formula: $d = 2r$
The radius is one-half the diameter.
  Formula: $r = \frac{d}{2}$

The circumference (C) is equal to pi times the diameter.
  Formula: $C = \pi d$
The circumference is equal to 2 times pi times the radius.
  Formula: $C = 2\pi r$

EXAMPLES:

Find the circumference of a circle when its diameter is 8 centimeters.

$C = \pi d$

$C = 3.14 \times 8$

$C = 25.12$ cm

$$\begin{array}{r} 3.14 \\ \times\quad 8 \\ \hline 25.12 \end{array}$$

$d = 8$ cm, $\pi = 3.14$; $C = ?$

ANSWER: 25.12 centimeters

Find the circumference of a circle when its radius is 21 inches.

$C = 2\pi r$

$C = 2 \times \frac{22}{7} \times 21$

$C = 132$ in.

$\frac{2}{1} \times \frac{22}{\underset{1}{7}} \times \frac{\overset{3}{\cancel{21}}}{1} = 132$

$r = 21$ in., $\pi = \frac{22}{7}$; $C = ?$

ANSWER: 132 inches

PRACTICE PROBLEMS

1 Find the diameter if the radius is:
    1. 7 cm   **2.** 23 m   **3.** 6.5 km   **4.** $10\frac{11}{16}$ in.   **5.** 2 ft 3 in.

2 Find the radius if the diameter is:
    1. 38 m   **2.** 5 mm   **3.** 8.9 km   **4.** $10\frac{2}{3}$ ft   **5.** 8 yd 2 ft

3 What is the circumference of a circle whose diameter is 60 meters?

4 Find the circumference of a circle having a diameter of:
    1. 5 mm   **2.** 35 cm   **3.** 260 m   **4.** 49 mi   **5.** 440 yd   **6.** 1.8 m   **7.** 8.4 km

5 What is the circumference of a circle whose radius is 26 meters?

6 Find the circumference of a circle having a radius of:
    1. 90 m   **2.** 7 mm   **3.** 56 km   **4.** 382 ft   **5.** 25 m   **6.** 8.4 km   **7.** $5\frac{1}{2}$ mi   **8.** $4\frac{3}{8}$ in.

7 What is the diameter of a circle whose circumference is 286 centimeters?

8 Find the diameter of a circle when its circumference is:
    1. 176 m   **2.** 198 cm   **3.** 330 yd   **4.** 40 mm   **5.** $6\frac{7}{8}$ in.

9 If the diameter of a circular table is 42 inches, what is the circumference of the table?

10 What distance in feet does the tip of a propeller travel in one revolution if its length (diameter) is 7 feet?

11 How long a metal bar do you need to make a basketball hoop with a diameter of 48 centimeters?

12 How much farther do you ride in one turn of a merry-go-round if you sit in the outside lane, 21 ft from the center, than if you sit in the inside lane, 14 ft from the center?

13 If the diameter of each wheel is 28 inches, how far does a bicycle go when the wheels revolve once? How many times do the wheels revolve in a distance of 1 mile?

Enrichment

The Greek letter pi (written π) stands for the number obtained when the circumference of any circle is divided by its diameter. $\pi = \frac{C}{d}$

The ancient Chinese used 3 as an approximation for pi. The Egyptians improved on this approximation around 1600 B.C. An astronomer, Ptolemy of Alexandria, calculated a value of pi to four decimal places. After, mathematicians attempted to calculate the exact value of pi. Their research showed that pi was an infinite decimal and impossible to calculate exactly.

The following are common values used for π: $\frac{22}{7}$ 3.14   3.1416   3.14159

15-5 Area of a Rectangle

Vocabulary

The **area** of any surface is the number of units of square measure contained in the surface. Area is measured in square units.

When computing area, express all the linear measurements with the same unit of measure.

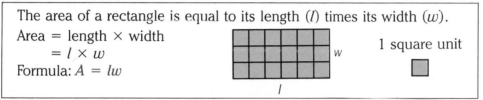

The area of a rectangle is equal to its length (l) times its width (w).

Area = length × width
   = l × w
Formula: $A = lw$

w 1 square unit

l

EXAMPLE:

Find the area of a rectangle 23 centimeters long and 16 centimeters wide.

16 cm

23 cm

l = 23 cm, w = 16 cm; A = ?

ANSWER: 368 cm²

$A = lw$
$A = 23 × 16$
$A = 368$ cm²

$$
\begin{array}{r}
23 \\
\times\ 16 \\
\hline
138 \\
23 \\
\hline
368
\end{array}
$$

PRACTICE PROBLEMS

1 What is the area of a rectangle if its length is 14 meters and its width is 9 meters?

2 Find the areas of the rectangles whose dimensions are:

  1. l = 23 mm   **2.** l = 12 in.   **3.** l = 6.8 m   **4.** $l = 1\frac{1}{3}$ yd   **5.** l = 3 ft 7 in.

     w = 17 mm w = 8 in. w = 1.625 m $w = 4\frac{1}{2}$ yd w = 1 ft 3 in.

3 Find the cost of each of the following:

  1. Resilvering a mirror 36 in. by 42 in. at $9.25 per square foot.
  2. Sodding a lawn 10 m by 6.2 m at $3.50 per square meter.

4 A schoolroom is 27 ft long and 20 ft wide. It has 6 windows, each measuring 3 ft by 9 ft. There are also 5 blackboards, each 5 ft by 3 ft; 6 blackboards, each $2\frac{1}{2}$ ft by 5 ft; and one bulletin board 6 ft by $3\frac{1}{2}$ ft. What is the ratio of the window area to the floor area? Find the ratio of the total blackboard area to the area of the bulletin board.

5 Find the area of a rectangular wing of an airplane if the span (length) is 47 ft 6 in. and chord (width) is 8 ft 3 in.

6 How many 4 cm by 6 cm tickets can be cut from 1.2 m by 1.5 m stock?

A square is a special kind of rectangle. We find the area of a square as we find the area of a rectangle. However, the length and width are equal.

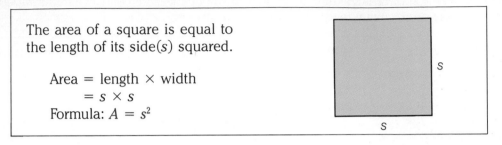

The area of a square is equal to the length of its side(s) squared.

Area = length × width
$= s \times s$
Formula: $A = s^2$

EXAMPLE:

Find the area of a square whose side is 37 meters long.

37 m

$s = 37$ m; $A = ?$

$A = s^2$
$A = (37)^2$
$A = 37 \times 37$
$A = 1{,}369$ m^2

```
   37
 × 37
  259
 1 11
1,369
```

ANSWER: 1,369 m^2

PRACTICE PROBLEMS

1 What is the area of a square whose side is 23 feet long?

2 Find the areas of squares whose sides measure:

   1. 10 cm   **2.** 42 mm   **3.** 32 in.

   4. 6,080 ft   **5.** 1,760 m **6.** 1.61 km

   7. 39.37 in. **8.** $18\frac{1}{2}$ ft   **9.** $4\frac{3}{8}$ in.

3 What is the cross sectional area of a square beam 14.5 cm on a side?

4 At $19.95 per square meter how much will a broadloom rug 5 m by 5 m cost?

5 The base of the Great Pyramid is 746 feet square. How many acres (to nearest hundredth) does it cover?

If we cut a right triangle off one end of a parallelogram (as shown) and move it to the other end, the resulting figure is a rectangle. Thus, the area of a parallelogram can be found by multiplying the length of a base times the height.

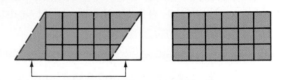

The area of a parallelogram is equal to the product of its base (b) and height (h).

Area = base × height

$= b \times h$

Formula: $A = bh$

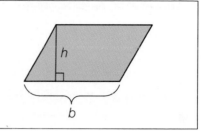

The height of a parallelogram is equal to the measure of its altitude.

EXAMPLE:

Find the area of a parallelogram having a height of 25 inches and a base 32 inches long.

25 in.

32 in.

$A = bh$
$A = 32 \times 25$
$A = 800 \text{ in.}^2$

$$
\begin{array}{r}
32 \\
\times\ 25 \\
\hline
160 \\
64 \\
\hline
800
\end{array}
$$

$b = 32 \text{ in.}, h = 25 \text{ in.}; A = ?$

ANSWER: 800 in.²

PRACTICE PROBLEMS

1 What is the area of a parallelogram if its height is 6 m and its base is 8 m long?

2 Find the areas of the parallelograms whose dimensions are:

1. $b = 14$ in.
$h = 26$ in.

2. $b = 98$ cm
$h = 75$ cm

3. $b = 4.7$ km
$h = 8.3$ km

4. $b = \frac{1}{2}$ in.
$h = 1\frac{1}{4}$ in.

5. $b = 4$ yd 1 ft
$h = 3$ yd 2 ft

3 Find the cost of seeding a lawn shaped like a parallelogram with a base length of 6 meters and a height of 5 meters. One kilogram of grass seed covers 30 square meters and costs $7.49.

15-8 Area of a Triangle

Any triangle is half of a corresponding parallelogram with the same base length and height. Thus, the area of a triangle is one-half that of the corresponding parallelogram.

> The area of a triangle is equal to one-half the base (b) times the height (h).
>
> Area $= \frac{1}{2} \times$ base $\times$ height
>
> $= \frac{1}{2} \times b \times h$
>
> Formula: $A = \frac{1}{2}bh$

The height of a triangle is equal to the measure of its altitude.

EXAMPLE:

Find the area of a triangle with a height of 26 millimeters and a base 17 millimeters long.

$b = 17$ mm, $h = 26$ mm
$A = ?$

$A = \frac{1}{2}bh$

$A = \frac{1}{2} \times 17 \times 26$

$A = 221$ mm²

$\frac{1}{\cancel{2}_1} \times \frac{17}{1} \times \frac{\cancel{26}^{13}}{1} = 221$

ANSWER: 221 mm²

PRACTICE PROBLEMS

1 What is the area of a triangle if its height is 10 inches and the base is 8 inches in length?

2 Find the areas of the triangles whose dimensions are:

1. $b = 12$ cm	**2.** $b = 10$ m	**3.** $b = 7$ km	**4.** $b = 16\frac{1}{2}$ ft	**5.** $b = 2$ ft
$h = 18$ cm	$h = 13$ m	$h = 5$ km	$h = 27$ ft	$h = 1$ ft 4 in.
6. $b = 8$ km	**7.** $b = 4.8$ m	**8.** $b = 1\frac{7}{8}$ in.	**9.** $b = 9\frac{1}{3}$ yd	**10.** $b = 4$ yd 7 in.
$h = 11$ km	$h = 3.4$ m	$h = 2\frac{1}{4}$ in.	$h = 6\frac{5}{8}$ yd	$h = 1$ ft 11 in.

3 How many square feet of surface does one side of a triangular sail expose if it has a base length of 10 ft and a height of 12 ft 6 in.?

15-9 Area of a Trapezoid

A diagonal separates a trapezoid into two triangles which have a common height but different bases.

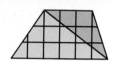

The area of a trapezoid is equal to one-half the height times the sum of the bases.

Area $= \frac{1}{2} \times$ height $\times$ sum of the two bases

$= \frac{1}{2} \times h \times (b_1 + b_2)$

Formula: $A = \frac{1}{2}h(b_1 + b_2)$ or $A = h \times \frac{b_1 + b_2}{2}$

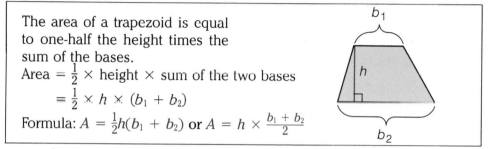

EXAMPLE:

Find the area of a trapezoid with bases 42 inches and 34 inches in length and a height of 29 inches.

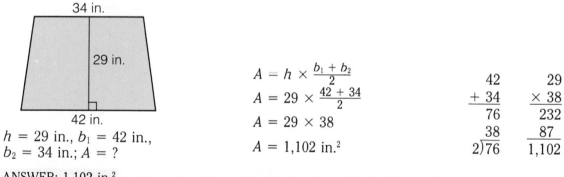

34 in.

29 in.

42 in.

$h = 29$ in., $b_1 = 42$ in., $b_2 = 34$ in.; $A = ?$

$A = h \times \frac{b_1 + b_2}{2}$

$A = 29 \times \frac{42 + 34}{2}$

$A = 29 \times 38$

$A = 1{,}102$ in.²

$$\begin{array}{r} 42 \\ +\ 34 \\ \hline 76 \end{array} \qquad \begin{array}{r} 29 \\ \times\ 38 \\ \hline 232 \\ 87 \\ \hline 1{,}102 \end{array}$$

$$2\overline{)76} \quad 38$$

ANSWER: 1,102 in.²

PRACTICE PROBLEMS

1 What is the area of a trapezoid if the height is 7 inches and the parallel sides are 8 inches and 14 inches long?

2 Find the areas of the trapezoids having the dimensions:

1. $h = 8$ cm
$b_1 = 4$ cm
$b_2 = 10$ cm

2. $h = 5$ mm
$b_1 = 9$ mm
$b_2 = 13$ mm

3. $h = 18$ m
$b_1 = 29$ m
$b_2 = 36$ m

4. $h = 6$ ft
$b_1 = 11\frac{3}{4}$ ft
$b_2 = 14\frac{1}{2}$ ft

5. $h = 10$ in.
$b_1 = 1$ ft
$b_2 = 1$ ft 4 in.

3 A section of a tapered airplane wing is shaped like a trapezoid. If the two parallel sides, measuring $3\frac{1}{2}$ feet and $5\frac{1}{2}$ feet, are 18 feet apart, find the area of the section.

The formula for the area of a circle also uses π. You may use $\frac{22}{7}$, 3.14, or—for greater accuracy—3.1416 for π. If we arrange the sectors of a circle approximately in the shape of a parallelogram, the areas are the same.

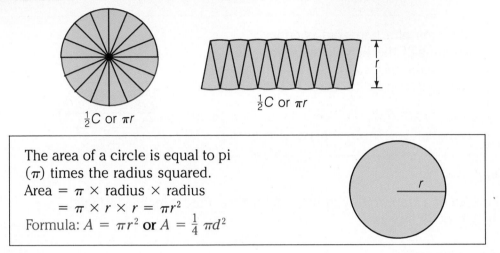

$\frac{1}{2}C$ or πr

$\frac{1}{2}C$ or πr

The area of a circle is equal to pi (π) times the radius squared.

Area $= \pi \times$ radius $\times$ radius

$= \pi \times r \times r = \pi r^2$

Formula: $A = \pi r^2$ or $A = \frac{1}{4} \pi d^2$

EXAMPLES:

Find the area of a circle having a radius of 5 meters.

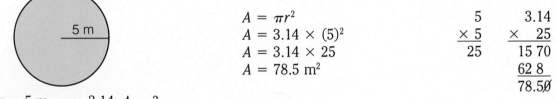

$A = \pi r^2$

$A = 3.14 \times (5)^2$

$A = 3.14 \times 25$

$A = 78.5 \text{ m}^2$

$$\begin{array}{r} 5 \\ \times\ 5 \\ \hline 25 \end{array} \qquad \begin{array}{r} 3.14 \\ \times\ \ \ 25 \\ \hline 15\ 70 \\ 62\ 8\ \ \\ \hline 78.5\cancel{0} \end{array}$$

$r = 5$ m, $\pi = 3.14$; $A = ?$

ANSWER: 78.5 m²

Find the area of a circle having a diameter of 14 feet.

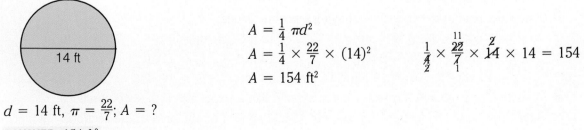

$A = \frac{1}{4} \pi d^2$

$A = \frac{1}{4} \times \frac{22}{7} \times (14)^2$

$A = 154 \text{ ft}^2$

$\frac{1}{\underset{2}{\cancel{4}}} \times \frac{\overset{11}{\cancel{22}}}{\underset{1}{\cancel{7}}} \times \overset{2}{\cancel{14}} \times 14 = 154$

$d = 14$ ft, $\pi = \frac{22}{7}$; $A = ?$

ANSWER: 154 ft²

PRACTICE PROBLEMS

1 What is the area of a circle whose radius is 6 centimeters?

2 Find the areas of circles having the following radii:

1. 13 m	**2.** 28 cm	**3.** 52 in.	**4.** 91 mm	**5.** 100 ft
6. 4.5 km	**7.** 1.375 mm	**8.** $5\frac{2}{3}$ yd	**9.** $\frac{1}{2}$ mi	**10.** 7 ft 4 in.

3 What is the area of a circle whose diameter is 24 feet?

4 Find the areas of circles having the following diameters:

1. 2 cm	**2.** 84 m	**3.** 19 yd	**4.** 63 mm	**5.** 220 yd
6. .75 km	**7.** 16.1 m	**8.** $\frac{5}{8}$ in.	**9.** $8\frac{1}{6}$ ft	**10.** 5 yd 1 ft

5 If a forest ranger can see from her tower for a distance of 42 km in all directions, how many square kilometers can she watch?

6 The dial of one of the world's largest clocks has a diameter of 50 ft. What is the area of the dial?

7 A revolving sprinkler sprays a lawn for a distance of 7 m. How many square meters does the sprinkler water in 1 revolution?

8 If a station can televise programs for a distance of 91 km, over what area may the programs be received?

9 Which is larger: the area of a circle 6 cm in diameter or the area of a square whose side is 6 cm? How much larger?

10 What is the area of one side of a washer if its diameter is $\frac{3}{4}$ in. and the diameter of the hole is $\frac{1}{4}$ in.?

Refresh Your Skills

1. Add: 1-4
39,627 + 46,858 + 90,377

2. Subtract: 1-5
50,602 − 43,575

3. Multiply: 1-6
806 × 450

4. Divide: 1-7
96)58,272

5. Add: 2-4
$8.79 + $13.63 + $.86

6. Subtract: 2-5
20 − 9.57

7. Multiply: 2-7
120 × 3.41

8. Divide: 2-8
.75)30

9. Add: 3-8
$8\frac{1}{2} + 5\frac{7}{8}$

10. Subtract: 3-9
$1\frac{1}{4} - \frac{4}{5}$

11. Multiply: 3-11
$\frac{3}{4} \times 1\frac{1}{2}$

12. Divide: 3-12
$2\frac{5}{8} \div 3$

13. Find 13% of $2,000
4-6

14. 25 is what percent of 20?
4-7

15. $12\frac{1}{2}$% of what number is 5?
4-8

15-11 Total Area of a Rectangular Solid

Vocabulary

The **total area** of the surface of a rectangular solid is the sum of the areas of all its faces and bases.

The faces of a rectangular solid are rectangles, as are the bases. Therefore, the total area of a rectangular solid is the sum of the areas of these rectangles.

Total area = 2 × length × width + 2 × length × height
+ 2 × width × height
= 2 × l × w + 2 × l × h + 2 × w × h

Formula: $A = 2lw + 2lh + 2wh$

EXAMPLE:

Find the total area of a rectangular solid 9 meters long, 6 meters wide, and 7 meters high.

$l = 9$ m, $w = 6$ m, $h = 7$ m; $A = ?$

ANSWER: 318 m²

$A = 2lw + 2lh + 2wh$
$A = 2 × 9 × 6 + 2 × 9 × 7 + 2 × 6 × 7$
$A = 108 + 126 + 84$
$A = 318$ m²

PRACTICE PROBLEMS

1 Find the total areas of rectangular solids with the following dimensions:

1. $l = 23$ cm
$w = 14$ cm
$h = 19$ cm

2. $l = 61$ mm
$w = 37$ mm
$h = 25$ mm

3. $l = 8.1$ m
$w = 2.7$ m
$h = 5.9$ m

4. $l = 9\frac{1}{2}$ in.
$w = 8$ in.
$h = 10\frac{3}{8}$ in.

5. $l = 5$ ft 4 in.
$w = 4$ ft 3 in.
$h = 7$ ft 6 in.

2 How many square meters of plywood will be needed to make a packing box with the dimensions 2.6 m by 2.5 m by 1.8 m?

3 A room is 18 ft long, 15 ft wide, and 9 ft high. Find the total area of the walls and ceiling, allowing a deduction of 64 ft² for the windows and doorway. How many gallons of paint are needed to cover the walls and ceiling with two coats if a gallon will cover 400 ft² with one coat? At $14.69 per gallon, how much will the paint cost?

15-12 Total Area of a Cube

The total area of the surface of a cube is equal to six times the area of one of its square faces.

Total area = 6 × edge × edge
= $6 \times e \times e = 6e^2$

Formula: $A = 6e^2$ or

$A = 6s^2$ (where s is the side of a square face).

EXAMPLE:

Find the total area of a cube whose edges measure 15 inches.

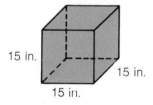

15 in.

15 in.

15 in.

$e = 15$ in.; $A = ?$
ANSWER: 1,350 in.^{2}

$A = 6e^2$
$A = 6 \times (15)^2$
$A = 6 \times 225$
$A = 1,350$ in.^{2}

$$\begin{array}{r} 15 \\ \times\ 15 \\ \hline 75 \\ 15 \\ \hline 225 \end{array} \qquad \begin{array}{r} 225 \\ \times\ \ 6 \\ \hline 1,350 \end{array}$$

PRACTICE PROBLEMS

1 Find the total areas of cubes whose edges measure:

1. 37 mm

2. 20 cm

3. 9.2 m

4. $3\frac{7}{8}$ in.

5. 1 ft 11 in.

2 A carton 1.5 m by 1.5 m by 1.5 m is made of cardboard. How many square meters of paper were used to make it if 10% extra was allowed for waste in cutting?

Lateral Area and Total Area of a Cylinder, and Area of a Sphere

Vocabulary

The **lateral area** of a cylinder is the area of the curved surface.

If you cut open a cylinder and lay the curved surface flat, it looks like a rectangle. Its area is equal to the product of the circumference of the base times the height.

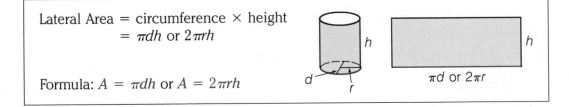

Lateral Area = circumference × height
 = πdh or $2\pi rh$

Formula: $A = \pi dh$ or $A = 2\pi rh$

EXAMPLE:

Find the lateral area of a cylinder having a diameter of 8 centimeters and a height of 9 centimeters.

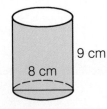

9 cm

8 cm

$A = \pi dh$

$A = 3.14 \times 8 \times 9$

$A = 226.08 \text{ cm}^2$

$$\begin{array}{r} 3.14 \\ \times\ \ \ 8 \\ \hline 25.12 \end{array}$$

$$\begin{array}{r} 25.12 \\ \times\ \ \ 9 \\ \hline 226.08 \end{array}$$

$d = 8$ cm
$h = 9$ cm
$\pi = 3.14$
$A = ?$

ANSWER: 226.08 cm²

To find the total area of a cylinder, add the lateral area to the area of the bases.

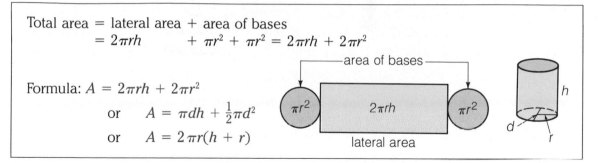

Total area = lateral area + area of bases
           = $2\pi rh$ + πr^2 + πr^2 = $2\pi rh + 2\pi r^2$

Formula: $A = 2\pi rh + 2\pi r^2$

    or $A = \pi dh + \frac{1}{2}\pi d^2$

    or $A = 2\pi r(h + r)$

EXAMPLE:

Find the total area of a cylinder when its radius is 21 feet and its height is 30 feet.

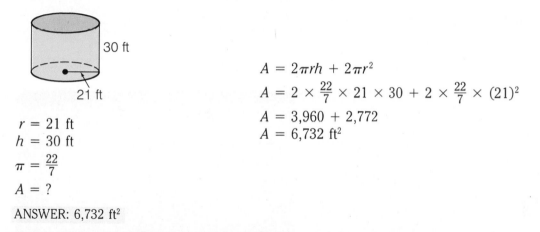

$r = 21$ ft

$h = 30$ ft

$\pi = \frac{22}{7}$

$A = ?$

ANSWER: 6,732 ft²

$A = 2\pi rh + 2\pi r^2$

$A = 2 \times \frac{22}{7} \times 21 \times 30 + 2 \times \frac{22}{7} \times (21)^2$

$A = 3,960 + 2,772$

$A = 6,732$ ft²

PRACTICE PROBLEMS

1 Find the lateral areas of cylinders with the following dimensions:

1. $d = 4$ m	**2.** $d = 60$ cm	**3.** $d = 28$ mm	**4.** $d = 3\frac{1}{2}$ in.	**5.** $d = 2$ ft 9 in.
$h = 10$ m	$h = 45$ cm	$h = 32$ mm	$h = 5\frac{1}{4}$ in.	$h = 4$ ft
6. $r = 3$ mm	**7.** $r = 14$ m	**8.** $r = 42$ cm	**9.** $r = 2\frac{3}{4}$ in.	**10.** $r = 8$ ft 6 in.
$h = 5$ mm	$h = 20$ m	$h = 18$ cm	$h = 8$ in.	$h = 5$ ft 4 in.

2 How many square inches of paper are needed to make a label on a can 4 inches in diameter and 5 inches high?

3 Find the total areas of cylinders with the following dimensions:

1. $r = 2$ cm	**2.** $r = 7$ m	**3.** $r = 30$ mm	**4.** $r = 4\frac{3}{8}$ in.	**5.** $r = 1$ ft 6 in.
$h = 4$ cm	$h = 12$ m	$h = 25$ mm	$h = 20$ in.	$h = 3$ ft
6. $d = 5$ m	**7.** $d = 4$ mm	**8.** $d = 35$ cm	**9.** $d = 2\frac{1}{2}$ in.	**10.** $d = 2$ ft 5 in.
$h = 8$ m	$h = 2$ mm	$h = 40$ cm	$h = 7$ in.	$h = 1$ ft 8 in.

4 How much asbestos paper covering is need to enclose the curved surface and the two ends of a hot water storage tank 35 centimeters in diameter and 1.5 meters high?

The area of the surface of a sphere is equal to 4 times pi (π) times the radius squared.

$$\text{Area} = 4 \times \pi \times r^2 \quad \text{Formula:} \ A = 4\pi r^2$$

EXAMPLE:

Find the area of the surface of a sphere having a radius of 6 cm.

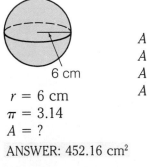

6 cm

$r = 6$ cm
$\pi = 3.14$
$A = ?$
ANSWER: 452.16 cm²

$A = 4\pi r^2$
$A = 4 \times 3.14 \times (6)^2$
$A = 12.56 \times 36$
$A = 452.16$ cm²

PRACTICE PROBLEMS

1 Find the surface areas of spheres having the following radii:

   1. 12 m   **2.** 28 mm   **3.** 6.3 cm   **4.** $9\frac{5}{8}$ in.   **5.** 5 ft 1 in.

2 Find the surface areas of spheres having the following diameters:

   1. 16 mm   **2.** 70 cm   **3.** 14.8 m   **4.** $6\frac{3}{4}$ yd   **5.** 4 ft 6 in.

3 What is the area of the earth's surface if its diameter is 7,900 miles?

15-14 Volume of a Rectangular Solid

Vocabulary

The **volume** of any solid is the number of units of cubic measure contained in the solid. Volume is measured in cubic units.

When computing volume, express all linear measurements with the same unit of measure.

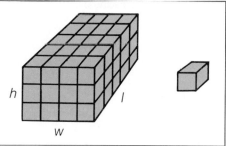

The volume of a rectangular solid is equal to the length (l) times the width (w) times the height (h).

Volume = length × width × height
$$= l \times w \times h$$
Formula: $V = lwh$

Sometimes the formula $V = Bh$ is used, where B is the area of the base ($l\,w$).

EXAMPLE:

Find the volume of a rectangular solid 8 meters long, 5 meters wide, and 7 meters high.

7 m
8 m
5 m
$l = 8$ m, $w = 5$ m, $h = 7$ m; $V = ?$
ANSWER: 280 m³

$V = lwh$
$V = 8 \times 5 \times 7$
$V = 280$ m³

$$\begin{array}{r} 8 \\ \times\,5 \\ \hline 40 \end{array} \qquad \begin{array}{r} 40 \\ \times\,7 \\ \hline 280 \end{array}$$

PRACTICE PROBLEMS

1 What is the volume of a rectangular solid if it is 7 in. long, 4 in. wide, and 9 in. high?

2 Find the volumes of rectangular solids having the following dimensions:

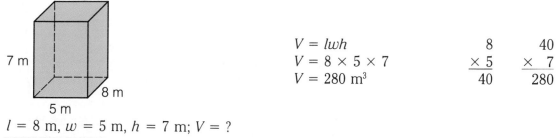

1. $l = 8$ mm
   $w = 3$ mm
   $h = 6$ mm

2. $l = 12$ m
   $w = 9$ m
   $h = 10$ m

3. $l = 17$ in.
   $w = 18$ in.
   $h = 14$ in.

4. $l = 36$ m
   $w = 36$ m
   $h = 13$ m

5. $l = 62$ cm
   $w = 40$ cm
   $h = 19$ cm

521

15-15 Volume of a Cube

A cube is also a rectangular solid, so its volume is also length times width times height. For a cube, these lengths of the three edges are all the same.

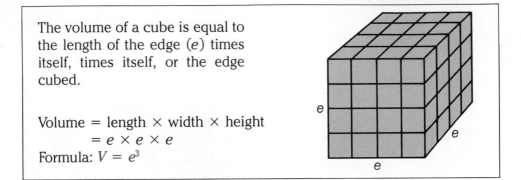

The volume of a cube is equal to the length of the edge (e) times itself, times itself, or the edge cubed.

Volume = length × width × height
   = $e \times e \times e$
Formula: $V = e^3$

EXAMPLE:

Find the volume of a cube whose edge measures 17 inches.

17 in.
17 in.
17 in.

$e = 17$ in.; $V = ?$

ANSWER: 4,913 in.³

$V = e^3$
$V = (17)^3$
$V = 17 \times 17 \times 17$
$V = 4{,}913$ in.³

```
   17        289
 × 17      ×  17
  119      2 023
   17      2 89
  289      4,913
```

PRACTICE PROBLEMS

1 What is the volume of a cube whose edge is 25 centimeters?

2 Find the volumes of cubes whose edges measure:

1. 9 m	**2.** 14 mm	**3.** 11 ft	**4.** 27 cm	**5.** 3.1 m
6. 4.5 cm	**7.** 1.09 m	**8.** $4\frac{3}{4}$ ft	**9.** 5 yd 2 ft	**10.** 1 ft 10 in.

3 How many cu. ft of space are in a bin 6 ft by 6 ft by 6 ft?

4 The volume of an 8-centimeter cube is how many times as large as the volume of a 2-centimeter cube?

Volume of a Right Circular Cylinder

The volume of a cylinder is equal to the area of the base times the height.
Volume = (area of the base) × height
$$= B \times h$$
$$= \pi r^2 \times h$$
Formula: $V = \pi r^2 h$ or $V = \frac{1}{4}\pi d^2 h$

Remember, $\frac{22}{7}$ or 3.14 can be used for π.

EXAMPLE:

Find the volume of a cylinder 75 cm high with a base that has a radius of 30 cm.

75 cm

30 cm

$V = \pi r^2 h$
$V = 3.14 \times (30)^2 \times 75$
$V = 3.14 \times 900 \times 75$
$V = 211,950 \text{ cm}^3$

$$\begin{array}{r} 30 \\ \times\ 30 \\ \hline 900 \end{array}$$

$$\begin{array}{r} 3.14 \\ \times\ 900 \\ \hline 2,826.00 \end{array}$$

$$\begin{array}{r} 2,826 \\ \times\ 75 \\ \hline 14\ 130 \\ 197\ 82 \\ \hline 211,950 \end{array}$$

$r = 30$ cm, $h = 75$ cm,
$\pi = 3.14;\quad V = ?$

ANSWER: 211,950 cm³

PRACTICE PROBLEMS

1 What is the volume of a cylinder if the radius of its base is 3 m and the height is 6 m?

2 What is the volume of a cylinder if the diameter of its base is 10 ft and the height is 16 ft?

3 Find the volumes of cylinders having the following dimensions:

    1. $r = 5$ mm   **2.** $r = 14$ m   **3.** $r = 19$ ft   **4.** $r = 2\frac{5}{8}$ in.   **5.** $r = 1$ ft 6 in.

        $h = 8$ mm $h = 10$ m $h = 25$ ft $h = 7\frac{1}{2}$ in. $h = 9$ ft 2 in.

4 Which container holds more, one that is 3 cm in diameter and 4 cm high or one that is 4 cm in diameter and 3 cm high? How many cubic centimeters more?

5 How many cubic yards of dirt must be dug to make a well 4 ft 6 in. in diameter and 42 ft deep?

6 What is the weight of a round steel rod 8 ft long and $\frac{3}{4}$ in. in diameter? A cubic foot of steel weighs 490 lb.

The volume of a sphere is equal to $\frac{4}{3}$ times pi (π) times the radius cubed.

Volume $= \frac{4}{3} \times \pi \times$ radius $\times$ radius $\times$ radius

$\qquad = \frac{4}{3} \times \pi \times r^3$

Formula: $V = \frac{4}{3}\pi r^3$ or $V = \frac{\pi d^3}{6}$

EXAMPLE:

Find the volume of a sphere with radius 1.2 cm.

1.2 cm

$\pi = 3.14$, $r = 1.2$ cm; $V = ?$

ANSWER: 7.23456 cm³

$V = \frac{4}{3}\pi r^3$

$V = \frac{4}{3} \times 3.14 \times (1.2)^3$

$V = \frac{4}{3} \times 5.42592$

$V = 7.23456$ cm³

The volume of a right circular cone is $\frac{1}{3}$ that of a right circular cylinder with the same base and height.

Volume $= \frac{1}{3} \times$ area of the base $\times$ height

$\qquad = \frac{1}{3} \times \pi r^2 \times h$

Formula: $V = \frac{1}{3}\pi r^2 h$

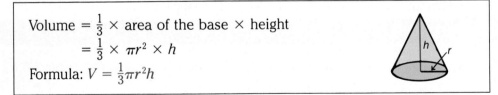

EXAMPLE:

Find the volume of a cone with radius 5 m and height 7 m.

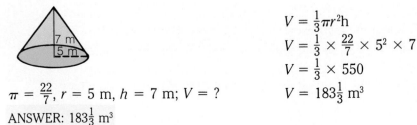

7 m
5 m

$\pi = \frac{22}{7}$, $r = 5$ m, $h = 7$ m; $V = ?$

ANSWER: $183\frac{1}{3}$ m³

$V = \frac{1}{3}\pi r^2 h$

$V = \frac{1}{3} \times \frac{22}{7} \times 5^2 \times 7$

$V = \frac{1}{3} \times 550$

$V = 183\frac{1}{3}$ m³

Similarly, the **volume of a regular pyramid** is $\frac{1}{3}$ that of a rectangular solid with the same base and height.

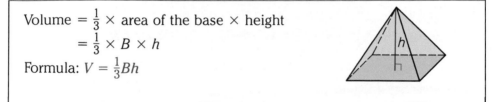

Volume $= \frac{1}{3} \times$ area of the base $\times$ height

$\qquad = \frac{1}{3} \times B \times h$

Formula: $V = \frac{1}{3}Bh$

EXAMPLE:

Find the volume of a regular triangular pyramid if the area of the base is 45 in.^{2} and the height is 9 in.

$V = \frac{1}{3}Bh$

$V = \frac{1}{3} \times 45 \times 9$

$V = 135$ in.^{3}

$h = 9$ in., $B = 45$ in.^{2}; $V = ?$

ANSWER: 135 in.^{3}

PRACTICE PROBLEMS

1 Find the volumes of spheres having the following radii:
  1. 30 m   **2.** 56 ft   **3.** 10.4 cm   **4.** $6\frac{5}{16}$ in.   **5.** 2 ft 8 in.

2 Find the volumes of spheres having the following diameters:
  1. 26 mm   **2.** 84 cm   **3.** 11.3 m   **4.** $12\frac{1}{2}$ in.   **5.** 4 ft 5 in.

3 How many cubic feet of air does a basketball contain if its diameter is 10 in.?

4 Find the volumes of cones having the following dimensions:
  1. $r = 8$ cm   **2.** $r = 14$ mm   **3.** $r = 6.5$ m   **4.** $r = 4$ ft 3 in.
     $h = 23$ cm $h = 12$ mm $h = 9$ m $h = 6$ ft

5 A conical pile of sand is 2.5 m in diameter and 9 m high. How many cubic meters of sand are in the pile?

6 How many bushels of grain are in a conical pile 16 ft in diameter and 5 ft high?

7 Find the volume of a pyramid when the side of its square base is 40 m and the height is 27 m.

8 What is the volume of a pyramid when its rectangular base is 23 m long and 18 m wide and the height is 35 m?

9 How many ft^3 of space are inside a tent in the shape of a square pyramid 15 ft on each side of the base and 16 ft high?

Many businesses depend on their air conditioning and refrigeration systems. Hospitals need to keep medicines cold. Food companies need to keep food fresh. When air conditioning or refrigeration equipment must be installed or repaired, *air-conditioning and refrigeration mechanics* do the job.

Air-conditioning and refrigeration mechanics read blueprints, install motors and other parts, and connect the equipment to ducts and electrical lines. When a system breaks down they check and repair it.

Air-conditioning and refrigeration mechanics often begin by helping experienced mechanics. Most have taken related courses in high school, vocational school, or junior college.

1. An air-conditioning and refrigeration mechanic needs to wrap some insulation around the perimeter of a rectangular cooling duct. If the duct is 8 inches by 16 inches, how long a piece of insulation is needed?

2. To install the correct size air conditioner, a mechanic needs to know the volume of a room. The room is 10 feet wide, 15 feet long, and 8 feet high. What is the volume?

3. How long a piece of insulation is needed to wrap around a cylindrical duct that is 7 inches in diameter?

4. The cylinder of a compressor has a radius of 1.2 centimeters. What is the area of the opening in the cylinder?

5. A tank for refrigerant is a cylinder 16 inches high. Its radius is 8 inches. What is its volume?

6. A grill has square openings $\frac{3}{4}$ inch on a side. What is the area of each opening?

CHAPTER REVIEW

Vocabulary

area p. 509   lateral area p. 518   total area p. 516
circumference p. 507   perimeter p. 504   volume p. 521

The numerals in the boxes indicate Lessons where help may be found.

1. What is the perimeter of a rectangle 135 meters long and 92 meters wide? $\boxed{15\text{-}1}$
2. Find the circumference of a circle whose diameter is 84 centimeters. $\boxed{15\text{-}4}$
3. What is the perimeter of a square whose side measures $7\frac{5}{8}$ inches? $\boxed{15\text{-}2}$
4. What is the circumference of a circle whose radius is 120 feet? $\boxed{15\text{-}4}$
5. Find the perimeter of a triangle whose sides measure 2.7 meters, 4.2 meters, and 3.5 meters. $\boxed{15\text{-}3}$

Find the area of:

6. A rectangle 65 meters long and 43 meters wide. $\boxed{15\text{-}5}$
7. A circle whose radius is 37 centimeters. $\boxed{15\text{-}10}$
8. A square whose side measures 21 feet. $\boxed{15\text{-}6}$
9. A parallelogram with an altitude of 159 mm and a base of 250 mm. $\boxed{15\text{-}7}$
10. A circle whose diameter is $8\frac{3}{4}$ inches. $\boxed{15\text{-}10}$
11. A triangle with a base of 60 km and a height of 82 km. $\boxed{15\text{-}8}$
12. A trapezoid with bases of 101 ft and 53 ft and a height of 64 ft. $\boxed{15\text{-}9}$

Find the total area of:

13. A cube whose edge measures 19 centimeters. $\boxed{15\text{-}12}$
14. A rectangular solid 31 yd long, 27 yd wide, and 16 yd high. $\boxed{15\text{-}11}$
15. A right circular cylinder 63 mm in diameter and 75 mm high. $\boxed{15\text{-}13}$
16. A sphere whose radius is 6 meters. $\boxed{15\text{-}13}$

Find the volume of:

17. A cube whose edge measures 38 millimeters. $\boxed{15\text{-}15}$
18. A cylinder 16 feet high with the radius of its base 10 feet. $\boxed{15\text{-}16}$
19. A rectangular solid 60 cm long, 43 cm wide, and 18 cm high. $\boxed{15\text{-}14}$
20. A sphere whose diameter is 9 meters. $\boxed{15\text{-}17}$
21. A right circular cone 40 ft high with the diameter of its base 28 ft. $\boxed{15\text{-}17}$
22. A square pyramid 12.5 m on each side and 15 m high. $\boxed{15\text{-}17}$
23. Which is longer, the perimeter of a bridge table measuring 76 centimeters on a side or the circumference of a circular table measuring 90 centimeters in diameter. $\boxed{15\text{-}2,\ 15\text{-}4}$
24. The cooking area of a circular grill having a diameter of 60 centimeters is how many times as large as the area of a grill having a diameter of 40 centimeters. $\boxed{15\text{-}10}$
25. A circular wading pool is 96 inches in diameter and 14 inches deep. How many cubic feet of water will it hold? $\boxed{15\text{-}16}$

COMPETENCY CHECK TEST

The numerals in the boxes indicate Lessons where help may be found.

Solve each problem and select the letter corresponding to your answer.

1. The perimeter of a rectangle 17 m long and 13 m wide is: 15-1
   a. 221 m   **b.** 30 m   **c.** 120 m   **d.** 60 m

2. The circumference of a circle whose diameter is 14 mm is: ($\pi = \frac{22}{7}$) 15-4
   a. 154 mm   **b.** 44 mm   **c.** 7 mm   **d.** Answer not given

3. The perimeter of a square whose side measures 31 inches is: 15-2
   a. 961 in.   **b.** 62 in.   **c.** 31 in.   **d.** 124 in.

4. The area of a triangle whose base is 27 in. and altitude is 16 in. is: 15-8
   a. 216 in.²   **b.** 43 in.²   **c.** 432 in.²   **d.** 22.5 in.²

5. The area of a rectangle 16 feet long and 7 feet wide is: 15-5
   a. 23 ft²   **b.** 112 ft²   **c.** 9 ft²   **d.** 46 ft²

6. The area of a circle whose radius is 35 cm is: ($\pi = \frac{22}{7}$) 15-10
   a. 3.5 m²   **b.** 110 cm²   **c.** 3,850 cm²   **d.** Answer not given

7. The area of a square whose side measures 1.5 m is: 15-6
   a. 22.5 m²   **b.** 2.25 m²   **c.** 3 m²   **d.** 6 m²

8. The volume of this rectangular box is: 15-14
   a. 14 in.³   **b.** 31 in.³

   c. 168 in.³   **d.** 84 in.³

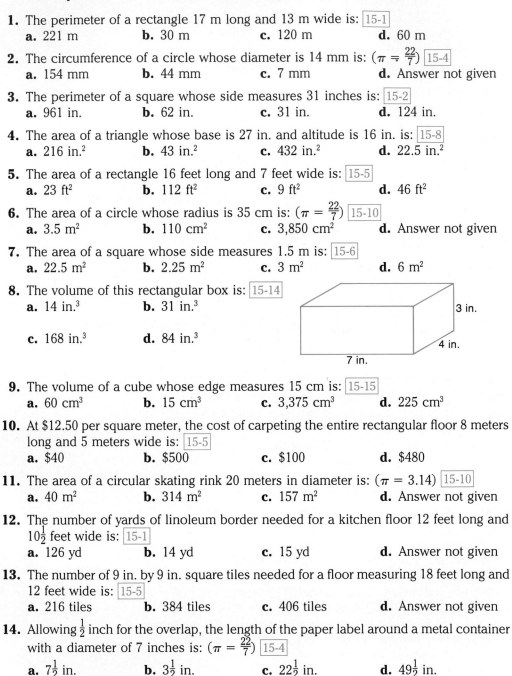

9. The volume of a cube whose edge measures 15 cm is: 15-15
   a. 60 cm³   **b.** 15 cm³   **c.** 3,375 cm³   **d.** 225 cm³

10. At $12.50 per square meter, the cost of carpeting the entire rectangular floor 8 meters long and 5 meters wide is: 15-5
    a. $40   **b.** $500   **c.** $100   **d.** $480

11. The area of a circular skating rink 20 meters in diameter is: ($\pi = 3.14$) 15-10
    a. 40 m²   **b.** 314 m²   **c.** 157 m²   **d.** Answer not given

12. The number of yards of linoleum border needed for a kitchen floor 12 feet long and $10\frac{1}{2}$ feet wide is: 15-1
    a. 126 yd   **b.** 14 yd   **c.** 15 yd   **d.** Answer not given

13. The number of 9 in. by 9 in. square tiles needed for a floor measuring 18 feet long and 12 feet wide is: 15-5
    a. 216 tiles   **b.** 384 tiles   **c.** 406 tiles   **d.** Answer not given

14. Allowing $\frac{1}{2}$ inch for the overlap, the length of the paper label around a metal container with a diameter of 7 inches is: ($\pi = \frac{22}{7}$) 15-4
    a. $7\frac{1}{2}$ in.   **b.** $3\frac{1}{2}$ in.   **c.** $22\frac{1}{2}$ in.   **d.** $49\frac{1}{2}$ in.

Achievement Test

12-1 Read, or write in words, each of the following:

  1. $\overline{EF}$ **2.** $\overleftrightarrow{MY}$ **3.** $\overrightarrow{LN}$ **4.** $\overleftrightarrow{CR}$ **5.** $\overline{BT}$ **6.** $\overrightarrow{HG}$

Name each of the following and express them symbolically:

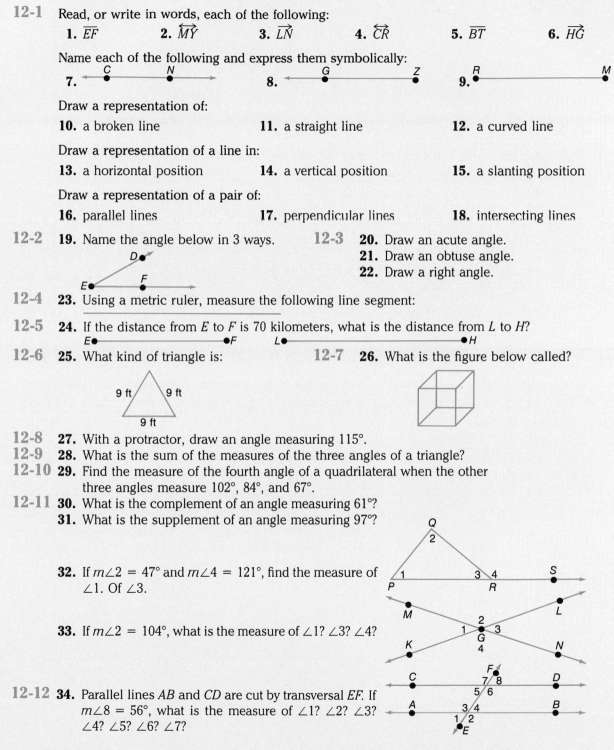

  7. **8.** **9.**

Draw a representation of:

10. a broken line   **11.** a straight line   **12.** a curved line

Draw a representation of a line in:

13. a horizontal position   **14.** a vertical position   **15.** a slanting position

Draw a representation of a pair of:

16. parallel lines   **17.** perpendicular lines   **18.** intersecting lines

12-2 **19.** Name the angle below in 3 ways.   **12-3** **20.** Draw an acute angle.
  21. Draw an obtuse angle.
  22. Draw a right angle.

12-4 **23.** Using a metric ruler, measure the following line segment:

12-5 **24.** If the distance from E to F is 70 kilometers, what is the distance from L to H?

12-6 **25.** What kind of triangle is: **12-7** **26.** What is the figure below called?

9 ft   9 ft

9 ft

12-8 **27.** With a protractor, draw an angle measuring 115°.
12-9 **28.** What is the sum of the measures of the three angles of a triangle?
12-10 **29.** Find the measure of the fourth angle of a quadrilateral when the other
  three angles measure 102°, 84°, and 67°.
12-11 **30.** What is the complement of an angle measuring 61°?
  31. What is the supplement of an angle measuring 97°?

32. If $m\angle 2 = 47°$ and $m\angle 4 = 121°$, find the measure of
  $\angle 1$. Of $\angle 3$.

33. If $m\angle 2 = 104°$, what is the measure of $\angle 1$? $\angle 3$? $\angle 4$?

12-12 **34.** Parallel lines AB and CD are cut by transversal EF. If
  $m\angle 8 = 56°$, what is the measure of $\angle 1$? $\angle 2$? $\angle 3$?
  $\angle 4$? $\angle 5$? $\angle 6$? $\angle 7$?

12-13 35. Find the indicated missing parts. **12-14 36.** Find the indicated missing parts:

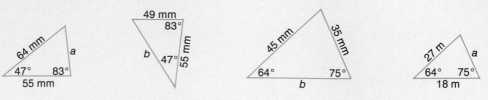

13-1 37. Make a copy of the following line segment: _____

13-2 38. Draw with a protractor an angle of 74°. Then construct with a compass an angle equal to it. Check with a protractor.

13-3 39. Construct a triangle with sides measuring 56 mm, 47 mm, and 63 mm.

40. Construct a triangle with sides measuring 7.2 cm and 5.1 cm and an included angle of 51°.

41. Construct a triangle with angles measuring 39° and 101° and an included side measuring $3\frac{1}{8}$ in.

13-4 42. Draw any line. Using a compass, construct a perpendicular to this line at a point on the line.

13-5 43. Draw any line. Using a compass, construct a perpendicular to this line from a point not on the line.

13-6 44. Draw a line segment 46 mm long. Using a compass, bisect this segment. Check with a ruler.

13-7 45. Draw an angle measuring 62°. Using a compass, bisect this angle. Check with a ruler.

13-8 46. Draw any line. Locate a point outside this line. Through this point construct a line parallel to the first line.

14-1 47. Find the square of 1.6

14-2 48. Find $\sqrt{169}$

14-3 49. Find $\sqrt{32}$ by the estimate, divide, and average method.

14-4 50. Find the hypotenuse of a right triangle if the altitude is 63 m and the base is 60 m.

51. Find the base of a right triangle if the hypotenuse is 120 cm and the altitude is 72 cm.

52. Find the altitude of a right triangle if the hypotenuse is 73 ft and the base is 55 ft.

14-5 53. Find the height of a TV antenna tower that casts a shadow 6 m long when a nearby light pole, 2.4 m high, casts a shadow .4 m long.

In right triangle *ABC*:

14-6 54. Find side *a*: **55.** Find side *b*: **56.** Find side *b*:

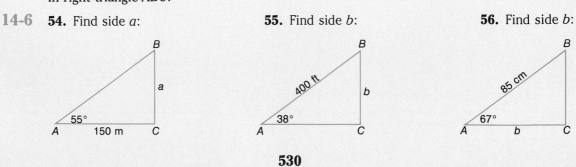

15-1   **57.** Find the perimeter of a rectangle 41 m long and 17 m wide.

15-2   **58.** Find the perimeter of a square whose side measures 26 cm.

15-3   **59.** Find the perimeter of a triangle with sides measuring $4\frac{3}{4}$ in., $3\frac{5}{8}$ in., and $5\frac{11}{16}$ in.

15-4   Find the circumference of:

     60. A circle whose diameter is 28 mm.

     61. A circle whose radius is 80 ft.

15-5   **62.** Find the area of a rectangle 51 cm long and 18 cm wide.

15-6   **63.** Find the area of a square whose side measures 47 ft.

15-7   **64.** Find the area of a parallelogram with an altitude of 24 m and a base of 49 m.

15-8   **65.** Find the area of a triangle whose altitude is 39 cm and base is 28 cm.

15-9   **66.** Find the area of a trapezoid with bases of 43 mm and 37 mm and a height of 25 mm.

15-10 Find the area of:

     67. A circle whose radius is 9 in.   **68.** A circle whose diameter is 56 in.

15-11 **69.** Find the total area of the outside surface of a rectangular solid 2.5 meters long, 1.7 meters wide and 2 meters high.

15-12 **70.** Find the total area of the outside surface of a cube whose edge measures $4\frac{1}{2}$ ft.

15-13 Find the total area of the outside surface of:

     71. A right circular cylinder 54 cm in diameter and 75 cm in height.

     72. A sphere whose diameter is 49 centimeters.

15-14 **73.** Find the volume of a rectangular solid 6.9 m long, 1.8 m wide, and 4.7 m high.

15-15 **74.** Find the volume of a cube whose edge measures 79 mm.

15-16 **75.** Find the volume of a right circular cylinder with a diameter of 70 ft and a height of 46 ft.

15-17 Find the volume of:

     76. A sphere whose diameter is 84 cm.

     77. A right circular cone 14 inches in diameter and 10 inches high.

     78. A square pyramid 10 meters on each side of the base and 15 meters high.

UNIT 5

Algebra

Inventory Test

16-1 Which of the following numerals name:

1. Positive numbers?

2. Negative numbers?

$$-\tfrac{2}{3} \qquad +86 \qquad -.625 \qquad -5\tfrac{1}{2} \qquad +1.7 \qquad -300$$

16-2 Which of the following numerals name:

3. Integers?　　**4.** Rational numbers?　　**5.** Irrational numbers?　　**6.** Real numbers?

$$-.3 \qquad +8\tfrac{3}{4} \qquad -69 \qquad -\sqrt{57} \qquad -\tfrac{7}{10} \qquad +\tfrac{36}{6}$$

16-3　**7.** Find the value of $|-63|$

16-4　**8.** Draw the graph of -3, -1, 0, and 2 on the number line.

9. Write the coordinates of which the following is the graph:

16-5　**10.** If -5 kilograms means 5 kilograms underweight, what does $+3$ kilograms mean?

16-6 Which of the following are true?

11. $-3 > -1$ 　　　　**12.** $+2 \not< -2$ 　　　　**13.** $0 < -4$ 　　　　**14.** $+4 \not> +6$

16-7 Using the number line as the scale, draw the vector and write the numeral represented by this vector that illustrates a movement from:

15. 0 to -4 　　　　**16.** -3 to $+5$ 　　　　**17.** $+1$ to -6 　　　　**18.** $+7$ to $+2$

16-8　**19.** What is the opposite of -18?

20. What is the additive inverse of $+23$?

21. Write symbolically: The opposite of negative ten is positive ten.

22. Which has the greater opposite number: $+2$ or $+5$?

16-9 Add:

23. $\begin{array}{r} -5 \\ +9 \\ \hline \end{array}$ 　　**24.** $\begin{array}{r} +6 \\ +7 \\ \hline \end{array}$ 　　**25.** $\begin{array}{r} -12 \\ +8 \\ \hline \end{array}$ 　　**26.** $\begin{array}{r} -9 \\ -7 \\ \hline \end{array}$ 　　**27.** $\begin{array}{r} -3 \\ +3 \\ \hline \end{array}$

28. $(-7) + (+4) + (-9) + (+11)$ 　　　　**29.** Simplify: $-4 + 6 - 3 - 7 + 5$

16-10 Subtract:

30. $\begin{array}{r} +6 \\ -8 \\ \hline \end{array}$ 　　**31.** $\begin{array}{r} +9 \\ +11 \\ \hline \end{array}$ 　　**32.** $\begin{array}{r} -5 \\ -4 \\ \hline \end{array}$ 　　**33.** $\begin{array}{r} -14 \\ +5 \\ \hline \end{array}$ 　　**34.** $\begin{array}{r} 0 \\ -4 \\ \hline \end{array}$

35. $(+5) - (-5)$ 　　　　**36.** Take 6 from 2.

16-11 Multiply:

37. $\begin{array}{r} +9 \\ -7 \\ \hline \end{array}$ 　　**38.** $\begin{array}{r} -4 \\ +8 \\ \hline \end{array}$ 　　**39.** $\begin{array}{r} -10 \\ -10 \\ \hline \end{array}$ 　　**40.** $\begin{array}{r} +7 \\ +3 \\ \hline \end{array}$ 　　**41.** $\begin{array}{r} 0 \\ -5 \\ \hline \end{array}$

42. $(-1)(-5)(-2)(+7)$ 　　　　**43.** Find the value of $(-2)^4$.

16-12 Divide:

44. $\dfrac{-40}{-8}$ 　　**45.** $\dfrac{-72}{+9}$ 　　**46.** $\dfrac{+24}{-1}$ 　　**47.** $\dfrac{+60}{+15}$ 　　**48.** $\dfrac{-16}{+16}$

49. $(-56) \div (-7)$ 　　　　**50.** Simplify: $\dfrac{6(-2) + 5(-3)}{-3(4 - 7)}$

17-1 Write as an algebraic expression:

51. The sum of l and w. **52.** The difference between ten and four.

53. The quotient of six divided by r. **54.** The product of a and nine.

55. Six times the square of the side (s).

17-2 Write each of the following as a sentence symbolically:

56. Some number y decreased by five is equal to twelve.

57. Each number x increased by two is greater than eight.

58. Ten times each number n is less than or equal to fourteen.

59. Each number s is greater than negative three and less than positive nine.

17-3 **60.** Express as a formula: Centripetal force (F) equals the product of the weight of the body (w) and the square of the velocity (v) divided by the product of the acceleration of gravity (g) and the radius of the circle (r).

17-4 Find the value of:

61. $m + 9n$ when $m = 8$ and $n = 3$.

62. $a + b(a - b)$ when $a = 6$ and $b = 2$.

17-5 **63.** Find the value of E when $I = 5$, $r = 16$, and $R = 20$.
Formula: $E = Ir + IR$

17-6 Which of the following equations:

64. Have 4 as the solution? **65.** Are equivalent? **66.** Is in simplest form?

$n - 3 = 7$	$n = 4$	$6n - 7 = 17$
$n + 9 = 13$	$\frac{n}{8} = 2$	$9n + n = 20$

17-7 What operation with what number do you use on both sides of each of the following equations to get an equivalent equation in simplest form? Also write this equivalent equation.

67. $7x = 56$ **68.** $m + 8 = 14$ **69.** $y - 11 = 6$ **70.** $\frac{a}{15} = 9$

Solve and check:

17-8 **71.** $x + 6 = 13$ **17-9** **72.** $b - 7 = 4$

17-10 **73.** $-7y = 63$ **74.** $4m = 76$

17-11 **75.** $\frac{a}{9} = 18$ **76.** $\frac{z}{-3} = 15$

17-12 **77.** $6n + 17 = 35$ **78.** $8x - 3x = 60$

17-13 **79.** Find the value of t when $v = 217$, $V = 25$, and $g = 32$. Formula: $v = V + gt$

17-14 80. Solve by the equation method:

Pedro bought a football at a 40% reduction sale paying $10.65 for it. What was the regular price?

17-15 Find the solutions of each of the following inequalities when the replacements for the variable are all the real numbers:

81. $8x > 136$ **82.** $n + 3 < 11$

83. $t - 5 \not> -7$ **84.** $4a + 13 \geq 1$

85. $6a - 7a < 0$ **86.** $\frac{d}{3} \neq -1$

87. $-7y \not< -21$ **88.** $2n - 5n > -15$

89. $\frac{3}{5} x \leq 42$

18-1 90. Write the ordered pair of numbers which has -6 as the first component and 4 as the second component.

18-2 What are the coordinates of:

91. Point A?   **92.** Point B?

93. Point C?   **94.** Point D?

95. Point E?   **96.** Point F?

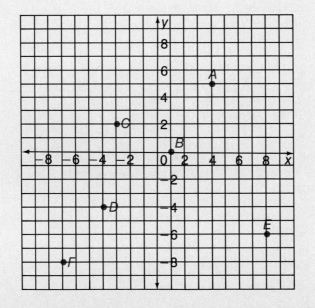

18-3 On graph paper, draw axes and plot the points with the following coordinates:

97. $G(-5, -2)$ **98.** $H(0, 6)$ **99.** $I(4, -3)$

18-4 100. Draw the graph of the equation: $x + y = 7$.

INTRODUCTION TO ALGEBRA

In algebra we continue to use the operational $(+, -, \times, \div$ or $\overline{)}$) of arithmetic. The exponent (p. 42), the square root symbol $\sqrt{\ }$, the parentheses ( ), the absolute value symbol | |, and new meanings for the $+$ and $-$ symbols (positive and negative) are also used.

In the lessons in reading, writing, and evaluation of algebraic expressions and sentences (equations, formulas, and inequalities) where the language of algebra is developed, the symbols $=, \neq, <, \not<, >, \not>, \leq,$ and $\geq$ are used in addition to the above symbols.

The number system is enlarged first by the introduction of negative numbers and then by the real numbers. Correspondingly the number line is extended to include points associated first with negative integers and then with real numbers.

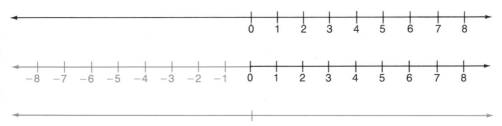

Knowledge of how to solve an equation provides a very essential mathematical tool both to solve problems and to deal with the important subject of the formula.

The graphing of the equation and the inequality on the number line and of the equation on the number plane are also included.

PROPERTIES

Operations in mathematics have certain characteristics or properties.

Commutative Property of Addition

The order in adding two numbers does not affect the sum.

Since $6 + 5 = 11$ and $5 + 6 = 11$, then $6 + 5 = 5 + 6$

Note: The operation of subtraction does not have this property.

$$6 - 5 \neq 5 - 6$$

Commutative Property of Multiplication

The order in multiplying two numbers does not affect the product.

Since $6 \times 3 = 18$ and $3 \times 6 = 18$, then $6 \times 3 = 3 \times 6$

Note: The operation of division does not have this property.

$$\frac{6}{3} \neq \frac{3}{6}$$

Associative Property of Addition

The order in adding three numbers does not affect the sum.

Since $5 + (6 + 4) = 15$ and $(5 + 6) + 4 = 15$,

then $5 + (6 + 4) = (5 + 6) + 4$

Note: The operation of subtraction does not have this property.

$$10 - 6 - 2 \neq 10 - (6 - 2)$$

Associative Property of Multiplication

The order in multiplying three numbers does not affect the product.

Since $6 \times (3 \times 4) = 72$ and $(6 \times 3) \times 4 = 72$,

then $6 \times (3 \times 4) = (6 \times 3) \times 4$

Note: The operation of division does not have this property.

$$24 \div 4 \div 2 \neq 24 \div (4 \div 2)$$

Distributive Property of Multiplication over Addition

Since $5 \times (7 + 3) = 5(10) = 50$ and $5 \times 7 + 5 \times 3 = 50$,

then $5 \times (7 + 3) = 5 \times 7 + 5 \times 3$

We say multiplication is distributed over addition.

Note: The operation of division is not distributed over addition.

$$12 \div (2 + 4) \neq 12 \div 2 + 12 \div 4$$

The same properties hold for all real numbers which include positive and negative numbers.

CHAPTER 16 Positive and Negative Numbers

16-1 Positive and Negative Numbers

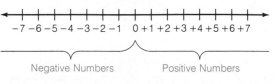

Positive numbers are designated by a centered or raised positive sign or by no sign at all. The numbers of arithmetic are all positive numbers.

$$+8, \; ^{+}11, \; 7, \; \tfrac{3}{5}, \text{ and } +\sqrt{11} \text{ are positive numbers.}$$

Negative numbers are designated by a centered or raised negative sign.

$$-4, \; ^{-}10, \; -\tfrac{5}{8}, \text{ and } -\sqrt{7} \text{ are negative numbers.}$$

We read $+8$ as "positive 8." We read -4 as "negative 4." Positive and negative numbers are sometimes called signed numbers or directed numbers. 0 is neither positive nor negative.

Look at the number line. Find $+5$ and -5. They are on opposite sides of 0 but are the same distance from 0. Such numbers are called opposites of each other. The opposite of 0 is 0.

The set of numbers consisting of all the whole numbers and their opposites is called the integers.

Integers: . . . , $-3, -2, -1, 0, 1, 2, 3, \ldots$
Positive integers: $1, 2, 3, \ldots$
Nonnegative integers: $0, 1, 2, 3, \ldots$
Negative integers: . . . , $-3, -2, -1$
Nonpositive integers: . . . , $-3, -2, -1, 0$

540

Even integers are integers divisible by 2. Odd integers are integers that are not divisible by 2.

> Even integers: . . . , $-4, -2, 0, 2, 4, . . .$
> Odd integers: . . . , $-5, -3, -1, 1, 3, 5, . . .$

EXAMPLES:

What number is the opposite of -8?

Look 8 units to the right of zero on the number line.

ANSWER: $+8$

List all the numbers described by the following:

$-5, -4, -3, . . . , 6$

ANSWER: $-5, -4, -3, -2, -1, 0, 1, 2, 3, 4,$
      $5, 6$

PRACTICE EXERCISES

1 Read, or write in words, each of the following:

1. -6	**2.** $+10$	**3.** -7	**4.** 15
5. -33	**6.** $+50$	**7.** $+63$	**8.** -147
9. -74	**10.** 27		

2 Write using symbols:

1. Negative twelve   **2.** Positive nine
3. Positive sixteen   **4.** Negative four
5. Positive twenty-eight   **6.** Negative seventy
7. Positive forty-one   **8.** Negative ninety-five

3 What number is the opposite of each of the following?

1. $+7$	**2.** -3	**3.** 0	**4.** -19
5. 46	**6.** -22	**7.** -100	**8.** $+79$
9. -61	**10.** 47		

4 **1.** Does -8 name a whole number? An integer?
   2. Does 17 name a whole number? An integer?

5 List all the numbers described by the following:

1. $-11, -10, -9, . . . , 3$
2. $-9, -7, -5, . . . , 9$
3. $-12, -10, -8, . . . , 10$

16-2 Rational and Real Numbers

Vocabulary

A **rational number** is a number that can be expressed as a quotient of two integers, with division by zero excluded.

An **irrational number** is a number that cannot be expressed as a quotient of two integers.

The **real numbers** are all the rational and irrational numbers.

Fractional numbers can be arranged in a definite order on the number line. There is a point corresponding to each fractional number. Each fractional number has an opposite.

The numeral $+\frac{2}{5}$ is read "positive two-fifths."

The numeral $-\frac{1}{4}$ is read "negative one-fourth."

All the fractional numbers and their opposites are rational numbers. Integers are also rational numbers, since an integer can be named in fraction form.

$-7 = -\frac{7}{1}$, $0 = \frac{0}{7}$, and $3\frac{1}{2} = \frac{7}{2}$ are all rational numbers.

A number that is both a non-terminating and a non-repeating decimal, such as the square root of any positive number other than a perfect square, is an irrational number.

$\sqrt{43} = 6.557 \ldots$ is an irrational number.

Together the rational numbers and the irrational numbers make up the real numbers. There are an infinite number of real numbers.

EXAMPLE:

To what sets of numbers does -8 belong?

ANSWER: -8 is an integer, a rational number, and a real number.

PRACTICE EXERCISES

1 Read, or write in words, each of the following:

1. $+\frac{5}{6}$ **2.** $-\frac{13}{4}$ **3.** $-6\frac{1}{3}$

4. $+.89$ **5.** -17.3 **6.** $2\frac{2}{5}$

7. $-\frac{11}{16}$ **8.** $-.004$

2 Write using symbols:

1. Negative one-fifth

2. Positive seven-twelfths

3. Positive nine-hundredths

4. Negative four and three-tenths

5. Negative nine and forty-eight hundredths

6. Positive one hundred sixty-one thousandths

3 What number is the opposite of each of the following?

1. $+\frac{5}{8}$ **2.** $-\frac{9}{4}$ **3.** $.07$

4. -2.75 **5.** $-1\frac{3}{5}$ **6.** $+.102$

7. $6\frac{4}{7}$ **8.** -58.3

4 Is $-\frac{3}{5}$ **1.** An integer?   **2.** A rational number?   **3.** A real number?

5 Is $+7\frac{2}{3}$ **1.** An integer?   **2.** A rational number?   **3.** A real number?

6 Is $.45$ **1.** An integer?   **2.** A rational number?   **3.** A real number?

7 Is -8.9 **1.** An integer?   **2.** A rational number?   **3.** A real number?

8 Is -6 **1.** An integer?   **2.** A rational number?   **3.** A real number?

9 Is $-\sqrt{31}$ **1.** An integer?   **2.** A rational number?   **3.** A real number?

Which of the following numerals name integers? Which name rational numbers? Which name irrational numbers? Which name real numbers?

10 **1.** -17 **2.** $+\frac{3}{4}$ **3.** $-.05$

     4. $-\sqrt{29}$ **5.** $\frac{16}{2}$ **6.** $-7\frac{1}{3}$

11 **1.** $-\frac{4}{7}$ **2.** $+100$ **3.** -5.08

     4. $-2\frac{3}{4}$ **5.** $\sqrt{13}$ **6.** $-\frac{18}{6}$

12 **1.** $-.375$ **2.** $-\frac{5}{8}$ **3.** $-\sqrt{16}$

     4. -69 **5.** $+\frac{28}{4}$ **6.** $3\frac{5}{6}$

13 **1.** $-\sqrt{5}$ **2.** $-\frac{15}{16}$ **3.** $+97$

     4. $.18$ **5.** $-4\frac{7}{12}$ **6.** $+\frac{21}{4}$

16-3 Absolute Value

Vocabulary

For any pair of opposite non-zero real numbers, the greater number is called the **absolute value** (symbol | |) for each of the numbers.

The absolute value of any number is the value of the corresponding arithmetic number which has no sign.

EXAMPLES:

What is the absolute value of -4?

ANSWER: 4

What is the absolute value of 0?

ANSWER: 0

What is the absolute value of $+4$?

ANSWER: 4

How is $|-9|$ read?

ANSWER: the absolute value of negative nine

PRACTICE EXERCISES

1 Read, or write in words, each of the following:

1. $|+11|$ **2.** $|-3|$ **3.** $|+35|$ **4.** $|-\frac{7}{10}|$ **5.** $|-.09|$ **6.** $|5\frac{2}{3}|$

2 Write, using the absolute value symbol:

1. The absolute value of negative thirteen.
2. The absolute value of positive eight-fifteenths.
3. The absolute value of positive three and nine-tenths.
4. The absolute value of negative fourteen-thousandths.

3 Find the absolute value of each of the following:

1. $|-7|$ **2.** $|+3|$ **3.** $|-6|$ **4.** $|0|$ **5.** $|-.8|$ **6.** $|+\frac{2}{3}|$
7. $|-2\frac{1}{2}|$ **8.** $|-27|$ **9.** $|19|$ **10.** $|-.82|$ **11.** $|-\frac{9}{4}|$ **12.** $|-1.75|$

In each of the following, first find the absolute value of each number as required; then apply the necessary operation to obtain the answer.

4   **1.** $|-4| + |-8|$ **2.** $|-\frac{3}{4}| + |-\frac{3}{4}|$
    3. $|+2\frac{4}{5}| + |-4\frac{1}{4}|$ **4.** $|-.91| + |+1.8|$

5   **1.** $|+13| - |-7|$ **2.** $|-\frac{4}{5}| - |+\frac{7}{10}|$
    3. $|-1\frac{1}{2}| + |-3\frac{2}{3}|$ **4.** $|-2.5| - |+.15|$

6   **1.** $|-9| \times |3|$ **2.** $|-\frac{2}{3}| \times |-\frac{3}{8}|$
    3. $|-3\frac{2}{5}| \times |+20|$ **4.** $|-6.08| \times |-.07|$

7   **1.** $|-54| \div |-9|$ **2.** $|-\frac{5}{8}| \div |+\frac{2}{5}|$
    3. $|3\frac{5}{6}| \div |-1\frac{7}{12}|$

16-4 Graphing on the Real Number Line

Vocabulary

The **real number line** is the complete collection of points corresponding to all the real numbers.

The real number line is endless in both directions. There is one and only one point corresponding to each real number and one and only one real number corresponding to each point.

Usually the real number line is labeled only with the numerals naming the integers. Only a part of it is shown at any one time.

Each point on the number line is called the graph of the number to which it corresponds. Each number is called the coordinate of the corresponding point on the line. Capital letters are generally used to identify these points.

> The graph of a number is a point on the number line whose coordinate is the number.
>
> *Point B* is the graph of -2.
> -2 is the coordinate of point *B*.

> The graph of a set of numbers is made up of the points on the number line whose coordinates are the numbers.
> The points *B, C, E,* and *H* are the graph of the numbers -3, -2, 0, and 3.

PROCEDURE

To draw the graph of a set of numbers on the number line:

(1) Draw an appropriate number line.
(2) Locate the point or points whose coordinates are the given numbers.

EXAMPLES:

Draw the graph of the following set of numbers on the number line: $-4, -2, 6, 8$

ANSWER:
$$-8\;-7\;-6\;-5\;-4\;-3\;-2\;-1\;\;0\;\;1\;\;2\;\;3\;\;4\;\;5\;\;6\;\;7\;\;8$$

Write the coordinates of which the following is the graph.

$$-7\;-6\;-5\;-4\;-3\;-2\;-1\;\;0\;\;1\;\;2\;\;3\;\;4\;\;5\;\;6\;\;7$$

ANSWER: $-2, -1, 0, 4, 6$

PRACTICE EXERCISES

$$\begin{array}{ccccccccccccccccccccc}
A & L & B & M & C & N & D & O & E & P & F & Q & G & R & H & S & I & T & J & U & V \\
\end{array}$$
$$-10\;-9\;-8\;-7\;-6\;-5\;-4\;-3\;-2\;-1\;\;0\;\;1\;\;2\;\;3\;\;4\;\;5\;\;6\;\;7\;\;8\;\;9\;\;10$$

1 What number corresponds to each of the following points?

1. E   **2.** G   **3.** M   **4.** S   **5.** F
6. L   **7.** J   **8.** R   **9.** C   **10.** V

2 What letter labels the point corresponding to each of the following numbers?

1. -3 **2.** $+9$ **3.** -8 **4.** -4 **5.** 0
6. $+6$ **7.** -5 **8.** $+1$ **9.** -10 **10.** $+7$

3 Draw the graph of each of the following sets of numbers on a number line:

1. -3 **2.** $-5, +2$
3. $-7, -4, 0, +1$ **4.** $-4, -2, 0, 2, 4$
5. $-6, -3, -1, 3, 5, 7$ **6.** $4, -2, +1$
7. $+4, -6, +1, -1, +7, -5, -3$ **8.** $-8, 4, 0, -6, -2, 5, 9, -1$
9. $-5, -4, -3, \ldots, 4$ **10.** $-9, -8, -7, \ldots, 2$

4 Draw the graph of each of the following on a number line:

1. all integers less than $+1$ and greater than -1.
2. all integers greater than -4 and less than 0.
3. all integers less than $+5$ and greater than -5.

5 Write the coordinates of which each of the following is the graph:

1.
$$-6\;-5\;-4\;-3\;-2\;-1\;\;0\;\;1\;\;2\;\;3\;\;4\;\;5\;\;6$$

2.
$$-6\;-5\;-4\;-3\;-2\;-1\;\;0\;\;1\;\;2\;\;3\;\;4\;\;5\;\;6$$

3.
$$-6\;-5\;-4\;-3\;-2\;-1\;\;0\;\;1\;\;2\;\;3\;\;4\;\;5\;\;6$$

4.
$$-6\;-5\;-4\;-3\;-2\;-1\;\;0\;\;1\;\;2\;\;3\;\;4\;\;5\;\;6$$

5.
$$-9\;-8\;-7\;-6\;-5\;-4\;-3\;-2\;-1\;\;0\;\;1\;\;2\;\;3$$

6.
$$-8\;-7\;-6\;-5\;-4\;-3\;-2\;-1\;\;0\;\;1\;\;2\;\;3\;\;4\;\;5\;\;6\;\;7\;\;8$$

7.
$$-9\;-8\;-7\;-6\;-5\;-4\;-3\;-2\;-1\;\;0\;\;1\;\;2\;\;3\;\;4\;\;5\;\;6\;\;7\;\;8\;\;9$$

16-5 Opposite Meanings

Positive and negative numbers are used in science, statistics, weather reports, stock reports, sports, and many other fields to express opposite meanings or directions.

EXAMPLE:

If $+1\frac{1}{4}$ points indicates a $1\frac{1}{4}$-point *gain* in a stock price, what does -2 points indicate?

ANSWER: 2-point *loss* in a stock price.

PRACTICE EXERCISES

1 If +250 meters represents 250 meters above sea level, what does −60 meters represent?

2 If −30 kilograms represents a downward force of 30 kilograms, what does +75 kilograms represent?

3 If an increase of 6% in the cost of living is indicated by +6%, how can a 4% decrease in the cost of living be indicated?

4 If 85 degrees west longitude is indicated by −85°, how can 54 degrees east longitude be indicated?

5 If +18 m.p.h. indicates a tail wind of 18 m.p.h., what does −23 m.p.h. represent?

6 If a deficiency of 47 millimeters of rainfall is indicated by −47 millimeters, how can an excess of 39 millimeters be indicated?

7 If −14° means 14 degrees below zero, what does +72° mean?

8 If $70 deposited in the bank is represented by +$70, how can $25 withdrawn from the bank be represented?

9 If an inventory shortage of 120 items is represented by −120 items, how can 58 items over be represented?

10 If a charge of 15 amperes of electricity is indicated by +15 amperes, how can a discharge of 17 amperes of electricity be indicated?

11 If +6 pounds means 6 pounds overweight, what does −9 pounds mean?

12 If −10° means 10 degrees south latitude, what does +15° mean?

13 If 8 yards lost in a football game is indicated by −8 yards, how can 5 yards gained be indicated?

14 If 4 degrees above normal temperature is represented by +4°, how can 7 degrees below normal temperature be represented?

15 If 25 meters to the left is indicated by −25 meters, how can 18 meters to the right be indicated?

16-6 Comparing Integers

We can compare integers just as we compare whole numbers. We can use the number line to help.

The number line may be drawn either horizontally or vertically.

```
    L   M   N   P   Q   R   S   T   U   V   W
 ◄──●───●───●───●───●───●───●───●───●───●───●──►
   -5  -4  -3  -2  -1   0  +1  +2  +3  +4  +5
```

On the horizontal number line, the number corresponding to the point farther to the right is greater. On the vertical number line, any number corresponding to a point is greater than any number corresponding to a point below it.

A	+5
B	+4
C	+3
D	+2
E	+1
F	0
G	-1
H	-2
I	-3
J	-4
K	-5

> Use the symbols >, <, and = to compare integers.

PROCEDURE

To compare two integers on the number line:

(1) Determine the position of each integer on a number line.
(2) If the first number is to the right or above the second number, use the symbol >. If the first number is to the left or below the second number, use the symbol <. If the two numbers are represented by the same point, use the symbol = .

EXAMPLE:

Insert >, <, or = to make a true sentence.

-3 ▨ -7

Since -3 is to the right of -7 on the number line, -3 is greater than -7.

ANSWER: $-3 > -7$

PRACTICE EXERCISES

1 On the horizontal number line, which point corresponds to the greater number:

1. Point T or point M?   **2.** Point P or point V?
3. Point Q or point N?   **4.** Point U or point L?

2 On the vertical number line, which point corresponds to the greater number:

1. Point C or point H?   **2.** Point G or point B?
3. Point F or point J?   **4.** Point K or point D?

3 Which number is greater?

1. +9 or +6   2. −8 or −5   3. 0 or +3   4. −7 or 0
5. +8 or −8   6. −1 or +4   7. −2 or +2   8. +5 or −10

4 Which number is less?

1. +3 or +7   2. −9 or −10   3. +4 or 0   4. 0 or −6
5. −8 or +5   6. +2 or −1   7. −4 or −7   8. −9 or +11

5 Which of the following sentences are true?

1. $+3 > -4$ 2. $0 < -5$ 3. $+2 < +8$ 4. $-12 > +10$
5. $-6 < -3$ 6. $-7 < -7$ 7. $+6 > +1$ 8. $+5 < -4$
9. $0 > +4$ 10. $-5 > -2$ 11. $+8 \not< +2$ 12. $-1 \not> +1$
13. $-9 < +9$ 14. $-2 \not> 0$ 15. $+15 \not< -21$ 16. $-3 < -7$

6 Rewrite each of the following and insert between the two numerals the symbol = , < , or > which will make the sentence true:

1. -8 _____ -1 2. $+6$ _____ $+2$ 3. $+9$ _____ -4
4. -7 _____ -9 5. $+11$ _____ -14 6. -5 _____ -5
7. -3 _____ 0 8. -12 _____ -2 9. 0 _____ -1

7 Name the following numbers in order of size (least first):

1. +4, −3, +7, −9, −2, +10, −4, 0, −6, +5
2. −1, +2, 0, −3, +8, −5, −7, +11, −10, +10

8 Name the following numbers in order of size (greatest first):

1. −3, +8, −7, +4, −8, 0, +3, −12, +9, −1
2. +5, −6, +1, −1, 0, +10, −9, −11, +8, −4

9 Which is greater:

1. The absolute value of −8 or the absolute value of +3?
2. The absolute value of +9 or the absolute value of −9?

10 Which has the greater opposite number?

1. +7 or +4   2. −3 or −8   3. −1 or +6   4. +5 or −10

Refresh Your Skills

1. Add: 1-4
   $658 + 594 + 769 + 945 + 268$

2. Subtract: 1-5
   $45,006 - 40,509$

3. Multiply: 1-6
   693×487

4. Divide: 1-7
   $108\overline{)3,132}$

5. Add: 2-4
   $.32 + 5.4$

6. Subtract: 2-5
   $.98 - .876$

7. Multiply: 2-7
   $3.2 \times .059$

8. Divide: 2-8
   $4.2\overline{).0126}$

9. Add: 3-8
   $7\frac{5}{12} + 4\frac{1}{3}$

10. Subtract: 3-9
    $6 - 1\frac{3}{10}$

11. Multiply: 3-11
    $2\frac{2}{3} \times 4\frac{4}{5}$

12. Divide: 3-12
    $8\frac{1}{2} \div \frac{3}{4}$

13. Find $\frac{1}{2}$% of 3,600
    4-6

14. What percent of 90 is 15?
    4-7

15. 3.90 is 25% of what amount?
    4-8

16-7 Opposite Directions and Vectors

Vocabulary

A **vector** is an arrow that represents a directed line segment and can be used to show directed distances.

Positive and negative numbers, used as directed numbers, show movements in opposite directions.

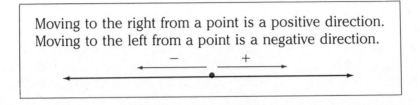

Moving to the right from a point is a positive direction.
Moving to the left from a point is a negative direction.

The sign in a numeral naming a directed number indicates the direction. The absolute value of the number represents the magnitude, or distance in units, of the movement.

EXAMPLES:

Give the meaning of +6.

ANSWER: +6 means moving 6 units to the right.

Represent +5 by a vector.

ANSWER:

Give the meaning of −7.

ANSWER: −7 means moving 7 units to the left.

Represent −3 by a vector.

ANSWER:

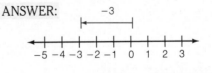

PROCEDURE

To determine the numeral that describes the movement from one point to another:

(1) Draw the vector between the points with the arrowhead pointing in the direction of the movement.
(2) Use the number of units of length of the vector as the absolute value of the number. Select the sign according to the direction of the vector: positive (+) if to the right, negative (−) if to the left.

EXAMPLE:

Picture the movement from +3 to −2.

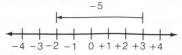

ANSWER: The numeral −5 describes this movement.

PRACTICE EXERCISES

1 A movement of how many units of distance and in what direction is represented by each of the following numbers?

1. −7   **2.** +9   **3.** −2   **4.** +12
5. +1   **6.** −4.25   **7.** $-2\frac{1}{4}$ **8.** $+\frac{2}{3}$

2 Represent each of the following movements by a numeral naming a signed number:

1. Moving 4 units to the left.   **2.** Moving 8 units to the right.
3. Moving 7.5 units to the right.   **4.** Moving $12\frac{1}{4}$ units to the left. −

3 Write the numeral that is represented by each of the following vectors:

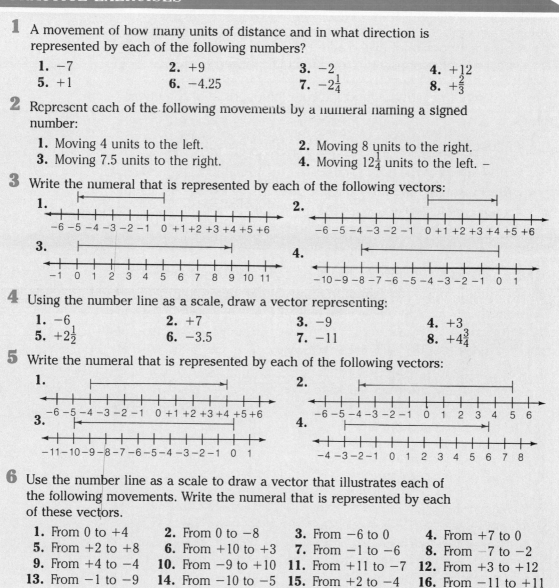

4 Using the number line as a scale, draw a vector representing:

1. −6   **2.** +7   **3.** −9   **4.** +3
5. $+2\frac{1}{2}$ **6.** −3.5   **7.** −11   **8.** $+4\frac{3}{4}$

5 Write the numeral that is represented by each of the following vectors:

6 Use the number line as a scale to draw a vector that illustrates each of the following movements. Write the numeral that is represented by each of these vectors.

1. From 0 to +4   **2.** From 0 to −8   **3.** From −6 to 0   **4.** From +7 to 0
5. From +2 to +8   **6.** From +10 to +3   **7.** From −1 to −6   **8.** From −7 to −2
9. From +4 to −4   **10.** From −9 to +10   **11.** From +11 to −7   **12.** From +3 to +12
13. From −1 to −9   **14.** From −10 to −5   **15.** From +2 to −4   **16.** From −11 to +11

16-8 Additive Inverse of a Number

Vocabulary

The opposite of a number is sometimes called the **additive inverse** of the number.

If the sum of two numbers is zero, each addend is the additive inverse of the other.

> -4 is the opposite of or the additive inverse of $+4$ (or 4).
> $+4$ (or 4) is the opposite of or the additive inverse of -4.

We use a centered negative sign to indicate the opposite of a number.

EXAMPLES:

Find the additive inverse of -8.

ANSWER: 8 or $(+8)$

Read, or write in words, $-(-13)$.

ANSWER: The opposite of negative thirteen.

Find the value of $-(+4)$.

The opposite of positive four is negative four.

ANSWER: -4

PRACTICE EXERCISES

1 What number is the additive inverse of each of the following numbers?

   1. -7 **2.** $+26$ **3.** 0 **4.** $-\frac{5}{8}$

2 What number is the opposite of each of the following numbers?

   1. $+10$ **2.** -6 **3.** $-.85$ **4.** 0

3 Read, or write in words, each of the following:

   1. -9 **2.** $-(+2)$ **3.** $-(-\frac{5}{6})$ **4.** $-(15) = -15$

4 Write symbolically, using the "opposite" symbol:

   1. The opposite of thirty-six.   **2.** The opposite of positive eight.
   3. The opposite of negative five.
   4. The opposite of negative twenty-three.
   5. The opposite of forty is negative forty.
   6. The opposite of negative one is positive one.
   7. The opposite of positive fifty-six is negative fifty-six.
   8. The opposite of negative sixty-seven is positive sixty-seven.

5 **1.** Find the opposite of the opposite of: -7; $+6$; $+\frac{3}{8}$; $-.9$; $-3\frac{1}{2}$
   2. What is always true about the opposite of the opposite of a number?

6 Find the value of each of the following:

   1. $-(+9) = ?$ **2.** $-(-3\frac{2}{5}) = ?$ **3.** $-(54) = ?$ **4.** $-(\frac{11}{4}) = ?$

16-9 Addition of Integers and Rational Numbers

We can use a number line to learn how to add integers and rational numbers.

PROCEDURE

To add on a number line:
(1) Draw a vector for the first addend, starting at zero.
(2) Draw a vector for the second addend, starting from the point reached by the first vector.
(3) If there are more than 2 addends, continue in this way for each addend.
(4) The coordinate of the final point reached is the sum.

> NOTE: If both addends are positive, the direction is positive. If both addends are negative, the direction is negative. If one addend is positive and the other negative, the direction is determined by the number with the greater absolute value.

EXAMPLES:

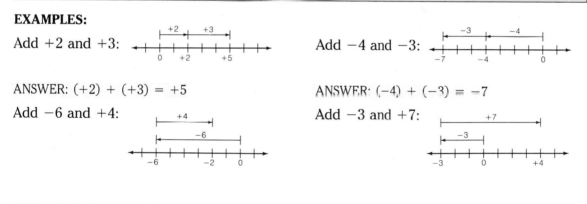

Add +2 and +3:

ANSWER: $(+2) + (+3) = +5$

Add −6 and +4:

ANSWER: $(-6) + (+4) = -2$

Add −4 and −3:

ANSWER: $(-4) + (-3) = -7$

Add −3 and +7:

ANSWER: $(-3) + (+7) = +4$

We can also add signed numbers algebraically.

PROCEDURE

To find the sum of two signed numbers:
(1) If both numbers are positive, find the sum of their absolute values and write a positive sign before its numeral.
(2) If both numbers are negative, find the sum of their absolute values and write a negative sign before its numeral.
(3) If one number is positive and one negative, subtract the smaller absolute value from the greater and write the sign of the number with the greater absolute value before the numeral.

EXAMPLES:

Add: +2	Add: −6	Add: −4	Add: −3
+ +3	+ +4	+ −3	+ +7
+5	−2	−7	+4

ANSWER: +5   ANSWER: −2   ANSWER: −7   ANSWER: +4

The sum of any number and its opposite is zero.

Add: +7	Add: −5
+ −7	+ +5
0	0

ANSWER: 0   ANSWER: 0

To add three or more numbers, first add the positive numbers. Then add the negative numbers. Finally, find the sum.

EXAMPLE: Add: $(+2) + (-9) + (+6) + (-3)$.

$$(+2) + (-9) + (+6) + (-3)$$
$$= (+8) + (-12)$$
$$= -4$$

ANSWER: −4

To simplify an algebraic expression such as $6 - 9 + 4 - 3$, think of it as $(+6) + (-9) + (+4) + (-3)$ where the given signs are part of the numerals for the signed numbers but the operation is considered to be addition.

EXAMPLE: Simplify: $6 - 9 + 4 - 3$.

$$6 - 9 + 4 - 3 = (+6) + (-9) + (+4) + (-3)$$
$$= (+10) + (-12)$$
$$= -2$$

ANSWER: −2

PRACTICE EXERCISES

1 Add on the number line, using vectors:

1. +5	**2.** −4	**3.** +7	**4.** +2	**5.** −8	**6.** −1	**7.** +4	**8.** 0
+3	−1	−3	−6	+2	+5	−4	−9

2 Add:

1. $+3$	**2.** $+9$	**3.** $+24$	**4.** $+8$	**5.** -5	**6.** -9	**7.** -27	**8.** -83
$\underline{+6}$	$\underline{+7}$	$\underline{+19}$	$\underline{+15}$	$\underline{-4}$	$\underline{-14}$	$\underline{-35}$	$\underline{-49}$

3

1. $+5$	**2.** $+11$	**3.** $+25$	**4.** $+91$	**5.** -9	**6.** -8	**7.** -23	**8.** -54
$\underline{-2}$	$\underline{-3}$	$\underline{-14}$	$\underline{-37}$	$\underline{+1}$	$\underline{+6}$	$\underline{+12}$	$\underline{+39}$

4

1. $+1$	**2.** $+3$	**3.** $+12$	**4.** $+20$	**5.** -6	**6.** -2	**7.** -18	**8.** -45
$\underline{-8}$	$\underline{-10}$	$\underline{-36}$	$\underline{-41}$	$\underline{+8}$	$\underline{+15}$	$\underline{+27}$	$\underline{+68}$

5

1. $+6$	**2.** -12	**3.** 0	**4.** -5
$\underline{-6}$	$\underline{+12}$	$\underline{+7}$	$\underline{0}$

6

1. $+\frac{3}{4}$	**2.** $-\frac{7}{8}$	**3.** $-.4$	**4.** -5.28
$\underline{-\frac{1}{4}}$	$\underline{-\frac{2}{3}}$	$\underline{+1.5}$	$\underline{-2.64}$

7

1. $+8$	**2.** -3	**3.** -3	**4.** $+9$
$+5$	-8	$+6$	-15
$\underline{+7}$	$\underline{-9}$	$\underline{-5}$	$\underline{+3}$

5. $+2$	**6.** -3	**7.** -4	**8.** -6
$+1$	-1	$+7$	-9
$+4$	-5	-8	$+7$
$\underline{+6}$	$\underline{-2}$	$\underline{+5}$	$\underline{-2}$

8 Add as indicated:

1. $(-3) + (-4)$
2. $(+8) + (-15)$
3. $(-1) + (+11)$
4. $(+5) + (-14) + (+7)$
5. $(-3) + (-9) + (-5)$
6. $[(+8) + (-11)] + (-4)$
7. $(-10) + [(-8) + (-2)]$
8. $(+7) + (-5) + (-10) + (+9)$
9. $(-7) + (+4) + (-1) + (+6)$
10. $(-12) + (+9) + (-6) + (+9)$

9 Simplify:

1. $2 + 6$
2. $4 - 8$
3. $-6 + 7$
4. $-5 - 3$
5. $18 - 9 + 5$
6. $9 - 7 + 2$
7. $5 - 6 - 8$
8. $-3 + 8 - 7$
9. $14 - 5 + 3 - 11$
10. $-9 - 6 + 7 - 2$
11. $6 - 8 + 2 - 4 + 5$
12. $-2 - 3 + 9 - 1 + 3$

10 Simplify by performing the indicated operations and finding the absolute values:

1. $|7 + 4|$
2. $|9 - 11|$
3. $|-4 - 5|$
4. $|-8 + 2|$
5. $|-10| + |+7|$
6. $|(-6) + (-2)|$
7. $-|14 - 5|$
8. $-|16-24|$
9. $-9 + |-8|$
10. $-|-(-4) + (-6)|$

WIND CHILL FACTOR

On a cold, windy day, have you ever felt a lot colder than the actual temperature would seem to indicate? You were right—because wind does affect how cold we feel. If the still air temperature on Mt. McKinley is 0°F but there is a wind of 40 miles per hour, the wind-chill temperature is −53°F.

People who work outdoors or climb mountains or ski in cold weather must take the wind-chill factor into account.

Here is a table that shows wind-chill factors.

Wind-Chill Chart								
Wind Speed mph	Actual Temperature °F							
	40	30	20	10	0	−10	−20	−30
	Wind-Chill Temperature °F							
10	28	16	4	−9	−21	−33	−46	−58
20	18	4	−10	−25	−39	−53	−67	−82
30	13	−2	−18	−33	−48	−63	−79	−94
40	10	−6	−21	−37	−53	−69	−85	−100
Greater wind speeds have little added effect.								

EXERCISES

1. What is the wind-chill temperature if the actual temperature is 10°F and the wind speed is 20 miles per hour?

2. Find the wind-chill air temperature if the actual temperature is 20°F and the wind speed is 30 miles per hour?

3. When the air temperature is 20°F, what is the change in wind-chill air temperature if the wind increases from 20 to 30 miles per hour?

4. Find the wind-chill air temperature if the actual temperature is 40°F and the wind speed is 20 miles per hour.

16-10 Subtraction of Integers and Rational Numbers

Subtraction is the inverse of addition. To find $(+3) - (-2) = ?$, we must find the addend which, when added to the given subtrahend, gives the minuend. For $(+3) - (-2) = ?$, we must find ? so that

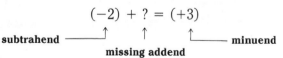

$$(-2) + ? = (+3)$$

subtrahend — missing addend — minuend

PROCEDURE

To subtract using vectors:
(1) Draw the vectors for the given subtrahend and minuend.
(2) Draw the vector from the point associated with the subtrahend to the point associated with the minuend.

EXAMPLE:

Subtract: $(+3) - (-2) = ?$

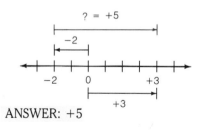

The distance from -2 to $+3$ is 5 units. The direction is to the right or the positive direction.

ANSWER: $+5$

When we compare

$(+3) - (-2) = +5$ and
$(+3) + (+2) = +5$

we see that subtracting -2 from $+3$ gives the same answer as adding $+2$ and $+3$. That is *subtracting a number* gives the same answer as *adding its opposite.*

PROCEDURE

To subtract signed numbers algebraically:
Add to the minuend the additive inverse of the number being subtracted.

EXAMPLES:

Subtract:	Add:		Subtract:	Add:		Subtract:	Add:
+2	+2		+2	+2		−2	−2
− +8 →	−8		− −8 →	+8		− +8 →	−8
	−6			+10			−10

ANSWER: −6   ANSWER: +10   ANSWER: −10

Subtract:	Add:		Subtract:	Add:		Subtract:	Add:
−2	−2		−8	−8		0	0
− −8 →	+8		− −2 →	+2		− +8 →	−8
	+6			−6			−8

ANSWER: .+6   ANSWER: −6   ANSWER: −8

PRACTICE EXERCISES

1 Subtract on the number line, using vectors:

1. +6
+3

2. +2
+9

3. −5
−1

4. −3
−7

5. +4
−6

6. −2
+5

7. 0
−8

8. −4
−4

Subtract:

2 **1.** 9
3

2. +11
+6

3. +15
+9

4. +17
+8

5. 5
7

6. 6
15

7. +2
+14
−

8. +28
+45

3 **1.** −6
−4

2. −10
−5

3. −8
−3

4. −12
−7

5. −5
−8

6. −7
−9

7. −6
−13

8. −51
−70

4 **1.** +7
−2

2. +9
−11

3. +6
−5

4. +18
−27

5. −12
+3

6. −6
+19

7. −5
+1

8. −16
+25

5
1. $+6$
   $\underline{+6}$
2. -5
   $\underline{-5}$
3. 0
   $\underline{-1}$
4. $+8$
   $\underline{0}$

5. $+11$
   $\underline{-11}$
6. $-.6$
   $\underline{-.7}$
7. $-\frac{7}{8}$
   $\underline{+\frac{3}{4}}$
8. $-2\frac{1}{2}$
   $\underline{-1\frac{2}{3}}$

6
1. $+5$
   $\underline{+9}$
2. -6
   $\underline{-2}$
3. -13
   $\underline{+7}$
4. $+20$
   $\underline{-15}$

5. -3
   $\underline{0}$
6. -14
   $\underline{+14}$
7. $+.9$
   $\underline{-1.5}$
8. 0
   $\underline{-17}$

7 Subtract as indicated:

1. $(-8) - (+3)$
2. $(+6) - (-11)$
3. $(2) - (-2)$
4. $(4 - 3) - (6 - 10)$
5. $(2 - 6) - (7 - 8)$
6. $(8 + 7) - (6 - 13)$
7. $[(-2) + (-7)] - (-9)$
8. $(+4) - [(-5) - (+6)]$
9. $[(-5) - (+8)] - (-12)$
10. $(-6) - [(-10) + (+3)]$

8
1. From -5 subtract 5.
2. Take 4 from 1.
3. From -3 take -7.
4. Subtract 10 from 3.
5. Take -9 from $+9$.
6. From 0 subtract -8.

9 Simplify by performing the indicated operations and finding the absolute values:

1. $|-4| - |-9|$
2. $|6 - 2| - |-5|$
3. $|1 - 3| - |4 - 7|$
4. $|-3 - 8| - 6$
5. $8 - |5 + 9|$
6. $|(0 - 2) - (4 - 2)|$

Refresh Your Skills

1. Add: 1-4
   $7{,}687 + 80{,}547 + 3{,}396$
2. Subtract: 1-5
   $140{,}000 - 69{,}076$
3. Multiply: 1-6
   $6{,}080 \times 319$
4. Divide: 1-7
   $360\overline{)225{,}720}$

5. Add: 2-4
   $.395 + .68 + .7$
6. Subtract: 2-5
   $7.8 - .65$
7. Multiply: 2-7
   $.09 \times 1.02$
8. Divide: 2-8
   $15.3\overline{)6.12}$

9. Add: 3-8
   $\frac{5}{6} + \frac{1}{2} + \frac{4}{5}$
10. Subtract: 3-9
    $2\frac{5}{8} - 1\frac{13}{16}$
11. Multiply: 3-11
    $6\frac{3}{4} \times 9\frac{1}{3}$
12. Divide: 3-12
    $6 \div 1\frac{1}{3}$

13. Find 9.2% of 10,000
    4-6
14. What percent of 95 is 95?
    4-7
15. .10 is 1% of what amount?
    4-8

Multiplication can be thought of as repeated addition.

$$(3)(4) = (+3)(+4) = (+4) + (+4) + (+4) = +12$$
$$(+3)(+4) = +12$$

The product of two positive numbers is a positive number.

Similarly,

$$(+3)(-4) = (-4) + (-4) + (-4) = -12$$
$$(+3)(-4) = -12$$

The product of a positive number and a negative number is a negative number.

Since $(-3)(+4)$ can be written as $(+4)(-3)$ by using the commutative property of multiplication, then,

$$(-3)(+4) = (+4)(-3) = (-3) + (-3) + (-3) = -12$$
$$(-3)(+4) = -12$$

The product of a negative number and a positive number is a negative number.

To determine what $(-3)(-4)$ equals, observe the following steps:

1. $(-3)(0) = 0$ — The product of any number and zero is zero.

2. But $[(+4) + (-4)] = 0$ — The sum of a number and its inverse is zero.

3. Then $(-3)[(+4) + (-4)] = 0$ — By substituting $[(+4) + (-4)]$ for 0 in step 1.

4. $[(-3)(+4)] + [(-3)(-4)] = 0$ — By the distributive property of multiplication over addition.

5. $-12 + [(-3)(-4)] = 0$ — Since $(-3)(+4) = -12$.

6. Therefore $[(-3)(-4)] = +12$ — In order for $-12 + [(-3)(-4)] = 0$ to be true, the product of $[(-3)(-4)]$ must be the additive inverse of -12, which is $+12$.

$$(-3)(-4) = +12$$

The product of two negative numbers is a positive number.

PROCEDURE

To find the product of two signed numbers:
(1) If the numbers are both positive or both negative, find the product of their absolute values and write a positive sign before this numeral.
(2) If one number is positive and the other negative, find the product of their absolute values and write a negative sign before this numeral.

(3) If there are more than two factors, find the product of their absolute values and write a positive sign before the numeral when the number of negative factors is an even number or a negative sign if the number of negative factors is an odd number.

EXAMPLES:

Multiply: $\times\ \begin{array}{r} +5 \\ +6 \\ \hline 30 \end{array}$ **The sign is positive.**

ANSWER: +30

Multiply: $\times\ \begin{array}{r} -5 \\ -6 \\ \hline 30 \end{array}$ **The sign is positive.**

ANSWER: +30

Multiply: $\times\ \begin{array}{r} +5 \\ -6 \\ \hline 30 \end{array}$ **The sign is negative.**

ANSWER: −30

Multiply: $\times\ \begin{array}{r} -5 \\ +6 \\ \hline 30 \end{array}$ **The sign is negative.**

ANSWER: −30

Multiply: $(-2)(+1)(-4)(-2)$ **3 negative factors. The sign is negative.**

$2 \times 1 \times 4 \times 2 = 16$

ANSWER: −16

If the product of two numbers is one (1), then each factor is said to be the multiplicative inverse or the reciprocal of the other.

EXAMPLES:

What is the multiplicative inverse of $\frac{2}{3}$?

ANSWER: $\frac{3}{2}$

What is the multiplicative inverse of $-\frac{3}{2}$?

ANSWER: $-\frac{2}{3}$

PRACTICE EXERCISES

Multiply:

1
1. $\begin{array}{r} +5 \\ +9 \end{array}$
2. $\begin{array}{r} +8 \\ +6 \end{array}$
3. $\begin{array}{r} +7 \\ +7 \end{array}$
4. $\begin{array}{r} +6 \\ +4 \end{array}$
5. $\begin{array}{r} -2 \\ -3 \end{array}$
6. $\begin{array}{r} -9 \\ -4 \end{array}$
7. $\begin{array}{r} -6 \\ -7 \end{array}$
8. $\begin{array}{r} -8 \\ -5 \end{array}$

2
1. $\begin{array}{r} -7 \\ +8 \end{array}$
2. $\begin{array}{r} -2 \\ +9 \end{array}$
3. $\begin{array}{r} -4 \\ +12 \end{array}$
4. $\begin{array}{r} -13 \\ +5 \end{array}$
5. $\begin{array}{r} +6 \\ -6 \end{array}$
6. $\begin{array}{r} +12 \\ -8 \end{array}$
7. $\begin{array}{r} +4 \\ -5 \end{array}$
8. $\begin{array}{r} +6 \\ -9 \end{array}$

3
1. $\begin{array}{r} 0 \\ +3 \end{array}$
2. $\begin{array}{r} -5 \\ 0 \end{array}$
3. $\begin{array}{r} 0 \\ -8 \end{array}$
4. $\begin{array}{r} +4 \\ 0 \end{array}$
5. $\begin{array}{r} -.4 \\ -.2 \end{array}$
6. $\begin{array}{r} +2.5 \\ -1.4 \end{array}$
7. $\begin{array}{r} -3\frac{1}{2} \\ +2\frac{1}{4} \end{array}$
8. $\begin{array}{r} -12 \\ -1\frac{2}{3} \end{array}$

4
1. $(-3)(-9)$
2. $(+4)(-7)$
3. $11(3-14)$
4. $-8(-7-5)$
5. $(+6)(-2)(-4)$
6. $(-2) \times [(-6)(-7)]$

Find the value of each of the following:

5
1. $(-3)^2$
2. $(+6)^2$
3. $(-5)^3$
4. $(+2)^4$
5. $(-4)^5$
6. $(-1)^6$
7. $(-3)^4$
8. $(-1)^5 \cdot (-3)^3$

6
1. $|(-1)(-5)|$
2. $|-4| \times |-7|$
3. $-10 \times |-6|$
4. $|12-3| \times |6-8|$
5. $(-6) \times |(-2)(-9)(-1)|$
6. $|-5|^2 \times |(-4)(-11)|$

7 Name the multiplicative inverse of each of the following:
1. -6
2. $+14$
3. $-\frac{2}{5}$
4. $-\frac{17}{4}$
5. $+\frac{5}{6}$
6. $-3\frac{1}{2}$

Just as subtraction is the inverse of addition, division is the inverse of multiplication.

> Since $(+3)(+4) = +12$, then $(+12) \div (+3) = +4$
>
> $$\frac{+12}{+3} = +4$$
>
> The quotient of two positive numbers is a positive number.

Similarly,

> Since $(-3)(+4) = -12$, then $(-12) \div (-3) = +4$
>
> $$\frac{-12}{-3} = +4$$
>
> The quotient of two negative numbers is a positive number.

> Since $(+3)(-4) = -12$,   Since $(-3)(-4) = +12$,
> then $(-12) \div (+3) = -4$ then $(+12) \div (-3) = -4$
>
> $$\frac{-12}{+3} = -4 \qquad\qquad \frac{+12}{-3} = -4$$
>
> The quotient of a negative number divided by a positive number or a positive number divided by a negative number is a negative number.

PROCEDURE

To divide two signed numbers:
(1) If the two numbers are both positive or both negative, find the quotient of their absolute values and write a positive sign before the quotient.
(2) If one number is positive and one number is negative, find the quotient of their absolute values and write a negative sign before the quotients.

EXAMPLES:

Divide: $(-8) \div (-4)$

$\frac{-8}{-4} = +2$ **The sign of the quotient is positive.**

ANSWER: $+2$

Divide: $(+6) \div (-2)$

$\frac{+6}{-2} = -3$ **The sign of the quotient is negative.**

ANSWER: -3

PRACTICE EXERCISES

Divide.

1 **1.** $\frac{+30}{+5}$ **2.** $\frac{+16}{+4}$ **3.** $\frac{+54}{+9}$ **4.** $\frac{+60}{+10}$ **5.** $\frac{-8}{-2}$ **6.** $\frac{-15}{-5}$ **7.** $\frac{-90}{-18}$ **8.** $\frac{-72}{-12}$

2 **1.** $\frac{+56}{-7}$ **2.** $\frac{+18}{-3}$ **3.** $\frac{+45}{-5}$ **4.** $\frac{+36}{-6}$ **5.** $\frac{-48}{+8}$ **6.** $\frac{-63}{+7}$ **7.** $\frac{-80}{+10}$ **8.** $\frac{-144}{+16}$

3 **1.** $\frac{-9}{-9}$ **2.** $\frac{+7}{-7}$ **3.** $\frac{-15}{-1}$ **4.** $\frac{-40}{+1}$ **5.** $\frac{0}{+6}$ **6.** $\frac{0}{-3}$ **7.** $\frac{0}{+15}$ **8.** $\frac{0}{-20}$

4 **1.** $(+36) \div (-9)$ **2.** $(-20) \div (-4)$
     3. $(-42) \div (+7)$ **4.** $(+75) \div (+5)$
     5. $(0) \div (-7)$ **6.** $(-8\frac{1}{2}) \div (-\frac{3}{4})$
     7. $(+.24) \div (-.3)$ **8.** $(-40) \div (-40)$
     9. $(+19) \div (-1)$ **10.** $(-100) \div (+25)$

5 **1.** $+2\overline{)-14}$ **2.** $-8\overline{)-64}$ **3.** $-3\overline{)-27}$
     4. $-5\overline{)+50}$ **5.** $.4\overline{)-3.2}$ **6.** $-7\overline{)+21}$

6 Simplify by performing the indicated operations:

1. $\dfrac{8 - 17}{3}$ **2.** $\dfrac{-10 - 4}{-7}$ **3.** $\dfrac{12 - 3(4)}{-2}$

4. $\dfrac{8(-1) + 7(-6)}{-5(5)}$ **5.** $\dfrac{(-4)^3 - (-2)^4}{(-10)^2}$ **6.** $\dfrac{6(4 - 7) - 5(6 - 12)}{-3(1 - 5)}$

7 Find the value of each of the following:

1. $\dfrac{|-16|}{8}$ **2.** $\dfrac{|-24|}{|-2|}$ **3.** $\dfrac{|-30|}{|-5|}$

4. $\dfrac{|-8| \times |-6|}{|-9| - |-5|}$ **5.** $\dfrac{|(-12)(+7)|}{|8 - 9| - |1 - 5|}$ **6.** $\dfrac{|-3|^3 - |1|^4}{|(-6) + (-7)|}$

Computer Activities

The computer program below asks for two numbers and computes their *sum* and *difference*.

```
10 PRINT "WHAT IS THE FIRST NUMBER";
20 INPUT X
30 PRINT "WHAT IS THE SECOND NUMBER";
40 INPUT Y
50 PRINT "THEIR SUM IS: " X + Y
60 PRINT "THEIR DIFFERENCE IS: "; X − Y
```

a) Enter the program and then run it to make sure it works.
b) Extend the program so that it *also* computes and prints out the *product* and the *quotient* of X and Y.

Almost everyone is affected by the weather. For farmers and pilots, an unexpected change in the weather can be a disaster. Thousands of workers study weather conditions so they can make accurate predictions. These workers are called *meteorologists*. They record temperatures, wind velocities, air pressure, and humidity. They watch for types of clouds that signal changes in the weather. They use weather balloons to measure conditions in the upper air. Some meteorologists even help with research on air pollution. Others develop better instruments for measuring weather conditions.

Many colleges and technical schools offer two-year programs. Meteorologists learn some of their skills on the job. They may work for government agencies or for private companies such as airlines.

1. On Monday night the lowest temperature was −8°C. On Tuesday night the lowest temperature was −11°C. Which night was colder?

2. A weather balloon at an altitude of 70,000 feet measured the temperature as −67°F. The temperature at ground level was 74°F. How much warmer was it at ground level?

3. A meteorologist measured the air pressure every hour. Each hour the change in pressure was −5 millibars. What was the total change after three hours?

4. The temperature dropped from 0°C to −24°C in six hours. What was the average change per hour?

5. At noon the temperature was 12°F. At 6 P.M. it was 19 degrees lower. What was the temperature at 6 P.M.?

6. The wind velocity changed from 31 mi/h to 23 mi/h. Use a signed number to express the amount of the change.

CHAPTER REVIEW

Vocabulary

additive inverse p. 552	negative integer p. 540	positive number p. 540
absolute value p. 544	negative number p. 540	rational number p. 542
directed number p. 540	nonnegative integer p. 540	real number p. 542
even integer p. 541	nonpositive integer p. 540	real number line p. 545
integer p. 540	odd integer p. 541	reciprocal p. 561
irrational number p. 542	opposites p. 540	signed number p. 540
multiplicative inverse p. 561	positive integer p. 540	vector p. 550

EXERCISES

The numerals in boxes indicate Lessons where help may be found.

Which of the following numerals name: 16-1

1. Negative numbers?   **2.** Positive numbers?

$$-3.4 \qquad +\tfrac{7}{8} \qquad -350 \qquad 0 \qquad -4\tfrac{3}{5} \qquad 91 \qquad +.852$$

Which of the following numerals name: 16-2

3. Integers?   **4.** Rational numbers?   **5.** Irrational numbers? **6.** Real numbers?

$$+4\tfrac{1}{4} \qquad -.6 \qquad 29 \qquad -\tfrac{32}{8} \qquad -47 \qquad \tfrac{5}{6} \qquad -63$$

7. Find the value of: $|-13|$. 16-3

8. Draw the graph of $-4, -2, -1, 0, 3, 6$ on a number line. 16-4

9. Write the coordinates of which the following is a graph: 16-4

10. If -175 feet represents 175 feet below sea level, what does $+825$ feet represent? 16-5

Which of the following sentences are true? 16-6

11. $-8 > +2$ **12.** $-6 < -3$ **13.** $-5 \not> 0$ **14.** $+1 \not< -7$

15. Name the following numbers in order of size (smallest first): $-1, +8, -10, 0, +5, -4$ 16-6

Using the number line as a scale, draw the vector for each of the following and write the numeral represented by it that indicates a movement from: 16-7

16. 0 to $+6$ **17.** -2 to 3   **18.** $+3$ to -5 **19.** $+6$ to $+1$

20. What is the opposite of -50? 16-8   **21.** What is the additive inverse of $+.8$? 16-8

22. Find the value of $-(-25)$. 16-8   **23.** Find the multiplicative inverse of $\tfrac{9}{16}$. 16-11

Add: 16-9   Subtract: 16-10

24. $\begin{array}{r} -6 \\ -5 \end{array}$ **25.** $\begin{array}{r} +8 \\ -11 \end{array}$ **26.** $(-4) + (+7)$ **28.** $\begin{array}{r} -7 \\ -9 \end{array}$ **29.** $\begin{array}{r} -4 \\ +6 \end{array}$ **30.** $(3) - (9)$

                                  27. $(10) + (-10)$ **31.** $(+4) - (-1)$

Multiply: 16-11   Divide: 16-12

32. $\begin{array}{r} +6 \\ -9 \end{array}$ **33.** $\begin{array}{r} -8 \\ -3 \end{array}$ **34.** $(5)(-11)$ **36.** $\tfrac{-18}{+3}$ **37.** $\tfrac{-9}{-9}$ **38.** $(30) \div (-1)$

                                  35. $(-3)(-12)$ **39.** $(-63) \div (-7)$

Find the value of each of the following numerical expressions:

40. $-6 + 2 - 7 + 14 - 5$ 16-9   **41.** $\tfrac{6-15}{-3}$ 16-12

COMPETENCY CHECK TEST

The numerals in the boxes indicate Lessons where help may be found.

Solve each problem and select the letter corresponding to your answer.

1. Which of the following names a negative number? 16-1
   a. $+9$ **b.** 0 **c.** $-\frac{3}{4}$ **d.** $+4\frac{1}{6}$

2. Which of the following names an integer? 16-2
   a. $-\sqrt{53}$ **b.** $+7\frac{2}{3}$ **c.** $-.89$ **d.** -3

3. The value of $|-16|$ is: 16-3
   a. 4 **b.** -16 **c.** 16 **d.** -4

4. The coordinates of which this is the graph are: 16-4

   a. $-3,2$ **b.** $-2,2$ **c.** $-2,3$ **d.** Answer not given

5. Which number is greatest? 16-6
   a. -6 **b.** 0 **c.** $+2$ **d.** -14

6. Which has the greatest opposite number? 16-8
   a. -10 **b.** $+14$ **c.** 0 **d.** -2

7. The value of $-(-8)$ is: 16-8
   a. -8 **b.** $+8$ **c.** 0 **d.** Answer not given

8. The sum of $+2$ and -7 is: 16-9
   a. -9 **b.** -5 **c.** $+9$ **d.** $+5$

9. The product of -8 and -3 is: 16-11
   a. $+24$ **b.** -11 **c.** -24 **d.** -5

10. When -4 is subtracted from -2, the answer is: 16-10
    a. -2 **b.** -6 **c.** $+2$ **d.** $+6$

11. When -18 is divided by $+6$, the quotient is: 16-12
    a. -12 **b.** -3 **c.** $+3$ **d.** $-\frac{1}{3}$

12. The number that is neither positive nor negative is: 16-1
    a. 0 **b.** 10 **c.** 100 **d.** $1,000$

13. The numbers $-3, +7, -5, 0, +2, -1$ arranged according to size, least first, are: 16-6
    a. $0, -1, +2, -3, -5, +7$ **b.** $-1, -3, -5, 0, +2, +7$
    c. $-5, -3, -1, 0, +2, +7$ **d.** Answer not given

14. The simplified answer for the numerical expression $5 - 11 + 2 - 4 + 9$ is: 16-9
    a. -2 **b.** -1 **c.** 0 **d.** 1

15. The multiplicative inverse of 6 is: 16-11
    a. -6 **b.** 6 **c.** $\frac{1}{6}$ **d.** Answer not given

CHAPTER **17** **Equations, Formulas, and Inequalities**

17-1 Algebraic Expressions

A numerical expression consists of a single numeral, or two or more numerals joined by operational symbols.

> 18; $7 + 5$; $28 - 14$; 5×43; $63 \div 7$; and $8 \times (3 + 11)$ are examples of numerical expressions.

A symbol that holds a place open for a number is called a variable.

> A; c; b_1; S; and $\triangle$ are examples of variables.

An algebraic expression may be a numerical expression or an expression containing one or more variables joined by operational symbols.

> b; $a - 5$; and $3y^2 - 7x + 9$ are examples of algebraic expressions.

A numeral is sometimes called a constant because it stands for a definite value. In an algebraic expression the multiplication symbol need not be used when the two factors are both variables ($b \times d$ may be expressed as bd) or a numeral and a variable. When a numeral and a variable are expressed as a product, the numeral always precedes the variable and the numeral is called the numerical coefficient of the variable.

$$5x \qquad\qquad\qquad 23a$$
↑———— **numerical coefficients** ———↑

EXAMPLES:

Write the product of n and 4 as an algebraic expression.

ANSWER: $4n$

Write ten times the sum of m and 6 as an algebraic expression.

ANSWER: $10(m + 6)$

PRACTICE EXERCISES

1 Name the variable in each of the following expressions:

1. $14 - x$ **2.** $3y + 36$ **3.** $\frac{z}{5} + 2$

4. $b_1 - 70$ **5.** $n(n + 4)$ **6.** $c^3 \times 5$

2 Write each of the following as an algebraic expression:

1. Six added to four.
2. Twelve times nineteen.
3. From ten subtract two.
4. Fifteen divided by three.
5. The square of nine.
6. The square root of twenty.
7. The sum of eight and three.
8. The difference between six and two.
9. The product of nine and five.
10. The quotient of twelve divided by four.
11. The sum of b and x.
12. a times y.
13. The product of g and eight.
14. The difference between c and g.
15. The quotient of d divided by r.
16. Two times the sum of l and w.
17. The cube of r.
18. The square root of s.
19. The product of b and h.
20. Three times the difference between t and nine.

3 Read, or write in words, each of the following:

1. $5n$
2. $19 - 7$
3. $y + 25$
4. $\frac{m}{x}$
5. cd^2x^3
6. $6a - 9b$
7. $(m + n)(m - n)$
8. $3x^2 \quad 4xy \mid 7y^2$

4 Write each of the following as an algebraic expression:

1. Twice the radius (r).
2. The circumference (C) divided by pi (π).
3. The principal (p) plus the interest (i).
4. 90° decreased by angle B.
5. Three times the side (s).
6. The sum of m∠A, m∠B, and m∠C.
7. The length (l) times the width (w) times the height (h).
8. The profit (p) added to the cost (c).
9. One half the product of the altitude (a) and base (b).
10. Angle B subtracted from 180°.
11. The square of the side (s).
12. The base (b) multiplied by the height (h) divided by two.
13. The sum of twice the length (l) and twice the width (w).
14. The cube of the edge (e).
15. The quotient of the interest (i) for one year divided by the principal (p).
16. The sum of bases b_1 and b_2 divided by two.
17. Four thirds the product of pi (π) and the cube of the radius (r).
18. The square of the altitude (a) added to the square of the base (b).
19. Twice pi (π) times the radius (r) times the sum of the height (h) and radius.
20. The square root of the difference between the square of the hypotenuse (h) and the square of the base (b).

5 Write an algebraic expression for each of the following:

1. Marilyn has n cents. She spends 95 cents for lunch. How many cents does she have left?
2. Scott is x years old. How old will he be in fifteen years?
3. Steve has d dollars in the bank. If he deposits y dollars, how many dollars will he then have in the bank?
4. How long does it take a train averaging r m.p.h. to travel m miles?
5. How many notebooks can be bought for x cents if one notebook costs y cents?

17-2 Algebraic Sentences

Mathematical Sentences

$7 + 9 = 16$	"Seven plus nine is equal to sixteen."
$x - 4 = 9$	"Some number x decreased by 4 is equal to nine."
$8a > 20$	"Eight times each number a is greater than twenty."
$n + 10 < 14$	"Each number n increased by ten is less than fourteen."

An open sentence is a mathematical sentence that contains a variable. An equation is an open sentence that has the equal sign ($=$) as its verb. An inequality is an open sentence that uses an inequality sign ($<$, $>$, $\leq$, or $\geq$) as its verb. An open sentence is neither true nor false.

Open Sentences

$x - 4 = 9$ Equation $n + 10 < 14$ Inequality

A line drawn through symbols like $\neq$, $\not<$, and $\not>$ reverses their meanings.

$x \neq 45$ "Some number x is *not* equal to forty-five."

A compound sentence is formed by joining two simple sentences with the connective "or" or "and." A disjunction is a compound sentence that is read with the connective "or." A conjunction is a compound sentence that is read with the connective "and."

Compound Sentences

$6n \leq 30$ means $6n < 30$ or $6n = 30$.   Disjunction

$8 < x < 12$ means $8 < x$ and $x < 12$.   Conjunction

EXAMPLES:

Read, or write in words, $b - 6 > 25$.

ANSWER: Some number b decreased by 6 is greater than 25.

Write "some number x increased by 12 is less than 15" using symbols.

ANSWER: $x + 12 < 15$

Explain the meaning of $4y \geq 12$ and determine if it is a conjunction or disjunction.

ANSWER: $4y > 12$ or $4y = 12$; disjunction

PRACTICE EXERCISES

1 Read, or write in words, each of the following:

  1. $n < 16$ **2.** $x > 29$
  3. $3y \neq 17$ **4.** $a - 5 < 53$
  5. $7t \not< -14$ **6.** $9m + 4 = 15$
  7. $8c - 9 \not> c + 7$ **8.** $2n + 6 < 3n - 8$
  9. $\frac{b}{5} > 20$ **10.** $15x - 4 \not> 18 - 7x$

2 Write each of the following as an open sentence using symbols:

  1. Some number x increased by ten is equal to forty-one.

  2. Each number n decreased by four is less than eighteen.

  3. Four times each number t is greater than twelve.

  4. Nine times each number y plus two is not equal to fifty.

  5. Each number a increased by five is not less than nine.

  6. Each number m divided by seven is not greater than fourteen.

  7. Seven times each number x is greater than or equal to thirty.

  8. Each number c plus nine is less than or equal to ten.

  9. Each number t is less than negative one and greater than negative five.

  10. Twelve times each number y is greater than or equal to four and less than or equal to fifteen.

3 Read, or write in words each of the following and determine if it is a conjunction or disjunction.

  1. $h \geq -6$ **2.** $19w \leq 38$
  3. $5d + 7 \leq 12$ **4.** $n - 16 \geq 4n + 9$
  5. $18 > x > 0$ **6.** $4 < b < 27$
  7. $12 \geq 7r \geq -6$ **8.** $2 < a \leq 45$

17-3 Formulas

A formula is a special kind of equation. It is a mathematical rule expressing the relationship of two or more quantities by means of numerals, variables, and symbols of operation. Often quantities are represented by the first letter of a key word.

$$p = 3s, \quad E = IR, \quad F = 1.8C + 32 \quad \text{are formulas.}$$

PROCEDURE

To express mathematical and scientific principles as formulas:
Write numerals, symbols of operation, and variables to show the relationships between the quantities.

EXAMPLE: Write as a formula: The circumference (C) of a circle is equal to pi (π) times the diameter (d).

ANSWER: $C = \pi d$

PROCEDURE

To translate a formula to a word statement:
Write a word rule stating the relationship expressed by the formula.

EXAMPLE: Translate: $a = \frac{360°}{n}$ where a = the measure of a central angle of a regular polygon and n = the number of sides.

ANSWER: The measure of a central angle of a regular polygon equals 360° divided by the number of sides.

PRACTICE EXERCISES

1 Express each of the following as a formula:

1. The perimeter of a square (p) is equal to four times the length of the side (s).

2. The area of a rectangle (A) is equal to the length (l) multiplied by the width (w).

3. The circumference of a circle (C) is equal to twice pi (π) times the radius (r).

4. The interest (i) is equal to the principal (p) times the rate of interest per year (r) times the time in years (t).

5. The radius of a circle (r) is equal to the diameter (d) divided by two.

6. The total amount (A) is equal to the sum of the principal (p) and the interest (i).

7. The area of a parallelogram (A) is equal to the product of the base (b) and height (h).

8. The volume of a cube (V) is equal to the cube of the edge (e).

9. The area of a circle (A) is equal to pi (π) times the square of the radius (r).

10. The area of a triangle (A) is equal to one half the product of the altitude (a) and base (b).

11. The sum of two complementary angles ($\angle A$ and $\angle B$) equals 90.°

12. The net price (n) is equal to the list price (l) less the discount (d).

13. The volume of a cylinder (V) is equal to one fourth pi (π) times the square of a diameter (d) times height (h).

14. The perimeter of a rectangle (p) is equal to twice the sum of the length (l) and width (w).

15. The rate of commission (r) is equal to the commission (c) divided by the sales (s).

16. The area of a trapezoid (A) is equal to the height (h) times the sum of the two parallel bases, b_1 and b_2, divided by two.

17. The perimeter of an isosceles triangle (p) is equal to the base (b) added to twice the length of the equal side (e).

18. The area of the surface of a rectangular solid (A) is equal to the sum of twice the product of the length (l) and width (w), twice the product of the length and height (h), and twice the product of the width and height.

19. The hypotenuse of a right triangle (h) is equal to the square root of the sum of the square of the altitude (a) and the square of the base (b).

20. The total area of a cylinder (A) is equal to twice pi (π) times the radius (r) times the sum of the height (h) and the radius.

2 Express each of the following formulas as a word statement:

1. $p = 3s$ where p = perimeter of an equilateral triangle and s = length of a side of an equilateral triangle.

2. $E = IR$ where E = electromotive force in volts, I = current in amperes, and R = resistance in ohms.

3. $t = \dfrac{d}{r}$ where t = time of travel, d = distance traveled, and r = average rate of speed.

4. $F = 1.8C + 32$ where F = Fahrenheit temperature reading and C = Celsius temperature reading.

5. $d = \frac{1}{2} gt^2$ where d = distance a freely falling body drops, g = acceleration of gravity, t = time of falling.

To evaluate an expression means to find the value of the expression for given numerical replacements of the variables. The value of an algebraic expression depends on the numerical values chosen for each variable in the expression. If these values change, the value of the expression usually changes.

PROCEDURE

To evaluate an algebraic expression:
(1) Copy the expression.
(2) Substitute the given numerical value for each variable.
(3) Perform the necessary operations indicated in the expression.

EXAMPLES:

Find the value of $2d + x$ when $d = 7$ and $x = 9$.

$$2d + x$$
$$= (2 \cdot 7) + 9$$
$$= 14 + 9$$
$$= 23$$

ANSWER: 23

Find the value of $2(l + w)$ when $l = 17$ and $w = 14$.

$$2(l + w)$$
$$= 2(17 + 14)$$
$$= 2(31)$$
$$= 62$$

ANSWER: 62

Find the value of $\frac{A}{b}$ when $A = 28$ and $b = 4$.

$$\frac{A}{b}$$
$$= \frac{28}{4}$$
$$= 7$$

ANSWER: 7

Find the value of $c^2 + d^2$ when $c = 8$ and $d = 3$.

$$c^2 + d^2$$
$$= (8)^2 + (3)^2$$
$$= 64 + 9$$
$$= 73$$

ANSWER: 73

PRACTICE EXERCISES

Find the value of each of the following algebraic expressions:

1 When $a = 12$ and $b = 6$:

1. $a + b$	2. $a - b$	3. ab	4. $\frac{a}{b}$
5. a^2	6. $7a$	7. $4(a - b)$	8. $a^2 - b^2$

2 When $x = 4$ and $y = 3$:

1. $x + 6y$	2. $5xy$	3. $4x - 8y$	4. $x(x - y)$
5. $9x^2y$	6. $x^2 - y^2$	7. $(x + y)^2$	8. $\frac{x + y}{2x - y}$

3 When $m = -8$, $n = -2$, and $x = 4$:

1. $3mnx$ **2.** $10mn - 7nx$ **3.** $(m + n)(m - n)$

4. $\frac{(n - x)^2}{n^2 - x^2}$ **5.** $m - n(m + n)$ **6.** $5m^2 + 2mx - 3x^2$

4 **1.** $4s$ when $s = 6$ **2.** $2r$ when $r = 12$

3. bh when $b = 10$ and $h = 7$ **4.** πd when $\pi = 3.14$ and $d = 26$

5. ab when $a = 45$ and $b = 51$ **6.** lwh when $l = 11$, $w = 7$, and $h = 15$

7. $2\pi r$ when $\pi = \frac{22}{7}$ and $r = 21$ **8.** πdh when $\pi = \frac{22}{7}$, $d = 63$, and $h = 20$

9. prt when $p = 400$, $r = .06$, and $t = 8$ **10.** $2\pi rh$ when $\pi = 3.14$, $r = 48$, and $h = 85$

5 **1.** $p + i$ when $p = 500$ and $i = 120$ **2.** $b + 2e$ when $b = 32$ and $e = 29$

3. $b_1 + b_2$ when $b_1 = 47$ and $b_2 = 35$ **4.** $2l + 2w$ when $l = 63$ and $w = 54$

5. $2lw + 2lh + 2wh$ when $l = 18$, $w = 14$, and $h = 17$

6 **1.** $90 - B$ when $B = 72$ **2.** $A - p$ when $A = 214$ and $p = 185$

3. $l - d$ when $l = 67$ and $d = 29$ **4.** $180 - A$ when $A = 104$

7 **1.** $\frac{d}{2}$ when $d = 46$ **2.** $\frac{ab}{2}$ when $a = 25$ and $b = 30$

3. $\frac{c}{\pi}$ when $c = 12.56$ and $\pi = 3.14$ **4.** $\frac{Bh}{3}$ when $B = 216$ and $h = 17$

8 **1.** a^2 when $a = 6$ **2.** s^2 when $s = 27$

3. b^2 when $b = 105$ **4.** h^2 when $h = 300$

5. πr^2 when $\pi = \frac{22}{7}$ and $r = 28$ **6.** $.7854d^2$ when $d = 50$

7. $6s^2$ when $s = 41$ **8.** e^3 when $e = 19$

9. $\frac{4}{3}\pi r^3$ when $\pi = 3.14$ and $r = 20$ **10.** $\pi r^2 h$ when $\pi = \frac{22}{7}$, $r = 70$, and $h = 45$

9 **1.** $\frac{1}{2}ab$ when $a = 72$ and $b = 66$ **2.** $\frac{1}{3}Bh$ when $B = 135$ and $h = 20$

3. $\frac{1}{4}\pi d^2$ when $\pi = 3.14$ and $d = 55$ **4.** $\frac{4}{3}\pi r^3$ when $\pi = \frac{22}{7}$ and $r = 42$

10 **1.** $a^2 + b^2$ when $a = 11$ and $b = 9$ **2.** $h^2 - a^2$ when $h = 36$ and $a = 27$

3. $V + gt$ when $V = 150$, $g = 32$, and $t = 7$ **4.** $p + prt$ when $p = 300$, $r = .06$, and $t = 8$

5. $2\pi rh + 2\pi r^2$ when $\pi = 3.14$, $r = 20$, and $h = 75$

6. $\pi dh + \frac{1}{2}\pi d^2$ when $\pi = \frac{22}{7}$, $d = 14$, and $h = 39$

11 **1.** $2(l + w)$ when $l = 37$ and $w = 28$ **2.** $\frac{5}{9}(F - 32)$ when $F = -13$

3. $I(r + R)$ when $I = 8$, $r = 3$, and $R = 10$ **4.** $(n - 1)d$ when $n = 10$ and $d = 6$

5. $\frac{h}{2}(b_1 + b_2)$ when $h = 34$, $b_1 = 27$, and $b_2 = 19$ **6.** $2\pi r(h + r)$ when $\pi = \frac{22}{7}$, $r = 7$, and $h = 13$

12 **1.** $\sqrt{A}$ when $A = 49$ **2.** $\sqrt{\frac{V}{\pi h}}$ when $V = 1{,}100$, $\pi = \frac{22}{7}$, and $h = 14$

3. $\sqrt{a^2 + b^2}$ when $a = 18$ and $b = 24$ **4.** $\sqrt{h^2 - a^2}$ when $h = 78$ and $a = 72$

5. $\sqrt{h^2 - b^2}$ when $h = 145$ and $b = 116$

17-5 Evaluating Formulas

Formulas are evaluated in the same way as expressions.

PROCEDURE

To determine the required value when evaluating a formula:
(1) Copy the formula and substitute the given values for the variables.
(2) Perform the necessary operations.

EXAMPLES:

Find the value of p when $s = 16$, using the formula $p = 4s$.

$$p = 4s$$
$$p = 4 \cdot 16$$
$$p = 64$$

ANSWER: 64

Find the value of I when $W = 100$ and $E = 25$, using $I = \frac{W}{E}$.

$$I = \frac{W}{E}$$
$$I = \frac{100}{25}$$
$$I = 4$$

ANSWER: 4

PRACTICE EXERCISES

Find the value of:

1
  1. d when $r = 85$. Formula: $d = 2r$ **2.** A when $B = 47$. Formula: $A = 90 - B$
  3. A when $b = 32$ and $h = 49$. Formula: $A = bh$
  4. c when $\pi = \frac{22}{7}$ and $d = 84$. Formula: $c = \pi d$
  5. A when $p = 620$ and $i = 53$. Formula: $A = p + i$
  6. p when $l = 73$ and $w = 28$. Formula: $p = 2l + 2w$
  7. i when $p = 650$, $r = .04$, and $t = 9$. Formula: $i = prt$
  8. V when $l = 41$, $w = 27$, and $h = 30$. Formula: $V = lwh$
  9. A when $\pi = 3.14$, $r = 48$, and $h = 75$. Formula: $A = 2\pi rh$
  10. A when $s = 50$. Formula: $A = s^2$ **11.** V when $e = 8$. Formula: $V = e^3$

2
  1. p when $l = 427$ and $w = 393$. Formula: $p = 2l + 2w$
  2. B when $A = 75$ and $C = 48$. Formula: $B = 180 - (A + C)$
  3. l when $a = 6$, $n = 12$, and $d = 5$. Formula: $l = a + (n - 1)d$
  4. I when $E = 220$ and $R = 11$. Formula: $I = \frac{E}{R}$
  5. P when $F = 90$, $d = 10$, and $t = 3$. Formula: $P = \frac{FD}{t}$
  6. v when $V = 27$, $g = 32$, and $t = 5$. Formula: $v = V + gt$
  7. A when $p = 150$, $r = .08$, and $t = 4$. Formula: $A = p(1 + rt)$
  8. I when $E = 220$, $n = 4$, $R = 8$, and $r = 3$. Formula: $I = \frac{nE}{R + nr}$

17-6 Solving Equations by Replacement

Solving an equation means finding the number represented by the variable which makes the sentence true. Any number that makes a sentence true is called a root or solution of the open sentence and is said to satisfy the sentence. Checking a solution is important. To check whether a number is a solution of a sentence, substitute the number for the variable and simplify. If the resulting sentence is true, the number is a solution.

> In the equation $n + 4 = 20$, $n + 4$ is the left member or left side and 20 is the right member or right side.

Equations which have exactly the same solutions are called equivalent equations.

> The equations $n + 4 = 20$ and $n = 16$ are equivalent equations because they both have the same solution, 16.

An equation is in simplest form when one member contains only the variable itself and the other member is a constant.

> The equation $n = 16$ is an equation in simplest form.

PROCEDURE

To solve an equation by replacement:

(1) Copy the equation.
(2) Substitute the given or chosen number for the variable in the equation.
(3) Simplify. If the resulting sentence is true, the number is a solution. If it is false, the number is not a solution.

EXAMPLES:

Is $n = 18$ a solution for $n - 2 = 16$?

$$n - 2 = 16$$
$$18 - 2 = 16?$$
$$16 = 16 \checkmark$$

ANSWER: $n = 18$ is a solution.

Is $n = 12$ a solution for $\frac{n}{6} = 4$?

$$\frac{n}{6} = 4$$
$$\frac{12}{6} = 4?$$
$$2 \neq 4$$

ANSWER: $n = 12$ is not a solution.

PRACTICE EXERCISES

1 Which value for the variable will make the given sentence true?

1. $x + 4 = 8$ **a.** $x = 12$ **b.** $x = 2$ **c.** $x = -4$ **d.** $x = \frac{1}{2}$ **e.** $x = 4$

2. $n - 2 = 10$ **a.** $n = 8$ **b.** $n = 5$ **c.** $n = 12$ **d.** $n = -8$ **e.** $n = -\frac{1}{5}$

3. $9r = 63$ **a.** $r = 72$ **b.** $r = 54$ **c.** $r = 0$ **d.** $r = 7$ **e.** $r = -6$

4. $\frac{b}{3} = 18$ **a.** $b = 6$ **b.** $b = \frac{1}{6}$ **c.** $b = 15$ **d.** $b = 54$ **e.** $b = 21$

2 Which value is the root of the given equation?

1. $m + 6 = 6$ **a.** $m = 12$ **b.** $m = -12$ **c.** $m = 0$ **d.** $m = 1$ **e.** $m = 6$

2. $8s = 24$ **a.** $s = 16$ **b.** $s = 32$ **c.** $s = -\frac{1}{3}$ **d.** $s = 0$ **e.** $s = 3$

3. $d - 9 = 3$ **a.** $d = 6$ **b.** $d = 3$ **c.** $d = -6$ **d.** $d = 12$ **e.** $d = -12$

4. $\frac{a}{8} = 4$ **a.** $a = 2$ **b.** $a = \frac{1}{2}$ **c.** $a = 32$ **d.** $a = 12$ **e.** $a = 4$

3 Which value is a solution of the given equation?

1. $12y = 6$ **a.** 6   **b.** 2   **c.** 18   **d.** $\frac{1}{2}$ **e.** -2

2. $n + 7 = 21$ **a.** 3   **b.** 28   **c.** $\frac{1}{3}$ **d.** 14   **e.** -3

3. $\frac{x}{12} = 8$ **a.** 20   **b.** 4   **c.** $\frac{2}{3}$ **d.** 96   **e.** $1\frac{1}{2}$

4. $w - 12 = 12$ **a.** -24 **b.** 0   **c.** 1   **d.** 24   **e.** 144

4 1. Write the left member of the equation $7x + 9 = 37$.
   2. Write the right member of the equation $5n - 3 = 42$.

5 1. Which of the following equations have 6 as a root?
   2. Which are equivalent equations?

   a. $y - 2 = 8$ **b.** $\frac{y}{3} = 2$ **c.** $9y = 54$ **d.** $y = 6$ **e.** $y + 17 = 23$

6 1. Which of the following equations have 7 as a root?
   2. Which are equivalent equations?
   3. Which one of the equivalent equations is in the simplest form?

   a. $9x = 63$ **b.** $x - 4 = 11$ **c.** $x = 7$ **d.** $\frac{x}{7} = 1$ **e.** $x + 15 = 8$

Inverse operations are operations that undo each other. Recall that addition undoes subtraction, subtraction undoes addition, multiplication undoes division, and division undoes multiplication.

$$(5 + 2) - 2 = 5 \qquad (8 - 6) + 6 = 8$$
$$(4 \times 3) \div 3 = 4 \qquad (14 \div 7) \times 7 = 14$$

We can use inverse operations to help solve equations. Before we use them, we need to learn about some properties of equality. The properties of equality allow us to obtain equations in simplest form, which give us the solutions to the equations.

Properties of Equality

I When we subtract the same number from both sides of an equation, the result is an equivalent equation.

II When we add the same number to both sides of an equation, the result is an equivalent equation.

III When we multiply both sides of an equation by the same nonzero number, the result is an equivalent equation.

IV When we divide both sides of an equation by the same nonzero number, the result is an equivalent equation.

These properties of equality tell us that the results are equal when equals are increased, decreased, multiplied, or divided by the same number. Division by zero is excluded.

PROCEDURE

To use a property of equality to solve an equation:
(1) Decide what operation will undo the operation in the equation.
(2) Choose the number that must be used with the operation. Operate with this number on both sides of the equation.

EXAMPLES:

For each given equation, decide what operation with what number should be used to solve the equation. Then write the equation in simplest form.

Given: $n + 2 = 12$
Subtract 2 from each side: $(n + 2) - 2 = 12 - 2$
Simplest form: $n = 10$

Given: $n - 2 = 12$
Add 2 to each side: $(n - 2) + 2 = 12 + 2$
Simplest form: $n = 14$

Given: $2n = 12$
Divide each side by 2: $\frac{2n}{2} = \frac{12}{2}$
Simplest form: $n = 6$

Given: $\frac{n}{2} = 12$
Multiply each side by 2: $2 \cdot \frac{n}{2} = 2 \cdot 12$
Simplest form: $n = 24$

PRACTICE EXERCISES

Find the missing number (▨) and, where required, the missing operation (?).

1 1. $(11 + 8) - ▨ = 11$ 2. $(16 + 22) \, ? \, ▨ = 16$ 3. $(n + 5) \, ? \, ▨ = n$

2 1. $(9 \times 7) \div ▨ = 9$ 2. $(54 \times 31) \, ? \, ▨ = 54$ 3. $(h \times 6) \, ? \, ▨ = h$

3 1. $(26 - 5) + ▨ = 26$ 2. $(40 - 18) \, ? \, ▨ = 40$ 3. $(x - 12) \, ? \, ▨ = x$

4 1. $(3 \times 8) \div ▨ = 8$ 2. $(23 \times 43) \, ? \, ▨ = 43$ 3. $(7m) \, ? \, ▨ = m$

5 1. $(36 \div 4) \times ▨ = 36$ 2. $(19 \div 6) \, ? \, ▨ = 19$ 3. $(y \div 8) \, ? \, ▨ = y$

6 1. $\frac{7}{12} \times ▨ = 7$ 2. $\frac{13}{5} \, ? \, ▨ = 13$ 3. $\frac{n}{4} \, ? \, ▨ = n$

7 1. $(r - 10) \, ? \, ▨ = r$ 2. $(16n) \, ? \, ▨ = n$ 3. $(w + 25) \, ? \, ▨ = w$

   4. $\left(\frac{b}{21}\right) \, ? \, ▨ = b$ 5. $(11x) \, ? \, ▨ = x$ 6. $(t - 37) \, ? \, ▨ = t$

8 In each of the following, what number do you subtract from both sides of the given equation to get an equivalent equation in simplest form? Also, write this equivalent equation.

   1. $n + 3 = 19$ 2. $x + 12 = 27$ 3. $y + 9 = 33$ 4. $r + 24 = 72$

9 In each of the following, what number do you add to both sides of the given equation to get an equivalent equation in simplest form? Also write this equivalent equation.

1. $w - 4 = 11$ **2.** $s - 19 = 6$ **3.** $h - 28 = 28$ **4.** $x - 15 = 0$

10 In each of the following, by what number do you divide both sides of the given equation to get an equivalent equation in simplest form? Also write this equivalent equation.

1. $9m = 72$ **2.** $15y = 90$ **3.** $20b = 35$ **4.** $12p = 7$

11 In each of the following, by what number do you multiply both sides of the given equation to get an equivalent equation in simplest form? Also write this equivalent equation.

1. $\frac{t}{3} = 6$ **2.** $\frac{m}{8} = 2$ **3.** $\frac{x}{14} = 9$ **4.** $\frac{a}{10} = 25$

12 What operation with what number do you use on both sides of the given equation to get an equivalent equation in simplest form? Also write this equivalent equation.

1. $x - 5 = 13$ **2.** $m + 11 = 21$

3. $\frac{n}{8} = 7$ **4.** $\frac{t}{6} = 12\cdot$

5. $8h = 56$ **6.** $4 = \frac{n}{10}$

7. $d - 110 = 40$ **8.** $42 = x - 28$

9. $6a = 42$ **10.** $12y = 9$

11. $b + 9 = 0$ **12.** $s - 31 = 31$

13. $x + 45 = 108$ **14.** $2 = 8n$

15. $75 = a + 17$ **16.** $18g = 144$

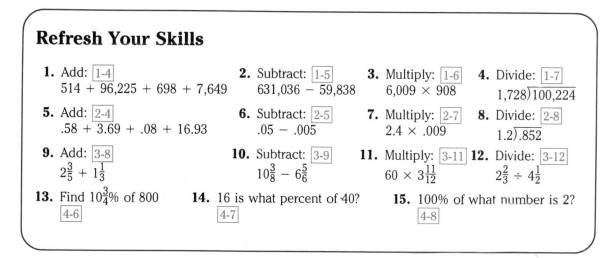

Refresh Your Skills

1. Add: 1-4
514 + 96,225 + 698 + 7,649

2. Subtract: 1-5
631,036 − 59,838

3. Multiply: 1-6
6,009 × 908

4. Divide: 1-7
1,728)100,224

5. Add: 2-4
.58 + 3.69 + .08 + 16.93

6. Subtract: 2-5
.05 − .005

7. Multiply: 2-7
2.4 × .009

8. Divide: 2-8
1.2).852

9. Add: 3-8
$2\frac{3}{5} + 1\frac{1}{3}$

10. Subtract: 3-9
$10\frac{3}{8} - 6\frac{5}{6}$

11. Multiply: 3-11
$60 \times 3\frac{11}{12}$

12. Divide: 3-12
$2\frac{2}{3} \div 4\frac{1}{2}$

13. Find $10\frac{3}{4}\%$ of 800
4-6

14. 16 is what percent of 40?
4-7

15. 100% of what number is 2?
4-8

17-8 Solving Equations by Subtraction

One way of solving equations is to use the properties of equalities and inverse operations. It is important to check your answer when you finish. When the operation indicated in the equation is addition, the inverse operation is subtraction.

PROCEDURE

To solve equations involving addition of the type $n + 4 = 20$:
(1) The inverse operation is subtraction, so subtract from both sides the number that is on the same side of the equation as the variable.
(2) Check your answer by substituting it for the variable in the equation.

EXAMPLE:

Solve and check: $n + 4 = 20$

$$n + 4 = 20$$
$$n + 4 - 4 = 20 - 4$$
$$n = 16$$

Check: $n + 4 = 20$
$16 + 4 = 20$?
$20 = 20$ ✔

The indicated operation is addition. Subtract 4 to undo adding 4.

ANSWER: $n = 16$

The equations $4 + n = 20$, $20 = n + 4$, and $20 = 4 + n$ are all solved by subtracting 4 from both sides of the equation. The solution, $16 = n$, can be rewritten as $n = 16$.

PRACTICE EXERCISES

Solve and check:

1
 1. $c + 6 = 14$
 2. $b + 18 = 31$
 3. $i + 37 = 63$
 4. $48 + w = 75$
 5. $29 + p = 106$
 6. $54 + n = 162$
 7. $67 = a + 28$
 8. $50 = m + 33$
 9. $105 = y + 79$
 10. $87 = 40 + x$
 11. $36 = 19 + d$
 12. $210 = 96 + r$
 13. $n + 80 = 80$
 14. $c + \$.09 = \$.23$
 15. $\$50 = i + \$42.50$

2
 1. $t + \frac{3}{4} = 4$
 2. $1\frac{1}{2} + x = 3\frac{5}{8}$
 3. $6.5 = .48 + n$
 4. $a + \$9 = \$10.50$
 5. $8\frac{1}{3} = 4\frac{1}{4} + d$
 6. $b + 9 = 4$
 7. $3 + y = 1$
 8. $a + 5 = -3$
 9. $-9 = r + 10$
 10. $g + 15 = 0$
 11. $62 = 62 + x$
 12. $n + 7 = 3.8$
 13. $25 + a = 97$
 14. $7 = s + .4$
 15. $\frac{3}{8} = \frac{1}{4} + x$

582

17-9 Solving Equations by Addition

When the operation indicated in the equation is subtraction, the inverse operation is addition.

PROCEDURE

To solve equations involving subtraction of the type $n - 4 = 20$:
(1) The inverse operation is addition, so add to both sides the number that is on the same side of the equation as the variable.
(2) Check your answer by substituting it for the variable in the given equation.

EXAMPLE:

Solve and check: $n - 4 = 20$.

$$n - 4 = 20$$
$$n - 4 + 4 = 20 + 4$$
$$n = 24$$

Check: $n - 4 = 20$
$24 - 4 = 20?$
$20 = 20$ ✔

The indicated operation is subtraction. Add 4 to undo subtracting 4.

ANSWER: 24

To solve $20 = n - 4$, you still add 4 to both sides. The solution, $24 = n$, can also be written as $n = 24$.

PRACTICE EXERCISES

Solve and check:

1
1. $c - 7 = 21$
2. $l - 18 = 5$
3. $A - 42 = 124$
4. $35 = n - 19$
5. $16 = b - 47$
6. $28 = y - 16$
7. $t - 9 = 0$
8. $0 = y - 26$
9. $c - 40 = 40$
10. $52 = m - 52$
11. $s - \$51 = \$120$
12. $9 = b - 2\frac{1}{2}$
13. $y - \$4.75 = \$3.25$
14. $d - \frac{7}{8} = 1\frac{1}{4}$
15. $9.6 = n - 7$

2
1. $a - \$.08 = \$.75$
2. $c - 4 = -9$
3. $r - 23 = -45$
4. $x - 18 = -10$
5. $b - 6 = -6$
6. $x - 2\frac{1}{2} = 2\frac{1}{2}$
7. $18 = n - 1.3$
8. $0 = x - 75$
9. $-5 = r - 5$
10. $51 = t - 51$
11. $w - 34 = -2$
12. $a - 29 = 76$
13. $\frac{1}{2} = d - 1\frac{3}{4}$
14. $x - \$.75 = \$.35$
15. $y - 1 = -9$

17-10 Solving Equations by Division

When the operation indicated in the equation is multiplication, the inverse operation is division.

PROCEDURE

To solve equations involving multiplication of the type $4n = 20$:
(1) The inverse operation is division, so divide both sides by the number by which the variable is multiplied in the given equation.
(2) Check your answer by substituting it for the variable in the given equation.

EXAMPLE:

Solve and check: $4n = 20$

$$4n = 20$$
$$\frac{4n}{4} = \frac{20}{4}$$
$$n = 5$$

Check: $4n = 20$
$$(4)(5) = 20?$$
$$20 = 20 \; \checkmark$$

The indicated operation is multiplication. Divide by 4 to undo multiplying by 4.

ANSWER: 5

To solve $20 = 4n$, you still divide both sides by 4. The solution, $5 = n$, can be written as $n = 5$.

PRACTICE EXERCISES

Solve and check:

1
1. $5a = 45$
2. $9c = 54$
3. $14n = 42$
4. $91 = 7y$
5. $60 = 12x$
6. $105 = 15b$
7. $2r = 18$
8. $48 = 4s$
9. $10w = 10$
10. $19 = 19h$
11. $8y = 0$
12. $0 = 16x$
13. $5c = 8$
14. $18w = 27$
15. $4y = 1$
16. $12d = 10$
17. $23 = 9n$
18. $4 = 5b$
19. $54 = 8r$
20. $12 = 16x$

2
1. $7t = 2.8$
2. $.4n = 36$
3. $10c = \$.90$
4. $\$.05a = \$1.45$
5. $\frac{1}{4}c = 20$
6. $\frac{7}{8}n = 42$
7. $24 = \frac{2}{3}n$
8. $3\frac{1}{2}t = 14$
9. $.06p = \$240$
10. $\frac{9}{16}r = 72$
11. $8x = -32$
12. $-5m = 40$
13. $-7y = -56$
14. $-x = -10$
15. $-64 = -4c$
16. $-y = 19$
17. $44 = -18b$
18. $-20s = -5$
19. $10c = -50$
20. $10 = -50c$
21. $12m = 300$
22. $140 = 15y$
23. $3.14d = 12.56$
24. $2\frac{1}{4}n = 13\frac{1}{2}$

17-11 Solving Equations by Multiplication

When the operation indicated in the equation is division, the inverse operation is multiplication.

PROCEDURE

To solve equations involving division of the type $\frac{n}{4} = 20$:
(1) The inverse operation is multiplication, so multiply both sides by the number by which the variable is divided in the given equation.
(2) Check your answer by substituting it for the variable in the given equation.

EXAMPLE:

Solve and check: $\frac{n}{4} = 20$

$$\frac{n}{4} = 20$$
$$4 \cdot \frac{n}{4} = 4 \cdot 20$$
$$n = 80$$

Check: $\frac{n}{4} = 20$
$$\frac{80}{4} = 20?$$
$$20 = 20 \checkmark$$

The indicated operation is division. Multiply by 4 to undo dividing by 4.

ANSWER: 80

To solve $20 = \frac{n}{4}$, you still multiply both sides by 4. The solution, $80 = n$, can also be written as $n = 80$.

PRACTICE EXERCISES

Solve and check:

1

1. $\frac{d}{2} = 19$
2. $\frac{s}{4} = 24$
3. $\frac{b}{7} = 3$
4. $\frac{n}{5} = 0$

5. $\frac{t}{8} = 8$
6. $12 = \frac{b}{3}$
7. $45 = \frac{m}{5}$
8. $1 = \frac{s}{9}$

9. $\frac{x}{3} = 9$
10. $\frac{c}{3.14} = 100$
11. $\frac{b}{5} = 1.4$
12. $\frac{1}{2}a = 17$

2

1. $\frac{3}{4}c = \$.36$
2. $0 = \frac{g}{8}$
3. $\frac{w}{5} = 30$
4. $\frac{k}{12} = 3$

5. $\frac{n}{4} = -5$
6. $\frac{r}{9} - -7$
7. $\frac{y}{-3} = 12$
8. $\frac{a}{-6} = -18$

9. $-8 = \frac{d}{-11}$
10. $\frac{x}{6} = 1$
11. $-5 = \frac{z}{7}$
12. $\frac{b}{16} = 16$

17-12 Solving Equations Using Two Operations

The solution of more difficult equations may require more than one operation.

PROCEDURE

To solve an equation using two operations:
(1) Simplify the equation by combining like terms (two or more terms containing the variable) where possible.
(2) Undo any additions or subtractions by the inverse operation.
(3) Undo any multiplications or divisions by the inverse operation.
(4) Check the answer in the original equation.

EXAMPLES:

Solve and check: $7x + 2 = 72$

$7x + 2 = 72$
$7x + 2 - 2 = 72 - 2$ **Subtract 2.**
$7x = 70$
$\frac{7x}{7} = \frac{70}{7}$ **Divide by 7.**
$x = 10$

ANSWER: 10

Check:

$7x + 2 = 72$
$(7 \cdot 10) + 2 = 72$
$70 + 2 = 72 \text{ ?}$
$72 = 72 \, \checkmark$

Solve and check: $7x + 2x = 72$

$7x + 2x = 72$ **Combine like terms.**
$9x = 72$
$\frac{9x}{9} = \frac{72}{9}$ **Divide by 9.**
$x = 8$

ANSWER: 8

Check:

$7x + 2x = 72$
$(7 \cdot 8) + (2 \cdot 8) = 72$
$56 + 16 = 72 \text{ ?}$
$72 = 72 \, \checkmark$

PRACTICE EXERCISES

Solve and check:

1
1. $2b + 18 = 46$
2. $3 + 5n = 62$
3. $51 = 9y + 6$
4. $120 = 76 + 2w$
5. $22h + 154 = 374$
6. $1.8c + 32 = 113$
7. $7t - 11 = 38$
8. $0 = 10n - 30$
9. $6a - 7 = 29$
10. $37 = 14b - 19$
11. $n + n = 50$
12. $9x + 3x = 84$

2
1. $p + .06p = 689$
2. $7y - y = 18$
3. $a - .05a = 760$
4. $6b = 91 - 15$
5. $82 = 5a + 9a$
6. $1.8c + 32 = 50$
7. $l - .35l = 195$
8. $\frac{3}{5}b = 21$
9. $\frac{5}{8}d = \$.75$
10. $t + \frac{1}{4}t = 6\frac{1}{2}$
11. $\frac{7}{8}n + 18 = 46$
12. $\frac{5}{9}b - 13 = 27$

17-13 Using Formulas

Often when using a formula to solve a problem, the result is an equation that must be solved for the unknown quantity.

PROCEDURE

To use a formula to solve a problem:
(1) Copy the formula.
(2) Substitute the given values for the variables.
(3) Perform the necessary operations.
(4) Solve the resulting equation for the value of the required variable.
(5) Check in the formula.

EXAMPLE:

Find the value of w when $A = 54$ and $l = 9$, using the formula $A = lw$.

$A = lw$ Check: $A = lw$
$54 = 9w$ $54 = 9(6)$?
$\frac{54}{9} = \frac{9w}{9}$ $54 = 54$ ✔
$6 = w$ or $w = 6$

ANSWER: $w = 6$

PRACTICE EXERCISES

Find the value of:

1
  1. s when $p = 18$. Formula: $p = 3s$ **2.** r when $d = 56$. Formula: $d = 2r$
  3. d when $C = 15.7$ and $\pi = 3.14$. Formula: $C = \pi d$
  4. t when $i = 140$, $p = 400$, and $r = 5\%$. Formula: $i = prt$
  5. h when $V = 6,280$, $\pi = 3.14$, and $r = 10$. Formula: $V = \pi r^2 h$
  6. b when $A = 72$ and $h = 18$. Formula: $A = bh$
  7. w when $V = 1,536$, $l = 16$, and $h = 12$. Formula: $V = lwh$

2
  1. b when $p = 91$ and $e = 29$. Formula: $p = b + 2e$
  2. A when $C = 42$. Formula: $C = A - 273$
  3. l when $p = 162$ and $w = 34$. Formula: $p = 2l + 2w$
  4. n when $S = 900$. Formula: $S = 180n - 360$
  5. V when $v = 323$, $g = 32$, and $t = 9$. Formula: $v = V + gt$
  6. r when $A = 640$, $p = 500$, $t = 7$. Formula: $A = p + prt$
  7. V when $P = 75$, $V' = 38$, $P' = 25$. Formula: $PV = P'V'$
  8. C when $d = 15$ and $\pi = 3.14$. Formula: $d = \frac{C}{\pi}$

GRAPHING AN EQUATION
ON THE NUMBER LINE

Once you have solved an equation, you can list its solution(s) or show them on a graph.

> The graph of an equation in one variable on a number line is the collection of all points on the number line whose coordinates are solutions of the equation.

EXAMPLE:

Draw the graph of $n + 2 = 8$.

$$n + 2 = 8$$
$$n = 6$$

First, solve the given equation.

Then, locate the point on the number line that corresponds to the solution and indicate the point by a solid dot.

EXERCISES

For each of the following equations draw an appropriate number line, then graph its solution. The replacements for the variable are all the real numbers.

1. $x = 5$ **2.** $b = -3$ **3.** $x + 4 = 7$

4. $a - 5 = 2$ **5.** $8a = 32$ **6.** $3y + 8 = 5$

7. $6n - n = 15$ **8.** $\frac{b}{3} = 2$ **9.** $5b - 6 = 29$

10. $|x| = 3$ **11.** $z = z + 2$ **12.** $5(x + 2) = 5x + 10$

Write a corresponding equation which is pictured by each of the following graphs:

13.

14.

15.

16.

17-14 Problem Solving by the Equation Method

> **Equations may be used to find:**
>
> **I** A fractional part or decimal part or percent of a number.
> **II** What fractional part or decimal part or percent one number is of another.
> **III** A number when a fractional part or decimal part or percent of it is known.

PROCEDURE

To solve problems using the equation method:

1 READ the problem carefully to find the facts related to the number you wish to determine.

2 PLAN
    a. Represent the unknown number by a variable, usually a letter.
    b. Form an equation by translating two equal facts, with at least one containing the unknown, into algebraic expressions and writing one expression on each side of the equal sign.

3 SOLVE the equation.

4 CHECK the answer directly with the facts in the given problem.

EXAMPLES:

I Find $\frac{3}{4}$ of 20.

$$\frac{3}{4} \times 20 = n$$
$$15 = n \text{ or } n = 15$$

ANSWER: 15

Find .75 of 20.

$$.75 \times 20 = n$$
$$15 = n \text{ or } n = 15$$

ANSWER: 15

Find 75% of 20.

$$75\% \times 20 = n$$
$$.75 \times 20 = n$$
$$15 = n \text{ or } n = 15$$

ANSWER: 15

II What fractional part of 20 is 15?

$$n \times 20 = 15$$
$$20n = 15$$
$$\frac{20n}{20} = \frac{15}{20}$$
$$n = \frac{3}{4}$$

ANSWER: $\frac{3}{4}$

What decimal part of 20 is 15?

$$n \times 20 = 15$$
$$20n = 15$$
$$\frac{20n}{20} = \frac{15}{20}$$
$$n = \frac{3}{4} = .75$$

ANSWER: .75

What percent of 20 is 15?

$$n\% \times 20 = 15$$
$$\frac{n}{100} \times 20 = 15$$
$$\frac{n}{5} = 15$$
$$n = 75$$

ANSWER: 75%

III $\frac{3}{4}$ of what number is 15?   .75 of what number is 15?   75% of what number is 15?

$\frac{3}{4} \times n = 15$ $.75 \times n = 15$ $75\% \times n = 15$

$\frac{3}{4}n = 15$ $.75n = 15$ $.75n = 15$

$\frac{4}{3} \cdot \frac{3}{4}n = \frac{4}{3} \cdot 15$ $\frac{.75n}{.75} = \frac{15}{.75}$ $\frac{.75n}{.75} = \frac{15}{.75}$

$n = 20$ $n = 20$ $n = 20$

ANSWER: 20   ANSWER: 20   ANSWER: 20

PRACTICE EXERCISES

By the equation method find the required number in each of these:

1 **1.** $\frac{1}{3}$ of 54   **2** **1.** .4 of 7   **3** **1.** 8% of 50

   2. $\frac{3}{8}$ of 72   **2.** .75 of 92   **2.** 60% of 18

   3. $\frac{2}{5}$ of 140   **3.** .36 of 250   **3.** 25% of 392

   4. $\frac{5}{6}$ of 258   **4.** .625 of 496   **4.** 140% of 475

   5. $\frac{7}{10}$ of 530   **5.** .19$\frac{3}{4}$ of 1,000   **5.** 87$\frac{1}{2}$% of 280

4 Express each as a fraction: **5** Express each as a decimal:

   1. What part of 25 is 15?   **1.** What part of 10 is 3?

   2. What part of 63 is 36?   **2.** What part of 25 is 17?

   3. 27 is what part of 81?   **3.** 12 is what part of 16?

   4. 45 is what part of 54?   **4.** 49 is what part of 56?

6 **1.** What percent of 50 is 9?   **7** **1.** $\frac{1}{3}$ of what number is 29?

   2. 30 is what percent of 48?   **2.** $\frac{4}{5}$ of what number is 56?

   3. 28 is what percent of 40?   **3.** 63 is $\frac{7}{12}$ of what number?

   4. What percent of 300 is 57?   **4.** $1\frac{3}{4}$ times what number is 84?

8 **1.** .06 of what number is 300?   **9** **1.** 25% of what number is 18?

   2. .375 of what number is 96?   **2.** 8% of what number is 52?

   3. 54 is .9 of what number?   **3.** 37 is 10% of what number?

   4. 1.04 of what number is 364?   **4.** 477 is 1.06% of what number?

10 **1.** If the school basketball team won 14 games, or $\frac{2}{3}$ of the games played, how many games were played?

   2. José received .54 of all votes cast in the election for school president. If he received 405 votes, how many students voted?

   3. If 817 students, or 95% of the school enrollment, were promoted to the next grade, how many students were enrolled in the school?

   4. There are 18 boys in a certain mathematics class. If this represents $\frac{3}{5}$ of the class enrollment, how many girls are in the class?

17-15   Solving Inequalities

To solve an inequality in one variable means to find all the numbers that will make the inequality a true sentence. The solution depends on the replacements allowed for the variable.

- When the replacements for the variable are all the *natural numbers*, the solutions of $n < 4$ are the natural numbers less than 4.

  Therefore, $n = 1$, 2, or 3.
- When the replacements for the variable are all the *whole numbers*, the solutions of $x > 11$ are the whole numbers greater than 11.

  Therefore, $x = 12, 13, 14, \ldots$.
- When the replacements for the variable are all the *real numbers*, the solutions of $y \neq 6$ are all the real numbers except 6.

  Therefore, $y =$ every real number except 6.
- When the replacements for the variable are all the *integers*, the solutions of $b \geq -3$ are -3 and all integers greater than -3.

  Therefore, $b = -3, -2, -1, 0, 1, 2, \ldots$.
- The solutions of $c \leq 5$ do not exist when the replacements for the variable are 6, 7, 8, 9, and 10.

  Therefore, c has no solution.
- When the replacements for the variable are all the *integers*, the solutions of $x \not> 2$ are the same as the solutions of $x \leq 2$, which are 2 and all the integers less than 2.

  Therefore $x = \ldots, -2, -1, 0, 1, 2$.
- When the replacements for the variable are all the *real numbers*, the solutions of $n \not< 4$ are the same as the solutions of $n \geq 4$, which are 4 and all the real numbers greater than 4.

  Therefore, $n = 4$ or any real number greater than 4.

Properties of Inequalities

I The same number can be added to or subtracted from both sides of an inequality without changing the order of the inequality.

II Both sides of an inequality can be multiplied or divided by the same positive number without changing the order of the inequality.

III When both sides of an inequality are multiplied or divided by the same negative number, an inequality of the reverse order results.

I EXAMPLES:

Solve $n + 2 < 6$ when the replacements for the variable are all the whole numbers.

$$n + 2 < 6$$
$$n + 2 - 2 < 6 - 2$$
$$n < 4$$
$$n = 0, 1, 2, \text{ or } 3$$

ANSWER: 0, 1, 2, or 3

Solve $x - 2 > 3$ when the replacements for the variable are all the real numbers.

$$x - 2 > 3$$
$$x - 2 + 2 > 3 + 2$$
$$x > 5$$
$$x = \text{all real numbers}$$
$$\text{greater than 5}$$

ANSWER: All real numbers greater than 5.

PRACTICE EXERCISES

Find the solutions of each of the following inequalities when the replacements for the variable are all the real numbers:

1 **1.** $n + 7 > 10$ **2.** $a + 6 > 6$ **3.** $23 + d > 12$

2 **1.** $c - 5 > 18$ **2.** $x - 3 > -4$ **3.** $r - \frac{1}{2} > 3\frac{1}{2}$

3 **1.** $b + 1 < 7$ **2.** $g + 9 < 2$ **3.** $m + .5 < 2.4$

4 **1.** $d - 9 < 6$ **2.** $t - 10 < 18$ **3.** $h - 3 < -5$

5 **1.** $m + 8 \geq 12$ **2.** $f + 17 \geq 0$ **3.** $w + 8 \nleq -8$

6 **1.** $x - 11 \geq 4$ **2.** $k - 3.5 \geq 8.7$ **3.** $a - 7 \nleq -4$

7 **1.** $y + 2 \leq 9$ **2.** $e + \frac{1}{4} \ngtr \frac{3}{4}$ **3.** $v + 16 \leq 13$

8 **1.** $r - 4 \leq 10$ **2.** $c - 6 \ngtr -8$ **3.** $x - 9 \leq 9$

9 **1.** $s + 9 \neq 17$ **2.** $n + .3 \neq 1.6$ **3.** $b + 12 \neq 1$

10 **1.** $w - 11 \neq 6$ **2.** $y - \frac{7}{8} \neq 1\frac{1}{2}$ **3.** $r - 2 \neq -2$

11 **1.** $x - 5 > 4$ **2.** $y + 8 < 3$ **3.** $n + 7 \geq 1$
    4. $a - 9 \leq 11$ **5.** $t + 3 \neq 3$ **6.** $b - 5 < -9$
    7. $r + 4 > 0$ **8.** $z - 12 \nleq 7$ **9.** $w + 6 \leq -4$

II EXAMPLES:

Solve $3c < 27$ when the replacements for the variable are all the natural numbers.

$$3c < 27$$
$$\frac{3c}{3} < \frac{27}{3}$$
$$c < 9$$
$$c = 1, 2, 3, \ldots, 8$$

ANSWER: 1, 2, 3, . . . , 8

Solve $4y + 7 \geq 7$ when the replacements for the variable are $-2, -1, 0, 1,$ and 2.

$$4y + 7 \geq 7$$
$$4y + 7 - 7 \geq 7 - 7$$
$$4y \geq 0$$
$$y \geq 0$$
$$y = 0, 1, \text{ or } 2$$

ANSWER: 0, 1, or 2

PRACTICE EXERCISES

Find the solutions of each of the following inequalities when the replacements for the variable are all the real numbers:

1 **1.** $7x > 14$ **2.** $9y > -72$ **3.** $8a > 2$

2 **1.** $\frac{m}{4} > 2$ **2.** $\frac{c}{10} > 7$ **3.** $\frac{r}{9} > -6$

3 **1.** $6z < 30$ **2.** $12s < -96$ **3.** $16b < 24$

4 **1.** $\frac{h}{8} < 3$ **2.** $\frac{p}{15} < -1$ **3.** $\frac{w}{7} < 5$

5 **1.** $10n \geq 50$ **2.** $18a \geq 12$ **3.** $\frac{3}{4}x \not< 24$

6 **1.** $\frac{b}{30} \geq 0$ **2.** $\frac{r}{4} \geq -6$ **3.** $\frac{z}{5} \not< 3$

7 **1.** $8s \leq 72$ **2.** $48h \not> 96$ **3.** $.05n \leq 20$

8 **1.** $\frac{y}{6} \leq 7$ **2.** $\frac{t}{14} \not> 2$ **3.** $\frac{b}{3} \leq -10$

9 **1.** $9c \neq 144$ **2.** $\frac{5}{8}x \neq 80$ **3.** $24f \neq 0$

10 **1.** $\frac{n}{12} \neq 5$ **2.** $\frac{d}{6} \neq -2$ **3.** $\frac{y}{10} \neq 9$

11 **1.** $15b < 60$ **2.** $\frac{a}{8} > -2$ **3.** $20x \geq -15$

       4. $9y \leq 39$ **5.** $6m \neq -48$ **6.** $\frac{t}{7} < 7$

       7. $\frac{n}{12} > 0$ **8.** $\frac{z}{5} \not< 15$ **9.** $\frac{7}{16}x \not> 56$

III EXAMPLES:

Solve $-5x > 30$ when the replacements for the variable are all the integers.

$$-5x > 30$$
$$\frac{-5x}{-5} < \frac{30}{-5} \quad \text{reverse order}$$
$$x < -6$$
$$x = \ldots, -9, -8, -7$$

ANSWER: $\ldots, -9, -8, -7$

Solve $-2b < -6$ when the replacements for the variable are all the integers.

$$-2b < -6$$
$$\frac{-2b}{-2} > \frac{-6}{-2}$$
$$b > 3$$
$$b = 4, 5, 6, \ldots$$

ANSWER: $4, 5, 6, \ldots$

PRACTICE EXERCISES

Find the solutions of each of the following inequalities when the replacements for the variable are all the real numbers:

1 **1.** $-5d > 30$ **2.** $-7c > 98$ **3.** $-.4h > 1.2$

2 **1.** $-4b > -28$ **2.** $-3m > -40$ **3.** $-\frac{3}{8}z > -15$

3 **1.** $-9x < 81$ **2.** $-16r < 96$ **3.** $-8a < 36$

4 **1.** $-2t < -16$ **2.** $-17y < -51$ **3.** $-1.5m < -9$

5 **1.** $-3y \geq 21$ **2.** $-10r \geq 0$ **3.** $-x \not< 2$

6 **1.** $-7n \geq -63$ **2.** $-\frac{4}{9}g \geq -72$ **3.** $-4t \not< -17$

7 **1.** $-9s \leq 180$ **2.** $-3y \not> 0$ **3.** $-18x \leq 90$

8 **1.** $-6z \leq -216$ **2.** $-7n \not> 63$ **3.** $-\frac{2}{3}y \leq -60$

9 **1.** $\frac{n}{-3} < 12$ **2.** $\frac{b}{-6} > -2$ **3.** $\frac{y}{-8} \leq 6$

    4. $\frac{c}{-10} \geq 9$ **5.** $\frac{a}{-3} < -7$ **6.** $\frac{x}{-5} > 8$

    7. $\frac{s}{-12} \not< -5$ **8.** $\frac{m}{-9} > 4$ **9.** $\frac{d}{-4} \not> -1$

10 **1.** $-7m < 56$ **2.** $-12z > -132$ **3.** $-4x \leq 92$

    4. $-10a \geq 45$ **5.** $-3r < 0$ **6.** $\frac{b}{-7} > -8$

    7. $-15y > 105$ **8.** $-2n \not< -1$ **9.** $-13g \not> -91$

Enrichment

Find the solutions of each of the following inequalities when the replacements for the variable are all the real numbers:

1. $2x + 7 < 15$ **2.** $5a - 3 > -28$ **3.** $6m + 1 \leq 13$

4. $7s - 5 \geq 9$ **5.** $8n + 3 \neq -61$ **6.** $3y - 7 < 20$

7. $10b + 5 > 0$ **8.** $17r - 9 \not< 42$ **9.** $9t + 6 < -57$

10. $14z - 8 \not> 69$ **11.** $7w + 15 > 113$ **12.** $4x - 11 \neq -9$

13. $3x + 2x > 15$ **14.** $9c - 3c < -72$ **15.** $5d - 7d \geq 14$

16. $y + y \neq -30$ **17.** $8m - 3m \leq 0$ **18.** $6t - 11t > -85$

19. $12b + 13b < -100$ **20.** $4x - 8x \not< 36$ **21.** $n - 2n \not> -5$

22. $10r + 3r \neq 65$ **23.** $12w - 5w > -84$ **24.** $7a - 15a < 144$

Find the solutions of each of the following inequalities.

Select only the solutions that are included in the set of replacements.

25. When replacements for the variable are 0, 1, 2, 3, 4, 5, 6, 7, 8, 9:
    a. $x + 5 > 8$ **b.** $n - 6 < 1$ **c.** $2b \leq 6$ **d.** $-3t \not< -12$

26. When replacements for the variable are $-3, -2, -1, 0, 1, 2, 3$:
    a. $b - 2 \neq 0$ **b.** $5n + 3 < 13$ **c.** $\frac{d}{6} > -1$ **d.** $2a - 5 \leq -1$

27. When replacements for the variable are all the prime numbers:
    a. $12d - 4d < 56$ **b.** $m - 7 \not> 16$ **c.** $6c - c \neq 55$ **d.** $-2n > -8$

28. When the replacements for the variable are all the integers:
    a. $d + 5 < -2$ **b.** $-7s \geq 35$ **c.** $\frac{y}{-4} \not> -5$ **d.** $4a - 7 \neq -15$

29. When the replacements are all the even whole numbers:
    a. $6b < 45$ **b.** $2a - 5 > 9$ **c.** $8a - 2a \not> 18$ **d.** $\frac{d}{9} \not< 4$

30. When the replacements are all the non-positive integers:
    a. $m - 1 > -6$ **b.** $-5x < -20$ **c.** $4t - 9 \not> -13$ **d.** $11y - y \geq -10$

GRAPHING AN INEQUALITY
ON THE NUMBER LINE

The graph of an inequality in one variable is the collection of all points whose coordinates are solutions of the given inequality.

> The graph of an inequality may be a line, a half-line, a ray, a line segment, or an interval. An *open dot* indicates that a point is not part of the graph. A *solid dot* indicates that it is part of the graph.
>
> Graph of $x > 2$
>
> Graph of $-3 \leq x < 2$

EXAMPLE:

Draw the graph of $4x - 9 < 3$.

$$4x - 9 < 3$$
$$4x < 12$$
$$x < 3$$

First, solve the inequality.

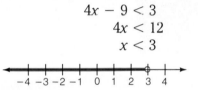

Then, draw the graph.

EXERCISES

For each of the following inequalities draw an appropriate number line, then graph its solutions. The replacements for the variable are all the real numbers.

1. $a > -1$ **2.** $a + 5 < 7$ **3.** $\frac{x}{2} \leq 2$

4. $2x - 1 \neq 1$ **5.** $-9x > 18$ **6.** $n + 2n \not> -12$

7. $-7y \not< -7$ **8.** $2b - 5b > -12$ **9.** $\frac{m}{3} \geq -1$

Draw the graph for each of the following on a number line:

10. $-3 < x < 4$ **11.** $-1 \leq x < 3$ **12.** $2 \geq x \geq -5$

Write a corresponding inequality which is pictured by each of the following graphs:

13.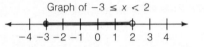

14.

15.

16.

17.

18.

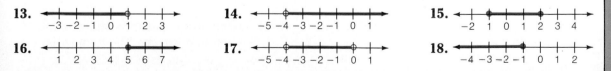

Car buyers are always looking for better and more efficient automobiles. *Automotive engineers* are constantly working towards building the better car. They do tests on engines and other equipment to see how well they work. Automotive engineers also design new parts to be safer or more efficient. They may do computer simulations (that is, models) or actually build models.

Automotive engineers work with design, production, and sales specialists. They may specialize in certain systems of an automobile, such as the electrical system or pollution control. Automotive engineers are college graduates. Many of them major in mechanical engineering.

1. An automotive engineer is studying how much a metal part expands when it is heated. The formula is $l_H - l_C = \Delta l$ where l_H is the length of the part when it is hot, l_C is the length when it is cold, and Δl is the difference. Solve for l_H if $l_C = 12$ mm and $\Delta l = 2$ mm.

2. An automotive engineer is studying the motion of a car. The formula is
$$v = \frac{d}{t}$$
where v is velocity, d is distance, and t is time. Solve for d if $v = 30$ ft/s and $t = 10$ s.

3. The displacement of an automobile engine is given by the formula
$$D = .784\, b^2 sn$$
where b is the bore, s is the stroke, and n is the number of cylinders. Solve the formula for s.

4. An automotive engineer is studying the acceleration of a test car. The formula is
$$v_2 = v_1 + at$$
where v_2 is the final velocity, v_1 is the initial velocity, a is the acceleration, and t is the time. Solve the formula for a.

CHAPTER REVIEW

Vocabulary

algebraic expression p. 568	formula p. 572	right member p. 577
compound sentence p. 570	inequality p. 570	root p. 577
conjunction p. 570	left number p. 577	simplest form p. 577
constant p. 568	numerical coefficient p. 568	solution p. 577
disjunction p. 570	numerical expression p. 568	variable p. 568
equation p. 570	open sentence p. 570	

The numerals in the boxes indicate where help may be found.

Write each of the following as an algebraic expression: 17-1

1. Thirty increased by twelve.
2. The quotient of r divided by d.
3. The difference in measures of angles S and T.
4. The square of t multiplied by sixteen.
5. The square root of the product of m and n.

Write each of the following sentences symbolically: 17-2

6. Some number n decreased by seven is equal to twenty-five.
7. Each number y increased by six is less than ten.
8. Twice each number w is greater than forty.

Express each of the following as a formula: 17-3

9. The circumference of a circle (c) is equal to the product of pi (π) and the diameter (d).
10. The sum of the measures of two supplementary angles E and F equals 180°.
11. Translate the following formula as a word statement: 17-3
   $p = 4s$ where p = perimeter of a square and s = length of side.

Find the value of: 17-4

12. $c + 3d$ when $c = 6$ and $d = -2$ 13. $x^2 - y^2$ when $x = -4$ and $y = 3$
14. $m - n(m + n)$ when $m = 8$ and $n = 2$
15. Find the value of E when $I = 6$, $R = 7$, and $r = 3$, using the formula $E = I$ ... 17-5
16. Which of the following equations have -4 as the solution? 17-6
   a. $x + 9 = 13$ **b.** $\frac{n}{2} = -2$ **c.** $y - 1 = 5$ **d.** $3c + 15 = $ ...

   ... $= 60$ 17-1
   ... $- 3n = 70$

Solve and check:

17. $n + 18 = 49$ 17-8   18. $x - 6 =$ ...
20. $\frac{t}{4} = 24$ 17-11   21. ...

... uction sale. 17-14
... integers. 17-15

23. Find the value of C when $F = 59$, ...

Solve by the equation method:

24. Find the regular price of a comput...
25. Find the solutions of $6n > -12$ w...

COMPETENCY CHECK TEST

The numerals in the boxes indicate Lessons where help may be found.

Solve each problem and select the letter corresponding to your answer.

1. "Twice the length of the equal side (e) subtracted from the perimeter (p)" may be written as the algebraic expression: 17-1

    a. $23 - p$ **b.** $p - 2e$ **c.** $3 - 2p$ **d.** Answer not given

2. The sentence "The average (A) of two numbers m and n equals the sum of these two numbers divided by two" expressed as a formula is: 17-3

    a. $A = 2(m + n)$ **b.** $A = \frac{2}{m + n}$ **c.** $A = \frac{m + n}{2}$ **d.** Answer not given

3. The value of $x^2 - 5y$ when $x = -6$ and $y = 2$ is: 17-4

    a. -12 **b.** -46 **c.** -22 **d.** 26

4. The value of v when $V = 45$, $g = 32$, and $t = 5$, using the formula $v = V + gt$ is: 17-5

    a. 205   **b.** 385   **c.** 82   **d.** Answer not given

5. Which of these equations has -3 as its solution? 17-6

    a. $y - 12 = 9$ **b.** $-y = -3$ **c.** $y + 5 = 2$ **d.** $4y - 9 = 3$

6. Which of these equations is equivalent to $x = 4$? 17-6

    a. $8x = 2$ **b.** $5x - 3 = 17$ **c.** $2x + 7x = 54$ **d.** $3x + 5 = -3$

7. The solution of $x - 5 = 10$ is: 17-9

    a. 15   **b.** 5   **c.** 2   **d.** -5

8. The solution of $\frac{n}{6} = 2$ is: 17-11

    a. 3   **b.** $\frac{1}{3}$ **c.** 4   **d.** 12

9. The solution of $2n + 11 = 25$ is: 17-12

    a. 18   **b.** 7   **c.** 14   **d.** Answer not given

10. The value of C when $F = 68$, using the formula $F = 1.8C + 32$ is: 17-13

    b. 33.8   **c.** 36   **d.** 20

11. of money that must be invested at the annual rate of 8% to earn \$10,000

    b. \$100,000   **c.** \$250,000   **d.** \$175,000

12. when the replacements for the variable are all the one-digit

    c. 2   **d.** 7

CHAPTER 18 Graphing in the Number Plane

18-1 Ordered Pairs

Vocabulary

An **ordered pair of numbers** is a pair of numbers expressed in a definite order so that one number is first (first component) and the other number is second (second component).

PROCEDURE

To express an ordered pair of numbers:
(1) Write the first component, followed by a comma.
(2) Write the second component.
(3) Enclose them within parentheses.

EXAMPLE:

Write an ordered pair of numbers which has 2 as the first component and 5 as the second component.

first component ⟶ ↓   ↓ ⟵ second component
$$(2 , 5)$$

ANSWER: (2,5)

PRACTICE EXERCISES

1 Write the ordered pair of numbers which has:

1. 1 as the first component and 3 as the second component.

2. 4 as the first component and 0 as the second component.

3. 7 as the first component and −3 as the second component.

4. −2 as the first component and 5 as the second component.

5. −8 as the first component and −9 as the second component.

2 Write all the ordered pairs of numbers which have:

1. 3, 4, or 5 as the first component and 4 as the second component.

2. 5 or 6 as the first component and 5 or 6 as the second component.

3. 1, 3, or 5 as the first component and 0 or 1 as the second component.

4. 2, 4, 6, or 8 as the first component and 2, 4, 6, or 8 as the second component.

5. −3, 0, or 3 as the first component and −2, −1, 1, or 2 as the second component.

Vocabulary

A **real number plane** is determined when a vertical real number line and a horizontal real number line are drawn perpendicular to each other.

The **coordinate axes** are the two perpendicular real number lines and are used to locate or plot points in the real number plane.

The **x-axis** is the horizontal number line.

The **y-axis** is the vertical number line.

The **x-coordinate** or **abscissa** is the number that indicates the horizontal distance, measured parallel to the x-axis, that a point is located to the left or right of the y-axis.

The **y-coordinate** or **ordinate** is the number that indicates the vertical distance, measured parallel to the y-axis, that a point is located above or below the x-axis.

The **coordinates** of a point are represented by an ordered pair of numbers in which the x-coordinate is first and y-coordinate is second.

The **origin** is the point where the x-axis and y-axis intersect.

The coordinate axes divide the real number plane into four regions, each of which is called a **quadrant**

The x- and y-coordinates are needed to locate a point on a number plane because the point is located with respect to both the x-axis and y-axis.

PROCEDURE

To locate a point on a number plane:
(1) Draw a perpendicular line from the point to the x-axis. Where the line intersects the x-axis is the x-coordinate of the point.
(2) Draw a perpendicular line from the point to the y-axis. Where the line intersects the y-axis is the y-coordinate of the point.
(3) Represent the x- and y-coordinates as an ordered pair of numbers in which the x-coordinate is the first component and the y-coordinate is the second component.

EXAMPLE:

What are the coordinates of the point on the number plane at the right?

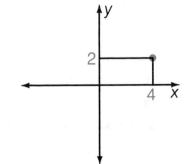

x-coordinate is 4
y-coordinate is 2

ANSWER: (4, 2)

In quadrant I, each point has a positive *x*-coordinate and a positive *y*-coordinate. $G(4, 4)$ $B(1, 1)$
In quadrant II, each point has a negative *x*-coordinate and a positive *y*-coordinate. $A(-2, 2)$ $D(-5, 1)$
In quadrant III, each point has a negative *x*-coordinate and a negative *y*-coordinate. $C(-2, -2)$ $H(-5, -4)$
In quadrant IV, each point has a positive *x*-coordinate and a negative *y*-coordinate. $E(3, -2)$ $F(1, -4)$
The origin has 0 for both the *x*-coordinate and *y*-coordinate. $I(0, 0)$
Any point on the *x*-axis has 0 for its *y*-coordinate. $K(-5, 0)$ $J(4, 0)$
Any point on the *y*-axis has 0 for its *x*-coordinate. $L(0, 5)$ $M(0, -3)$

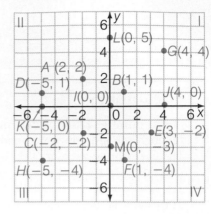

PRACTICE EXERCISES

What are the coordinates of each of the following indicated points?

1   **2**

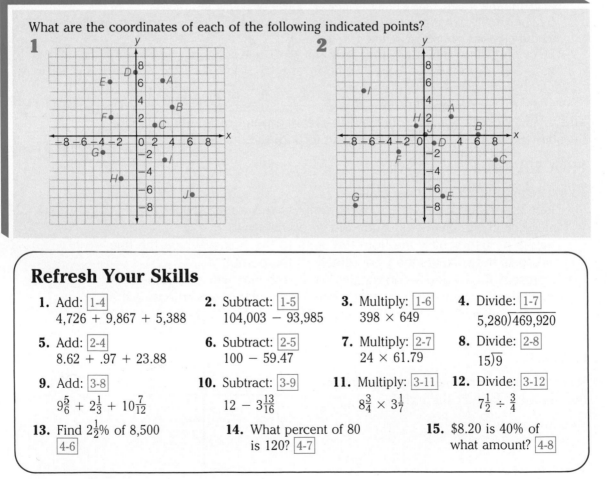

Refresh Your Skills

1. Add: `1-4`
4,726 + 9,867 + 5,388

2. Subtract: `1-5`
104,003 − 93,985

3. Multiply: `1-6`
398 × 649

4. Divide: `1-7`
$5{,}280\overline{)469{,}920}$

5. Add: `2-4`
8.62 + .97 + 23.88

6. Subtract: `2-5`
100 − 59.47

7. Multiply: `2-7`
24 × 61.79

8. Divide: `2-8`
$15\overline{)9}$

9. Add: `3-8`
$9\frac{5}{6} + 2\frac{1}{3} + 10\frac{7}{12}$

10. Subtract: `3-9`
$12 - 3\frac{13}{16}$

11. Multiply: `3-11`
$8\frac{3}{4} \times 3\frac{1}{7}$

12. Divide: `3-12`
$7\frac{1}{2} \div \frac{3}{4}$

13. Find $2\frac{1}{2}\%$ of 8,500
`4-6`

14. What percent of 80 is 120? `4-7`

15. $8.20 is 40% of what amount? `4-8`

18-3 Plotting Points in the Number Plane

PROCEDURE

To plot a point in the number plane:
(1) Locate the point by using the first component of the given ordered pair as the x-coordinate and the second component as the y-coordinate.
(2) Use a dot to indicate the position of the point.

EXAMPLES:

Plot point K whose coordinates are $(2, 3)$.

x-coordinate is 2.
y-coordinate is 3.

Plot point L whose coordinates are $(-3, 4)$.

x-coordinate is -3.
y-coordinate is 4.

Plot point M whose coordinates are $(-2, -6)$

x-coordinate is -2.
y-coordinate is -6.

Plot point N whose coordinates are $(5, 0)$.

x-coordinate is 5.
y-coordinate is 0.

ANSWERS:

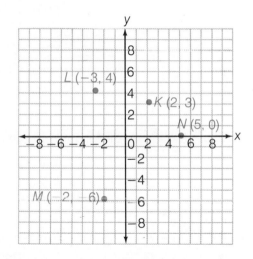

PRACTICE EXERCISES

On graph paper, draw axes and plot the points that have the following coordinates:

1
1. $A(6, 1)$
2. $B(-8, 4)$
3. $C(5, -3)$
4. $D(-3, -3)$
5. $E(-5, 0)$
6. $F(-3, 7)$
7. $G(4, 9)$
8. $H(0, 5)$
9. $I(1, -6)$

2
1. $J(-5, -4)$
2. $K(7, 3)$
3. $L(8, -8)$
4. $M(-6, -3)$
5. $N(0, 0)$
6. $O(-3, 2)$
7. $P(0, -1)$
8. $Q(-4, 4)$
9. $R(4, 0)$
10. $S(-7, 6)$
11. $T(6, -9)$

LONGITUDE AND LATITUDE

The position of any point on the earth's surface is determined by the intersection of its meridian of longitude and its parallel of latitude. Meridians of longitude are imaginary circles which pass through the North Pole and South Pole. Parallels of latitude are imaginary circles which are parallel to the equator.

The prime meridian from which longitude is calculated is the meridian that passes through Greenwich near London, England. West longitude extends from this prime meridian (0° longitude) westward halfway around the earth to the International Date Line (180° longitude). East longitude extends eastward from the prime meridian to the International Date Line.

The equator is 0° latitude. North latitude is measured north of the equator and south latitude is measured south of the equator. The North Pole is 90° north latitude and the South Pole is 90° south latitude. North latitude is indicated by the letter N, south latitude by S, east longitude by E, and west longitude by W.

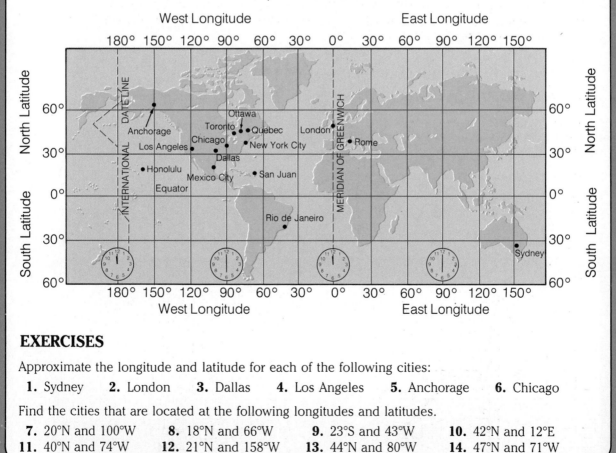

Map of the World

EXERCISES

Approximate the longitude and latitude for each of the following cities:

1. Sydney   **2.** London   **3.** Dallas   **4.** Los Angeles   **5.** Anchorage   **6.** Chicago

Find the cities that are located at the following longitudes and latitudes.

7. 20°N and 100°W   **8.** 18°N and 66°W   **9.** 23°S and 43°W   **10.** 42°N and 12°E
11. 40°N and 74°W   **12.** 21°N and 158°W   **13.** 44°N and 80°W   **14.** 47°N and 71°W

Graphing an Equation in the Number Plane

Vocabulary

The **graph of an equation** in two variables in the real number plane is the collection of all points in the real number plane whose coordinates are the ordered pairs of numbers which are the solutions of the equation.

PROCEDURE

To draw the graph of an equation:
1. Make a table.
2. Select three values for the x-coordinate.
3. Substitute each value of the x-coordinate for the x variable in the equation.
4. Find the corresponding value of the y-coordinate by solving the equation for the y variable.
5. Plot the three points in the number plane.
6. Draw a line through the three plotted points.

EXAMPLES:

Draw the graph of $x + y = 4$

x	y
0	4
2	2
3	1

$$0 + y = 4 \qquad 2 + y = 4$$
$$y = 4 \qquad y = 2$$
$$3 + y = 4$$
$$y = 1$$

ANSWER:

Draw the graph of $y = 3x$.

x	y
0	0
2	6
-2	-6

$$y = 3(0) \qquad y = 3(2)$$
$$y = 0 \qquad y = 6$$
$$y = 3(-2)$$
$$y = -6$$

ANSWER:

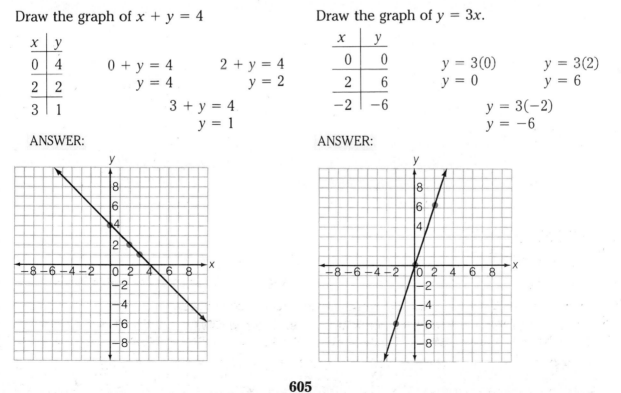

Sometimes one variable is missing in a given equation. In such cases no matter what value is substituted for the missing variable, the other variable remains constant.

When the variable for x is missing, the line is parallel to the x-axis.
When the variable for y is missing, the line is parallel to the y-axis.

EXAMPLES:

Draw the graph of $4y = 12$.

$4y = 12$
$\quad y = 3$ may be written as:
$\qquad 0x + y = 3$

x	y
0	3
3	3
-4	3

ANSWER:

Draw the graph of $3x = -15$.

$3x = -15$
$\quad x = -5$ may be written as:
$\qquad x + 0y = -5$

x	y
-5	0
-5	3
-5	-2

ANSWER:

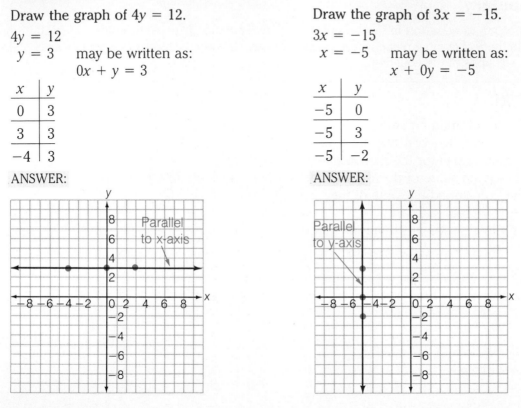

PRACTICE EXERCISES

Draw the graph of each of the following equations:

1
1. $y = x + 5$
2. $y = x - 3$
3. $x = y + 1$
4. $x = y - 6$
5. $x + y = 0$
6. $x + y = 2$
7. $x + y = -9$
8. $x - y = 4$
9. $x - y = -2$

2
1. $y = 5x$
2. $x = 2y$
3. $y = x$
4. $y = -3x$
5. $2y = 4x$
6. $2x = -2y$
7. $-3y = 12x$
8. $4y = 7x$
9. $-12y = -8x$

3
1. $y = 4$
2. $x = 2$
3. $y = -5$
4. $x = -6$
5. $y = 0$
6. $x = 0$
7. $7y = 21$
8. $12x = -60$
9. $-9y = -18$

Career Applications

Cloth, rugs, and needlework kits are available in many patterns. *Textile designers* create and plan these patterns. They use graph paper to draw a design. Each line on the graph paper may represent a thread. The graph paper is an enlarged scale model of the textile. A designer planning a woven fabric marks the paper to show which thread will go over which. A designer planning a needlepoint canvas marks what stitch and what color yarn will be used at each point. Many designers do their work on computers, using special graphics software.

Textile designers need to know about art and graphic design. They should also know the best styles and techniques for different types of textiles. Most textile designers learn their skills in technical schools or colleges.

1. A textile designer makes this graph for a needlepoint pattern. Some points will have blue stitches. What are the coordinates of each point?
   a. *A*   **b.** *B*
   c. *C*   **d.** *D*

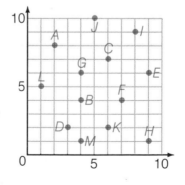

2. Some points will have red stitches. Name the point for each pair of coordinates.
   a. (6, 2)   **b.** (1, 5)
   c. (8, 9)   **d.** (4, 1)

3. A textile designer makes a pattern using rows of stitches. Graph these equations on the same set of coordinates.
   a. $y = 2x - 1$ **b.** $y = 2x + 2$ **c.** $y = 2x + 5$

607

CHAPTER REVIEW

Vocabulary

abscissa p. 601
coordinate p. 601
coordinate axis p. 601
graph of an
equation p. 605

ordered pair
of numbers p. 600
ordinate p. 601
origin p. 601
quadrant p. 601

real number plane p. 601
x-axis p. 601
x-coordinate p. 601
y-axis p. 601
y-coordinate p. 601

The numerals in boxes indicate the Lessons where help may be found.

1. Write the ordered pair of numbers which has 5 as the first component and -3 as the second component. $\boxed{18\text{-}1}$

What are the coordinates of: $\boxed{18\text{-}2}$

2. Point M?

3. Point N?

4. Point P?

5. Point Q?

6. Point R?

7. Point S?

On graph paper, draw axes and plot points with the following coordinates: $\boxed{18\text{-}3}$

  8. $T\ (3, -2)$ **9.** $V\ (-4, -4)$ **10.** $W\ (-7, 3)$ **11.** $Z\ (0, -5)$

12. $A\ (6, 7)$ **13.** $B\ (6, 0)$ **14.** $C\ (-5, -2)$ **15.** $D\ (0, 0)$

Draw the graph of each of the following equations: $\boxed{18\text{-}4}$

16. $y = x + 3$ **17.** $x + y = 8$ **18.** $y = -4x$ **19.** $y = -4$

20. Without drawing the graph, which of the following equations has a line that passes through the origin $(0, 0)$? $\boxed{18\text{-}4}$

  a. $x + y = 15$ **b.** $7y = 14$ **c.** $3x = 5y$ **d.** $x - y = 6$

21. Without drawing the graph, which of the following equations has as a graph a line that is parallel to the y-axis? $\boxed{18\text{-}4}$

  a. $8x = 3y$ **b.** $2x = 8$ **c.** $x + y = 11$ **d.** $y = 3x - 7$

22. Which of the following ordered pairs of numbers are solutions of the equation $x + y = 9$? $\boxed{18\text{-}4}$

  a. $(4, -5)$ **b.** $(11, -2)$ **c.** $(2, 7)$ **d.** $(-3, 6)$

23. Without drawing the graph, which number pairs are coordinates of points on the line forming the graph of $2x - y = 8$? $\boxed{18\text{-}4}$

  a. $(-5, 2)$ **b.** $(7, 6)$ **c.** $(3, -2)$ **d.** $(-1, -10)$

COMPETENCY CHECK TEST

The numerals in boxes indicate Lessons where help may be found.

Solve each problem and select the letter corresponding to your answer.

1. The ordered pair of numbers which has 8 as its first component and −5 as the second component is: $\boxed{18\text{-}1}$
  a. (−5, 8)   **b.** (8, −5)   **c.** (8, −8)   **d.** (5, −5)

2. The coordinates of point *P* are: $\boxed{18\text{-}2}$
  a. (4, −3)   **b.** (−4, −3)   **c.** (−3, 4)   **d.** (3, −4)

3. The coordinates of point *Q* are: $\boxed{18\text{-}2}$
  a. (−4, 0)   **b.** (0, 0)   **c.** (0, −4)   **d.** (0, 4)

4. The coordinates of point *R* are: $\boxed{18\text{-}2}$
  a. (−5, −6)   **b.** (−5, 6)   **c.** (6, −5)   **d.** (−6, 5)

5. The coordinates of point *S* are: $\boxed{18\text{-}2}$
  a. (−3, 6)   **b.** (−6, −3)   **c.** (6, −3)   **d.** (−3, −6)

6. In the quadrant below the *x*-axis and to the left of the *y*-axis all the points have: $\boxed{18\text{-}2}$
  a. Positive *x*-coordinates and negative *y*-coordinates
  b. Negative *x*-coordinates and positive *y*-coordinates
  c. Positive *x*-coordinates and positive *y*-coordinates
  d. Negative *x*-coordinates and negative *y*-coordinates

7. To draw the graph of the equation $x + y = 10$, use the coordinates: $\boxed{18\text{-}4}$

a.

x	8	−7	−6
y	2	−3	4

b.

x	3	−8	5
y	7	2	−5

c.

x	1	−4	−2
y	9	6	−8

d.

x	12	2	7
y	−2	8	3

8. Without drawing the graph, the graph of the equation $y = 5x$ is a line that: $\boxed{18\text{-}4}$
  a. Is parallel to *y*-axis   **b.** Is parallel to *x*-axis   **c.** Passes through origin (0, 0)   **d.** Answer not given

9. The ordered pair that is a solution of the equation $2x + y = 9$ is: $\boxed{18\text{-}4}$
  a. (3, 4)   **b.** (−1, 12)   **c.** (4, 1)   **d.** (−2, 5)

10. One of the points on the line forming the graph of the equation $3x - y = 10$ has the coordinates: $\boxed{18\text{-}4}$
  a. (3, 1)   **b.** (−5, 5)   **c.** (1, 7)   **d.** (4, 2)

Achievement Test

16-1 Which of the following numerals name:

    1. Positive numbers?   **2.** Negative numbers?

        -41 $+6\frac{2}{3}$ $-.084$ 195 0 $-\frac{5}{8}$

16-2 Which of the following numerals name:

    3. Integers   **4.** Rational numbers   **5.** Irrational numbers   **6.** Real numbers

        -48 $+6\frac{2}{5}$ $-.9$ $+\frac{20}{5}$ $-\sqrt{31}$ $-\frac{2}{3}$

16-3   **7.** Find the value of: $|-52|$

16-4   **8.** Draw a graph of -4, -3, 0, and 3 on a number line.

    9. Write the coordinates of which the following is the graph:

16-5   **10.** If a tailwind of 20 miles per hour is indicated by $+20$ m.p.h., how can a headwind of 32 miles per hour be indicated?

16-6 Which of the following are true?

    11. $-3 < -1$ **12.** $-10 > +6$ **13.** $+2 \not< -5$ **14.** $-4 \not> 0$

16-7 Using the number line as the scale, draw the vector and write the numeral represented by this vector that illustrates a movement from:

    15. $+2$ to $+8$ **16.** -7 to -3 **17.** $+3$ to -4 **18.** -1 to -5

16-8   **19.** What is the opposite of -20?

    20. What is the additive inverse of $+14$?

    21. Write symbolically: The opposite of negative twelve is positive twelve.

    22. Which has the greater opposite number: -3 or $+2$?

16-9 Add:

    23. $+8$ **24.** -4 **25.** -15 **26.** $+9$ **27.** -7
        $\underline{-10}$ $\underline{-5}$ $\underline{+7}$ $\underline{+6}$ $\underline{+7}$

    28. $(-6) + (+11) + (-7)$

    29. $-6 + 8 - 3 - 2 + 7$

16-10 Subtract:

    30. -2 **31.** -6 **32.** $+5$ **33.** $+3$ **34.** 0
        $\underline{-9}$ $\underline{+6}$ $\underline{+8}$ $\underline{-12}$ $\underline{-7}$

    35. $(-9) - (-12)$

    36. Take 9 from 5.

16-11 Multiply:

37. -8 $\underline{+3}$

38. -9 $\underline{-7}$

39. $+2$ $\underline{+11}$

40. $+5$ $\underline{-5}$

41. 0 $\underline{-6}$

42. $(-3)(+4)(-2)$

43. Find the value of: $(-3)^5$

16-12 Divide:

44. $\frac{-48}{6}$ **45.** $\frac{+32}{-8}$ **46.** $\frac{-90}{15}$ **47.** $\frac{+35}{+7}$ **48.** $\frac{+12}{-12}$

49. $(-54) \div (+9)$

50. Simplify: $\frac{6(-7+5)-4(-1-1)}{-2(9-13)}$

17-1 Write as an algebraic expression:

51. The product of b and four.   **52.** The quotient of d divided by t.

53. The sum of p and i.   **54.** Pi (π) times the square of the radius (r).

55. The difference between one hundred eighty and A.

17-2 Write each of the following sentences symbolically:

56. Some number n increased by seven is equal to thirty-six.

57. Each number x decreased by ten is less than sixteen.

58. Five times each number y is greater than or equal to forty.

59. Each number w is greater than negative nine and less than positive two.

17-3   **60.** Express as a formula: The distance (d) a freely falling body drops is one half the product of the acceleration due to gravity (g) and the square of the time of falling (t).

17-4 Find the value of:

61. $x - 4y$ when $x = 5$ and $y = 2$.

62. $3m^2 + 2mn - n^2$ when $m = 4$ and $n = -2$.

17-5   **63.** Find the value of A when $\pi = \frac{22}{7}, r = 14$, and $h = 10$, using the formula: $A = 2\pi r(h + r)$.

17-6 Which of the following equations:

64. Have 3 as the solution?   **65.** Are equivalent?   **66.** Is in simplest form?

$n - 3 = 4$ $n + 6 = 9$ $5n - 8 = 7$

$\frac{n}{6} = 2$ $n = 3$ $4n + 2n = 12$

17-7 What operation with what number do you use on both sides of each of the following equations to get an equivalent equation in simplest form? Also write this equivalent equation.

67. $9x = 72$ **68.** $y + 10 = 15$ **69.** $n - 14 = 3$ **70.** $\frac{r}{6} = 12$

Solve and check:

17-8   **71.** $n + 9 = 7$ 17-9   **72.** $x - 6 = 11$

17-10   **73.** $-12y = 84$ **74.** $8n = 6$

17-11   **75.** $\frac{c}{14} = 10$ **76.** $\frac{x}{-5} = 25$

17-12   **77.** $8b - 19 = 13$ **78.** $7x + 8x = 75$

17-13   **79.** Find the value of d when $l = 63$, $a = 7$, and $n = 9$, using the formula $l = a + (n - 1)d$

17-14   **80.** Solve by the equation method: If the school baseball team won 15 games, or $\frac{5}{8}$ of the games played, how many games were lost?

17-15   Find the solutions of each of the following inequalities when the replacements for the variable are all the real numbers:

     81. $5n < 60$ **82.** $b - 4 > 9$ **83.** $b + 2 \not< -3$

     84. $3x - 4 \neq -10$ **85.** $-7y \not< 14$ **86.** $\frac{n}{6} \geq -5$

     87. $3y - 5y > 0$ **88.** $\frac{2}{3}x < 18$ **89.** $n - 2n \leq -14$

18-1   **90.** Write the ordered pair of numbers that has 0 for the first component and -7 as the second component.

18-2   What are the coordinates of:

     91. Point G?   **92.** Point H?   **93.** Point I?

     94. Point J?   **95.** Point K?   **96.** Point L?

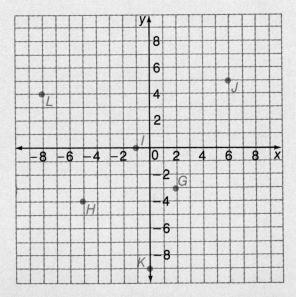

18-3   On graph paper, draw axes and plot the points with the following coordinates:

     97. $M (-4, -6)$ **98.** $N (5, 0)$ **99.** $P (-7, 2)$

18-4   **100.** Draw the graph of the equation $x - y = 5$.

Table of Trigonometric Values

Angle	Sine	Cosine	Tangent	Angle	Sine	Cosine	Tangent
0°	.0000	1.0000	.0000	46°	.7193	.6947	1.0355
1°	.0175	.9998	.0175	47°	.7314	.6820	1.0724
2°	.0349	.9994	.0349	48°	.7431	.6691	1.1106
3°	.0523	.9986	.0524	49°	.7547	.6561	1.1504
4°	.0698	.9976	.0699	50°	.7660	.6428	1.1918
5°	.0872	.9962	.0875	51°	.7771	.6293	1.2349
6°	.1045	.9945	.1051	52°	.7880	.6157	1.2799
7°	.1219	.9925	.1228	53°	.7986	.6018	1.3270
8°	.1392	.9903	.1405	54°	.8090	.5878	1.3764
9°	.1564	.9877	.1584	55°	.8192	.5736	1.4281
10°	.1736	.9848	.1763	56°	.8290	.5592	1.4826
11°	.1908	.9816	.1944	57°	.8387	.5446	1.5399
12°	.2079	.9781	.2126	58°	.8480	.5299	1.6003
13°	.2250	.9744	.2309	59°	.8572	.5150	1.6643
14°	.2419	.9703	.2493	60°	.8660	.5000	1.7321
15°	.2588	.9659	.2679	61°	.8746	.4848	1.8040
16°	.2756	.9613	.2867	62°	.8829	.4695	1.8807
17°	.2924	.9563	.3057	63°	.8910	.4540	1.9626
18°	.3090	.9511	.3249	64°	.8988	.4384	2.0503
19°	.3256	.9455	.3443	65°	.9063	.4226	2.1445
20°	.3420	.9397	.3640	66°	.9135	.4067	2.2460
21°	.3584	.9336	.3839	67°	.9205	.3907	2.3559
22°	.3746	.9272	.4040	68°	.9272	.3746	2.4751
23°	.3907	.9205	.4245	69°	.9336	.3584	2.6051
24°	.4067	.9135	.4452	70°	.9397	.3420	2.7475
25°	.4226	.9063	.4663	71°	.9455	.3256	2.9042
26°	.4384	.8988	.4877	72°	.9511	.3090	3.0777
27°	.4540	.8910	.5095	73°	.9563	.2924	3.2709
28°	.4695	.8829	.5317	74°	.9613	.2756	3.4874
29°	.4848	.8746	.5543	75°	.9659	.2588	3.7321
30°	.5000	.8660	.5774	76°	.9703	.2419	4.0108
31°	.5150	.8572	.6009	77°	.9744	.2250	4.3315
32°	.5299	.8480	.6249	78°	.9781	.2079	4.7046
33°	.5446	.8387	.6494	79°	.9816	.1908	5.1446
34°	.5592	.8290	.6745	80°	.9848	.1736	5.6713
35°	.5736	.8192	.7002	81°	.9877	.1564	6.3138
36°	.5878	.8090	.7265	82°	.9903	.1392	7.1154
37°	.6018	.7986	.7536	83°	.9925	.1219	8.1443
38°	.6157	.7880	.7813	84°	.9945	.1045	9.5144
39°	.6293	.7771	.8098	85°	.9962	.0872	11.4301
40°	.6428	.7660	.8391	86°	.9976	.0698	14.3007
41°	.6561	.7547	.8693	87°	.9986	.0523	19.0811
42°	.6691	.7431	.9004	88°	.9994	.0349	28.6363
43°	.6820	.7314	.9325	89°	.9998	.0175	57.2900
44°	.6947	.7193	.9657	90°	1.0000	.0000	
45°	.7071	.7071	1.0000				

Index

Photo Credits

Selected Answers

Chapter 1 Related Practice Exercises

Lesson 1-1 **1** 1. 793,058 **2** 1. 61 437
3 1. ninety-six **4** 1. four hundred twenty-six
5 1. eight thousand, two hundred seventy-eight
6 1. six thousand, four **7** 1. twenty-seven thousand four hundred thirty-two **8** 1. one hundred fifty thousand **9** 1. three million **10** 1. nine million, two hundred fifty thousand **11** 1. seven million, one hundred twenty-two thousand, eight hundred forty-three **12** 1. thirty two million, four hundred twenty-nine thousand, seven hundred eighty-four
13 1. seven million **14** 1. five billion **15** 1. nine trillion **16** 1. seven million, five hundred ninety three thousand **17** 1. twenty-eight thousand, nine hundred **18** 1. six million, five hundred eighty-two thousand **19** 1. seventy-five billion **20** 1. nine trillion **21** 1. six thousand, nine hundred sixty-nine hundred
Lesson 1-2 **1** 1. 400 **2** 1. 4,610
3 1. 3,900,000 **4** 1. 48,018,007,000
5 1. 4,950,000,000,000 **6** 1. 14,600
Lesson 1-3 **1** 1. 30 **2** 1. 30 **3** 1. 200
4 1. 600 **5** 1. 9,000 **6** 1. 9,000 **7** 1. 90,000
8 1. 60,000 **9** 1. 300,000 **10** 1. 600,000
11 1. 5,000,000 **12** 1. 7,000,000
13 1. 6,000,000,000 **14** 1. 9,000,000,000
15 1. 6,000,000,000,000 **16** 1. 7,000,000,000,000
Lesson 1-4 **1** 1. 60 **2** 1. 67 **3** 1. 142
4 1. 20 **5** 1. 158 **6** 1. 19 **7** 1. 247
8 1. 20 **9** 1. 94 **10** 1. 688 **11** 1. 9,898
12 1. 78,799 **13** 1. 939 **14** 1. 8,769
15 1. 89,986 **16** 1. 1,596 **17** 1. 8,783
18 1. 79,906 **19** 1. 1,525 **20** 1. 9,302
21 1. 105,005 **22** 1. 27 **23** 1. 34,757
24 1. 32 **25** 1. 616 **26** 1. c **27** 1. 53
Lesson 1-5 **1** 1. 55 **2** 1. 15 **3** 1. 24
4 1. 17 **5** 1. 30 **6** 1. 3 **7** 1. 24 **8** 1. 336
9 1. 512 **10** 1. 484 **11** 1. 210 **12** 1. 217
13 1. 7 **14** 1. 227 **15** 1. 6,223 **16** 1. 3,217
17 1. 3,737 **18** 1. 6,168 **19** 1. 2,349
20 1. 4,815 **21** 1. 12,241 **22** 1. 33,427
23 1. 22,166 **24** 1. 35,895 **25** 1. 68,668
26 1. 13,955 **27** 1. 65,539 **28** 1. 331,722
29 1. 5,524,025 **30** 1. 7 **31** 1. 55
32 1. 148 **33** 1. 261 **34** 1. 574 **35** 1. c
36 1. 19
Lesson 1-6 (Multipliers-one digit) **1** 1. 64
2 1. 189 **3** 1. 92 **4** 1. 288 **5** 1. 963
6 1. 868 **7** 1. 752 **8** 1. 4,286 **9** 1. 96,956
10 1. 91 **11** 1. 160 **12** 1. 2,100 **13** 1. 208

14 1. 1,620 **15** 1. 42,240 **16** 1. 4,008
17 1. 21,280 **18** 1. 168 **19** 1. 104
20 1. 1,296 **21** 1. b **22** 1. 279
Lesson 1-6 (Multipliers-two or more digits)
1 1. 276 **2** 1. 1,296 **3** 1. 3,321 **4** 1. 86,373
5 1. 2,235,672 **6** 1. 7,488 **7** 1. 46,656
8 1. 643,566 **9** 1. 7,382,928 **10** 1. 3,049,212
11 1. 21,332,160 **12** 1. 393,478,722
13 1. 3,100 **14** 1. 1,770 **15** 1. 13,026
16 1. 373,474 **17** 1. 348,783 **18** 1. 32,800
19 1. 820,615 **20** 1. 1,234,240 **21** 1. 840
22 1. 35,742 **23** 1. 512 **24** 1. 952 **25** 1. b
26 1. 2,136
Lesson 1-7 (Divisors-1 digit) **1** 1. 31 **2** 1. 17
3 1. 231 **4** 1. 141 **5** 1. 236 **6** 1. 41
7 1. 72 **8** 1. 3,224 **9** 1. 1,651 **10** 1. 2,497
11 1. 749 **12** 1. 23,292 **13** 1. 9,122
14 1. 210 **15** 1. 304 **16** 1. 105 **17** 1. 3,001
18 1. 2,004 **19** 1. 2 r4 **20** 1. 10 r5 **21** 1. 3 r8 **22** 1. 19 r2 **23** 1. 60 r5 **24** 1. 146
25 1. 233 **26** 1. b **27** 1. 23
Lesson 1-7 (Divisors-two or more digits) **1** 1. 5
2 1. 7 **3** 1. 28 **4** 1. 76 **5** 1. 123 **6** 1. 548
7 1. 2,415 **8** 1. 5,219 **9** 1. 189 **10** 1. 705
11 1. 220 **12** 1. 2,004 **13** 1. 862 r25
14 1. 172 r4 **15** 1. 7 **16** 1. 52 **17** 1. 48
18 1. 629 **19** 1. 754 **20** 1. 37 **21** 1. 301
22 1. 320 **23** 1. 426 **24** 1. 528 r113
25 1. 467 r80 **26** 1. 8 **27** 1. 42 **28** 1. 467
29 1. 5,485 **30** 1. 654 **31** 1. 17 **32** 1. 39
33 1. c **34** 1. 3
Lesson 1-8 **1** 1. yes **2** 1. 3 **3** 1. 1, 2, 3, 4, 6, 8, 12, 16, 24, 48 **4** 1. 1, 4 **5** 1. 1, 2, 3, 4, 6, 12
6 1. 2 **7** 1. 2

Chapter 2 Related Practice Exercises

Lesson 2-1 **1** 1. eight tenths **2** 1. three hundredths **3** 1. twenty-four hundredths **4** 1. one and six tenths **5** 1. two and fifty-one hundredths
6 1. five thousandths **7** 1. twenty-four thousandths **8** 1. eight hundred thirty-two thousandths **9** 1. six and five thousandths
10 1. seven ten-thousandths **11** 1. six hundred-thousandths **12** 1. one millionth
13 1. eight and twenty-five ten-thousandths
14 1. point nine three; ninety-three hundredths
15 1. thirty-six point eight nine; thirty-six and eighty-nine hundredths
Lesson 2-2 **1** 1. .4 **2** 1. 6.5 **3** 1. .08
4 1. .36 **5** 1. 6.04 **6** 1. .003 **7** 1. .069
8 1. .274 **9** 1. 2.017 **10** 1. .008
11 1. .00003 **12** 1. .000006 **13** 1. 500.0058
14 1. .7 **15** 1. 61.2
Lesson 2-3 **1** 1. .3 **2** 1. .1 **3** 1. .52
4 1. .32 **5** 1. .215 **6** 1. .545 **7** 1. .2059
8 1. .3414 **9** 1. .00008 **10** 1. .00008

11 1. .000003   **12** 1. .00005   **13** 1. $.27
14 1. $.64   **15** 1. 6   **16** 1. 10   **17** 1. $6
18 1. $2
Lesson 2-4   **1** 1. .8   **2** 1. 1.6   **3** 1. .07
4 1. .15   **5** 1. .94   **6** 1. 1.32   **7** 1. 1.00
8 1. 5.9   **9** 1. 14.3   **10** 1. 12.13   **11** 1. 5.41
12 1. 10.85   **13** 1. 6.18   **14** 1. 4.6
15 1. 8.009   **16** 1. 34.83   **17** 1. 391.77
18 1. 2.69   **19** 1. $2.78   **20** 1. $17.24
21 1. .9   **22** 1. 1.08   **23** 1. $.98
24 1. 15.462   **25** 1. a   **26** 1. 1.5
Lesson 2-5   **1** 1. .4   **2** 1. .26   **3** 1. .18
4 1. .07   **5** 1. .09   **6** 1. .3   **7** 1. 4.2   **8** 1. 2.8
9 1. 5.0   **10** 1. .621   **11** 1. 4.11   **12** 1. 12.3
13 1. .0863   **14** 1. .10709   **15** 1. .3007
16 1. .46 or .4600   **17** 1. .0021   **18** 1. .8
19 1. 2.247   **20** 1. .5   **21** 1. .326   **22** 1. .7
23 1. 4.4   **24** 1. $2.25   **25** 1. .1   **26** 1. .05
27 1. .15   **28** 1. 3.36   **29** 1. 7.5   **30** 1. $.25
31 1. .45   **32** 1. c   **33** 1. .6
Lesson 2-6   **1** 1. .3   **2** 1. 1.47   **3** 1. .89
4 1. 3.010   **5** 1. .01, .01, .001, .0001   **6** 1. .19, .201,
.21, 1.2   **7** 1. True   **8** 1. False   **9** 1. c
Lesson 2-7   **1** 1. .4   **2** 1. 2.07   **3** 1. 1.235
4 1. 7.6176   **5** 1. 10   **6** 1. .24   **7** 1. .09
8 1. .018   **9** 1. .009   **10** 1. 52.2   **11** 1. .72
12 1. .06   **13** 1. .102   **14** 1. .048
15 1. .2072   **16** 1. .0134   **17** 1. .0012
18 1. .09568   **19** 1. .00004   **20** 1. .247632
21 1. .000048   **22** 1. 3.5   **23** 1. .06
24 1. $1.68   **25** 1. $14.91   **26** 1. $3.56
27 1. $525   **28** 1. 15   **29** 1. a   **30** 1. 4.8
Lesson 2-8 (divisors-whole numbers)   **1** 1. 2.3
2 1. .43   **3** 1. .121   **4** 1. .2234   **5** 1. .84
6 1. .038   **7** 1. .28   **8** 1. .4   **9** 1. .05
10 1. .6   **11** 1. .6   **12** 1. 4.24   **13** 1. .29
14 1. c   **15** 1. .19
Lesson 2-8 (divisors-tenths)   **1** 1. 19   **2** 1. 2.8
3 1. .42   **4** 1. .841   **5** 1. .002   **6** 1. 24.3
7 1. .35   **8** 1. 30   **9** 1. 5   **10** 1. 11.3
11 1. .05   **12** 1. 97
Lesson 2-8 (divisors-hundredths)   **1** 1. 17
2 1. 11.6   **3** 1. 234.77   **4** 1. .992   **5** 1. .03
6 1. 57   **7** 1. 1900   **8** 1. 720   **9** 1. 300
10 1. 717.4   **11** 1. 173   **12** 1. b   **13** 1. 30
Lesson 2-8 (divisors-thousandths)   **1** 1. 109
2 1. 1004.7   **3** 1. 14.91   **4** 1. .09   **5** 1. 3.2
6 1. 2,000   **7** 1. 270   **8** 1. 87,700   **9** 1. 3,000
10 1. 291.5   **11** 1. c   **12** 1. 200
Lesson 2-9   **1** 1. .7   **2** 1. .9   **3** 1. .07
4 1. .75   **5** 1. .375   **6** 1. .14   **7** 1. .444
8 1. .50   **9** 1. .83   **10** 1. .286   **11** 1. 2.5
12 1. .25   **13** 1. .05   **14** 1. .2
Lesson 2-10   **1** 1. 40   **2** 1. 50   **3** 1. 5
4 1. 5   **5** 1. 700   **6** 1. 650   **7** 1. 56
8 1. 4,000   **9** 1. 37.5   **10** 1. 200   **11** 1. 23
12 1. $3,500   **13** 1. 400   **14** 1. $20
Lesson 2-11   **1** 1. 50   **2** 1. 400   **3** 1. 3

4 1. 2.6   **5** 1. .3   **6** 1. 58   **7** 1. 700
8 1. 2,000   **9** 1. 42   **10** 1. 20   **11** 1. 72.1
12 1. 854   **13** 1. 3,000   **14** 1. 50,000
15 1. 657   **16** 1. 350   **17** 1. 265.3
18 1. 6,582   **19** 1. 952,000,000
20 1. 83,000,000,000   **21** 1. 17,000,000,000,000
Lesson 2-12   **1** 1. 2   **2** 1. 3.4   **3** 1. .5
4 1. .09   **5** 1. .35   **6** 1. 2   **7** 1. 3.82   **8** 1. .59
9 1. .04   **10** 1. .0021   **11** 1. .295   **12** 1. 8
13 1. 3.725   **14** 1. .628   **15** 1. .085
16 1. .000925   **17** 1. .2849   **18** 1. 617
19 1. 39   **20** 1. 8
Lesson 2-13   **1** 1. 83,000,000   **2** 1. 6,800,000
3 1. 17,450,000   **4** 1. 8,560,000   **5** 1. 4,500,000
6 1. 78,000,000,000   **7** 1. 9,400,000,000
8 1. 12,060,000,000   **9** 1. 7,568,000,000
10 1. 8,300,000,000   **11** 1. 4,000,000,000,000
12 1. 3,900,000,000,000   **13** 1. 5,920,000,000,000
14 1. 8,227,000,000,000   **15** 1. $9,600,000,000,000
16 1. 9,000   **17** 1. 8,700   **18** 1. $19,000
19 1. 3,800   **20** 1. 850
Lesson 2-14   **1** 1. 9 hundred   **2** 1. 13.4 hundred
3 1. 25.68 hundred   **4** 1. 6 thousand   **5** 1. 8.5
thousand   **6** 1. 7.13 thousand   **7** 1. 2.654
thousand   **8** 1. 8 million   **9** 1. 7.3 million
10 1. 11.47 million   **11** 1. 28.063 million   **12** 1. 9
billion   **13** 1. 7.5 billion   **14** 1. 16.82 billion
15 1. 51.006 billion   **16** 1. 6 trillion   **17** 1. 4.7
trillion   **18** 1. 32.08 trillion   **19** 1. 17.526 trillion
20 1. $63 thousand   **21** 1. $84 million   **22** 1. $58
billion   **23** 1. $6.5 trillion

Chapter 3   Related Practice Exercises
Lesson 3-1   **1** 1. $\frac{1}{2}$ **2** 1. $\frac{1}{2}$ **3** 1. $\frac{4}{5}$ **4** 1. $\frac{3}{5}$
5 1. $\frac{2}{3}$ **6** 1. $\frac{5}{8}$ **7** 1. $\frac{1}{4}$
Lesson 3-2   **1** 1. 1   **2** 1. $1\frac{1}{2}$ **3** 1. $1\frac{1}{2}$
4 1. $4\frac{1}{2}$ **5** 1. $2\frac{1}{2}$ **6** 1. 3   **7** 1. 7   **8** 1. $4\frac{2}{3}$
9 1. $7\frac{3}{5}$ **10** 1. $6\frac{1}{2}$
Lesson 3-3   **1** 1. 4   **2** 1. 8   **3** 1. 70   **4** 1. $\frac{8}{64}$
5 1. $\frac{20}{10}$
Lesson 3-4   **1** 1. $\frac{1}{2}, \frac{2}{4}, \frac{3}{6}, \frac{4}{8}, \ldots$ **2** 1. $\frac{1}{6}, \frac{2}{12}, \frac{3}{18}, \frac{4}{24}, \ldots$
3 1. Yes   **4** 1. No
Lesson 3-5   **1** 1. $\frac{1}{3}$ **2** 1. $\frac{2}{3}$ **3** 1. $\frac{1}{6}$ **4** 1. $\frac{1}{4}$
5 1. True   **6** 1. True   **7** 1. False   **8** 1. True
9 1. $\frac{1}{2}, \frac{1}{3}, \frac{1}{5}$ **10** 1. $\frac{1}{6}, \frac{1}{4}, \frac{1}{2}$ **11** 1. 9   **12** 1. 2
13 1. $3.28   **14** 1. $17.35
Lesson 3-6   **1** 1. b, c   **1** 1. 0, 5, 10, 15, 20,
25, . . .   **3** 1. 0, 15, 30   **4** 1. 20, 40, 60, 80
5 1. 24   **6** 1. 12   **7** 1. 48
Lesson 3-7   **1** 1. 4, $\frac{2}{4}, \frac{1}{4}$ **2** 1. 6, $\frac{2}{6}, \frac{3}{6}$ **3** 1. 12, $\frac{3}{12}$,
$\frac{2}{12}$ **4** 1. 12, $\frac{6}{12}, \frac{4}{12}, \frac{3}{12}$
Lesson 3-8   **1** 1. $\frac{4}{5}$ **2** 1. $\frac{1}{2}$ **3** 1. 1   **4** 1. $1\frac{1}{3}$
5 1. $1\frac{2}{3}$ **6** 1. $\frac{3}{4}$ **7** 1. $1\frac{5}{12}$ **8** 1. $\frac{43}{60}$ **9** 1. 1

10 1. $8\frac{7}{8}$ **11** 1. $13\frac{3}{5}$ **12** 1. $7\frac{3}{4}$ **13** 1. $13\frac{3}{4}$
14 1. 12 **15** 1. $4\frac{1}{3}$ **16** 1. $13\frac{1}{2}$ **17** 1. $6\frac{2}{5}$
18 1. $10\frac{11}{24}$ **19** 1. $6\frac{1}{2}$ **20** 1. $15\frac{1}{4}$ **21** 1. $11\frac{1}{2}$
22 1. $7\frac{11}{16}$ **23** 1. 4 **24** 1. $10\frac{7}{8}$ **25** 1. $\frac{3}{8}$
26 1. $3\frac{5}{8}$ **27** 1. $1\frac{1}{4}$ **28** 1. $1\frac{1}{2}$

Lesson 3-9 **1** 1. $\frac{1}{5}$ **2** 1. $\frac{1}{2}$ **3** 1. $\frac{11}{16}$ **4** 1. $\frac{1}{6}$
5 1. $\frac{1}{12}$ **6** 1. $4\frac{2}{5}$ **7** 1. $2\frac{1}{4}$ **8** 1. $4\frac{3}{5}$ **9** 1. $5\frac{1}{3}$
10 1. 1 **11** 1. $2\frac{1}{2}$ **12** 1. $6\frac{3}{4}$ **13** 1. $4\frac{5}{8}$
14 1. $2\frac{1}{2}$ **15** 1. $3\frac{2}{15}$ **16** 1. $3\frac{7}{12}$ **17** 1. $2\frac{3}{4}$
18 1. $5\frac{5}{6}$ **19** 1. $7\frac{17}{20}$ **20** 1. $\frac{1}{4}$ **21** 1. $4\frac{3}{4}$
22 1. $\frac{2}{5}$ **23** 1. $4\frac{1}{2}$ **24** 1. $\frac{9}{16}$ **25** 1. $\frac{1}{24}$

Lesson 3-10 **1** 1. $\frac{6}{5}$ **2** 1. $\frac{13}{8}$ **3** 1. $\frac{41}{5}$ **4** 1. $\frac{14}{3}$

Lesson 3-11 **1** 1. $\frac{1}{8}$ **2** 1. $\frac{3}{10}$ **3** 1. $\frac{3}{5}$ **4** 1. $\frac{5}{16}$
5 1. $\frac{1}{6}$ **6** 1. $\frac{2}{3}$ **7** 1. $\frac{5}{6}$ **8** 1. 5 **9** 1. 6
10 1. $1\frac{1}{2}$ **11** 1. $2\frac{1}{4}$ **12** 1. $2\frac{1}{3}$ **13** 1. 3
14 1. 10 **15** 1. $\frac{3}{4}$ **16** 1. $5\frac{1}{4}$ **17** 1. $2\frac{2}{3}$
18 1. 34 **19** 1. $8\frac{1}{2}$ **20** 1. $8\frac{4}{5}$ **21** 1. 42
22 1. $25\frac{1}{2}$ **23** 1. $12\frac{2}{3}$ **24** 1. 2 **25** 1. $1\frac{7}{8}$
26 1. $1\frac{3}{5}$ **27** 1. $1\frac{3}{32}$ **28** 1. 2 **29** $2\frac{4}{5}$
30 1. $3\frac{3}{16}$ **31** 1. $\frac{1}{12}$ **32** 1. $11\frac{55}{64}$ **33** 1. 126
34 1. 258 **35** 1. $29\frac{1}{4}$ **36** 1. 12 **37** 1. 1,760
38 1. $10 **39** 1. $82 **40** 1. $\frac{3}{4}$ **41** 1. 24
42 1. $.49 **43** 1. $.20

Lesson 3-12 **1** 1. $\frac{3}{4}$ **2** 1. $2\frac{1}{2}$ **3** 1. $\frac{1}{3}$ **4** 1. 2
5 1. $1\frac{1}{7}$ **6** 1. $\frac{1}{6}$ **7** 1. $\frac{3}{40}$ **8** 1. 18 **9** 1. 20
10 1. $6\frac{2}{3}$ **11** 1. $3\frac{1}{2}$ **12** 1. $1\frac{1}{2}$ **13** 1. $2\frac{13}{16}$
14 1. 4 **15** 1. $3\frac{1}{3}$ **16** 1. $\frac{3}{4}$ **17** 1. $3\frac{3}{5}$
18 1. $2\frac{2}{3}$ **19** 1. $2\frac{2}{5}$ **20** 1. $7\frac{1}{2}$ **21** 1. $\frac{1}{20}$
22 1. $\frac{8}{15}$ **23** 1. 5 **24** 1. $2\frac{2}{5}$ **25** 1. $\frac{2}{3}$
26 1. $\frac{21}{40}$ **27** 1. 1 **28** 1. $7\frac{1}{2}$ **29** 1. $1\frac{3}{5}$
30 1. $\frac{32}{45}$

Lesson 3-13 **1** 1. $\frac{1}{4}$ **2** 1. $\frac{3}{4}$ **3** 1. $\frac{7}{25}$ **4** 1. $\frac{1}{2}$
5 1. $\frac{1}{8}$ **6** 1. $\frac{3}{5}$ **7** 1. $1\frac{1}{6}$ **8** 1. $1\frac{1}{4}$ **9** 1. 2
10 1. 1

Lesson 3-14 **1** 1. 10 **2** 1. 12 **3** 1. 32
4 1. 72 **5** 1. 642

Lesson 3-15 **1** 1. .1 **2** 1. .8 **3** 1. .39
4 1. .75 **5** 1. $.37\frac{1}{2}$ **6** 1. $.33\frac{1}{3}$ **7** 1. 1.15
8 1. $.33\frac{1}{3}$ **9** 1. $.57\frac{1}{7}$ **10** 1. .5 or .50
11 1. $.37\frac{1}{2}$ **12** 1. $.77\frac{7}{9}$ **13** 1. 1.5 or 1.50
14 1. 1.6 or 1.60 **15** 1. $1.55\frac{5}{9}$ **16** 1. $2.37\frac{1}{2}$
17 1. .571 **18** 1. .9514

Lesson 3-16 **1** 1. $\frac{3}{5}$ **2** 1. $\frac{3}{4}$ **3** 1. $\frac{1}{50}$ **4** 1. $\frac{2}{5}$

5 1. $\frac{7}{8}$ **6** 1. $1\frac{1}{5}$ **7** 1. $1\frac{1}{4}$ **8** 1. $1\frac{1}{3}$ **9** 1. $\frac{1}{8}$
10 1. $\frac{9}{250}$ **11** 1. $1\frac{3}{8}$ **12** 1. $\frac{5}{16}$ **13** 1. $\frac{1}{400}$
14 1. $3\frac{9}{16}$

Lesson 3-17 **1** 1. $\frac{8}{15}$ **2** 1. 2:9 **3** 1. $\frac{5}{8}$ **4** 1. $\frac{1}{3}$
5 1. $\frac{11}{2}$ **6** 1. $\frac{8}{1}$ **7** 1. $\frac{3}{2}$ **8** 1. $\frac{n}{7}$ **9** 1. $\frac{4}{5}$
10 1. $\frac{7}{2}$ **11** 1. 20 or $\frac{20}{1}$ **12** 1. $\frac{1}{4}$ **13** 1. $\frac{88}{1}$ or
88 **14** 1. 1, 2, 4 **15** 1. $\frac{2}{3}, \frac{3}{2}, \frac{5}{2}, \frac{5}{5}$

Lesson 3-18 **1** 1. $\frac{6}{3}$ **2** 1. $\frac{x}{30} = \frac{9}{54}$ **3** 1. 2, 3
4 1. $t = 24$ **5** 1. $a = 5$ **6** 1. $n = 114$
7 1. $x = 63$ **8** 1. $y = 4$ **9** 1. 56 **10** 1. 15 items

Chapter 4 Related Practice Exercises

Lesson 4-1 **1** 1. 3% **2** 1. 47% **3** 1. 4
hundredths **4** 1. 6%, .06, $\frac{6}{100}$ **5** 1. $\frac{18}{100}$ or 18:100
Lesson 4-2 **1** 1. .06 **2** 1. .16 **3** 1. .40 or .4
4 1. 1.34 **5** 1. 1.30 or 1.3 **6** 1. 1 **7** 1. $.37\frac{1}{2}$ or
.375 **8** 1. $.04\frac{1}{2}$ or .45 **9** 1. $.60\frac{1}{2}$ or .605
10 1. $1.00\frac{7}{8}$ or 1.00875 **11** 1. .875 **12** 1. .1775
13 1. .035 **14** 1. .0125 **15** 1. .26375
16 1. .007 **17** 1. $.00\frac{1}{2}$ or .005
Lesson 4-3 **1** 1. 1% **2** 1. 28% **3** 1. 60%
4 1. 139% **5** 1. 120% **6** 1. $12\frac{1}{2}$% **7** 1. $1\frac{1}{2}$%
8 1. $10\frac{1}{2}$% **9** 1. $137\frac{1}{2}$% **10** 1. 87.5%
11 1. 26.25% **12** 1. 25% **13** 1. 124.5%
14 1. 100% **15** 1. $\frac{1}{4}$% **16** 1. .5%
Lesson 4-4 **1** 1. $\frac{1}{2}$ **2** 1. $\frac{1}{3}$ **3** 1. $\frac{3}{50}$ **4** 1. $1\frac{1}{10}$
Lesson 4-5 **1** 1. 25% **2** 1. $83\frac{1}{3}$% **3** 1. 7%
4 1. 54% **5** 1. 9% **6** 1. $66\frac{2}{3}$% **7** 1. $71\frac{3}{7}$%
8 1. $55\frac{5}{9}$% **9** 1. 100% **10** 1. 175%
11 1. $166\frac{2}{3}$% **12** 1. 150% **13** 1. $262\frac{1}{2}$%
14 1. $266\frac{2}{3}$%
Lesson 4-6 **1** 1. 12.48 **2** 1. .36 **3** 1. .412
4 1. 10 **5** 1. 3 **6** 1. 100 **7** 1. 32.48
8 1. 54 **9** 1. 72 **10** 1. .75 **11** 1. 4.9
12 1. .24 **13** 1. .9 **14** 1. .016 **15** 1. $.20
16 1. $.03 **17** 1. $9.80
Lesson 4-7 **1** 1. 60% **2** 1. 25% **3** 1. $83\frac{1}{3}$%
4 1. $33\frac{1}{3}$% **5** 1. $42\frac{6}{7}$% **6** 1. $28\frac{4}{7}$% **7** 1. 100%
8 1. 400% **9** 1. 120% **10** 1. $133\frac{1}{3}$%
11 1. 75% **12** 1. 25% **13** 1. 25%
14 1. 15% **15** 1. 35% **16** 1. 3%
Lesson 4-8 **1** 1. 200 **2** 1. 50 **3** 1. 15
4 1. 234 **5** 1. 500 **6** 1. 100 **7** 1. 59
8 1. 45 **9** 1. 78 **10** 1. 800 **11** 1. 400

12 1. 18   **13 1.** 500   **14 1.** 2,000   **15 1.** 200

Chapter 5   Related Practice Problems

Lesson 5-1   1 1. They drove 424 miles. It took 8 hours. What was the average rate of speed? (What is 424 divided by 8?)   **2 1.** Ribbon is 42 inches long. Ribbon is cut into 3 pieces.   **3 1.** How many miles did the train travel?   **4 1.** How much is his weekly salary?   **5 1.** The movie lasts $1\frac{1}{2}$ hours.   **6 1.** The number of items for sale.
Lesson 5-2   1 1. multiplication   **2 1.** addition
3 1. subtraction   **4 1.** division   **5 1.** 23 × .50
6 1. answers will vary   **7 1.** 12 − 7 = 5
8 1. 25 × 6 = n   **9 1.** .09 × a = $225
10 1. 60 ÷ 8 = n   **11 1.** s = 24 × 18
Lesson 5-3   1 1. 60,000; 59,290   **2 1.** 1,700; 1,705
3 1. 300; $299\frac{1}{7}$ **4 1.** 5,000; 4,827   **5 1.** 1.2; 1.19
6 1. .25; .2475   **7 1.** .16; .1554   **8 1.** .3; .263
Lesson 5-4   1 1. c   **2 1.** a
Lesson 5-5   1 1. How much more than Teresa does her brother weigh?   **2 1.** $18.46
Lesson 5-6   1 1. How much did Jill spend?
2 1. How many pairs of sneakers does he buy?
3 1. $1,799.82
Lesson 5-7   1 1. What are the two total scores? What are the two averages? What is the difference between the two averages?   **2 1.** Gloria's team; 8 points

Chapter 6   Practice Problems

Lesson 6-1 p. 276   1 answers will vary
2 1. $164   **3 1.** $193.20   **4 1.** $266.80   **5 1.** 39, $247.65   **6 1.** 39, $325.65   **p. 277   1** answers will vary   **2 1.** $4.75   **3 1.** $4,420   **4 1.** $8,400
5 1. $8,970   **6 1.** $180   **7 1.** $675
8 1. $188.31   **9 1.** $600   **10 1.** $295
11 1. $59   **12 1.** $180 per week   **p. 278**
1 1. $33.32   **2 1.** $198   **3 1.** 151, $105.70   **p. 279   1** $140   **2** $500   **3** $234   **4 1.** $560
5 1. $110   **6** $37.50, $150   **7** $2,687.50
Lesson 6-2   1 $77,655; $5,845   **2 1.** $59.64; $366.36   **3** 12%   **4 1.** 15%   **5** 3%   **6** $2,500
7 1. $2,300   **8 1.** $56.52; $684.52   **9** $225.28
10 $91.06   **11** $4,327.61   **12** 15%   **13** $8,567
Lesson 6-3   1 $796.50   **2 1.** $89.78   **3 1.** $28
4 $10.15   **5** $90   **6 1.** $48   **7 1.** 60%
8 1. 20%   **9** $12\frac{1}{2}$%   **10 1.** 20%   **11** 21%
12 1. 25%   **13 1.** $121.10   **14 1.** $33\frac{1}{3}$%
15 $105   **16** $6   **17** $93.31; $16.51   **18** $12\frac{1}{2}$%
Lesson 6-4   1 1. $4.14   **2 1.** $3.30   **3 1.** $208
4 1. $502.32
Lesson 6-5   1 1. $4.23   **2 1.** $27.85
Lesson 6-6   1 1. $1.70   **2 1.** $2.04   **3 1.** $4.55

4 1. $15.75   **5 1.** $139.50
Lesson 6-7   1 $9.70; $12.83; $159.47   **2 1.** $87.37
3 1. $83.51   **4** $140.25   **5** $15.60; $15.33; $3.26; $183.31
Lesson 6-8   1 1. $1,388   **2 1.** $209   **3 1.** refund, $64

Chapter 7   Practice Problems

Lesson 7-1   1 1. $.36   **2 1.** $.75   **3 1.** $.63
4 1. $.80   **5 1.** $6.66   **6 1.** $6.66; $3.75; $11.40
Lesson 7-2   1 1. $.37   **2 1.** 86¢   **3 1.** thirty-four and six tenths cents   **4 1.** two dollars and thirty-seven and four tenths cents   **5 1.** $149   **6 1.** 86¢
7 1. 29.7¢   **8 1.** 90¢   **9 1.** $1.3725; lowest
10 1. $1.85;   **11 1.** $2.987; highest   **12 1.** $.29
13 1. 5 for 49¢   **14 1.** 10¢   **15 1.** .916¢
16 1. a half-gallon at $1.12   **17 1.** 6¢
18 1. 23.8¢   **19 1.** 62.5¢   **20 1.** 2.26-kg bag at 95¢   **21 1.** $.45
Lesson 7-3   1 1. Tables $69.30; Lamps $38.15; Sofas $297.50; Bedroom sets $629.30; Bookcases $50.93 nearest cent; Mirrors $26.59 nearest cent   **2** $3.75; $11.25
3 $5,685.50   **4 1.** $2.10   **5** 20%   **6 1.** 25%
7 $199.80   **8 1.** $40   **9** $37.41   **10** $29.50
11 $8.40; $47.60   **12** $20.79; $109.16   **13 1.** $49.41; $133.59   **14** $286.71   **15 1.** $115.20   **16** 24%
17 15%   **18** $900   **19** $320   **20 1.** $56
21 $120
Lesson 7-4   1 1. $.21   **2 1.** $3.66   **3 1.** $3.59
4 1. $11.92   **5 1.** $2.95
Lesson 7-5   1 1. $.10   **2 1.** $.96   **3 1.** $.15
4 1. $7.28   **5 1.** $7   **6 1.** $.03   **7 1.** $.42
8 1. $20.50   **9 1.** $.02
Lesson 7-6   1 1. 8,352; 8,417   **2 1.** $55.54
3 1. 6,529; 6,774   **4 1.** $7.35   **5 1.** 8.6¢
6 1. 25,240   **7 1.** 1,040 gal.; $1,285.69
Lesson 7-7   1 1. $1,631.25   **2** 330 miles   **3** 9.2 h; 32.86 gal.   **4** $287.45
Lesson 7-8   1 $22   **2** $33.80; $33.80   **3 1.** $45
4 $38   **5** $31.80   **6 1.** $9.54   **7 1.** $10
Lesson 7-9 p. 322   1 1. 12%   **2 1.** 24%   **p. 324   1 1.** 14.25%   **2 1.** $6.35; $11.82
Lesson 7-10   1 1. $35   **2 1.** $198.72; $69.55
3 $30.60
Lesson 7-11   1 1. $294.20   **2 1.** $0   **3** $13,152
4 1. $163.40   **5 1.** $5,000   **6** $24,397.75
7 1. 20-payment life is $223 more.
Lesson 7-12   1 1. $.007   **2 1.** 2¢   **3 1.** 40 mills
4 1. 4¢ per $1   **5 1.** 50 mills per $1   **6 1.** $8 per $100   **7 1.** $45 per $1,000   **8 1.** 4%   **9 1.** 2%
10 $571.20   **11** $1,225.50   **12 1.** $464
13 1. $592   **14 1.** $520   **15 1.** $200
16 1. $150   **17** $779.10   **18 1.** $669.80
Lesson 7-13   1 1. $385.42   **2 1.** $68,125
3 $49,000   **4** $622   **5** $669.25   **6** $965.57
7 1. $5,200   **8** The cost of owning is $2,312.25 more.

Chapter 8 Practice Problems

Lesson 8-1 **1** See check stub and check register for position. **2** $406.75; balance $1,999.82
3 $1,491.42 **4** $1,249.47 **5** $727.34; balance $1,976.81 **6** $1,491.12 **7** $1,396.17
Lesson 8-2 **1** **1.** $30 **2** **1.** $1,624 **3** **1.** $315
4 **1.** $19.25 **5** **1.** $11.25 **6** **1.** $14 **7** **1.** $45
8 **1.** $8 **9** **1.** $4 **10** **1.** $5.95 **11** **1.** $45; $345
12 $64.75 **13** $588 **14** **1.** 12% **15** $250,000
Lesson 8-3 **1** **1.** 12% **2** **1.** 52% **3** **1.** 365%
4 **1.** 300% **5** **1.** $564; $50 **6** **1.** $636; $53
Lesson 8-4 **1** **1.** $45 **2** **1.** $1,980
3 **1.** $349.84 **4** **1.** $125.04 **5** $87.30
Lesson 8-5 **1** **1.** $451.14 **2** **1.** $563.92
3 **1.** $12,775.40 **4** **1.** $179,204.40 **5** $383.32
Lesson 8-6 **1** **1.** $219.50; $119.50 **2** **1.** $674.16; $4,455.51 **3** **1.** $443.37; 94.37 **4** **1.** $510
5 **1.** $20 **6** **1.** $788.77 **7** **1.** $31.71
8 **1.** $81.21 **9** **1.** $5.62 **10** **1.** $627.50
Lesson 8-7 p. 349 **1** **1.** $1,587.50 **2** **1.** $5,199.13
3 **1.** $3,750.19 **4** **1.** $5\frac{3}{4}$ points above low and $1\frac{3}{8}$ points below high **5** **1.** INA In-$.48 **6** **1.** INA IN. . .
7 **1.** $696 **8** **1.** $1,032.50 **p. 351** **1** **1.** 1989; 5% **1** **1.** $885 **3** **1.** $50.63 **4** **1.** See Bond Quotations. **p. 352 top** **1** **1.** $1,421
2 **1.** $1,573 **p. 352 bottom** **1** $1,612.50
2 $50.00 **3** $4,747.50

Chapter 9 Practice Exercises

Lesson 9-1 **1** **1.** kilometer **2** **1.** 2 cm **3** **1.** 4 cm **4** **1.** 7 cm **5** **1.** 66 mm **6** **1.** 80 mm
7 **1.** 14 mm; 1 cm 4 mm **8** 6.5 cm **9** **1.** 10
10 **1.** 1,000 **11** **1.** 7,583 **12** **1.** 340
13 **1.** 47 **14** **1.** 4,390 **15** **1.** 780 **16** **1.** 4.6
17 **1.** 600 **18** **1.** 8.1 **19** **1.** .7; 7; 70; 7,000; 70,000; 700,000 **20** **1.** 74 **21** **1.** 482 **22** **1.** 3.8
23 **1.** 6.094 **24** **1.** 601.8 **25** 10 meters
26 21.75 km **27** .306 m or 306 mm **28** **1.** 9,000 mm **29** **1.** 7 m **30** 6,800 m; 6.7 km; 526,000 cm; 692,400 mm **31** 84.5 mm; 845 m; 8.45 km; 8,450,000 cm
Lesson 9-2 **1** **1.** milligram **2** **1.** 10
3 **1.** 1,000 **4** **1.** 90 **5** **1.** 3.9 **6** **1.** 250
7 **1.** 600 **8** **1.** 300 **9** **1.** 53.3 **10** **1.** 7.864
11 **1.** 3 **12** **1.** 5; 500,000; 5,000,000 **13** **1.** 56
14 **1.** 485 **15** **1.** 5.2 **16** **1.** 8.052
Lesson 9-3 **1** **1.** kiloliter **2** **1.** 10 **3** **1.** 1,000
4 **1.** 70 **5** **1.** 8,500 **6** **1.** 310 **7** **1.** .9
8 **1.** 290 **9** **1.** 79 **10** **1.** 6.582 **11** **1.** 70; 700; 7,000 **12** **1.** 74 **13** **1.** 6,019 **14** **1.** 3.7
15 **1.** 8,037 **16** **1.** 401.4 **17** 5 liters
18 **1.** 65mL **19** .639 L and 6.39 dL or equal; 62.8 cL;

637 mL **20** 622 cL; 6,235 mL; 89.4 dL; 19 L
Lesson 9-4 **1** **1.** square centimeter **2** **1.** 10,000 cm^2 **3** **1.** 600 **4** **1.** 3,700 **5** **1.** 8.19
6 **1.** 45,000,000 **7** **1.** .80 **8** **1.** 7.5 **9** **1.** 5
10 **1.** 400,000 m^2; 4 km^2
Lesson 9-5 **1** **1.** cubic meter **2** **1.** 1,000,000 mm^3 **3** **1.** 8,000 mm^3 **4** **1.** 27,000,000 cm^3
5 **1.** 63,000 dm^3 **6** **1.** 17 m^3 **7** **1.** 21 cm^3
8 **1.** 7 **9** **1.** 13 **10** **1.** 5 **11** **1.** 6,000 **12** 800 dm^3 **13** 171,000 kg **14** 6.6 L; 6.6 kg **15** 7,540 L; 7,540 kg

Chapter 10 Practice Exercises

Lesson 10-1 **1** **1.** 96 **2** **1.** 324 **3** **1.** 67
4 **1.** 18 **5** **1.** 42,240 **6** **1.** 66 **7** **1.** 17
8 **1.** 2 **9** **1.** 12,320 **10** **1.** 49.5 **11** **1.** 7,240
12 **1.** 5 **13** **1.** 7 **14** **1.** 3,200 **15** **1.** 6
16 **1.** 3 **17** **1.** 5 **18** **1.** $\frac{1}{3}$ **19** **1.** $\frac{7}{12}$
20 **1.** $\frac{1}{4}$ **21** **1.** 13 ft 11 in. **22** **1.** 5 yd 2 in.
23 **1.** 20 ft 8 in. **24** **1.** 3 ft 2 in. **25** **1.** 36,480
26 **1.** 22,293$\frac{1}{3}$ **27** **1.** 8.06 **28** **1.** 5 **29** **1.** 21
30 4.34
Lesson 10-2 **1** **1.** 96 **2** **1.** 18 **3** **1.** 14,000
4 **1.** 132,160 **5** **1.** 12,400 **6** **1.** 4 **7** **1.** 3
8 **1.** 3 **9** **1.** $\frac{3}{4}$ **10** **1.** $\frac{3}{5}$ **11** **1.** $\frac{1}{4}$ **12** **1.** 4 lb 14 oz. **13** **1.** 3 lb 9 oz **14** **1.** 18 lb 12 oz **15** **1.** 2 lb 3 oz
Lesson 10-3 **1** **1.** 144 **2** **1.** 160 **3** **1.** 896
4 **1.** 40 **5** **1.** 6 **6** **1.** 48 **7** **1.** 11 **8** **1.** 3
9 **1.** 32 **10** **1.** 27 **11** **1.** 7 **12** **1.** 3
13 **1.** 5 **14** **1.** 3 **15** **1.** 2 **16** **1.** $\frac{1}{2}$
17 **1.** $\frac{1}{2}$ **18** **1.** $\frac{1}{2}$ **19** **1.** 8 gal. 3 qt **20** **1.** 3 qt **21** **1.** 3 gal. 1 pt **22** **1.** 4 gal. 1 qt
Lesson 10-4 **1** **1.** 8 **2** **1.** 120 **3** **1.** 36
4 **1.** 9 **5** **1.** $\frac{1}{2}$ **6** **1.** $\frac{3}{4}$ **7** 12 bu **8** 4 pk 6 qt
9 40 bu 2 pk **10** **1.** 3 pk 1 qt
Lesson 10-5 **1** **1.** 2,304 **1** **1.** 11 **3** **1.** 5
4 **1.** 3,680 **5** **1.** 5 **6** **1.** 17 **7** **1.** $\frac{1}{2}$ **8** **1.** $\frac{8}{9}$
9 **1.** $\frac{3}{8}$ **10** **1.** $\frac{3}{4}$
Lesson 10-6 **1** **1.** 65,664 **2** **1.** 793,152
3 **1.** 1,080 **4** **1.** 17 **5** **1.** 5 **6** **1.** 5 **7** **1.** $\frac{1}{4}$
8 **1.** $\frac{1}{3}$ **9** **1.** $\frac{1}{3}$ **10** **1.** $\frac{5}{9}$
Lesson 10-7 **1** **1.** 8 **2** **1.** 64 **3** **1.** 6
4 **1.** 1,617 **5** **1.** 75 **6** **1.** 4,687$\frac{1}{2}$
Lesson 10-8 **1** **1.** 1,440 **2** **1.** 45 **3** **1.** 720
4 **1.** 45 **5** **1.** 30 **6** **1.** 156 **7** **1.** 25 **8** **1.** $\frac{3}{4}$

9 1. $\frac{2}{3}$ **10 1.** $\frac{1}{6}$ **11 1.** $\frac{2}{7}$ **12 1.** $\frac{3}{4}$ **13** 23 h
14 9 yr 9 mo **15** 15 yr 9 mo **16** 1 da. 16 h 20
min **17 1.** a half-hour **18 1.** 16 seconds
19 1. 9 P.M. **20 1.** 5 min **21 1.** 16 h
22 1. Answers will vary. **23 1.** 31 **24 1.** 61
25 Answers will vary.
Lesson 10-9 p. 400 **1** 7:00 P.M.; 10:00 P.M.; 12:17 A.M.
2 6:25 P.M.; 8:17 P.M.; 10:00 P.M. **3** 280; 213 min or 3 h
33 min; 78.9 km/h **4** 21 min **p. 401 (top)** **1** 8:20
A.M., CST; 10:15 A.M., CST; 1 h. 55 min; no **2** 235; 12:40
P.M., CST; meal **3** 2 h 20 min **4** 25 min. **5** 3 hr
40 min **p. 402 (bottom)** **1** 9:30 A.M., EST; 11:59 A.M.,
CST; 7:20 P.M., CST; 12:15 A.M., CST **2** 3:45 P.M., MST;
10:30 A.M., CST **3** 18 h 20 min **4** 5:00 P.M., CST; 4:50
A.M., MST; 12 h 50 min **5** Bus 820 takes 31 h 45 min;
Bus 824 takes 32 h 25 min; Bus 824 takes 2 h 40 min
more
Lesson 10-10 **1 1.** 4 **2 1.** 4.5 **3 1.** 2,500
4 1. 108 **5 1.** 37.84 **6 1.** 286 **7 1.** 240
8 1. 16.4 **9 1.** 69.2 **10 1.** 5,280
Lesson 10-11 **1 1.** 323 K **2 1.** 7°C **3 1.** 30°C
4 1. 140°F

Chapter 11 Practice Exercises

Lesson 11-1 **1 1.** 400 miles; 100 miles **2 1.**

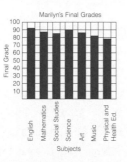

Lesson 11-2 **1 1.** 2 pupils; 1 pupil **2 1.**

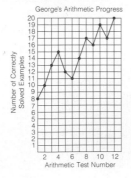

Lesson 11-3 **1 1.** $\frac{1}{4}$, $\frac{1}{5}$ **2 1.**

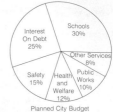

Planned City Budget

Lesson 11-4 **1, 2, 3, 1.**

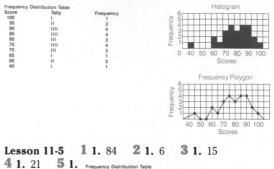

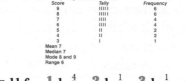

Lesson 11-5 **1 1.** 84 **2 1.** 6 **3 1.** 15
4 1. 21 **5 1.**

Frequency Distribution Table		
Score	Tally	Frequency
9	IIIII	6
8	IIIII	6
7	IIII	4
6	IIII	4
5	II	2
4	II	2
3	I	1

Mean 7
Median 7
Mode 8 and 9
Range 6

Lesson 11-6 **1 1.** $\frac{4}{9}$ **2 1.** $\frac{1}{3}$ **3 1.** $\frac{1}{5}$ **4** $\frac{1}{30}$; 29 to
1 **5 1.** (1) $\frac{26}{52}$ or $\frac{1}{2}$; (2) $\frac{4}{52}$ or $\frac{1}{13}$; (3) $\frac{12}{52}$ or $\frac{3}{13}$; (4) $\frac{1}{52}$

Chapter 12 Practice Exercises

Lesson 12-1 **1 1.** line AG or GA, $\overleftrightarrow{AG}$, $\overleftrightarrow{GA}$ **2 1.** line
segment EF or FE, $\overline{EF}$ $\overline{FE}$ **3 1.** line ST or TS, $\overleftrightarrow{ST}$, $\overleftrightarrow{TS}$
4 1. ray AB, $\overrightarrow{AB}$ **5** ray BG, ray BH, ray BD; $\overrightarrow{BG}$, $\overrightarrow{BH}$,
$\overrightarrow{BD}$ **6 1.** line CN **7** H **8** $\overline{RS}$, $\overline{ST}$ **9** $\overline{KM}$, $\overline{MN}$,
$\overline{NL}$ **10 1.** $\overline{FG}$, $\overline{GH}$, $\overline{HF}$ **11** $\overline{DF}$, $\overline{FG}$, $\overline{GE}$ **12** $\overline{AC}$
and $\overline{BC}$ **13 1.** curved **14 1.** horizontal
15 1. b, h, i **16 1.** $\overline{CE}$ **17 1.** D **18 1.** Yes
19 True **20** True **21** True **22** True
Lesson 12-2 **1** sides-$\overline{DE}$, $\overline{DF}$ vertex-D **2** $\angle F$, $\angle 2$
3 1. $\angle 4$ **4** $\angle 1 = \angle EFH$; $\angle 2 = \angle EFG$; $\angle 3 = \angle GFH$
5 $\angle MKL$ $\angle LKM$ $\angle K$ $\angle 1$; $\angle KML$ $\angle LMK$ $\angle M$ $\angle 2$; $\angle MLK$
$\angle L$ $\angle 3$ **6** $\angle ATM$ or $\angle MTA$: $\angle ATN$ or $\angle NTA$; $\angle NTB$ or
$\angle BTN$; $\angle BTM$ or $\angle MTB$ **7 1.** $\angle Q$
Lesson 12-3 **1** degrees; ° **2** 360° **3** 90°, $\frac{1}{4}$
4 0°, 90° **5** 90°, 180° **6** $\frac{1}{2}$, 180° **7** 1, 3 **8** 126°,
94° **9 1.** C, E, G **10 1.** Drawings will vary.
Lesson 12-4 **1** Show lengths indicated. **2 1.** $2\frac{3}{8}$
in. **3 1.** Show lengths indicated. **4** 3.7 cm.
5 1. 44 mm. **6** Drawings will vary. **7** 88 mm.

Lesson 12-5 **1 1.** 105 m **2 1.** 180 km **3 1.** 240 mi **4 1.** 12 ft **5 1.** 4 ft **6 1.** 2,400 m or 2.4 km **7 1.** 45 km **8 1.** 3 in. **9 1.** 6 in. **10 1.** 5 in. **11 1.** 3 cm **12 1.** 4 mm **13** 1:60,000,000 **14** 1:16 **15** 1:15,000,000 **16** 1:160 **17** 1:1,800 **18** 1:576 **19 1.** 3 cm **20 1.** 3 in. **21** 24 m by 18 m **22** $2\frac{3}{4}$ in. × $1\frac{1}{2}$ in. **23 1.** 72 mi **24 1.** 60 km **25** 112 mi **26** 130 km

Lesson 12-6 **1 1.** Plane *ABC* **2 1.** triangle **3 1.** equilateral **4** 1, 3, and 5 **5 1.** three **6** True **7** True **8** True

Lesson 12-7 **1 1.** rectangular solid **2 1.** 6 **3 1.** 5 **4** True **5** True **6** True

Lesson 12-8 **1 1.** 80° **2 1.** angle *S* **3 1.** 150° **4 1.** 180° **5-6** Drawings will vary. **7 1.** 60° **8** 90° **9** 180°

Lesson 12-9 **1 1.** 64° **2** 60° **3** 62° **4** 55° **5 1.** 42°

Lesson 12-10 **1** 360°, 360° **2 1.** 133° **3** 65°, 115°, 115° **4** 72° **5 1.** 900°

Lesson 12-11 **1** 1, 2, 3, 6 **2 1.** 69° **3 1.** 64° **4 1.** 35°, 55°, 35° **5** Each angle is less than 90°. **6** 1, 4, 5 **7 1.** 136° **8 1.** 148° **9** No, an acute angle. Since an obtuse angle is greater than 90°, its supplement must be less than 90°. **10** ∠3, yes, ∠2, yes **11 1.** 68° **12 1.** 110° **13 1.** 163° **14 1.** 35° **15** 111°, 69° **16** 52°, 127° **17** 61°, 99° **18** 84°, 70° **19** 49°, 86° **20** Yes. The measure of an exterior angle, when it is a right angle, will equal the measure of the adjacent interior angle. No. Since the measure of an exterior angle of a triangle is equal to the sum of the measures of the opposite two interior angles and each interior angle must have a measure greater than zero, the measures of an exterior angle and one opposite interior angle will not be the same.

Lesson 12-12 **1 1.** transversal **2** equal, ∠6, ∠4 **3** equal, ∠6, ∠3, ∠5, ∠4 **4** 180°, supplementary **5** 180°, supplementary **6** vertical or opposite **7 1.** ∠6 and ∠7, and ∠3 and ∠2 are opposite angles and hence equal. ∠3 and ∠6 are alternate-interior angles and hence are equal. Therefore, if ∠3 and ∠6 are equal, their opposite angles, ∠2 are ∠7 are equal. **8 1.** 130° **9** parallel **10 1.** Yes

Lesson 12-13 **1** Triangles *ABC* and *JKL* are congruent because three sides of triangle *ABC* are equal to three sides of triangle *JKL*. **2** Triangles *GHI* and *JKL* are congruent because two angles and an included side of triangle *GHI* are equal respectively to two angles and an included side of triangle *JKL*. **3** Triangles *DEF* and *JKL* are incongruent because two sides and an included angle of triangle *DEF* are equal respectively to two sides and an included angle of triangle *JKL*. **4 1.** 67° **5 1.** 25° **6** Given *AB* = *BC*, ∠*ABD* = ∠*DBC* and

BC = *BD*, triangles *ABD* and *DBC* are congruent because two sides and an included angle of triangle *ABD* are equal respectively to two sides and an included angle of triangle *DBC*. **7** It is given that side *EF* = side *HG* and side *EH* = side *FG*. Since both triangles share a common side *EG*, the two triangles *EGH* and *EFG* have three equal sides and are therefore congruent.

Lesson 12-14 **1** Triangles *DEF* and *JKL* are similar because two angles of triangle *DEF* are equal to two angles of triangle *JKL*. **2** Triangles *DEF* and *GHI* are similar because the ratios of the corresponding sides are equal. **3** Triangles *ABC* and *JKL* are similar because two sides of triangle *ABC* are proportional to two corresponding sides of triangle *JKL* and the included angles are equal. **4 1.** 10 mm **5 1.** 144 ft **6** 275 m **7** 144 ft

Chapter 13 Practice Exercises

Lesson 13-1 **1 1.** $1\frac{1}{4}$ in. **2 1.** $1\frac{3}{4}$ in. **3**

Lesson 13-2 **1 1.** **2** Angles less than 90°. **3** Angles greater than 90°, but less than 180°. **4** Angles of 90°. **5** Drawings will vary.

Lesson 13-3 p. 472 **1 1.** **2** Yes **3** Triangle with sides $2\frac{3}{16}$ in., $1\frac{5}{8}$ in., and $1\frac{5}{8}$ in. **4 1.** 2 in., $2\frac{3}{4}$ in., $1\frac{1}{8}$ in. **5 1.** 20 mm, 35 mm, 29 mm

p. 473 **1 1.**

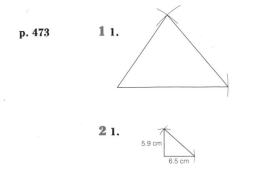

2 1.

5.9 cm

6.5 cm

3 Yes   **4 1.** 3 cm, 2 cm

p. 474   **1 1.**   **2**

5 cm

3   **4**

$2\frac{11}{16}$ in.

5 1. 4 in.

Lesson 13-4   **1**   **2**

35 mm

43 mm

3   **4** Scale: 1 in. = 8 ft

$2\frac{1}{2}$ in. $2\frac{5}{8}$ in.

$2\frac{1}{2}$ in. $4\frac{1}{8}$ in.

5 Yes

Lesson 13-5   **1**   **2**

3   **4**

5 1. Yes

Lesson 13-6   **1 1.**   **2**

3 1.   **4**

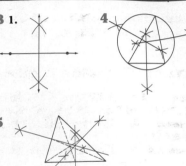

5

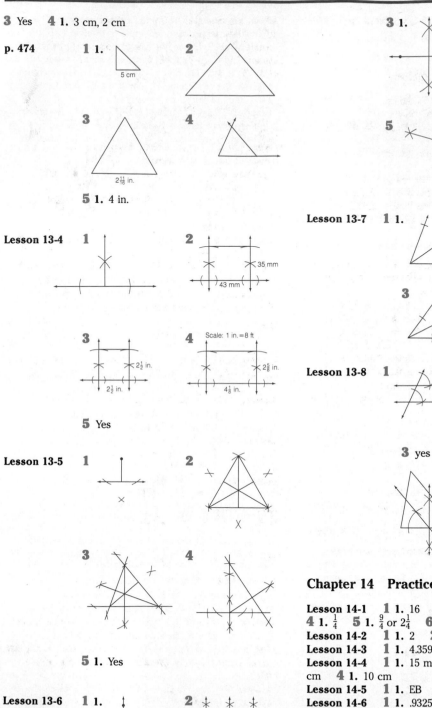

Lesson 13-7   **1 1.**   **2** Drawings will vary.

3   **4** Drawings will vary.

Lesson 13-8   **1**   **2**

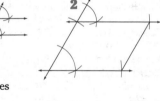

3 yes

Chapter 14   Practice Exercises

Lesson 14-1   **1 1.** 16   **2 1.** 256   **3 1.** .04
4 1. $\frac{1}{4}$ **5 1.** $\frac{9}{4}$ or $2\frac{1}{4}$ **6 1.** 36
Lesson 14-2   **1 1.** 2   **2 1.** 20   **3 1.** $\frac{3}{4}$
Lesson 14-3   **1 1.** 4.359   **2 1.** 2.449
Lesson 14-4   **1 1.** 15 m   **2 1.** 12 mm   **3 1.** 7
cm   **4 1.** 10 cm
Lesson 14-5   **1 1.** EB   **2 1.** AC
Lesson 14-6   **1 1.** .9325   **2 1.** .9816   **3 1.** .9511
4 1. 21°   **5 1.** 65°   **6 1.** 26°   **7 1.** 87°   **8 1.** 41°
9 1. 84°   **10 1.** 171.56 cm   **11 1.** 392.5 m

12 1. 64.95 m   **13 1.** 21.1 ft   **14 1.** 69.47 m
15 1. 223.7 yd   **16 1.** 45.1 cm

Chapter 15   Practice Exercises

Lesson 15-1   **1** 136 mm   **2 1.** 50 cm   **3** 196 m; $1,175.02   **4** 9 m   **5** 7 ft   **6** 68 ft
Lesson 15-2   **1** 32 km   **2 1.** 100 cm   **3** 120 yd **4** $104.07
Lesson 15-3   **1 1.** 42 m   **2 1.** 189 cm   **3** 38 cm   **4** 592 m
Lesson 15-4   **1 1.** 14 cm   **2 1.** 19 m   **3** 188.4 m   **4 1.** 15.7 mm   **5** 163.28 m   **6 1.** 565.2 m **7** 91 cm   **8 1.** 56 m   **9** 132 in.   **10** 21.98 ft or 22 ft   **11** 150.72 cm   **12** 44 ft   **13** 88 in., 720 times each
Lesson 15-5   **1** 126 m²   **2 1.** 391 mm²
3 1. $97.13   **4** $\frac{3}{10}, \frac{50}{7}$ **5** $391\frac{7}{8}$ ft²   **6** 750 tickets
Lesson 15-6   **1** 529 ft²   **2 1.** 100 cm²   **3** 210.25 cm²   **4** $498.75   **5** 12.78 acres
Lesson 15-7   **1** 48 m²   **2 1.** 364 in.²   **3** $7.49
Lesson 15-8   **1** 40 in.²   **2 1.** 108 cm²   **3** $62\frac{1}{2}$ ft²
Lesson 15-9   **1** 77 in.²   **2 1.** 56 cm²   **3** 81 ft²
Lesson 15-10   **1** 113.04 cm²   **2 1.** 530.66 m²
3 452.16 ft²   **4 1.** 3.14 cm²   **5** 5,538.96 km²
6 1,962.5 ft²   **7** 153.86 m²   **8** 26,002.34 km²
9 area of square; 7.74 cm²   **10** 11.393 in.²
Lesson 15-11   **1 1.** 2,050 cm²   **2** 31.36 m²   **3** 800 ft²,   **4** gal., $58.76
Lesson 15-12   **1 1.** 8,214 mm²   **2** 14.85 m²
Lesson 15-13 p. 519   **1 1.** 125.6 m²   **2** 62.8 in.²
3 1. 75.36 cm²   **4** 1.840825 m² or 18,408.25 cm²
p. 520   **1 1.** 1,808.64 m²   **2 1.** 803.84 mm²
3 195,967,400 mi²
Lesson 15-14   **1** 252 in.³   **2 1.** 144 mm³
Lesson 15-15   **1** 15,625 cm²   **2 1.** 729 m³   **3** 216 ft³   **4** 64 times as large.
Lesson 15-16   **1** 169.56 m³   **2** 1,256 ft³   **3 1.** 628 mm³   **4** Second container 9.42 cm² more or $9\frac{3}{7}$ cm²
5 $24\frac{3}{4}$ yd³   **6** 12 lb $\frac{1}{2}$ oz
Lesson 15-17   **1 1.** 113,040 m³   **2 1.** 9,198,107 mm³   **3** .3 ft³   **4 1.** 1,540.69 cm²   **5** 14.72 m³
6 267.95 bu   **7** 14,400 m³   **8** 4,830 m³   **9** 1,200 ft³

Chapter 16   Practice Exercises

Lesson 16-1   **1 1.** negative six   **2 1.** -12
3 1. -7   **4 1.** no, yes   **5 1.** -11, -10, -9, -8, -7, -6, -5, -4, -3, -2, -1, 0, 1, 2, 3
Lesson 16-2   **1 1.** positive five-sixths   **2 1.** $-\frac{1}{5}$
3 1. $-\frac{5}{8}$ **4 1.** no   **5 1.** no   **6 1.** no   **7 1.** no
8 1. yes   **9 1.** no   **10 1.** integer; rational; real

11 1. rational; real   **12 1.** rational; real
13 1. irrational; real
Lesson 16-3   **1 1.** The absolute value of positive eleven   **2 1.** $|-13|$ **3 1.** 7   **4 1.** 12   **5 1.** 6
6 1. 27   **7 1.** 6
Lesson 16-4   **1 1.** -2   **2 1.** 0
3 1. [number line: -3 -2 -1 0]   **4 1.** [number line: -1 0 1]
5 1. -3
Lesson 16-5   **1** 60 meters below sea level   **2** an upward force of 75 kilograms   **3** -4%   **4** +54°
5 a head wind of 23 m.p.h.   **6** +39 millimeters
7 72° above zero   **8** -$25   **9** +58 items
10 -17 amperes   **11** 9 pounds under weight
12 15° N latitude   **13** +5 yards   **14** -7°
15 +18 meters
Lesson 16-6   **1 1.** Point T **2 1.** Point C
3 1. +9   **4 1.** +3   **5 1.** T   **6 1.** <   **7 1.** -9, -6, -4, -3, 0, +4, +5, +7, +10   **8 1.** +9, +8, +4, +3, 0, -1, -3, -7, -8, -12   **9 1.** -8   **10 1.** -4
Lesson 16-7   **1 1.** 7 to the left   **2 1.** -4
3 1. -5   **4 1.** answers will vary   **5 1.** +8

6 1. [number line: -1 0 1 2 3 4 5 with +4 arrow]

Lesson 16-8   **1 1.** 7   **2 1.** -10   **3 1.** the opposite of +9 (or 9)   **4 1.** -(36)   **5 1.** -7; +6; $+\frac{3}{8}$; -.9; $-3\frac{1}{2}$ **6 1.** -9
Lesson 16-9   **1 1.** +8   **2 1.** +9   **3 1.** +3
4 1. -7   **5 1.** 0   **6 1.** $+\frac{1}{2}$ **7 1.** +20
8 1. -7   **9 1.** 8   **10 1.** 11
Lesson 16-10   **1 1.** +3   **2 1.** 6   **3 1.** -2
4 1. +9   **5 1.** 0   **6 1.** -4   **7 1.** -11
7 1. -10   **8 1.** -10   **9 1.** -5
Lesson 16-11   **1 1.** 45   **2 1.** -56   **3 1.** 0
4 1. +27   **5 1.** +9   **6 1.** 5   **7 1.** $-\frac{1}{6}$
Lesson 16-12   **1 1.** +6   **2 1.** -8   **3 1.** +1
4 1. -4   **5 1.** -7   **6 1.** -3   **7 1.** 2

Chapter 17   Practice Exercises

Lesson 17-1   **1 1.** X **2 1.** 4 + 6 or 6 + 4
3 1. five times n **4 1.** $2r$ **5 1.** $n - 95$
Lesson 17-2   **1 1.** Some number n is less than sixteen   **2 1.** $x + 10 = 41$ **3 1.** disjunction; Some number h is greater than or equal to negative six
Lesson 17-3   **1 1.** $p = 4s$ **2 1.** The perimeter of an equilateral triangle equals three times the length of a side.
Lesson 17-4   **1 1.** 18   **2 1.** 22   **3 1.** 192
4 1. 24   **5 1.** 620   **6 1.** 18   **7 1.** 23   **8 1.** 36
9 1. 2,376   **10 1.** 202   **11 1.** 130   **12 1.** 7
Lesson 17-5   **1 1.** 170   **2 1.** 1,640
Lesson 17-6   **1 1.** 4   **2 1.** 0   **3 1.** $\frac{1}{2}$
4 1. $7x + 9$ **5 1.** b, c, d, e   **6 1.** a, c, d

Lesson 17-7 **1** 1. 8 **2** 1. 7 **3** 1. 5 **4** 1. 3
5 1. 4 **6** 1. 12 **7** 1. +10 **8** 1. 3; $n = 16$
9 1. 4; 15 **10** 1. 9; $m = 8$ **11** 1. 3; $t = 18$
12 1. add 5; $x = 18$
Lesson 17-8 **1** 1. $c = 8$ **2** 1. $t = 3\frac{1}{4}$
Lesson 17-9 **1** 1. $c = 28$ **2** 1. $a = \$.83$
Lesson 17-10 **1** 1. $a = 9$ $t = .4$
Lesson 17-11 **1** 1. $d = 38$ **2** 1. $c = \$.48$
Lesson 17-12 **1** 1. $b = 14$ **2** 1. $b = 4$
Lesson 17-13 **1** 1. 6 **2** 1. 33
Lesson 17-14 **1** 1. 18 **2** 1. 2.8 **3** 1. 4
4 1. $\frac{3}{5}$ **5** 1. .3 **6** 1. 18% **7** 1. 87
8 1. 5,000 **9** 1. 72 **10** 1. 21
Lesson 17-15 p. 592 **1** 1. $n > 3$ **2** 1. $c > 23$
3 1. $b < 6$ **4** 1. $d < 15$ **5** 1. $m \geq 4$ **6** 1. $x \geq$
15 **7** 1. $y \leq 7$ **8** 1. $r \leq 14$ **9** 1. all real
numbers except 8 **10** 1. all real numbers except
17 **11** 1. $x > 9$ **p. 593** **1** 1. $x > 2$ **2** 1. $n >$
8 **3** 1. $z < 5$ **4** 1. $n < 24$ **5** 1. $n \geq 5$ **6** 1. b
≥ 0 **7** 1. $s \leq 9$ **8** 1. $y \leq 42$ **9** 1. all real
numbers except 16 **10** 1. all real numbers except
60 **11** 1. $b < 4$ **p. 593–594** **1** 1. $d < -6$
2 1. $b < 7$ **3** 1. $x > -9$ **4** 1. $t > 8$ **5** 1. $y \leq$
-7 **6** 1. $n \leq 9$ **7** 1. $s \geq 20$ **8** 1. $z \geq 36$
9 1. $n > -36$ **10** 1. $m > -8$

Chapter 18 Practice Exercises

Lesson 18-1 **1** 1. $(1, 3)$ **2** 1. $(3, 4); (4, 4); (5, 4)$

Lesson 18-2 **1** 1. $A (3,6)$ $B (4, 3)$ $C (2, 1)$ D
$(0, 7)$ $E (-3, 6)$ $F (-3, 2)$ $G (-4, -2)$ $H (-2,$
$-5)$ $I (3, -3)$ $J (6, -7)$

Lesson 18-3 **1** 1. **2** 1.

Lesson 18-4 **1** 1. **2** 1.

3 1.

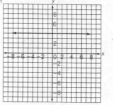

Erick Marroquin

Refresher Math
(new ed. 1986)

~~Jennifer Twine~~

~~Wendy Dew~~

~~Bobbie Leggett~~

C. f. Hus

ms. B Gaudette

T6133